Clinical Anatomy for Anesthesiologists

Clinical Anatomy for Anesthesiologists

Richard S. Snell, M.D., Ph.D.
Professor and Chairman
Department of Anatomy
George Washington University School
of Medicine and Health Sciences
Washington, D.C.

Jordan Katz, M.D.
Professor, Department of Anesthesiology
University of California, San Diego
Staff Anesthesiologist
University Hospital and
Veterans Administration Medical Center
San Diego, California

APPLETON & LANGE
Norwalk, Connecticut/San Mateo, California

0-8385-1258-5

Notice: Our knowledge in clinical sciences is constantly changing. As new information becomes available, changes in treatment and in the use of drugs become necessary. The authors and the publisher of this volume have taken care to make certain that the doses of drugs and schedules of treatment are correct and compatible with the standards generally accepted at the time of publication. The reader is advised to consult carefully the instruction and information material included in the package insert of each drug or therapeutic agent before administration. This advice is especially important when using new or infrequently used drugs.

88 89 90 91 92 / 10 9 8 7 6 5 4 3 2 1

Prentice-Hall International (UK) Limited, *London*
Prentice-Hall of Australia, Pty. Ltd., *Sydney*
Prentice-Hall Canada, Inc., *Toronto*
Prentice-Hall Hispanoamericana, S.A., *Mexico*
Prentice-Hall of India Private Limited, *New Delhi*
Prentice-Hall of Japan, Inc., *Tokyo*
Prentice-Hall of Southeast Asia (Pte.) Ltd., *Singapore*
Editora Prentice-Hall do Brasil Ltda., *Rio de Janeiro*
Prentice-Hall, *Englewood Cliffs, New Jersey*

LC: 88-071850
ISBN: 0-8385-1258-5

Designer: M. Chandler Martylewski

PRINTED IN THE UNITED STATES OF AMERICA

For Maureen and Ruby

Contents

Preface

This book is written to provide the anesthesiologist with the basic knowledge of anatomy necessary for practicing anesthesiology. It is recognized that this medical subspecialty requires a detailed knowledge of the anatomy of the respiratory system, and the cardiovascular system, and the nervous system, including the autonomic nervous system, while other areas of the body may be covered more superficially.

Surface anatomy and surface landmarks have been emphasized and photographs of living subjects have been added to assist in this study. Normal radiographs of the head, neck, thorax, abdomen, and pelvis, have been included. Examples of computed tomography scans of the thorax have also been added.

The practical application of anatomical facts to anesthesiology has been emphasized throughout this book, in the form of clinical notes. Clinical problems requiring anatomical knowledge for their solution are presented at the end of each chapter.

The majority of the illustrations have been kept simple and many are in color. Overview drawings of the distribution of many spinal nerves and cranial nerves have been included.

RSS and JK
1988

Acknowledgments

We thank the many anesthesiology residents, clinical colleagues, and friends who have made valuable suggestions regarding the preparation of the manuscript. We are greatly indebted to the late Dr. Alvin Wyman, Clinical Professor of Radiology at the George Washington University School of Medicine, for the loan of radiographs and to Dr. David O. Davis, Professor of Radiology, for examples of computed tomography scans. In this connection, we also thank the Audio Visual Department for their fine photographic work.

We wish to express our sincere thanks to Ira Alan Grunther for the original illustrations. To the librarians of the George Washington University School of Medicine our thanks are due for continued help in procuring reference material. Special thanks go to Mrs. Uma Gupta for her patience and skill in typing the manuscript.

To the staff of Appleton & Lange we wish to express our sincere thanks for guiding this book through every phase of its production.

We wish to thank Little, Brown and Co. for granting us permission to reproduce figures taken from the following textbooks written by Dr. Snell: *Clinical Embryology for Medical Students*, Third Edition, 1983; *Clinical and Functional Histology for Medical Students*, First Edition, 1984; *Clinical Anatomy for Medical Students*, Third Edition, 1986; *Clinical Neuroanatomy for Medical Students*, Second Edition, 1986.

Clinical Anatomy for Anesthesiologists

CHAPTER 1

Respiratory System
Upper Airway

Airway management is one of the prime responsibilities of the anesthesiologist either in the context of operating room anesthesia or resuscitation. All techniques of airway management from manual manipulation through endotracheal intubation require a detailed knowledge of the anatomy of the respiratory system.

If attention is paid to anatomic structures during *routine* case management, the problems that can occur during unusual situations are more easily handled. It is the purpose of this chapter to present in some detail those anatomic features that are significant for airway management. Because the relative anatomy may change as the person matures, comparisons, when necessary, between anatomic structures in infancy, childhood, and adulthood are presented.

The airway extends from the nostrils of the nose and the lips of the mouth to the alveoli of the lungs. The importance of keeping a clear airway applies to any patient who is unconscious, but also to those who are receiving a general anesthetic by any means.

NOSE

The nose consists of the external nose and the nasal cavity.

EXTERNAL NOSE

The external nose has a free tip and is attached to the forehead by the root. The external orifices are two elliptical apertures called the **external nares or nostrils,** which are separated from each other by the nasal septum (*Fig. 1–1*). In the adult, each measures about 1.5 to 2 cm anteroposteriorly and 0.5 to 1 cm transversely; they are, of course, much smaller in the child. The lateral margin, the **ala nasi** is rounded and mobile.

The framework of the external nose is made up above by the nasal bones, the frontal processes of the maxillae, and the nasal part of the frontal bone. Below, the framework is formed of plates of hyaline cartilage, which include the upper and lower lateral nasal cartilages and the septal cartilage (*Fig. 1–1*).

Muscles of the External Nose

Nasalis

Origin. The transverse part is attached to the maxilla just lateral to the nose; the alar part arises from the maxilla below the transverse part.

Insertion. The transverse part expands into an aponeurosis, which crosses the bridge of the nose to join the muscle of the opposite side. The alar part is attached to the cartilage of the ala nasi.

Action. The transverse part compresses the nasal cavity at the vestibule. The alar part pulls the ala nasi laterally and widens the nostril; the latter action is seen on deep inspiration.

Nerve Supply. Buccal branch of facial nerve.

Depressor Septi

Origin. The maxilla above the central incisor tooth.

Inserton. Lower edge of nasal septum.

Action. Steadies the nasal septum during deep inspiration and assists alar part of nasalis muscle in widening or keeping open the nostril.

Nerve Supply. Buccal branch of facial nerve.

Blood Supply of the External Nose

The skin of the ala nasi and the lower part of the septum are supplied by branches from the facial artery. The remainder of the skin of the external nose is supplied by branches of the ophthalmic and the maxillary arteries.

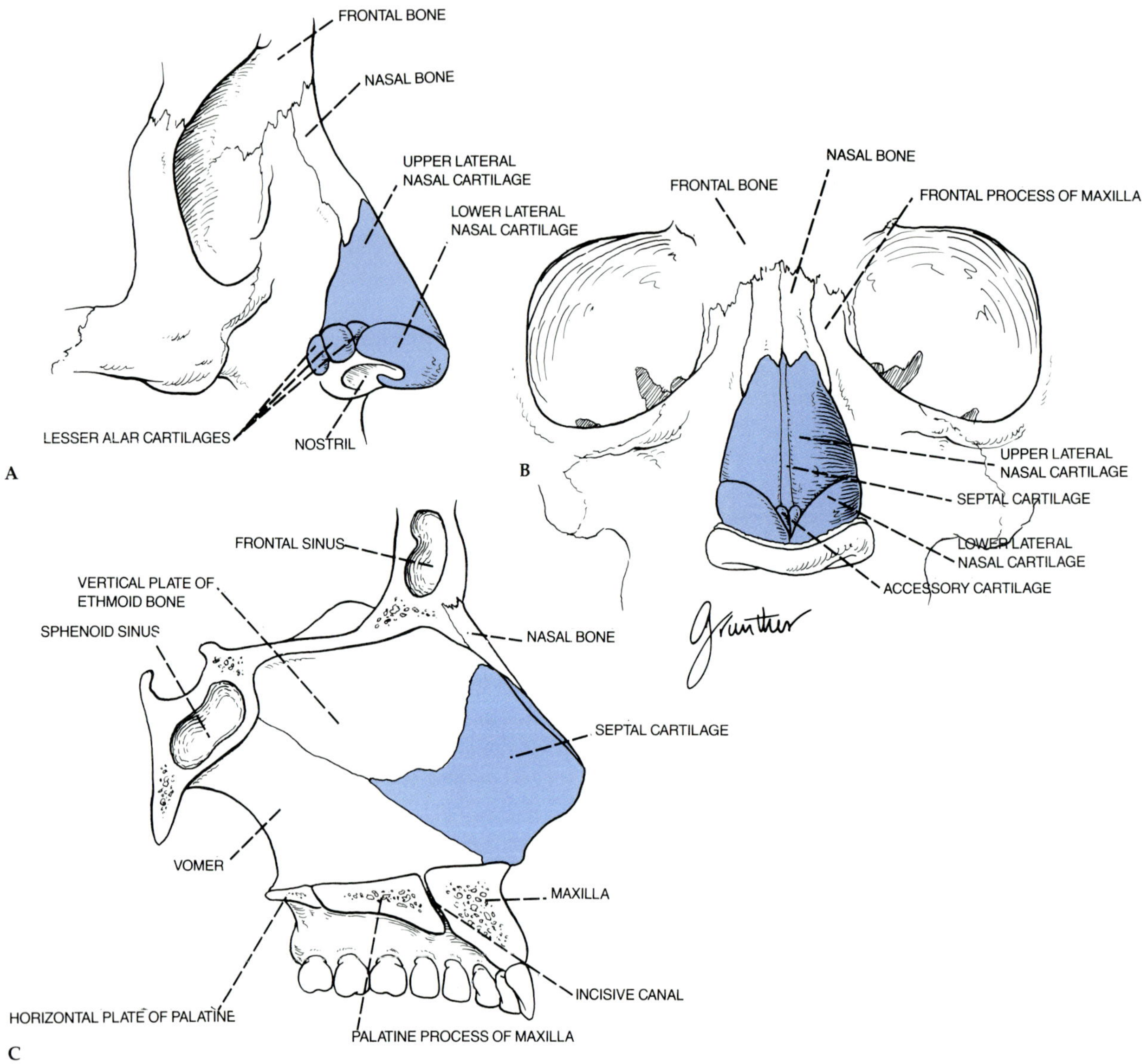

Figure 1–1. External nose and nasal septum. **A.** Lateral view of bony and cartilaginous skeleton of external nose. **B.** Anterior view of bony and cartilaginous skeleton of external nose. **C.** Bony and cartilaginous skeleton of nasal septum.

Sensory Nerve Supply of the External Nose

The skin of the external nose is supplied by the infratrochlear and external nasal branches of the ophthalmic nerve and the infraorbital branch of the maxillary nerve.

NASAL CAVITY

The nasal cavity extends from the external nares or nostrils in front to the **posterior nasal apertures or choanae** behind. Each choana is oval-shaped and measures in the adult about 2.5 cm vertically and 1.25 cm transversely.

The nasal cavity is divided into right and left halves by the nasal septum. Each half consists of three regions, the vestibular, the olfactory, and the respiratory.

The **nasal vestibule,** is the slightly expanded area of the nasal cavity lying just inside the nostril (*Fig. 1–2*). It is lined with skin and has coarse stiff hairs that curve downward and forward to the nostril. The hairs are particularly long in the male after middle age.

The vestibule extends superiorly to a curved ridge called the **limen nasi** that coincides with the upper margin of the lower nasal cartilage. Here the skin of the vestibule becomes continuous with the mucous membrane of the remainder of the nasal cavity.

The **olfactory region** is confined to the superior part of the nasal cavity. The olfactory mucous membrane lines the upper surface of the superior concha and the sphenoethmoidal recess. It also lines a corresponding area on the nasal septum and lines the roof.

The **respiratory region** is the larger region and occupies the remainder of the nasal cavity below the superior concha.

Walls of the Nasal Cavity

Each half of the nasal cavity has a floor, a roof, a lateral wall, and a medial or septal wall.

The **floor** is formed by the palatine process of the maxilla and the horizontal plate of the palatine bone (*Fig. 1–2*). It is gently concave transversely and flat anteroposteriorly.

The **roof** is narrow and is formed anteriorly by the nasal and frontal bones, in the middle by the cribriform plate of the ethmoid, and posteriorly by the sphenoid (*Fig. 1–2*). The highest point on the roof is formed by the ethmoid bone.

The **lateral wall** is marked by three projections called the **superior, middle,** and **inferior nasal conchae** (*Fig. 1–2*). The area below each concha is referred to as a **meatus.** The **sphenoethmoidal recess** is a small area of the nose that lies above the superior concha and in front of the body of the sphenoid bone. It receives the opening of the **sphenoidal air sinus.**

The **superior meatus** lies below and lateral to the superior concha (*Fig. 1–2*). It receives the openings of the **posterior ethmoidal sinuses.** The **middle meatus** lies below and lateral to the middle concha. It has on its lateral wall a rounded prominence, the **bulla ethmoidalis,** caused by the bulging of the underlying middle ethmoidal air sinuses, which open on its upper border. A curved cleft, the **hiatus semilunaris,** lies immediately below the bulla. The anterior end of the hiatus leads into a funnel-shaped channel called the **infundibulum.** The **maxillary sinus** opens into the middle meatus via the hiatus semilunaris. The **frontal sinus** opens into and is continuous with the infundibulum. The **anterior ethmoidal sinuses** also open into the infundibulum. The middle meatus is continuous in front with a depression called the **atrium.** The atrium is limited above by a ridge the **agger nasi** (*Fig. 1–2*). Below and in front of the atrium is the vestibule.

The **inferior meatus** lies below and lateral to the inferior concha and receives the opening of the **nasolacrimal duct** (*Fig. 1–2*). A fold of mucous membrane forms an imperfect valve, which guards the opening of the duct.

The **medial wall or nasal septum,** is an osteocartilaginous partition covered by adherent mucous membrane. The upper part is formed by the vertical plate of the ethmoid and the vomer (*Fig. 1–1*). The anterior part is formed by the septal cartilage. The septum rarely lies in the median plane thus increasing the size of one-half of the nasal cavity and decreasing the size of the other. Ridges or spurs of bone often project from the septum.

Mucous Membrane of the Nasal Cavity

The mucous membrane lines the nasal cavities with the exception of the vestibules, which are lined by skin. The mucous membrane is closely adherent to the underlying periosteum or perichondrium and is thickest over the conchae and nasal septum and is thinnest in the meatuses. There are two types of mucous membrane: (1) olfactory and (2) respiratory

The **olfactory mucous membrane** lines the upper surface of the superior concha and the sphenoethmoidal recess. It also lines the corresponding area on the nasal septum and lines the roof. The olfactory nerve cells send their nerve fibers through the openings in the cribriform plate of the ethmoid and end in the olfactory bulbs (*Fig. 1–3*). The surface of the mucous membrane is kept moist by the secretions of numerous serous glands.

The **respiratory mucous membrane** lines the lower part of the nasal cavities. The surface is kept moist by the production of mucus secreted by glands and goblet cells. The surface is covered with ciliated columnar cells. There is a large plexus of veins in the submucous connective tissue. Congestion of the veins causes swelling of the mucous membrane, which considerably reduces the size of the nasal cavity and the openings communicating with it.

Nerve Supply of the Nasal Cavity

The olfactory nerves arise from the special olfactory cells in the olfactory mucous membrane described previously.

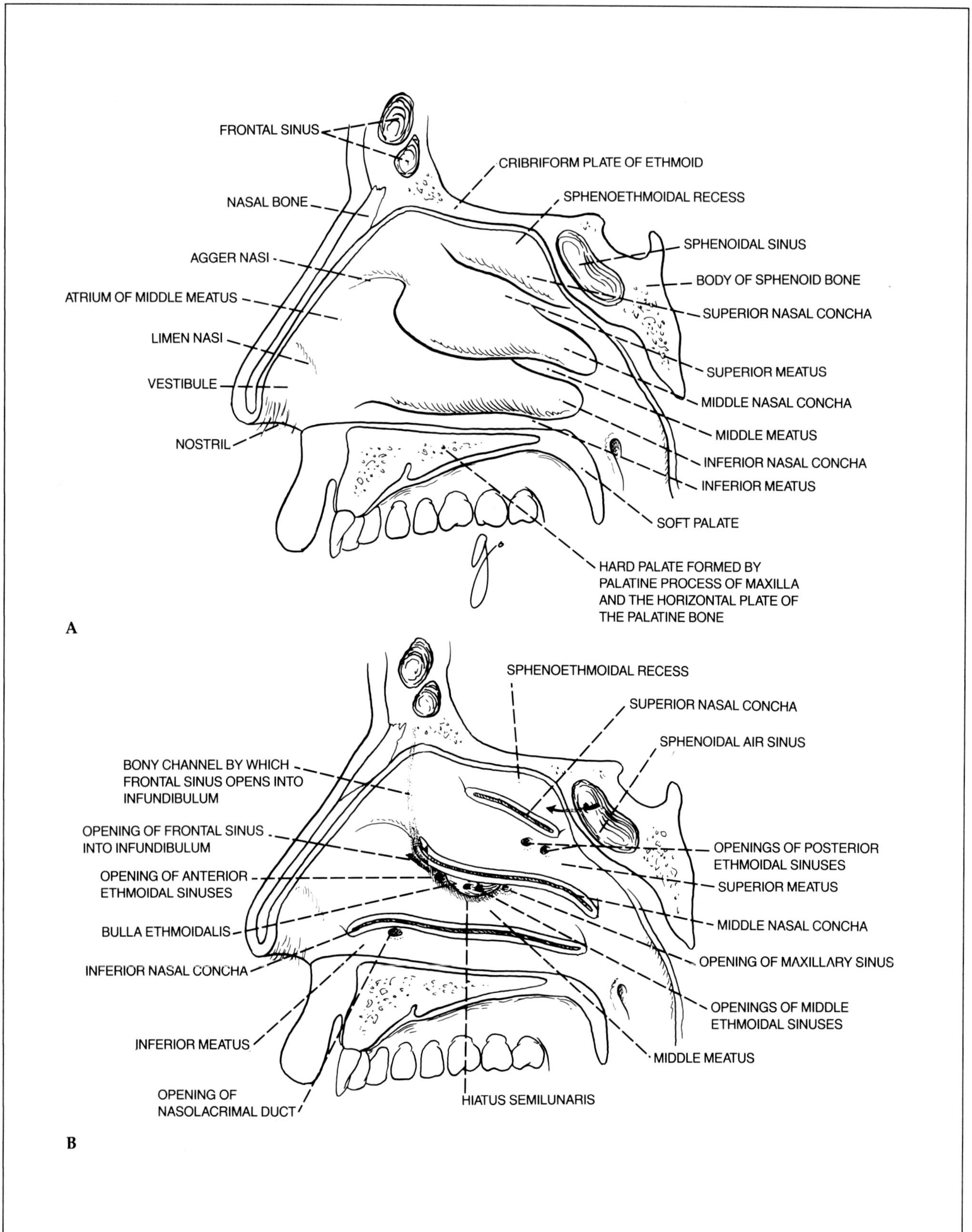

Figure 1–2. A. Lateral wall of the right nasal cavity. **B.** Lateral wall of the right nasal cavity; the superior, middle, and inferior conchae have been partially removed to show openings of the paranasal sinuses and the nasolacrimal duct into the meati.

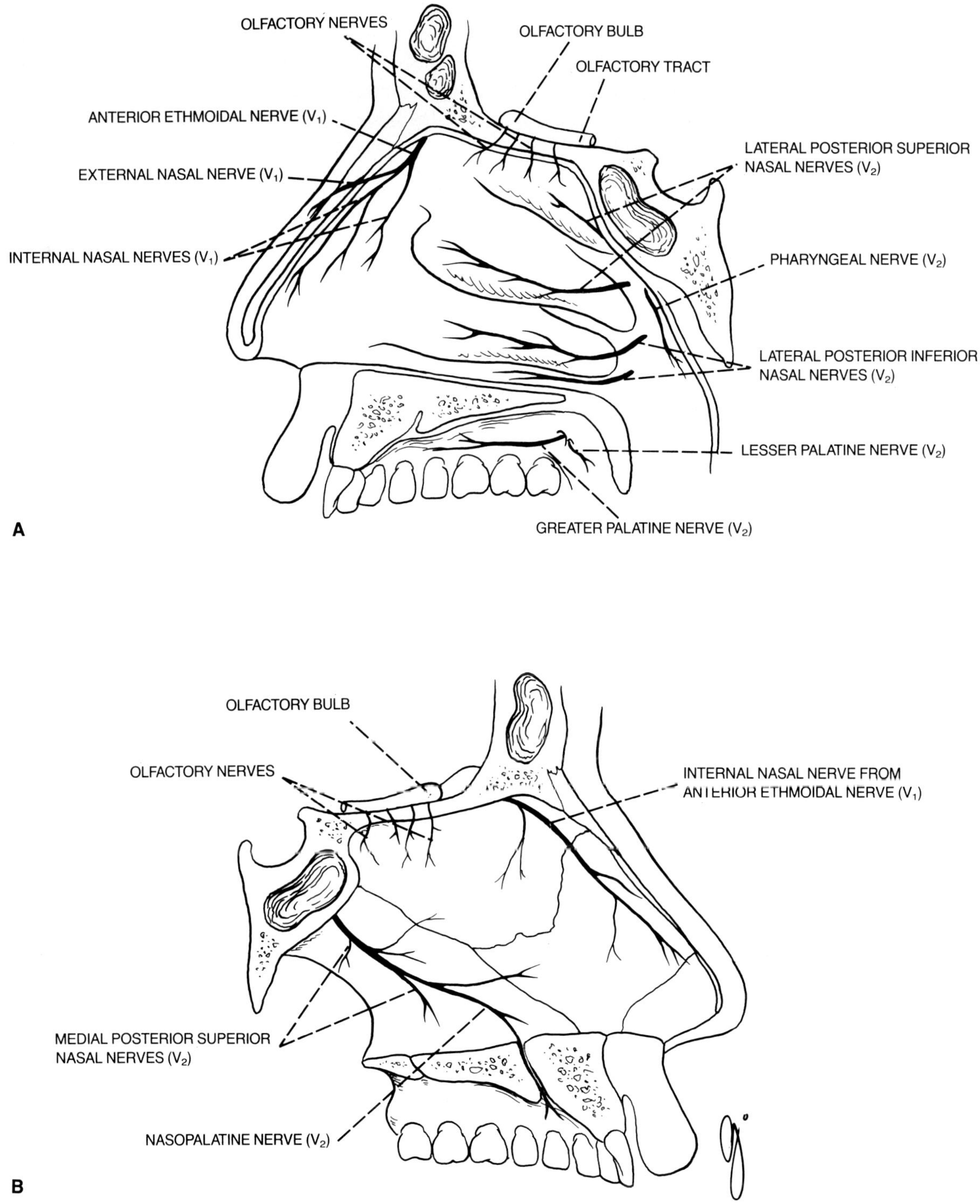

Figure 1–3. A. Lateral wall of nasal cavity showing sensory innervation of mucous membrane. **B.** Nasal septum showing sensory innervation of mucous membrane.

They ascend through the cribriform plate to the olfactory bulbs (*Fig. 1–3*).

The nerves of ordinary sensation are branches of the nasociliary nerve, a branch of the ophthalmic division (V_1) of the trigeminal nerve and branches from the maxillary division (V_2) of the trigeminal nerve. The distribution of these nerves is shown in Figure 1–3. The nasal mucous membrane is extremely sensitive and may be a site of reflex responses. Plugging a patient's nose with gauze may produce intense restlessness during the recovery period.

Sympathetic postganglionic vasoconstrictor fibers from the superior cervical ganglion supply the nasal arteries and veins. Increased sympathetic activity causes the mucous membrane to shrink and decreased activity causes the mucous membrane to swell. Parasympathetic postganglionic secretomotor fibers from the pterygopalatine ganglion supply the nasal glands.

Blood Supply to the Nasal Cavity

The arterial supply to the nasal cavity is derived mainly from branches of the maxillary artery. The most important branch is the sphenopalatine artery, which enters the nasal cavity through the pterygopalatine foramen (*Fig. 1–4*). The sphenopalatine artery anastomoses with the septal branch of the superior labial branch of the facial artery in the region of the vestibule. This is a very common site of bleeding from the nose (Kiesselbach's area).

The submucous venous plexus is drained by veins that accompany the arteries.

The **lymphatic vessels** draining the vestibule end in the submandibular nodes. The remainder of the nasal cavity is drained by vessels that pass to the upper deep cervical nodes.

Functions of the Nose

1. The nose forms part of the airway to the lungs. Normally adults breath through the nose rather than through the mouth in quiet respiration. It is interesting to note that the resistance to nose breathing is about 1.5 times greater than mouth breathing.
2. The inspired air is warmed by the blood in the venous plexuses present in the submucosa. It has been estimated that the nose can, irrespective of ambient temperature, raise the temperature of the inspired air to within 2 to 3 percent of the body temperature.
3. The inspired air is humidified by the mucus secreted by the glands and goblet cells of the respiratory mucous membrane. The daily volume of nasal secretions is about 1 liter. By the time the inspired air has left the nose it is totally humidified. Humidified air is necessary for the normal action of the cilia present on the columnar cells lining the upper part of the respiratory pathway. It is also necessary to keep the cells lining the alveoli moist.
4. The olfactory mucous membrane detects odors in the inspired air.
5. The respiratory mucous membrane cleans the inspired air. Large foreign bodies may be trapped by the stiff hairs of the vestibular skin. Small dust particles may pass further into the nose where they are removed from the air by the moist sticky surface of the mucous membrane. The extensive surface area provided by the folded conchae greatly facilitates the process. The contaminated mucus contains bactericidal agents that are a powerful defense against invading organisms. The mucus is continually being moved backward by the ciliary action of the columnar ciliated epithelium that covers the surface. On reaching the pharynx, the mucus is swallowed. Reflex sneezing also helps to rid the airway of foreign materials.
6. Vocal resonance is provided by the volume of air in the nasal passages.

PARANASAL SINUSES

The paranasal sinuses are cavities found in the maxilla, frontal, sphenoid, and ethmoid bones (*Fig. 1–5*). They are lined with mucoperiosteum and communicate with the nasal cavity through small openings.

The **maxillary sinus** is pyramidal in shape and on each side lies within the body of the maxilla (*Fig. 1–5*). The roof is formed by the floor of the orbit and the floor is formed by the alveolar process. The maxillary sinus opens into the middle meatus of the nose through the hiatus semilunaris (*Fig. 1–2*).

The **frontal sinuses,** two in number, are contained within the frontal bone (*Fig. 1–5*). Rudimentary or absent at birth, they are well developed by the eighth year. They are situated above and behind the medial end of the eyebrow. Each frontal sinus opens into the middle meatus of the nose through the infundibulum (*Fig. 1–2*).

The **sphenoidal sinuses,** two in number, lie within the body of the sphenoid bone (*Fig. 1–5*). Each sinus opens into the sphenoethmoidal recess above the superior concha (*Fig. 1–2*).

The **ethmoidal sinuses** are contained within the ethmoid bone on each side (*Fig. 1–5*) between the nose and the orbit. The anterior sinus opens into the infundibulum, the middle sinus into the middle meatus on or above the bulla ethmoidalis, and the posterior sinus into the superior meatus (*Fig. 1–2*).

The mucus produced by the glands in the mucous membrane is moved into the nose by the ciliary action of the columnar cells. Drainage of the mucus is also achieved by the syphon action created during blowing of the nose.

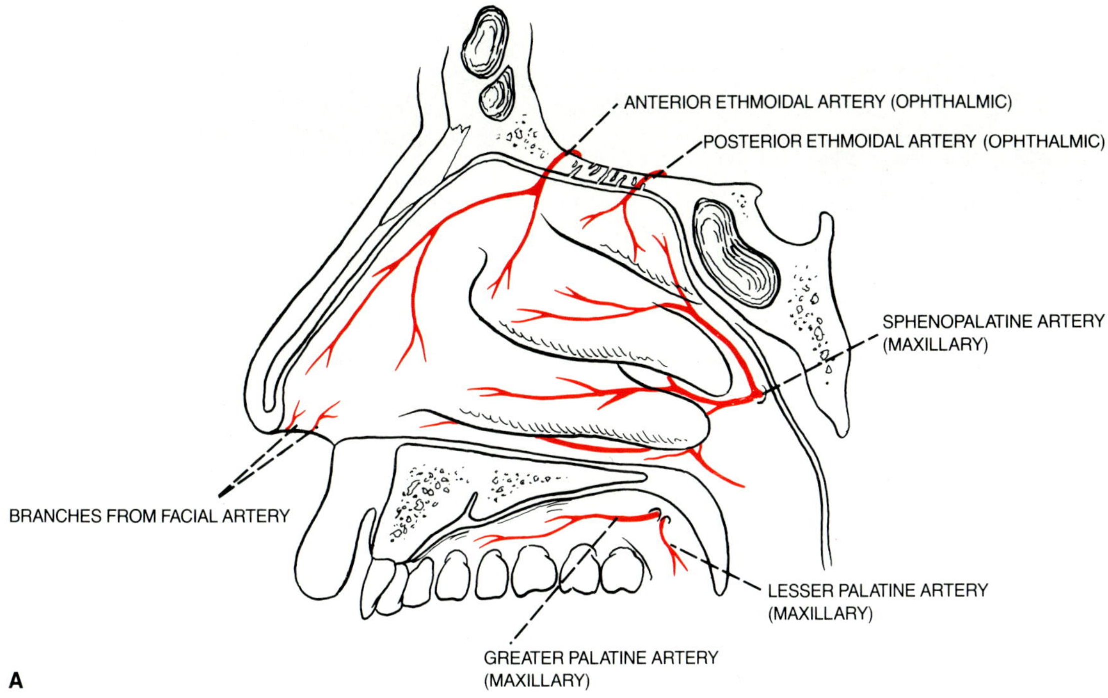

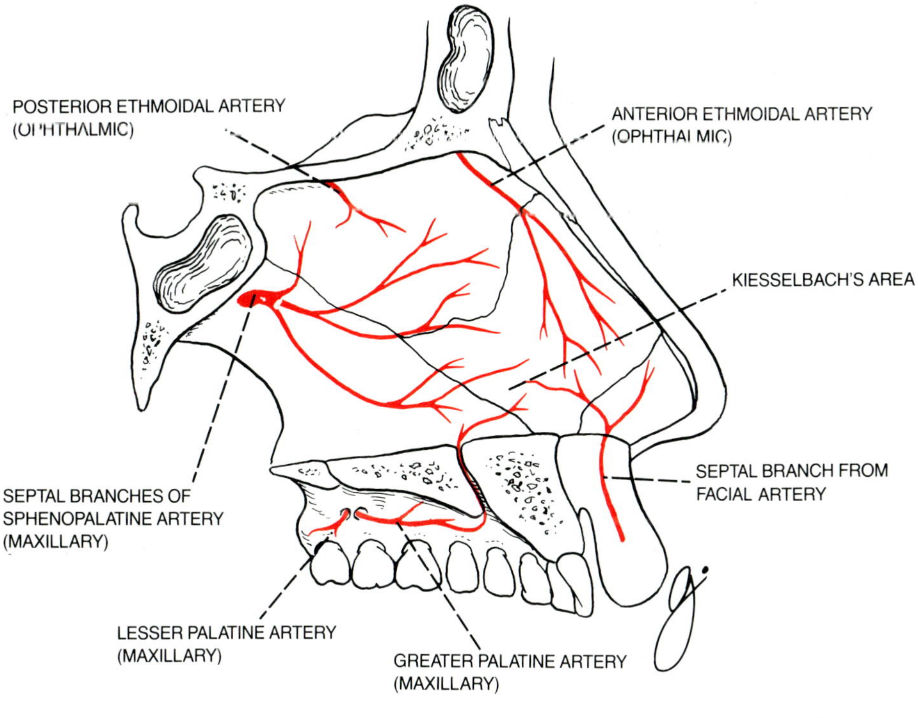

Figure 1–4. A. Lateral wall of nasal cavity showing the arterial supply of the mucous membrane. **B.** Nasal septum showing the arterial supply of the mucous membrane.

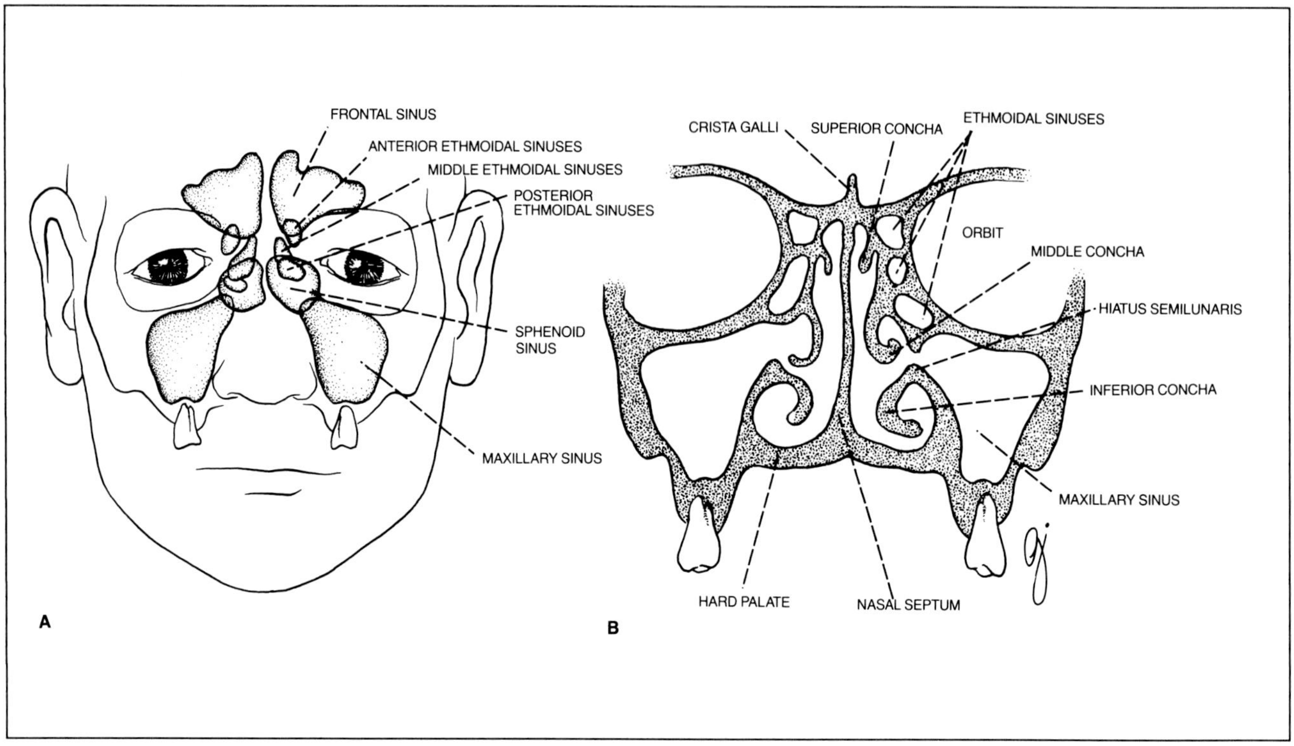

Figure 1–5. A. The position of the paranasal sinuses as related to the face. **B.** Coronal section through the nasal cavity showing the ethmoidal and the maxillary sinuses.

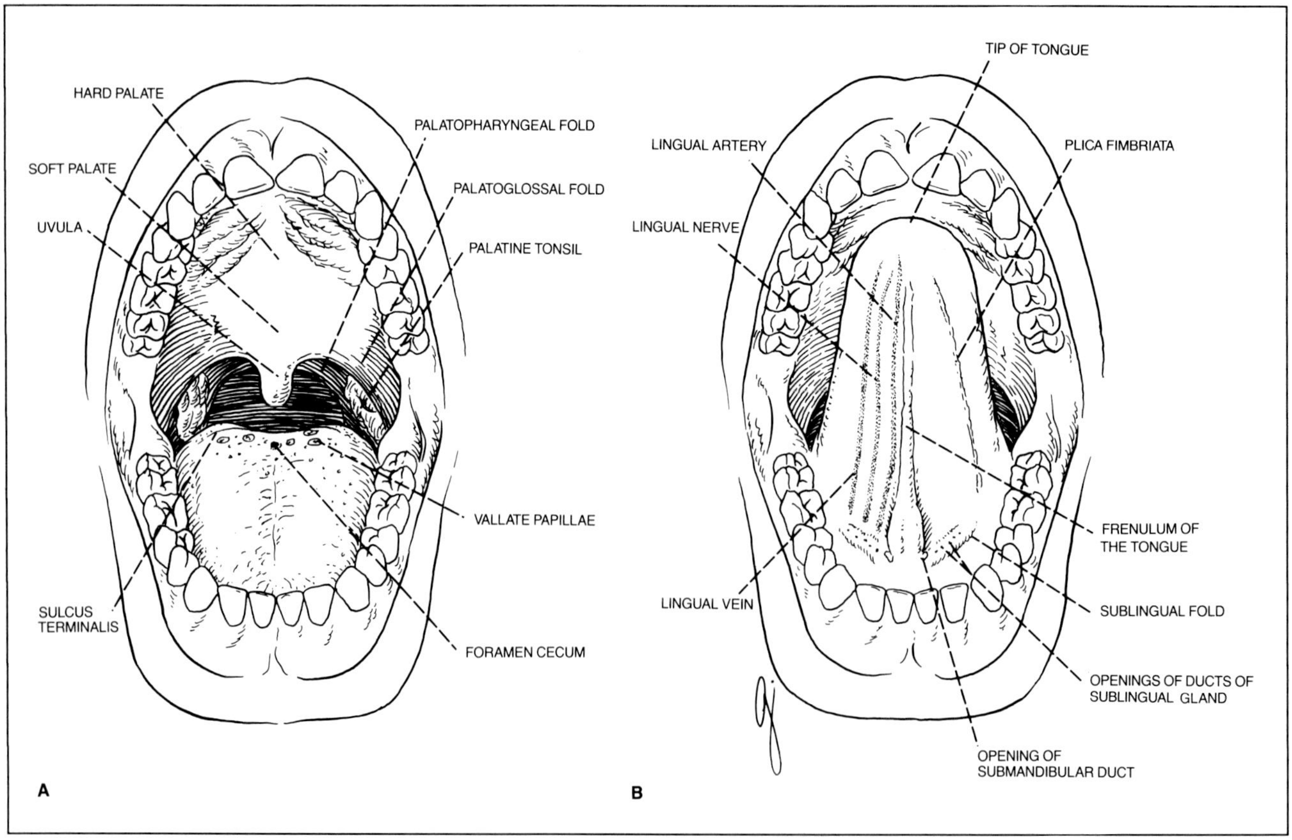

Figure 1–6. A. Mouth open to display cavity of mouth and oral pharynx. **B.** Undersurface of tongue.

Function of the Paranasal Sinuses

The function of the sinuses is to act as resonators to the voice; as they are less dense than solid bone, their presence decreases the weight of the skull.

CLINICAL NOTES

NASOTRACHEAL INTUBATION

Nasotracheal intubation is an indicated technique in several circumstances. Among the more common circumstances are in patients who have limited jaw mobility, cervical arthritis, certain airway tumors, and conditions in which prolonged postoperative ventilation might be required. The technique of nasotracheal intubation can be accomplished with the patient awake using topical anesthesia or after general anesthesia is induced. It is usually routine but, as is the experience of all practicing anesthesiologists, sometimes nasotracheal intubation can be very difficult or impossible. Reasons for difficulties include a deflected nasal septum or presence of bony spurs in the nasal cavity.

Before any attempt of nasal intubation, examination of the nares and nasal cavities is required. Because it is often difficult to determine which half of the nasal cavity is most patent, a history and physical examination, including direct inspection, may have to be complemented by appropriate x-ray studies.

Deflected Nasal Septum

Deflection of the nasal septum is very common in the male adult and is believed to be caused in many cases by trauma in childhood. Congestion of the respiratory mucous membrane, hypertrophic rhinitis, or polyps may also reduce the size of the nasal airway. The most voluminous part of the nasal cavity is close to the floor and it is usually possible to pass a well-lubricated tube through the nostril along the inferior meatus into the nasopharynx.

MUCOUS MEMBRANE

The great vascularity and delicacy of the nasal mucous membrane dictates that extreme care should always be exerted in passing a nasal tube; a severe hemorrhage in this area is difficult to treat. To avoid causing unnecessary bleeding, good topicalization with the addition of a vasoconstrictor is often indicated. When obstructed the nasal cavity can be expanded by shrinking the mucous membranes with a vasoconstrictor and then further dilating the orifice with progressively larger soft nasal airways.

MOUTH

LIPS

The lips are two fleshy folds that surround the oral orifice (*Fig. 1–6*). They are covered on the outside by skin and are lined on the inside by mucous membrane. The substance of the lips is made up by the orbicularis oris muscle and the muscles that radiate out from the lips into the face (*Fig. 1–7*). Also included are the labial blood vessels and nerves, connective tissue, and many small salivary glands. The **philtrum** is the shallow vertical groove seen in the midline on the outer surface of the upper lip. There are median folds of mucous membrane connecting the inner surface of the lips to the gums; these are the **labial frenula.**

MOUTH CAVITY

The mouth extends from the lips anteriorly to the oropharyngeal isthmus posteriorly. The **oropharyngeal isthmus** is formed on each side by the palatoglossal fold (*Fig. 1–6*). The **vestibule of the mouth** lies between the lips and the cheeks externally and the gums and the teeth internally. It is a slit-like space that communicates with the exterior through the **oral fissure.** When the jaws are closed, it communicates with the mouth proper behind the third molar tooth on each side. The vestibule is limited above and below by the reflection of the mucous membrane from the lips and cheeks to the gums.

The lateral wall of the vestibule is formed by the **cheek,** which is made up by the buccinator muscle and is lined with mucous membrane. The tone of the buccinator muscle and that of the muscles of the lips keeps the walls of the vestibule in contact with one another. The **duct of the parotid salivary gland** opens on a small papilla into the vestibule opposite the upper second molar tooth.

The **mouth proper** has a roof formed by the hard palate in front and the soft palate behind. The floor is formed largely by the anterior two-thirds of the tongue and by the reflection of the mucous membrane from the sides of the tongue to the gum of the mandible. A fold of mucous membrane called the **frenulum of the tongue** connects the undersurface of the tongue in the midline to the floor of the mouth (*Fig. 1–6*). Lateral to the frenulum can be seen the **deep lingual vein** through the mucous membrane. Lateral to the lingual vein, the mucous membrane forms a fringed fold, the **plica fimbriata** (*Fig. 1–6*).

The **orifice of the submandibular duct** opens onto the floor of the mouth on the summit of a small papilla on either side of the frenulum of the tongue (*Fig. 1–6*).

The **sublingual gland** projects up into the mouth producing a low rounded fold of mucous membrane that extends backward and laterally from the submandibular papilla (*Fig. 1–6*).

Mucous Membrane of the Mouth Cavity

The mucous membrane of the mouth consists of stratified squamous nonkeratinized epithelium lying on connective

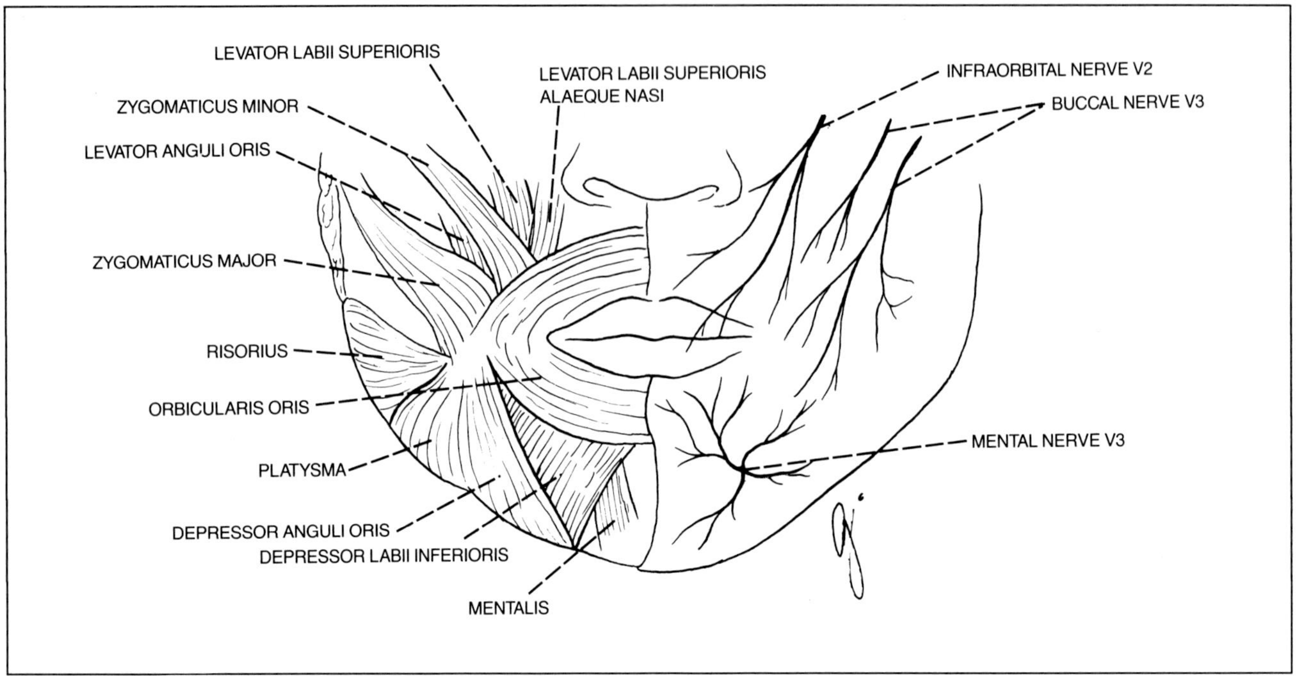

Figure 1–7. Arrangement of the facial muscles around the lips; the sensory nerve supply of the lips is also shown.

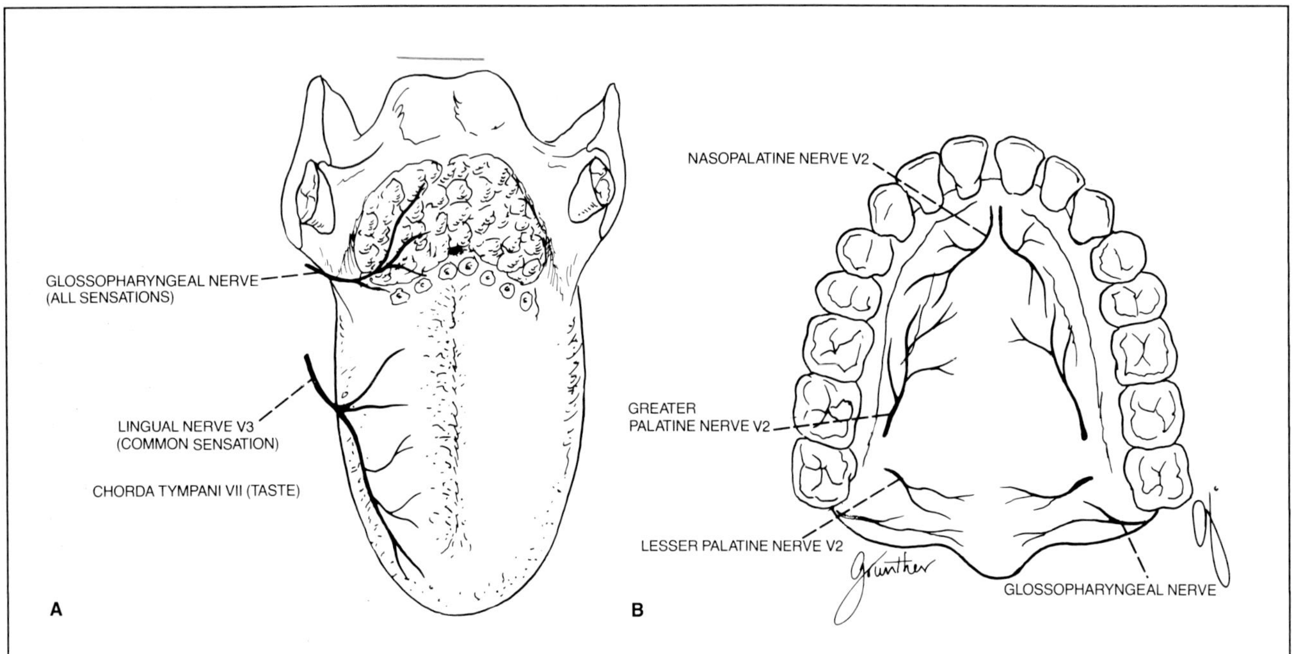

Figure 1–8. A. Sensory nerve supply to the mucous membrane of the tongue. **B.** Sensory nerve supply to the mucous membrane of the hard and soft palate; taste fibers run with the branches of the maxillary nerve (V_2) and join the greater petrosal branch of the facial nerve.

tissue. Small groups of mucous glands pour their secretions onto the surface. In the vestibule, the mucous membrane is tethered to the buccinator muscle by elastic fibers in the submucosa that prevent redundant folds of mucous membrane from being bitten between the teeth when the jaws are closed. The gingiva, or gum, is the part of the mucous membrane that surrounds the teeth. Its surface is covered with keratinized stratified squamous epithelium, and is strongly attached to the alveolar periosteum.

Sensory Nerve Supply of the Mucous Membrane of the Mouth Cavity

The roof of the mouth is supplied by the greater palatine and nasopalatine nerves (*Fig. 1–8*). The nerve fibers are derived from the maxillary division of the trigeminal nerve.

The floor of the mouth is supplied by the lingual nerve, a branch of the mandibular division of the trigeminal nerve. The taste fibers travel in the chorda tympani nerve, a branch of the facial nerve.

The cheek is supplied by the buccal nerve, a branch of the mandibular division of the trigeminal nerve. (Note that the buccinator muscle is innervated by the buccal branch of the facial nerve.)

TEETH

The **deciduous teeth** are 20 in number; 4 incisors, 2 canine, and 4 molars in each jaw. They begin to erupt at about the sixth month after birth and have all erupted by the end of the second year. The teeth of the lower jaw usually appear before those of the upper jaw.

The **permanent teeth** are 32 in number, including 4 incisors, 2 canines, 4 premolars, and 6 molars in each jaw (*Fig. 1–6*). They begin to erupt at the sixth year. The last tooth to erupt is the third molar, however, and this may take place between the 17th and 30th years. The teeth of the lower jaw usually appear before those of the upper jaw.

TONGUE

The tongue is shaped like an inverted boot with its toe projecting forward (*Fig. 1–9*). It is composed of a mass of striated muscle covered with mucous membrane. The muscles of the tongue are intrinsic and extrinsic. The intrinsic muscles are confined to the tongue and are arranged in longitudinal, transverse, and vertical directions, and they alter the shape of the tongue. The extrinsic muscles of the tongue extend from the tongue to the soft palate and the styloid process of the temporal bone above, and the mandible and the hyoid bone below. They alter the position of the tongue. The tongue is divided into right and left halves by a median **fibrous septum.** The origin, insertion, nerve supply, and action of the tongue muscles are summarized in Table 1–1.

The mucous membrane of the upper surface of the tongue may be divided into anterior and posterior parts by a V-shaped sulcus, the **sulcus terminalis** (*Fig. 1–6*). The apex of the sulcus projects backward and is marked by a small pit, the **foramen cecum.** The sulcus serves to divide the tongue into the anterior two-thirds, or oral part, and the posterior third, or pharyngeal part. Three types of papillae are present on the upper surface of the anterior two-thirds of the tongue: (1) the filiform papillae, (2) the fungiform papillae, and (3) the vallate papillae.

TABLE 1–1. MUSCLES OF THE TONGUE

Name of Muscle	Origin	Insertion	Nerve Supply	Action
Intrinsic Muscles				
Longitudinal Transverse Vertical	Median septum and submucosa	Mucous membrane	Hypoglossal nerve	Alters shape of tongue
Extrinsic Muscles				
Genioglossus	Superior genial spine of mandible	Blends with other muscles of tongue and body of hyoid bone	Hypoglossal nerve	Protrudes apex through mouth
Hyoglossus	Body and greater cornu of hyoid bone	Blends with other muscles of tongue	Hypoglossal nerve	Depresses tongue
Styloglossus	Styloid process of temporal bone	Blends with other muscles of tongue	Hypoglossal nerve	Draws tongue upward and backward
Palatoglossus muscle of soft palate see Table 1–2				

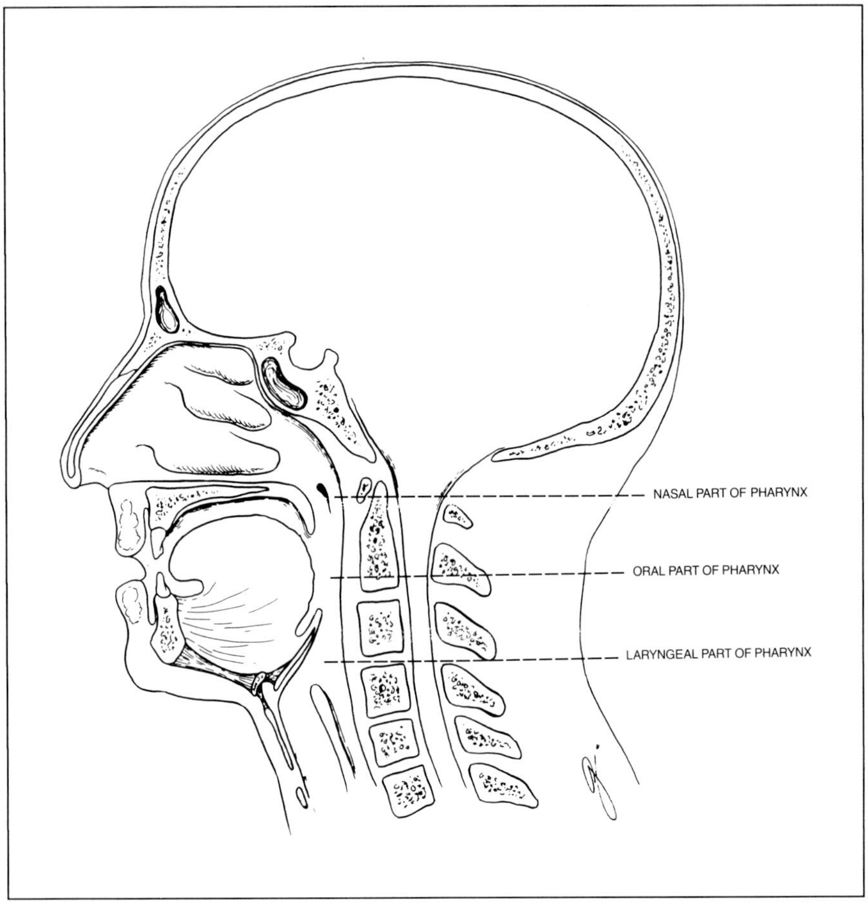

Figure 1–9. Sagittal section through the nose, mouth, pharynx, and larynx to show the subdivisions of the pharynx.

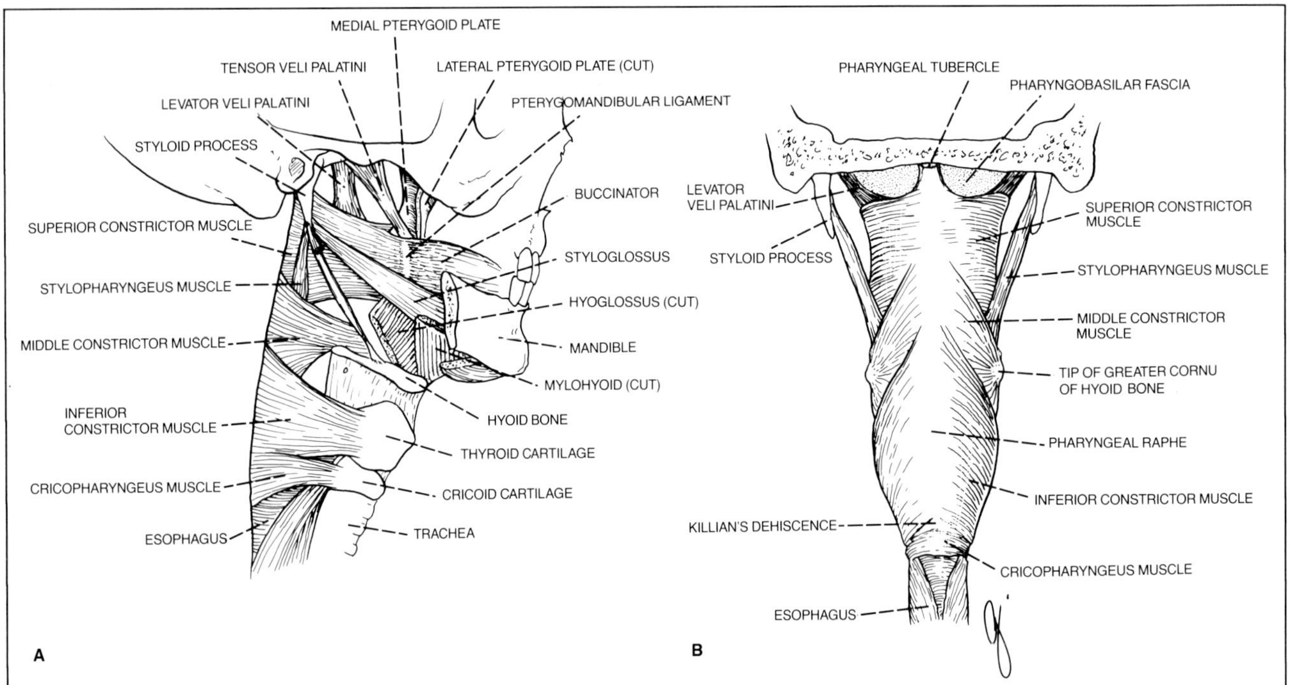

Figure 1–10. A. Pharynx, lateral view, showing the three constrictor muscles. **B.** Pharynx, posterior view, showing the three constrictor muscles and the stylopharyngeus.

The **filiform papillae** are very numerous and cover the anterior two-thirds of the tongue on its upper surface. They form small conical projections and are whitish in color due to the thickness of the cornified epithelium.

The **fungiform papillae** are much less numerous than the filiform papillae and are scattered on the sides and apex of the tongue. They are mushroom-shaped and possess a vascular connective tissue core, which imparts a reddish tinge to the papillae.

The **vallate papillae** are 10 to 12 in number and are situated in a row immediately in front of the sulcus terminalis. Each papilla protrudes slightly from the surface and is surrounded by a circular furrow in the walls of which lie the tastebuds. In the midline anteriorly, the underface of the tongue is connected to the floor of the mouth by a fold of mucous membrane called the **frenulum of the tongue** (*Fig. 1–6*).

The mucous membrane covering the posterior third of the tongue is devoid of papillae but has an irregular surface produced by underlying lymphatic nodules, the **lingual tonsil.**

Blood Supply. Lingual artery, tonsillar branch of facial artery, ascending pharyngeal artery.

Lymphatic Drainage. Tip drains into submental lymph nodes, sides of anterior two-thirds drain into submandibular and deep cervical nodes. Posterior one-third drains into deep cervical nodes.

Sensory Innervation. Anterior two-thirds: lingual nerve (general sensation) and chorda tympani (taste). Posterior third: glossopharyngeal for general sensation and taste (*Fig. 1–8*).

Movements of Tongue

Protrusion. Genioglossus muscles on both sides.

Retraction. Styloglossus and hyoglossus muscles on both sides.

Depression. Hyoglossus muscles on both sides.

Retraction and Elevation of Posterior Third. Styloglossus and palatoglossus on both sides.

Shape Changes These are produced by the intrinsic muscles.

PALATE

The palate forms the roof of the mouth and the floor of the nasal cavity. It may be divided into two parts: (1) the hard palate in front and (2) the soft palate behind.

The **hard palate** is formed by the palatine processes of the maxillae and the horizontal plates of the palatine bones (*Fig. 1–1*). The upper and lower surfaces of the hard palate are covered with mucous membrane. The **soft palate** is a mobile fold attached to the posterior border of the hard palate (*Fig. 1–11*). It is covered on its upper and lower surfaces by mucous membrane and contains an aponeurosis of dense fibrous tissue, muscle fibers, lymphoid tissue, glands, vessels, and nerves. Its free posterior border has in the midline a conical projection called the **uvula.** The soft palate is continuous at the sides wih the lateral wall of the pharynx.

From the undersurface of the soft palate, a muscular fold extends downward and forward on each side and runs to the side of the base of the tongue. It is covered with mucous membrane and is called the **palatoglossal arch or fold** (*Fig. 1–6*). The muscle contained within the arch is the **palatoglossus muscle.** Another similar arch extends from the undersurface of the soft palate downward and laterally to join the pharyngeal wall. It is also covered with mucous membrane and is called the **palatopharyngeal arch or fold** (Fig. 1–6). The muscle contained within the arch is the palatopharyngeus muscle. The **palatine tonsils,** which are masses of lymphoid tissue, are situated between the palatoglossal and palatopharyngeal arches (*Fig. 1–11*).

The palatoglossal arch on each side serves as a landmark and is the place where the mouth cavity becomes continuous with the oral part of the pharynx. At the posterior border of the soft palate the oral part of the pharynx becomes continuous with the nasal part of the pharynx (*Fig. 1–9*).

The muscles of the soft palate are the tensor veli palatini, the levator veli palatini, the palatoglossus, the palatopharyngeus, and the musculus uvulae. The origins, insertions, nerve supply, and actions of these muscles are summarized in Table 1–2.

Blood Supply. Greater palatine branch of maxillary artery, ascending palatine branch of facial artery, ascending pharyngeal artery.

Lymphatic Drainage. Deep cervical lymph nodes.

Sensory Innervation. Greater and lesser palatine nerves, nasopalatine and glossopharyngeal nerves (*Fig. 1–8*).

Movements of the Soft Palate

The channel between the oral part of the pharynx and the nasal part of the pharynx can be closed by raising the soft palate. Closure occurs during swallowing to prevent food from entering the nasal part of the pharynx and the back of the nose; closure also occurs during speech. The soft palate is raised by the contraction of the **levator veli palatini** muscle on each side. At the same time, the upper fibers of the **superior constrictor** muscle contract and pull the posterior pharyngeal wall forward. The **palatopharyngeus** muscles on both sides also contract, so that the palatopharyngeal arches are pulled medially, like side curtains.

TABLE 1–2. MUSCLES OF THE SOFT PALATE

Name of Muscle	Origin	Insertion	Nerve Supply	Action
Tensor veli palatini	Spine of sphenoid, auditory tube	Forms palatine aponeurosis with muscle of opposite side	Nerve to medial pterygoid from mandibular nerve	Tenses soft palate
Levator veli palatini	Petrous part of temporal bone, auditory tube	Palatine aponeurosis	Pharyngeal plexus	Raises soft palate
Palatoglossus	Palatine aponeurosis	Side of tongue	Pharyngeal plexus	Pulls root of tongue upward and backward, narrows oropharyngeal isthmus
Palatopharyngeus	Palatine aponeurosis and border of hard palate	Posterior border of thyroid cartilage	Pharyngeal plexus	Elevates wall of pharynx, pulls palatopharyngeal arches medially
Uvulae	Posterior border of hard palate	Mucous member of uvula	Pharyngeal plexus	Elevates uvula

SALIVARY GLANDS

There are three main pairs of salivary glands called the **parotid, submandibular,** and **sublingual glands,** and in addition there are many small **buccal glands** scattered throughout the mouth.

Parotid Gland

The parotid gland lies in a deep hollow below the external auditory meatus, behind the ramus of the mandible and in front of the sternocleidomastoid muscle. The **parotid duct** emerges from the anterior border of the gland and passes forward over the lateral surface of the masseter muscle. It enters the vestibule of the mouth on a small papilla opposite the upper second molar tooth.

Submandibular Gland

The submandibular gland lies beneath the lower border of the body of the mandible. It can be divided into superficial and deep parts, which are continuous with each other around the posterior border of the mylohyoid muscle. The **submandibular duct** emerges from the anterior end of the deep part of the gland and runs forward beneath the mucous membrane of the mouth. It opens into the mouth on a small papilla, which is situated at the side of the frenulum of the tongue (*Fig. 1–6*).

Sublingual Gland

The sublingual gland lies beneath the mucous membrane (sublingual fold) of the floor of the mouth close to the frenulum of the tongue. The **sublingual ducts** (8 to 20 in number) open into the mouth on the summit of the sublingual fold, but a few may open into the submandibular duct (*Fig. 1–6*).

CLINICAL NOTES

Lips and Vestibule

Asymmetry of the lips and paralysis of the buccinator muscle, with a tendency to accumulate saliva and food in the vestibule, would indicate a lesion of the facial nerve on that side.

Tongue

In an unconscious patient, there is a tendency for the tongue to fall backward and obstruct the laryngeal opening. This is caused by the loss of tone of the extrinsic muscles and unless quickly corrected, will lead to all of the sequelae of airway obstruction.

The following corrective measures need be undertaken. The neck should be extended at the atlanto-occipital joint. Stretching of the fascia and muscles of the front of the neck causes a forward and downward movement of the mandible that is correctable by placing a finger below the symphysis menti and pulling the mandible forward and up. Sometimes this is inadequate to relieve the obstruction and should be supplemented by placing the fingers behind the angles of the mandible and exerting forward pressure. In case of a fat or heavily muscled face it is often best to rest the forearms against the firm surface of the bed to be able to exert the necessary pressure to do the above maneuver. This moves the mandible forward causing the tongue to be displaced away from the laryngeal opening, as the mandible is attached to the tongue by the genioglossus muscle.

In the immediate postoperative period, if airway obstruction is caused by a lax tongue keeping the patient in a semilateral position during the recovery is often all that is required.

There are additional anatomic problems that may interfere with the maintenance of a normal airway or endotracheal intubation. Among the most annoying clinically is the presence of protruding incisor teeth; the normal route for endotracheal intubation is blocked by the tooth. It is, therefore, necessary to put the endotracheal tube in an extreme lateral position to approach the glottis.

Patients with a receding jaw secondary to a small mandible are often intubation challenges. Frequently oral intubation is impossible and the nasal route is resorted to. Because this anatomic configuration approaches the picture seen in younger children, however, a small straight blade, such as a Miller #2 or Miller #3, can overcome the visual difficulties noted when a curved blade of the MacIntosh type is used.

It must be remembered that total visualization of the glottis is not necessary for endotracheal intubation. If one is able to see the epiglottis, the following maneuvers can often result in a successful intubation. The tube is laid on the laryngeal side of the epiglottis and advanced while kept in contact with this surface. Often this procedure alone will allow the tube to go into the trachea. Occasionally when the tube is caught at the anterior glottic constriction the head should be flexed slightly allowing the pressure of the tongue to displace the endotracheal tube somewhat posteriorly and hence into the opening of the glottis. Frequently this maneuver has to be supplemented by turning the head slightly to one side or another. The use of styleted endotracheal tubes also may be of some benefit in this situation.

A short muscular neck is another difficulty that all clinical anesthetists are familiar with. Again, in this situation, a variety of laryngoscope blades might provide better visualization and the basic principles as outlined previously can also be used.

When oral endotracheal intubation is impossible in these situations, nasotracheal intubation may be successful, as the tube approaches the glottis slightly more posteriorly and is directed more toward it.

PHARYNX

The pharynx is situated behind the nasal cavities, the mouth, and the larynx (*Fig. 1–9*) and may be divided into nasal, oral, and laryngeal parts. The pharynx is funnel-shaped, its upper, wider end lying under the skull, and its lower, narrow end becoming continuous with the esophagus opposite the sixth cervical vertebra. The pharynx has a musculomembranous wall, which is deficient anteriorly. Here, it is replaced by the posterior openings into the nose (choanae), the opening into the mouth, and the inlet of the larynx. The mucous membrane is continuous with that of the nasal cavities, the mouth, and the larynx. By means of the auditory tube, the mucous membrane is also continuous with that of the tympanic cavity.

The muscles in the wall of the pharynx consist of the superior, middle, and inferior constrictor muscles (*Fig. 1–10*), whose fibers run in a more or less circular direction, and the stylopharyngeus and salpingopharyngeus muscles, whose fibers run in a more or less longitudinal direction. The origin, insertion, nerve supply, and action of these muscles are summarized in Table 1–3.

The three constrictor muscles extend around the pharyngeal wall and insert into a fibrous raphe, which extends from the pharyngeal tubercle on the basilar part of the occipital bone down to the esophagus. It should be noted that the three constrictor muscles overlap one another so that the middle constrictor lies on the outside of the lower part of the superior constrictor and the inferior constrictor lies outside the lower part of the middle constrictor (*Fig. 1–10*).

The lower part of the inferior constrictor, which arises from the cricoid cartilage, is called the **cricopharyngeus muscle** (*Fig. 1–10*). The fibers of the cricopharyngeus pass horizontally around the lowest and narrowest part of the pharynx and act as a sphincter. **Killian's dehiscence** is the area on the posterior pharyngeal wall between the upper propulsive part of the inferior constrictor and the lower sphincteric part, the cricopharyngeus (*Fig. 1–10*).

A part of the superior constrictor, which arises from the soft palate, sweeps around the pharyngeal wall close to its upper border. This band of muscle is called the **palatopharyngeal sphincter** and produces a rounded ridge called the **ridge of Passavant.**

Interior of Pharynx

The pharynx is divided into three parts: the nasal pharynx, the oral pharynx, and the laryngeal pharynx.

Nasal Pharynx. The nasal pharynx lies above the soft palate and behind the nasal cavities (*Fig. 1–11*). In the submucosa of the roof is a collection of lymphoid tissue called the **pharyngeal tonsil.** The **pharyngeal** isthmus is the opening in the floor between the soft palate and the posterior pharyngeal wall. On the lateral wall is the opening of the **auditory tube,** the elevated edge of which is called the **tubal elevation.** The **pharyngeal recess** is a depression in the pharyngeal wall behind the tubal elevation. The **salpingopharyngeal fold** is a vertical fold of mucous membrane covering the salpingopharyngeus muscle.

Oral Pharynx. The oral pharynx lies behind the oral cavity (*Fig. 1–11*). The floor is formed by the posterior

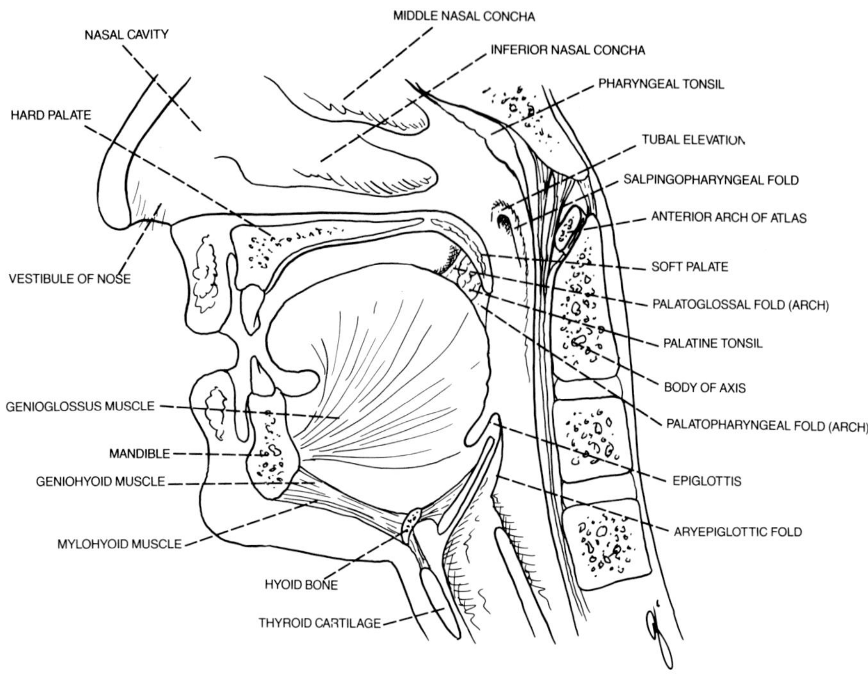

Figure 1–11. Sagittal section of the head and neck showing the relations of the nasal cavity, mouth, pharynx, and larynx.

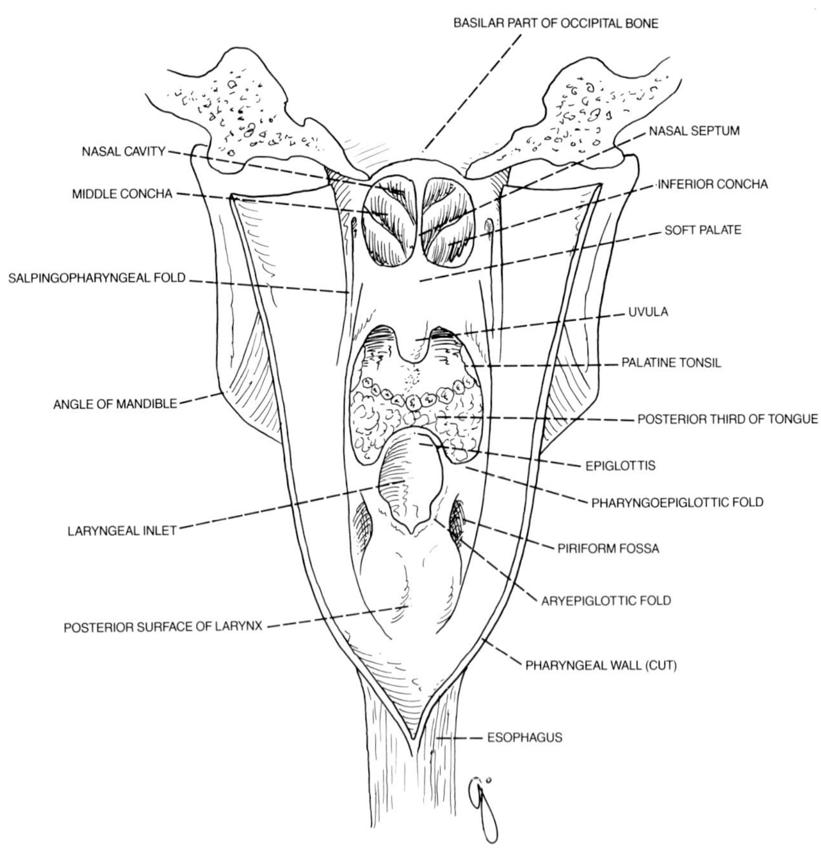

Figure 1–12. Interior of the pharynx as seen on posterior view. The posterior wall has been cut in the midline, and the right and left halves have been turned laterally.

TABLE 1–3. MUSCLES OF THE PHARYNX

Name of Muscle	Origin	Insertion	Nerve Supply	Action
Superior constrictor	Medial pterygoid plate, pterygoid hamulus, pterygomandibular ligament, mylohyoid line of mandible	Pharyngeal tubercle of occipital bone; pharyngeal raphe midline posteriorly	Pharyngeal plexus	Aids soft palate in closing off nasal pharynx; propels bolus downward
Middle constrictor	Lower part of stylohyoid ligament, lesser and greater cornu of hyoid bone	Pharyngeal raphe midline posteriorly	Pharyngeal plexus	Propels bolus downward
Inferior constrictor	Lamina of thyroid cartilage, cricoid cartilage	Pharyngeal raphe midline posteriorly	Pharyngeal plexus	Propels bolus downward
Cricopharyngeus (lowest fibers of inferior constrictor)				Sphincter at lower end of pharynx
Stylopharyngeus	Styloid process of temporal bone	Posterior border of thyroid cartilage	Glossopharyngeal nerve	Elevates larynx during swallowing
Salpingopharyngeus	Auditory tube	Blends with palatopharyngeus	Pharyngeal plexus	Elevates pharynx
Palatopharyngeus (see muscles of palate Table 1–2)				

one third of the tongue and the interval between the tongue and the epiglottis. In the midline is the **median glossoepiglottic fold,** and on each side the **lateral glossoepiglottic fold.** The depression on each side of the median glossoepiglottic fold is called the **vallecula** (*Fig. 1–18*).

On the lateral wall on each side are the palatoglossal and the palatopharyngeal arches or folds and the **palatine tonsils** between them (*Fig. 1–11*). The **palatoglossal arch** is a fold of mucous membrane covering the palatoglossus muscle. The interval between the two palatoglossal arches is called the **oropharyngeal isthmus** and marks the boundary between the mouth and the pharynx. The **palatopharyngeal arch** is a fold of mucous membrane covering the palatopharyngeus muscle. The **tonsillar sinus** is a recess between the palatoglossal and palatopharyngeal arches, which is occupied by the palatine tonsil.

Laryngeal Pharynx. The laryngeal pharynx lies behind the opening into the larynx (*Fig. 1–11*). The lateral wall is formed by the thyroid cartilage and the thyrohyoid membrane. The **piriform fossa** is a depression in the mucous membrane situated on each side of the laryngeal inlet. (*Fig. 1–12*). It is a common site for the lodging of fish bones.

Sensory Nerve Supply of the Mucous Membrane of the Pharynx. The nasal part of the pharynx is mainly supplied by the maxillary nerve (V_2); the oral part by the glossopharyngeal nerve; the mucous membrane around the entrance into the larynx is supplied by the internal laryngeal branch of the vagus nerve (*Fig. 1–13*).

Blood Supply of Pharynx. Ascending pharyngeal, tonsillar branches of facial, and branches of maxillary and lingual arteries.

Lymphatic Drainage of Pharynx. Directly into the deep cervical lymph nodes or indirectly via the retropharyngeal or paratracheal nodes into the deep cervical nodes.

Palatine Tonsils

The palatine tonsils are two masses of lymphoid tissue located in the lateral walls of the oral part of the pharynx between the palatoglossal and palatopharyngeal arches (*Fig. 1–11*). Each tonsil is covered by mucous membrane, and its free medial surface projects into the pharynx. The surface is pitted by numerous small openings, which lead into the **tonsillar crypts.** The tonsil reaches its maximum size during early childhood, but after puberty it diminishes considerably in size. The lymphatic drainage of the tonsil is into the upper deep cervical lymph nodes, just below and behind the angle of the mandible.

Waldeyer's Ring. The lymphoid tissue that surrounds the opening into the respiratory and digestive systems forms a ring. The lateral part of the ring is formed by the palatine tonsils and tubal tonsils (lymphoid tissue around the opening of the auditory tube). The pharyngeal tonsil in the roof of the nasopharynx forms the upper part and the lingual tonsil on the posterior third of the tongue forms the lower part.

Swallowing (Deglutition). The masticated food is formed into a ball or bolus on the dorsum of the tongue

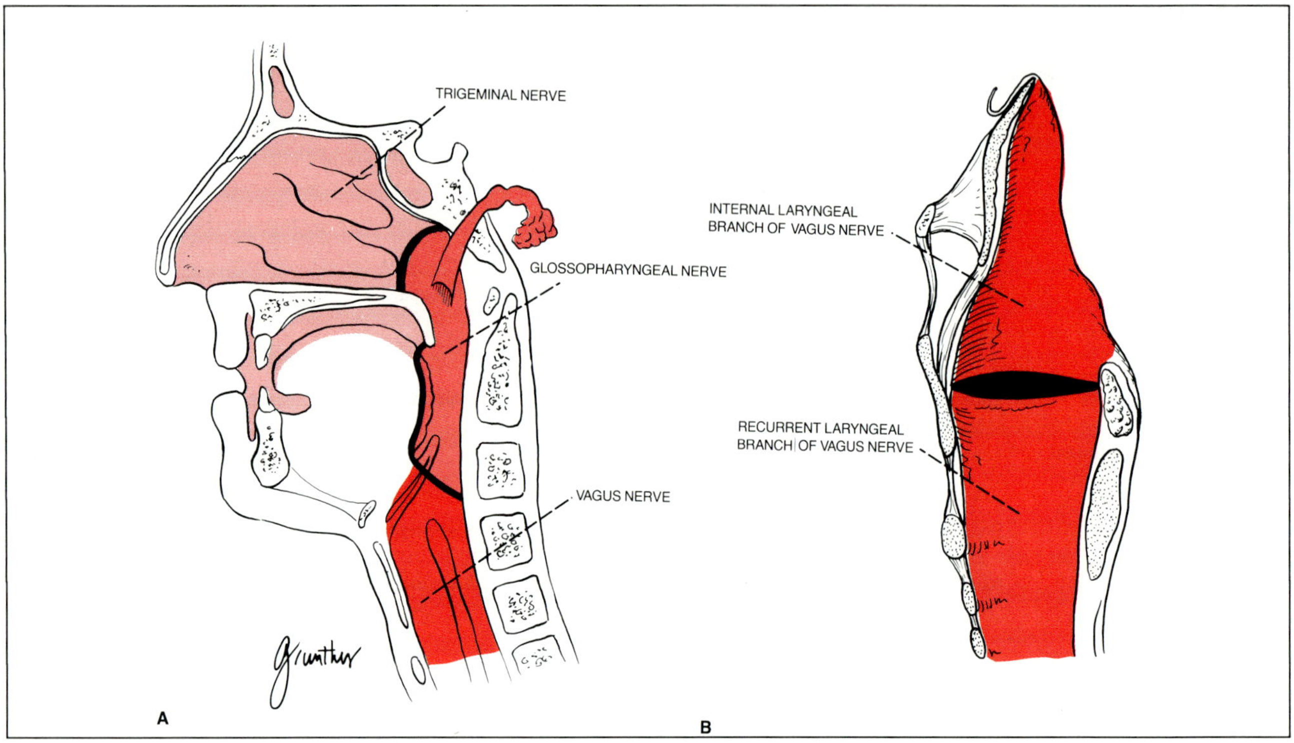

Figure 1–13. Sagittal section of the head and neck showing the sensory innervation of the mucous membrane of the different parts of the pharynx and larynx.

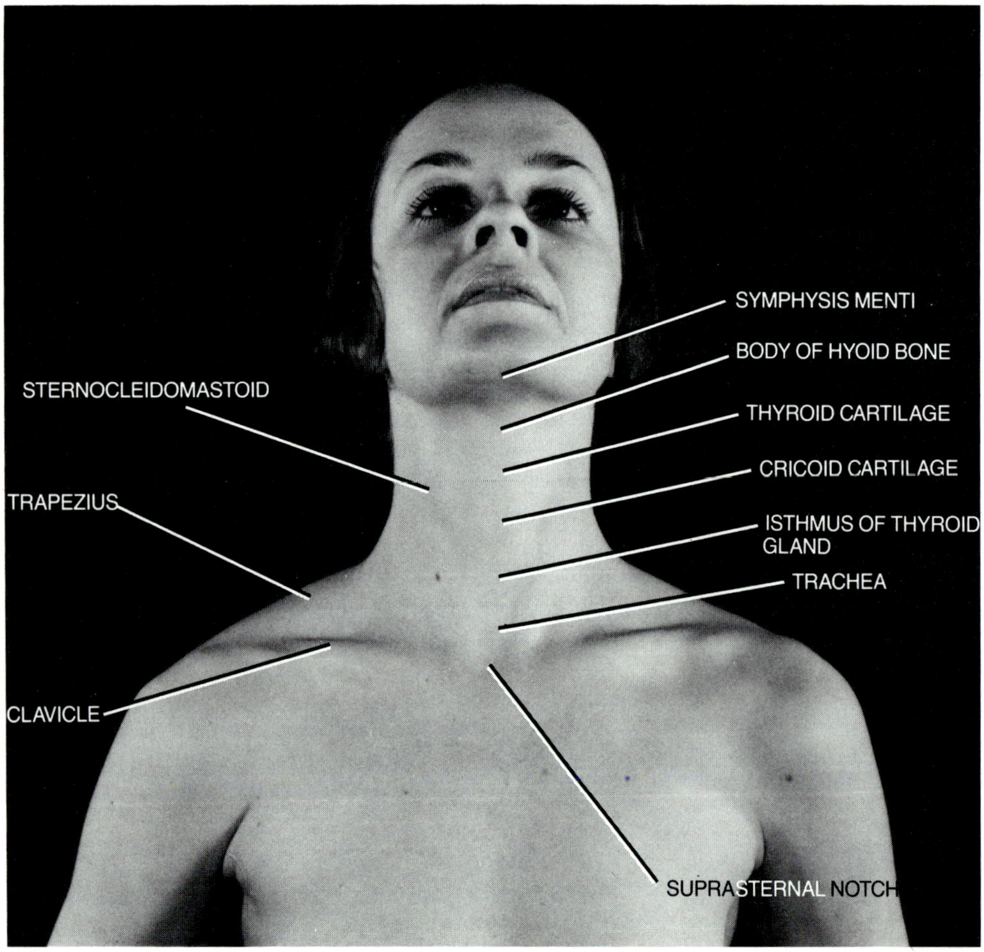

Figure 1–14. Anterior view of the head and the neck of a 29-year-old woman, showing important surface landmarks.

and voluntarily pushed upward and backward against the undersurface of the hard palate. This is brought about by the contraction of the styloglossus muscles on both sides, which pull the root of the tongue upward and backward. The palatoglossus muscles squeeze the bolus backward into the pharynx. From this point onward the process of swallowing becomes an involuntary act.

The nasal part of the pharynx is shut off from the oral part of the pharynx by the elevation of the soft palate, the pulling forward of the posterior wall of the pharynx by the palatopharyngeal sphincter, and the pulling medially of the palatopharyngeal arches. This prevents the passage of food and drink into the nasal cavities.

The larynx and the laryngeal part of the pharynx are pulled upward by the contraction of the stylopharyngeus, salpingopharyngeus, thyrohyoid, and palatopharyngeus muscles. The main part of the larynx is thus elevated to the posterior surface of the epiglottis, and the entrance into the larynx is closed. The laryngeal entrance is made smaller by the approximation of the aryepiglottic folds and the arytenoid cartilages are pulled forward by the contraction of the aryepiglottic, oblique arytenoid, and thyroarytenoid muscles.

The bolus moves downward over the epiglottis, the closed entrance into the larynx, and reaches the lower part of the pharynx as the result of the successive contraction of the superior, middle, and inferior constrictor muscles. Some of the food slides down the groove on either side of the entrance into the larynx, i.e., down through the **piriform fossae.** Finally, the lower part of the pharyngeal wall (the cricopharyngeus muscle) relaxes and the bolus enters the esophagus.

CLINICAL NOTES

Lymphoid Tissue of the Pharynx

Enlargement of the lymphoid tissue of the pharynx may seriously encroach on the airway, especially in children. The pharyngeal tonsil and the palatine tonsils reach their maximum normal size in early childhood. After puberty, together with other lymphoid tissues in the body, they gradually atrophy.

Mechanism of Swallowing

In the conscious individual, the crossing over of the food and fluid from the mouth to the esophagus, and the air from the nose to the larynx, is taken care of efficiently by the mechanism of swallowing, in which the act of respiration is momentarily suspended. In the unconscious individual, when the reflex mechanisms are not functioning, it is possible for food and fluid to enter the bronchial tree or air to enter the stomach. Moreover, should the unconscious individual vomit, the regurgitated gastric contents may be inhaled into the lungs.

Surface Anatomy of the Airway in the Neck

Preferably the neck should be palpated while the patient is supine, when the muscles overlying the deeper structures are relaxed and the structures become easier to feel.

Anteriorly, in the midline, the following structures may be palpated from above downward:

The **symphysis menti** may be felt where the two halves of the body of the mandible unite in the midline (*Fig. 1–14*).

The **body of the hyoid bone** lies opposite the body of the third cervical vertebra (*Figs. 1–14 and 1–15*). The hyoid bone is a horseshoe-shaped structure and the greater cornua can be felt on each side of the neck between the finger and the thumb. The hyoid bone moves superiorly when the patient swallows.

The **thyrohyoid membrane** occupies the interval between the hyoid bone and the thyroid cartilage (*Fig. 1–15*).

The notched **upper border of the thyroid** cartilage lies opposite the fourth cervical vertebra (*Figs. 1–14 and 1–15*). The anterior border of the thyroid cartilage is more prominent in the male adult than in the female adult.

The **cricothyroid ligament** fills in the interval between the cricoid cartilage and the thyroid cartilage (*Fig. 1–15*).

The **cricoid cartilage** lies at the level of the sixth cervical vertebra at the junction of the larynx with the trachea (*Figs. 1–14 and 1–15*). It is not as easy to feel as the thyroid cartilage, but it can be identified with gentle palpation when the patient is asked to swallow and the cartilage rises in the neck.

The **cricotracheal ligament** fills in the interval between the cricoid cartilage and the first ring of the trachea (*Fig. 1–15*). It can be recognized with gentle palpation.

The **first ring of the trachea** can usually be identified.

The **isthmus of the thyroid gland** can be recognized as a soft structure crossing in front of the second, third, and fourth rings of the trachea (*Figs. 1–14 and 1–15*).

The **inferior thyroid veins** and the **thyroidea ima artery,** (when present) although not palpable, lie in front of the fifth, sixth, and seventh rings of the trachea. It should be noted that the trachea is now receding from the surface as it approaches the root of the neck.

The **brachiocephalic artery,** the **left brachiocephalic vein,** the **thymus gland,** and even the upper margin of the **arch of the aorta,** are sometimes present in front of the trachea just above the suprasternal notch in the young child.

The **jugular arch** connects the two anterior jugular veins just above the suprasternal notch.

The **suprasternal notch,** which is the upper margin of the manubrium sterni can be felt between the anterior ends of the clavicles (*Fig. 1–15*). It lies opposite the

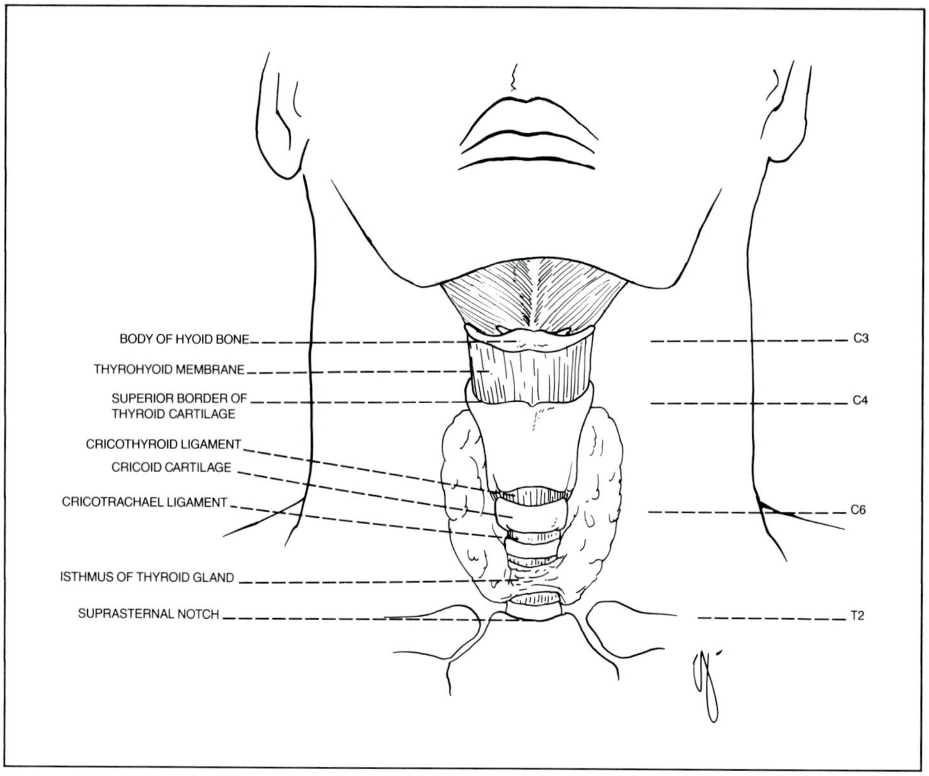

Figure 1–15. Head and neck of an adult showing the vertebral levels of different parts of the larynx.

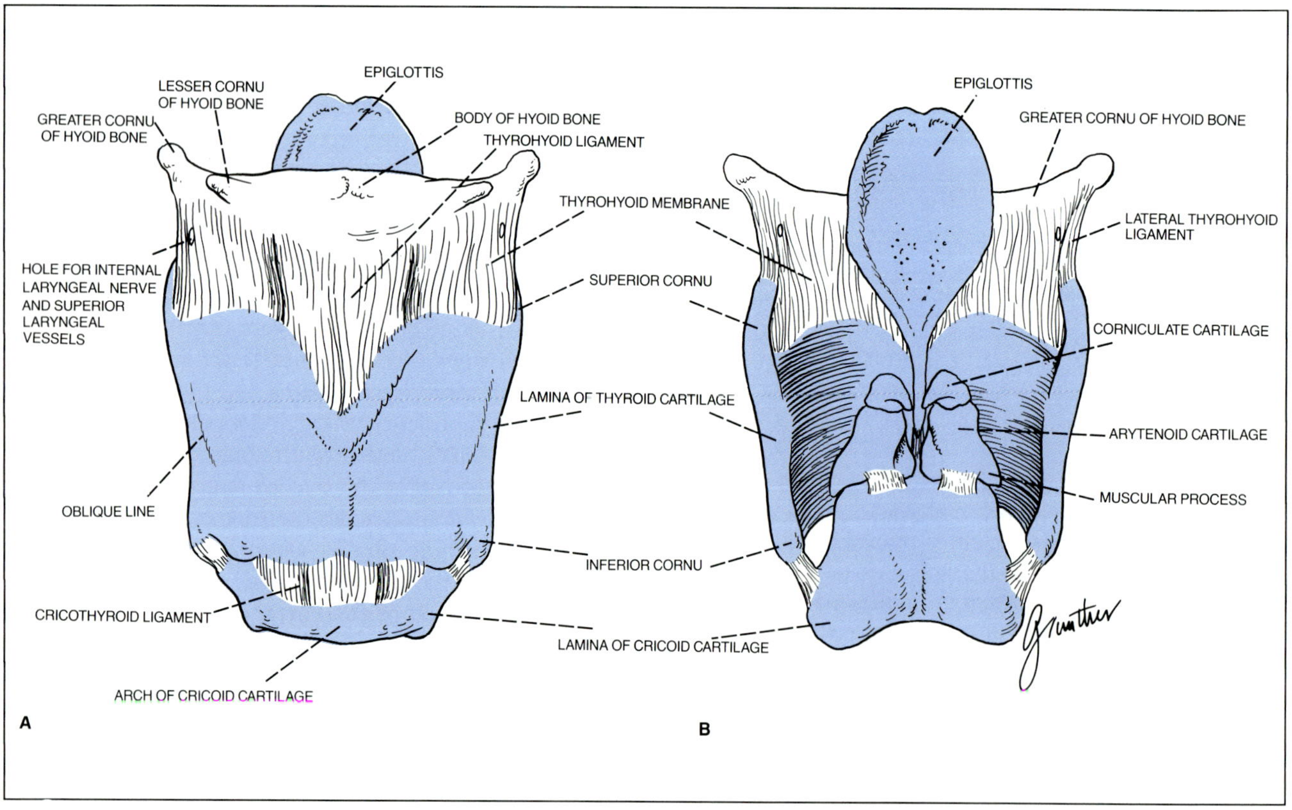

Figure 1–16. Larynx and its ligaments. **A.** Anterior view. **B.** Posterior view.

lower border of the body of the second thoracic vertebra in the midrespiratory position.

The trachea in the neck lies in the midline and it should be possible for the examiner to confirm this by inserting the index and middle finger into the grooves on either side of the trachea, between the trachea and the sternocleidomastoid muscles. It should be remembered that in the adult the trachea may measure as much as 1 inch (2.5 cm) in diameter, whereas in a baby it is much narrower (8 mm).

LARYNX

The larynx is an organ that provides a protective sphincter at the inlet of the air passages and is responsible for voice production. It is situated below the tongue and the hyoid bone, between the great blood vessels of the neck, and lies in front of the fourth, fifth, and sixth cervical vertebrae (*Fig. 1–15*). It opens above into the laryngeal part of the pharynx, and below, is continuous with the trachea. The larynx is covered in front by the infrahyoid strap muscles and at the sides by the thyroid gland.

The framework of the larynx is formed of cartilages that are held together by ligaments and membranes, moved by muscles, and lined by mucous membrane.

Cartilages of Larynx (*Fig. 1–16*)

Thyroid Cartilage. The thyroid cartilage is the largest cartilage of the larynx (*Fig. 1–16*). It consists of two laminae of hyaline cartilage that meet in the midline in the prominent V angle (the so-called Adam's apple). The posterior border extends upward into a **superior cornu** and downward into an **inferior cornu.** On the outer surface of each lamina is an **oblique line** for the attachment of the sternothyroid, the thyrohoid, and the inferior constrictor muscles.

Cricoid Cartilage. The cricoid cartilage is formed of hyaline cartilage and shaped like a signet ring having a broad lamina behind and a shallow arch in front (*Figs. 1–16 and 1–17*). The cricoid cartilage lies below the thyroid cartilage and on each side of the lateral surface there is a facet for articulation with the inferior cornu of the thyroid cartilage. Posteriorly, the lamina has on its upper border on each side a facet for articulation with the arytenoid cartilage. All these joints are synovial in nature.

Arytenoid Cartilages. The arytenoid cartilages are two in number, small, and pyramidal-shaped (*Fig. 1–17*). They are situated at the back of the larynx, on the lateral part of the upper border of the lamina of the cricoid cartilage. Each cartilage has an **apex** above that articulates with the small corniculate cartilage, a **base** below that articulates with the lamina of the cricoid cartilage, and a **vocal process** that projects forward and gives attachment to the vocal ligament. There is a **muscular process** that projects laterally and gives attachment to the posterior and lateral cricoarytenoid muscles.

Corniculate Cartilages. The corniculate cartilages are two small conical-shaped cartilages that articulate with the arytenoid cartilages (*Fig. 1–17*). They give attachment to the aryepiglottic folds.

Cuneiform Cartilages. The cuneiform cartilages are two small rod-shaped cartilages that are found in and serve to strengthen the aryepiglottic folds (*Fig. 1–17*).

Epiglottis. The epiglottis is a leaf-shaped lamina of elastic cartilage lying behind the root of the tongue (*Fig. 1–17*). It is attached by its stalk to the back of the thyroid cartilage. The sides of the epiglottis are attached to the arytenoid cartilages by the aryepiglottic folds of the mucous membrane. The upper edge of the epiglottis is free. The covering of mucous membrane passes forward on to the posterior surface of the tongue as the **median glossoepiglottic fold;** the depression on each side of the fold is called the **vallecula** (*Fig. 1–18*). Laterally, the mucous membrane passes onto the wall of the pharynx as the **lateral glossoepiglottic fold.**

Membranes and Ligaments of Larynx

Thyrohyoid Membrane. The thyrohoid membrane connects the upper margin of the thyroid cartilage to the upper margin of the posterior surface of the hyoid bone (*Fig. 1–16*). In the midline it is thickened to form the **median thyrohyoid ligament.** The membrane is pierced on each side by the superior laryngeal vessels and the internal laryngeal nerve, a branch of the superior laryngeal nerve (*Fig. 1–22*).

Cricotracheal Ligament. The cricotracheal ligament connects the cricoid cartilage to the first ring of the trachea (*Fig. 1–17*).

Fibroelastic Membrane. The fibroelastic membrane lies beneath the mucous membrane lining the larynx. The upper part is thin and called the **quadrangular membrane** (*Fig. 1–17*). It extends between the epiglottis and the arytenoid cartilages, its thickened inferior margin forming the **vestibular ligaments.** The vestibular ligaments form the interior of the vestibular folds (*Fig. 1–17*). The lower part of the fibroelastic membrane is called the **cricothyroid ligament.** The anterior part of the cricothyroid ligament is thick and connects the cricoid cartilage to the lower margin of the thyroid cartilage. The lateral part of the ligament is thin and is attached below to the upper margin of the cricoid cartilage. The superior margin of the ligament, instead of being attached to

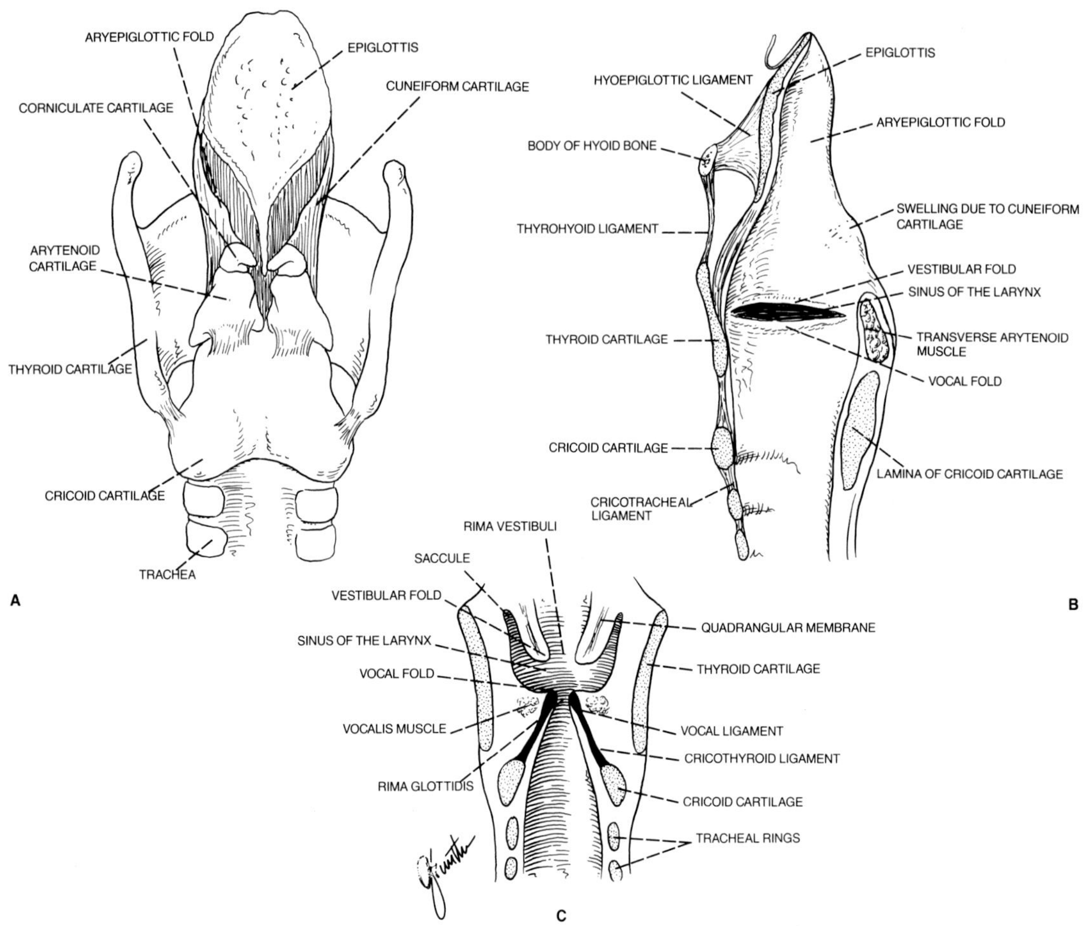

Figure 1–17. Larynx. **A.** Posterior view showing aryepiglottic fold in position. **B.** Sagittal section through larynx showing vestibular fold, sinus, and vocal fold. **C.** Coronal section through larynx showing vocal ligament and cricothyroid ligament.

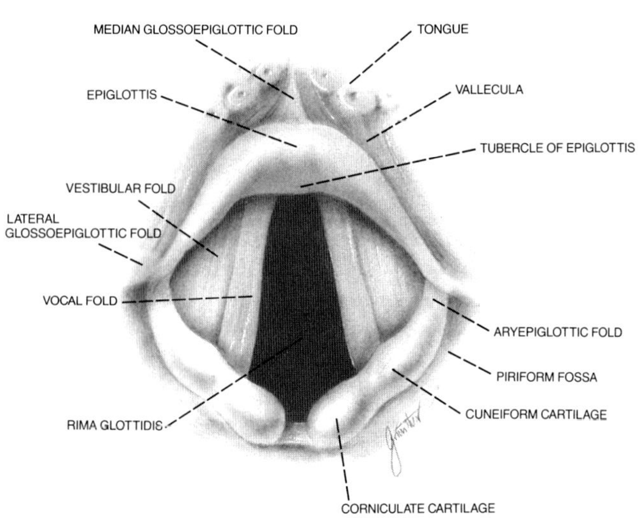

Figure 1–18. The laryngeal inlet as seen from above.

the lower margin of the thyroid cartilage, ascends on the medial surface of the thyroid cartilage. Its upper margin, composed almost entirely of elastic tissue, forms the important **vocal ligament** on each side. The vocal ligaments form the interior of the **vocal folds** (vocal cords) (*Fig. 1–17*). The anterior end of each vocal ligament is attached to the thyroid cartilage and the posterior end to the vocal process of the arytenoid cartilage.

Cavity of the Larynx

The **inlet** of the larynx looks backward and upward into the laryngeal part of the pharynx. The opening is wider in front than behind and is bounded in front by the epiglottis, laterally by the aryepiglottic fold of mucous membrane, and posteriorly by the arytenoid cartilages with the corniculate cartilages. The cuneiform cartilage lies within and strengthens the aryepiglottic fold and produces a small elevation on the upper border. The **piriform fossa** is a recess situated on either side of the fold and inlet (*Fig. 1–18*). It is bounded medially by the aryepiglottic fold and laterally by the thyroid cartilage and the thyrohyoid membrane. Beneath the mucosa lies the terminal branches of the internal laryngeal nerve, which may be blocked by the application of local analgesic solutions to the piriform fossa (*Fig. 1–22*).

The cavity of the larynx extends from the inlet to the lower border of the cricoid cartilage where it is continuous with the cavity of the trachea. It is divided into three regions by the presence of an upper and a lower pair of folds of mucous membrane that project into the cavity from the sides of the larynx. The upper pair of folds are fixed and called the **vestibular folds** and the gap between them is named the **rima vestibuli** (*Fig. 1–17*). The lower pair are mobile, concerned with voice production, and are called the **vocal folds;** the gap between them is named the **rima glottidis** or **glottis.** The glottis is bounded in front by the vocal folds and behind by the medial surfaces of the arytenoid cartilages. The glottis is the narrowest part of the larynx and measures from front to back about 2.5 cm in the male adult and less in the female. In children, the lower part of the larynx within the cricoid cartilage is the narrowest part.

The three regions of the cavity of the larynx are (1) the upper region called the **vestibule,** situated between the inlet and the vestibular folds, its anterior wall is formed by the epiglottis, the lower part of which projects backward as the **tubercule;** (2) the middle region, situated between the vestibular and vocal folds; and (3) the lower region, situated between the vocal folds and the lower border of the cricoid cartilage. The **vestibular fold** on each side is formed of mucous membrane and contains within it the vestibular ligament, which is the thickened lower edge of the quadrangular membrane. The fold is fixed (i.e., does not move with respiration) and is pink in color when viewed with a laryngoscope (*Fig. 1–18*).

The **vocal fold** on each side is formed of mucous membrane and contains within it the vocal ligament, which is the thickened upper edge of the cricothyroid ligament. The fold is mobile, moves with respiration, and is white in color when viewed with a laryngoscope (*Fig. 1–18*).

The **sinus of the larynx** is a small recess on each side of the larynx situated between the vestibular and vocal folds. It is lined with mucous membrane (*Fig. 1–17*).

The **saccule of the larynx** is a diverticulum of mucous membrane that ascends from the sinus of the larynx between the vestibular fold and the thyroid cartilage (*Fig. 1–17*). The submucosa contains large numbers of mucous glands that open onto the surface of the mucous membrane.

Muscles of the Larynx

The muscles of the larynx may be divided into two groups: (1) extrinsic and (2) intrinsic.

Extrinsic Muscles. The extrinsic muscles move the larynx up and down during the act of swallowing. Note that many of these muscles are attached to the hyoid bone and the hyoid bone is attached to the thyroid cartilage by the thyrohyoid membrane. It follows that movements of the hyoid bone are accompanied by movements of the larynx.

The **elevator muscles** of the larynx include the digastric, the stylohyoid, the mylohyoid, the geniohyoid, the stylopharyngeus, the salpinopharyngeus, and the palatopharyngeus muscles. The **depressor muscles** of the larynx include the sternothyroid, the sternohyoid, and the omohyoid muscles. The details of the origins, insertions, nerve supply, and action are given in Table 1–4.

Intrinsic Muscles. The intrinsic muscles modify the inlet of the larynx and move the vocal folds (*Figs. 1–19 and 1–20*). The details of the origins, insertions, nerve supply, and action are given in Table 1–5.

Movements of the Vocal Folds

The movements of the vocal folds are dependent on the movements of the arytenoid cartilages, which rotate and slide up and down on the sloping shoulders of the superior border of the cricoid cartilage.

The rima glottidis is opened by contraction of the posterior cricoarytenoid, which rotates the arytenoid cartilage and adducts the vocal process (*Fig. 1–21*). The elastic tissue in the capsules of the cricoarytenoid joints keeps the arytenoid cartilages apart so that the posterior part of the glottis is open.

The rima glottidis is closed by contraction of the lateral cricoarytenoid, which rotates the arytenoid cartilage and adducts the vocal process (*Fig. 1–21*). The posterior part of the glottis is narrowed by drawing the aryte-

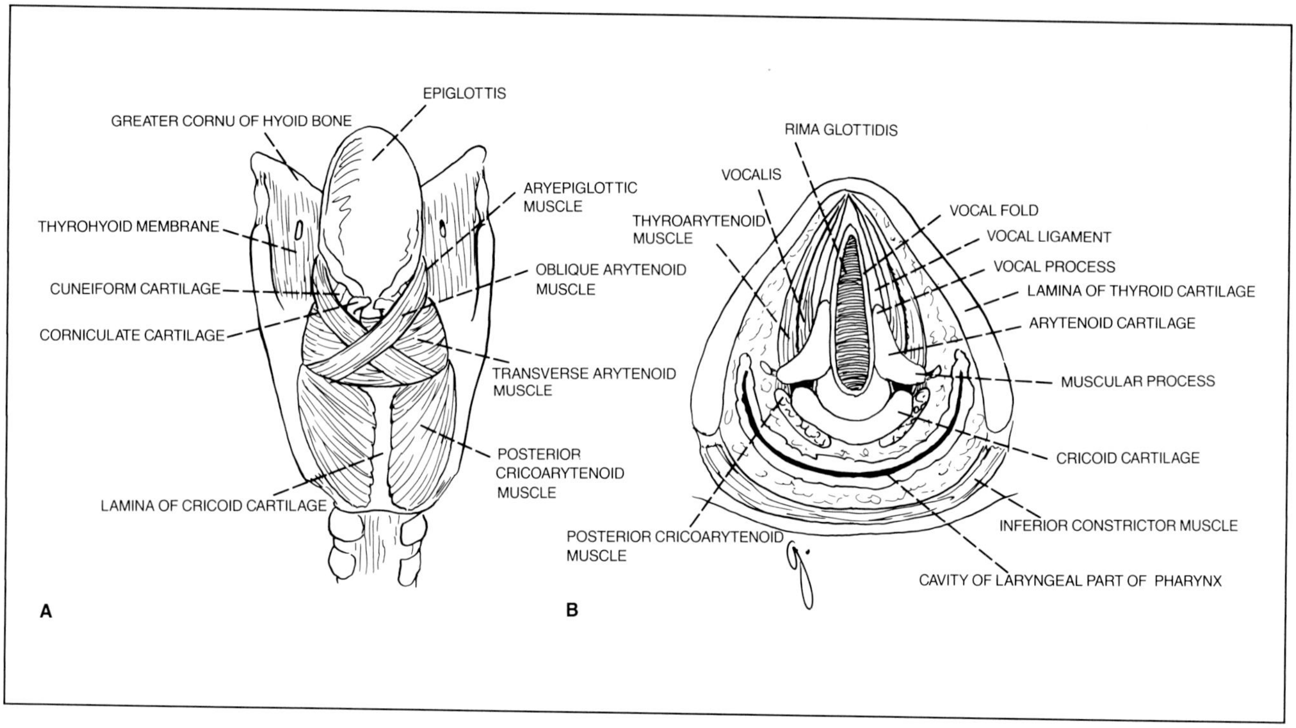

Figure 1–19. A. Intrinsic muscles of the larynx as seen on posterior view. **B.** Horizontal section through the larynx at the level of the rima glottidis.

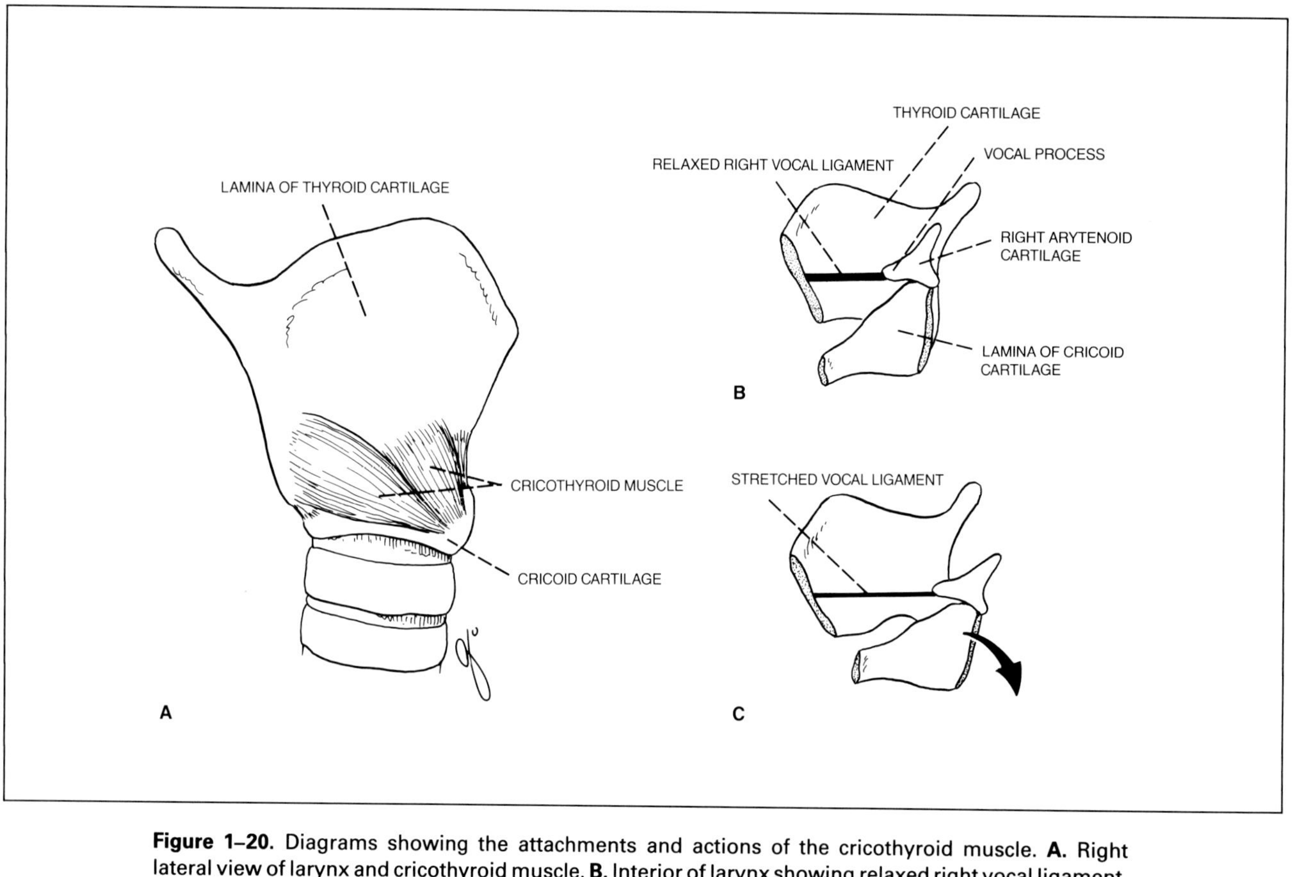

Figure 1–20. Diagrams showing the attachments and actions of the cricothyroid muscle. **A.** Right lateral view of larynx and cricothyroid muscle. **B.** Interior of larynx showing relaxed right vocal ligament. **C.** Interior of larynx showing right vocal ligament stretched as the result of the tilting backward of the cricoid and arytenoid cartilages by the contraction of the cricothyroid muscle.

TABLE 1–4. EXTRINSIC MUSCLES OF THE LARYNX

Name of Muscle	Origin	Insertion	Nerve Supply	Action
Elevator Muscles				
Digastric				
Posterior belly	Mastoid process of temporal bone	Intermediate tendon held to hyoid bone by fascial sling	Facial nerve	Depresses mandible or elevates hyoid bone
Anterior belly	Body of mandible		Nerve to mylohyoid	
Stylohyoid	Styloid process	Body of hyoid bone	Facial nerve	Elevates hyoid bone
Mylohyoid	Mylohyoid line of body of mandible	Body of hyoid bone and fibrous raphe	Inferior alveolar nerve	Elevates floor of mouth and hyoid bone or depresses mandible
Geniohyoid	Inferior mental spine of mandible	Body of hyoid bone	First cervical nerve	Elevates hyoid bone or depresses mandible
Stylopharyngeus	Styloid process	Posterior border of thyroid cartilage	Glossopharyngeal nerve	Elevates larynx and pharynx
Salpinogopharyngeus	Cartilaginous part of auditory tube	With palatopharyngeus, posterior border of thyroid cartilage	Pharyngeal plexus	Elevates larynx and pharynx; dilates auditory tube
Palatopharyngeus	Hard palate and pharyngeal aponeurosis	Posterior border of thyroid cartilage	Pharyngeal plexus	Elevates larynx and pharynx; pulls palatopharyngeal arch medially
Depressor Muscles				
Sternothyroid	Manubrium sterni	Oblique line lamina of thyroid cartilage	Ansa cervicalis; C1	Depresses larynx
Sternohyoid	Manubrium sterni and clavicle	Body of hyoid bone	Ansa cervicalis; C1, C2, and C3	Depresses hyoid bone
Omohyoid				
Inferior belly	Upper margin of scapula and suprascapular ligament	Held to clavicle and first rib by fascial sling		
Superior belly	Lower border body of hyoid bone		Ansa cervicalis; C1, C2, and C3	Depresses hyoid bone
Thyrohyoid	Oblique line lamina of thyroid cartilage	Lower border of body of hyoid bone	First cervical nerve	Depresses hyoid bone or elevates larynx

noid cartilages together by contracting the transverse arytenoid muscles.

The vocal folds are stretched by contraction of the cricothyroid muscle (*Fig. 1–20*). The vocal folds are slackened by contraction of the vocalis, a part of the thyroarytenoid muscle (*Fig. 1–19*).

Movements of the Vocal Folds with Respiration

On quiet inspiration, the vocal folds are abducted and the rima glottidis is triangular in shape with the apex in front (*Fig. 1–21*). On expiration, the vocal folds are adducted, leaving a small gap between them (*Fig. 1–21*).

On deep inspiration, the vocal folds are maximally abducted and the triangular shape of the glottis is converted into a diamond shape due to the maximal lateral rotation of the arytenoid cartilages (*Fig. 1–21*).

Sphincteric Function of the Larynx

There are two sphincters in the larynx: (1) at the inlet and (2) at the rima glottidis.

The sphincter at the inlet is used only during swallowing. As the bolus of food is passed backward between the tongue and the hard palate, the larynx is pulled up beneath the back of the tongue. The inlet of the larynx is narrowed by the action of the oblique arytenoid and aryepiglottic muscles. The epiglottis is pushed backward by the tongue and serves as a cap over the laryngeal inlet. The bolus of food, or fluids, now enters the esophagus by passing over the epiglottis or moving down the

TABLE 1–5. INTRINSIC MUSCLES OF THE LARYNX

Name of Muscle	Origin	Insertion	Nerve Supply	Action
Muscles Controlling Laryngeal Inlet				
Oblique arytenoid	Muscular process of arytenoid cartilage	Apex of opposite arytenoid cartilage	Recurrent laryngeal nerve	Brings aryepiglottic folds together and narrow inlet
Aryepiglottic	Part of oblique arytenoid, apex of arytenoid cartilage	Epiglottis	Recurrent laryngeal nerve	Narrows inlet
Thyroepiglottic	Medial surface of thyroid cartilage	Epiglottis and aryepiglottic fold	Recurrent laryngeal nerve	Widens inlet by pulling aryepiglottic folds apart
Muscles Controlling Movements of Vocal Folds				
Cricothyroid	Side of cricoid cartilage	Lower border of lamina and inferior cornu of thyroid cartilage	External laryngeal nerve	Tilts cricoid and arytenoid posteriorly; tenses vocal folds
Thyroarytenoid	Inner surface of angle of lamina of thyroid cartilage	Anterolateral surface of arytenoid cartilage	Recurrent laryngeal nerve	Pulls forward arytenoid and relaxes vocal folds
Vocalis	Part of thyroarytenoid muscle that lies on lateral side of vocal ligament			
Lateral cricoarytenoid	Upper border of arch of cricoid cartilage	Muscular process of arytenoid cartilage	Recurrent laryngeal nerve	Adducts vocal ligament
Posterior cricoarytenoid	Back of cricoid cartilage	Muscular process of arytenoid cartilage	Recurrent laryngeal nerve	Abducts vocal ligament
Transverse arytenoid	Back and medial surface of arytenoid cartilage	Back and medial surface of opposite arytenoid cartilage	Recurrent laryngeal nerve	Closes posterior part of rima glottidis

grooves on either side of the laryngeal inlet, the piriform fossae.

In coughing or sneezing, the rima glottidis serves as a sphincter. After inspiring, the vocal folds are adducted, and the muscles of expiration are made to contract strongly. As a result, the intrathoracic pressure rises, whereupon the vocal folds are suddenly abducted. The sudden release of the compressed air will often dislodge foreign particles or mucus from the respiratory tract and carry the material up into the pharynx. Here, they are either swallowed or expectorated.

In abdominal straining associated with micturition, defecation, and parturition, the air is often held temporarily in the respiratory tract by closing the rima glottidis. After deep inspiration the rima glottidis is closed. The muscles of the anterior abdominal wall now contract, and the upward movement of the diaphragm is prevented by the presence of compressed air within the respiratory tract. After a prolonged effort the person often releases some of the air by momentarily opening the rima glottidis, producing a grunting sound.

Voice Production in the Larynx

The intermittent release of expired air between the adducted vocal folds results in their vibration and in the production of sound. The **frequency, or pitch,** of the voice is determined by changes in the length and tension of the vocal ligaments. The **quality** of the voice depends on the resonators above the larynx, namely, the pharynx, the mouth, and the paranasal sinuses. The quality is controlled by the muscles of the soft palate, tongue, floor of the mouth, cheeks, lips, and jaws. Normal speech depends on the modification of the sound into recognizable consonants and vowels by the use of the tongue, the teeth, and the lips. Vowels sounds are usually purely oral with the soft palate raised; that is to say, the air is channeled through the mouth rather than the nose.

Speech involves the intermittent release of expired air between the adducted vocal folds. **Singing** a note requires a more prolonged release of the expired air between the adducted vocal folds. In **whispering,** the vocal folds are adducted, but the arytenoid cartilages are separated; the vibrations are given to a constant stream of

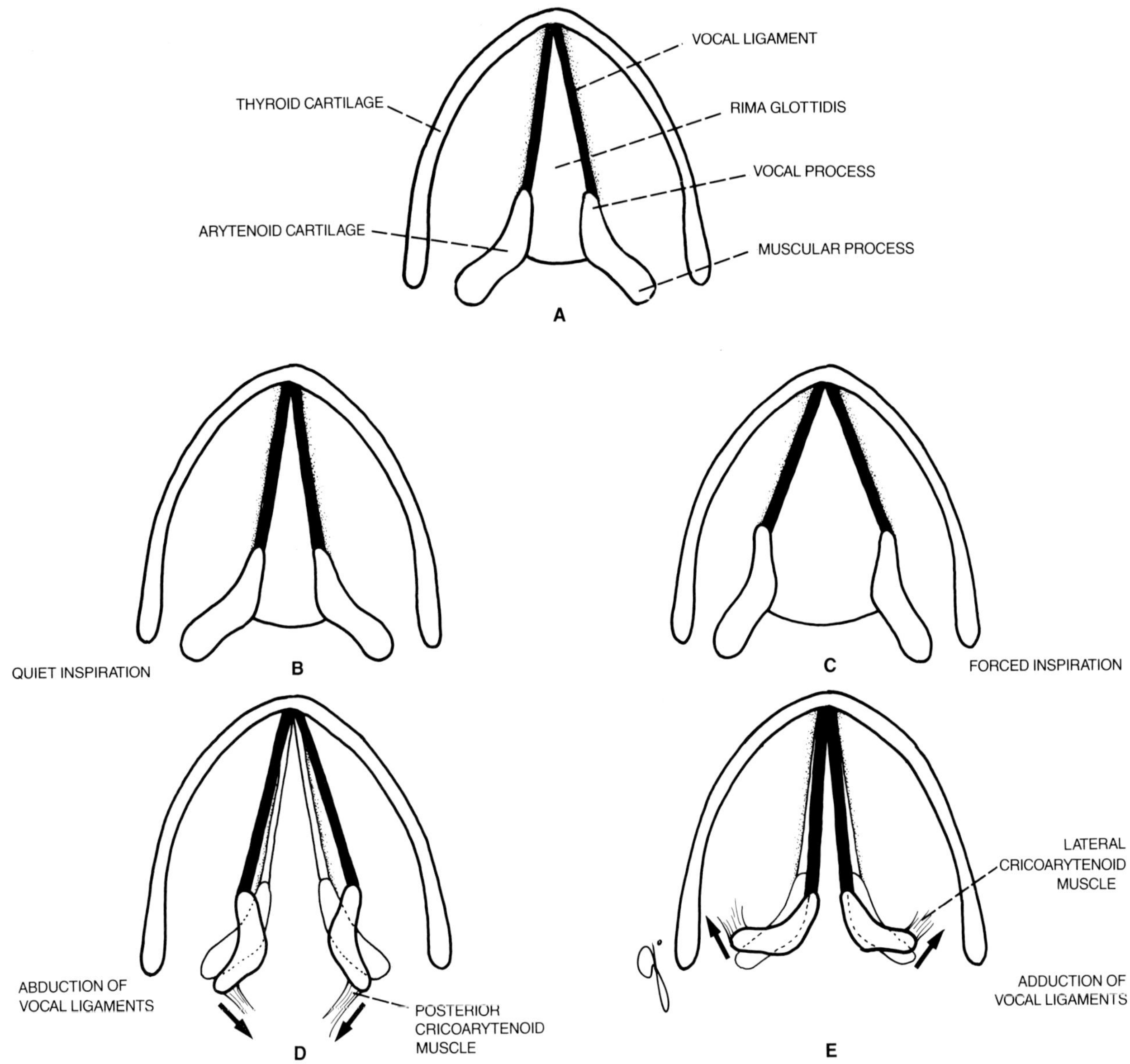

Figure 1–21. Movements of the vocal ligaments and the arytenoid cartilages. **A.** The attachments of the vocal ligaments. **B.** The position of the vocal ligaments in quiet inspiration; note that the anterior part of the rima glottidis is triangular in shape and the posterior part between the arytenoid cartilages is rectangular in shape. **C.** The position of the vocal ligaments in forced inspiration; the rima glottidis is now diamond-shaped. **D.** Abduction of the vocal ligaments caused by contraction of the posterior cricoarytenoid muscles. **E.** Adduction of the vocal ligaments caused by the contraction of the lateral cricoarytenoid muscles.

expired air that passes through the posterior part of the rima glottidis.

Mucous Membrane of the Larynx. The mucous membrane of the larynx lines the cavity and is covered with ciliated columnar epithelium. On the vocal folds, however, where the mucous membrane is subject to repeated trauma during phonation, the mucous membrane is covered with stratified squamous epithelium.

Nerve Supply of the Larynx. The **sensory nerve supply** to the mucous membrane above the vocal folds is from the internal laryngeal nerve. Below the level of the vocal folds the mucous membrane is supplied by the recurrent laryngeal nerve (*Fig. 1–22*).

The **motor nerve supply** to the intrinsic muscles is the recurrent laryngeal nerve, except for the cricothyroid muscle, which is supplied by the external laryngeal nerve.

Blood Supply and Lymphatic Drainage of the Larynx. The **arterial supply of the larynx** is from branches of the superior and inferior thyroid arteries. The **lymphatic drainage** is into the deep cervical lymph nodes.

CLINICAL NOTES

Larynx in Children

In children, the neck is shorter and the larynx is more cephalad than in adults (*Fig. 1–23*). At birth, the cricoid cartilage lies at the level of the fourth cervical vertebra and only at the age of six does it lie opposite the sixth cervical vertebra. The glottis at birth lies opposite the second cervical vertebra.

The epiglottis is U-shaped and less flexible in the child, which sometimes makes it difficult to line up the oral, the pharyngeal, and the tracheal axes when passing a laryngoscope.

The rima glottidis tends to be more anterior in the child compared with the adult. The vocal folds in the child have a thicker submucosa than in an adult, therefore edema of the folds is more likely to occlude the glottis in a child than in the adult.

As mentioned previously, the cavity of the larynx is narrowest within the cricoid ring in a child, whereas the glottis is the narrowest part of the cavity in an adult.

Paralysis of the Vocal Folds

The muscles of the larynx are innervated by the recurrent laryngeal nerves, with the exception of the cricothyroid muscle, which is supplied by the external laryngeal nerve. Both these nerves are vulnerable during operations on the thyroid gland because of the close relationship between them and the arteries of the gland. The external laryngeal nerve may be injured during ligation of the superior thyroid artery and the recurrent laryngeal nerve may be damaged during ligation of the inferior thyroid artery. The left recurrent laryngeal nerve may be involved in a bronchial or esophageal carcinoma or in secondary metastatic deposits in the mediastinal lymph nodes. The right and left recurrent laryngeal nerves may be damaged by malignant involvement of the deep cervical lymph nodes.

Section of the external laryngeal nerve produces weakness and huskiness of the voice, as the vocal fold cannot be tensed (*Fig. 1–24*). The cricothyroid muscle is paralyzed.

Unilateral complete section of the recurrent laryngeal nerve results in the vocal fold on the affected side assuming the position midway between abduction and adduction (*Fig. 1–24*). It lies just lateral to the midline. Speech is not greatly affected, as the other vocal fold compensates to some extent and moves toward the affected vocal fold.

Bilateral complete section of the recurrent laryngeal nerve results in both vocal folds assuming the position midway between abduction and adduction (*Fig. 1–24*). Breathing is impaired, as the rima glottidis is partially closed and speech is lost.

Unilateral partial section of the recurrent laryngeal nerve results in a greater degree of paralysis of the abductor muscles than of the adductor muscles. The affected vocal fold assumes the adducted midline position (*Fig. 1–24*). This phenomenon has not been explained satisfactorily. It must be assumed that the abductor muscles receive a greater number of nerves than the adductor muscles, and thus partial damage of the recurrent laryngeal nerve results in damage to relatively more nerve fibers to the abductor muscles. Another possibility is that the nerve fibers to the abductor muscles are traveling in a more exposed position in the recurrent laryngeal nerve and are, therefore, more prone to be damaged.

Bilateral partial section of the recurrent laryngeal nerve results in bilateral paralysis of the abductor muscles and the drawing together of the vocal folds (*Fig. 1–24*). Acute dyspnea and stridor follow, and tracheotomy may be necessary.

The interior of the larynx may be inspected indirectly with a laryngeal mirror passed through the open mouth into the oral pharynx. A more satisfactory method is the direct method using the laryngoscope. The neck is brought forward on a pillow and the head is fully extended at the atlanto-occipital joint. The illuminated instrument can then be introduced into the larynx over the back of the tongue. The valleculae, the piriform fossae, the epiglottis, and the aryepiglottic folds are clearly seen. The two elevations produced by the corniculate and cuneiform cartilages can be recognized. Within the larynx, the vestibular folds and the vocal folds can be seen. The former are fixed, widely separated, and reddish in color; the latter move with respiration and are white in color. With quiet breathing, the rima glottidis is triangular in shape, with the apex in front. With deep inspiration, the rima glottidis assumes

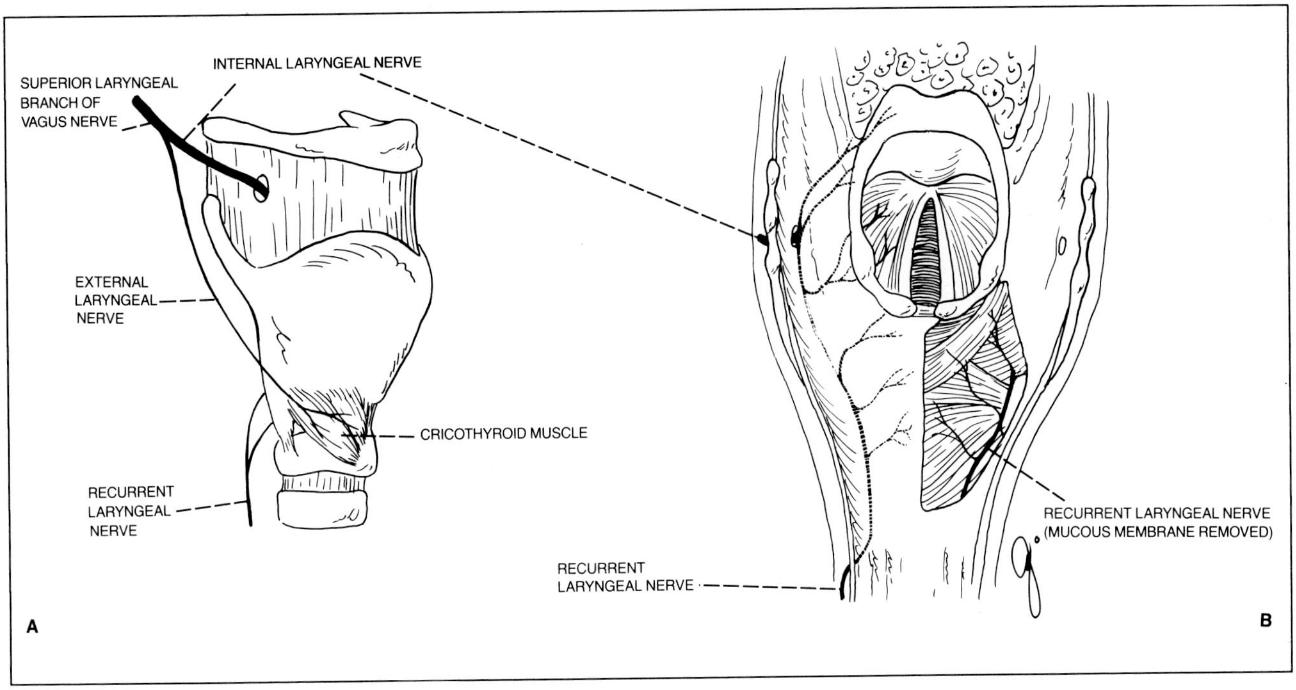

Figure 1–22. A. Lateral view of larynx showing the internal and external laryngeal branches of the superior laryngeal branch of the vagus nerve. **B.** The distribution of the terminal branches of the internal and recurrent laryngeal nerves. The larynx is viewed from above and posteriorly.

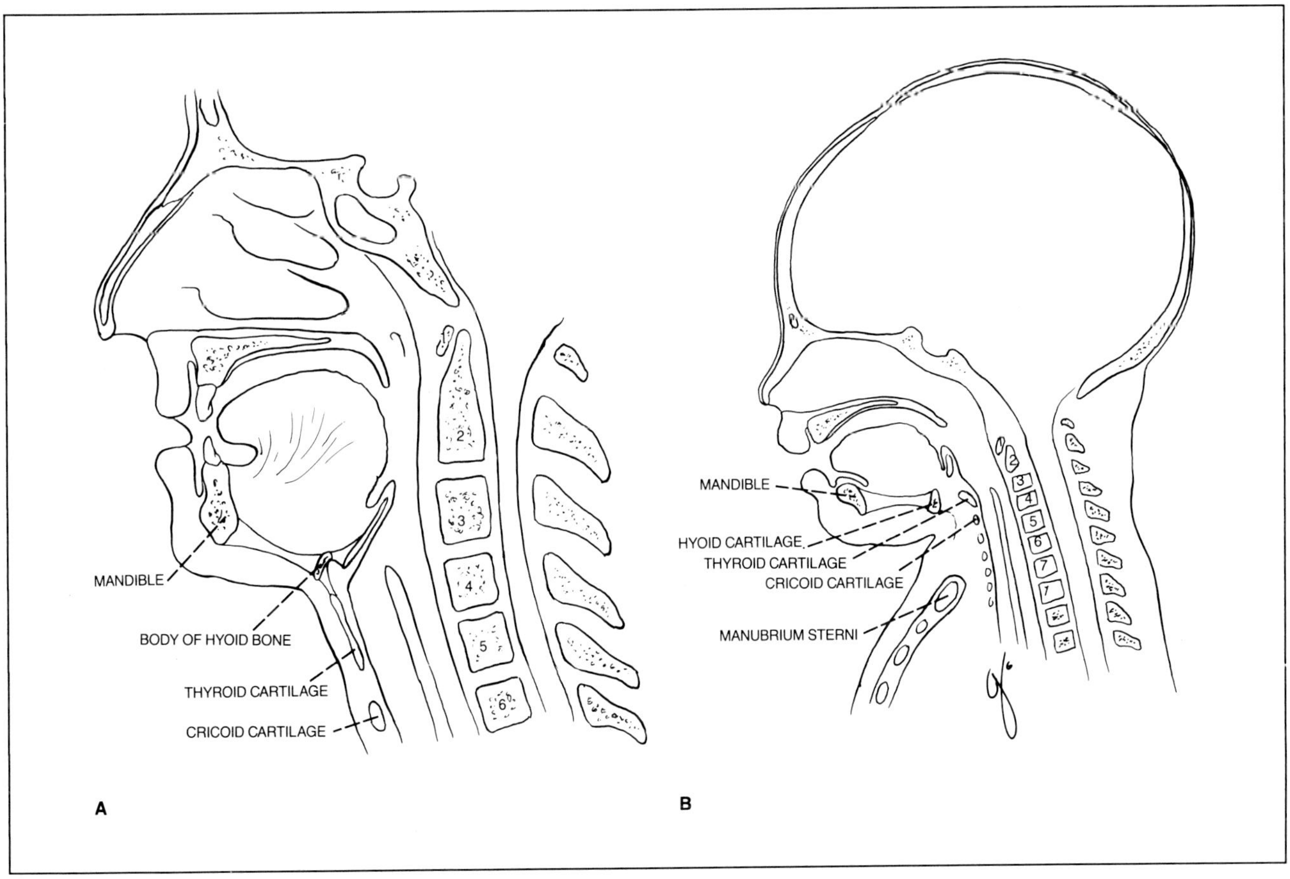

Figure 1–23. Sagittal sections of the neck of (**A**) an adult and (**B**) an infant shortly after birth. Shows the different vertebral levels in these age groups.

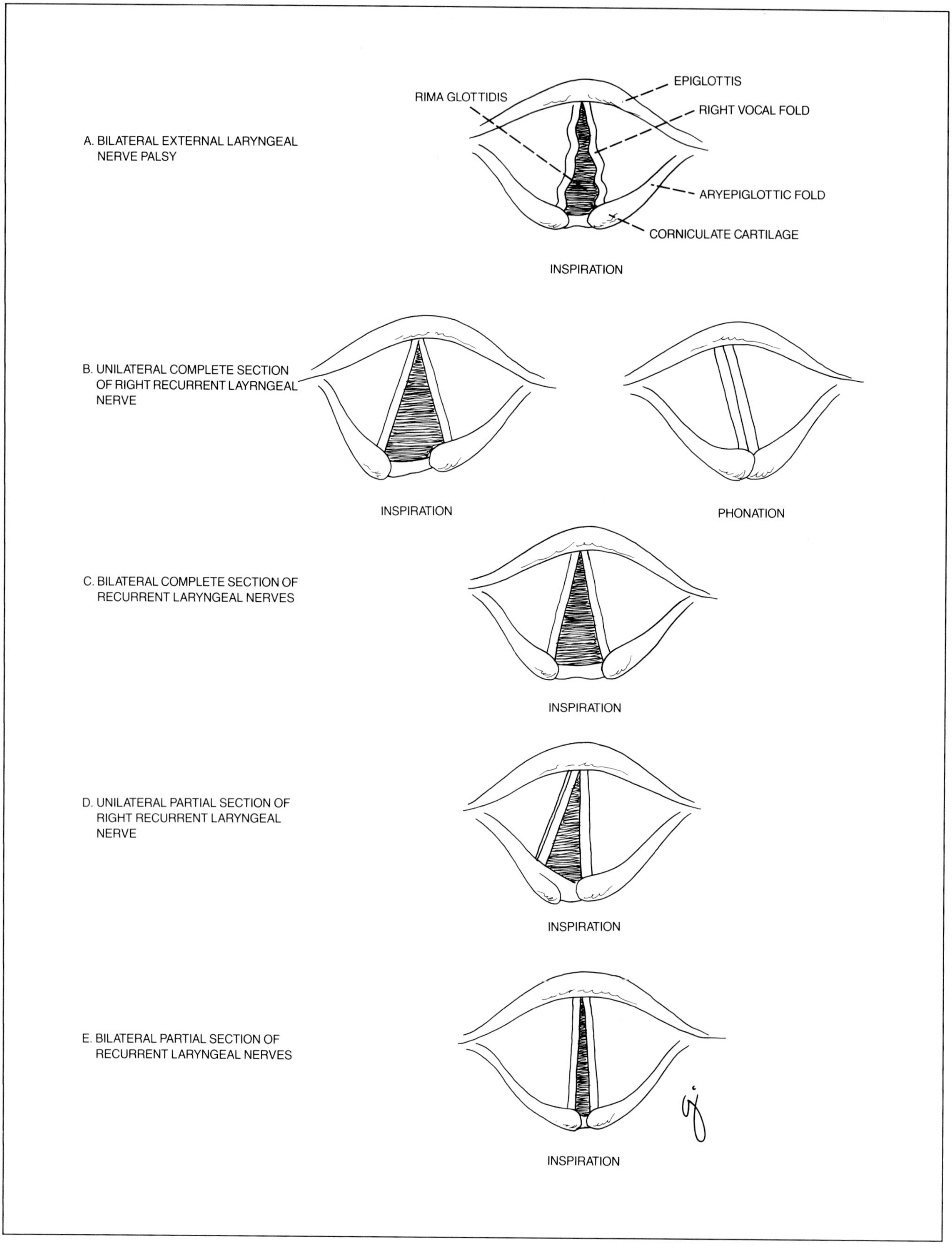

Figure 1–24. The position of the vocal folds after damage to the external and recurrent laryngeal nerves.

a diamond shape due to the lateral rotation of the arytenoid cartilages.

If the patient is asked to breathe deeply, the vocal folds become widely abducted, and the inside of the trachea can be seen.

TRACHEA (Fig. 1–25)

The trachea is a mobile cartilaginous and membranous tube. It begins as a continuation of the larynx at the lower border of the cricoid cartilage at the level of the sixth cervical vertebra. It descends in the midline of the neck and in the thorax ends at the **carina** by the dividing into right and left principal (main) bronchi at the level of the sternal angle (opposite the disk between the fourth and fifth thoracic vertebrae). Here it lies slightly to the right of the midline. During expiration the bifurcation rises by about one vertebral level, and during deep inspiration may be lowered as far as the sixth thoracic vertebra, a distance of about 3 cm.

In the adult the trachea is about 4.5 inches (11.25 cm) long and 1 inch (2.5 cm) in diameter. In the infant, the diameter may be as small as 8 mm during the first year and as the child grows the diameter in millimeters corresponds approximately to the age in years. The fibroelastic tube is kept patent by the presence of U-shaped cartilaginous bars (rings) or hyaline cartilage embedded in its wall. The posterior free ends of the cartilage are connected by smooth muscle, the **trachealis muscle.**

The mucous membrane of the trachea is lined with pseudostratified ciliated columnar epithelium (*Figs. 1–26 and 1–27*). There are many goblet cells and tubular mucous glands.

Relations in the Neck

Anterior. Skin, fascia, isthmus of thyroid gland (in front of the second, third, and fourth rings), inferior thyroid veins, jugular arch, thyroidea ima artery (if present), and left brachiocephalic vein in the child, overlapped by the sternothyroid and sternohyoid muscles.

Posterior. Right and left recurrent laryngeal nerves, esophagus.

Laterally. Lobes of the thyroid gland and the carotid sheath and contents.

Relations in Superior Mediastinum of Thorax

Anterior. Sternum, thymus, left brachiocephalic vein, origins of brachiocephalic and left common carotid arteries, arch of aorta.

Posterior. Esophagus and left recurrent laryngeal nerve.

Right Side. Azygos vein, right vagus nerve, and pleura.

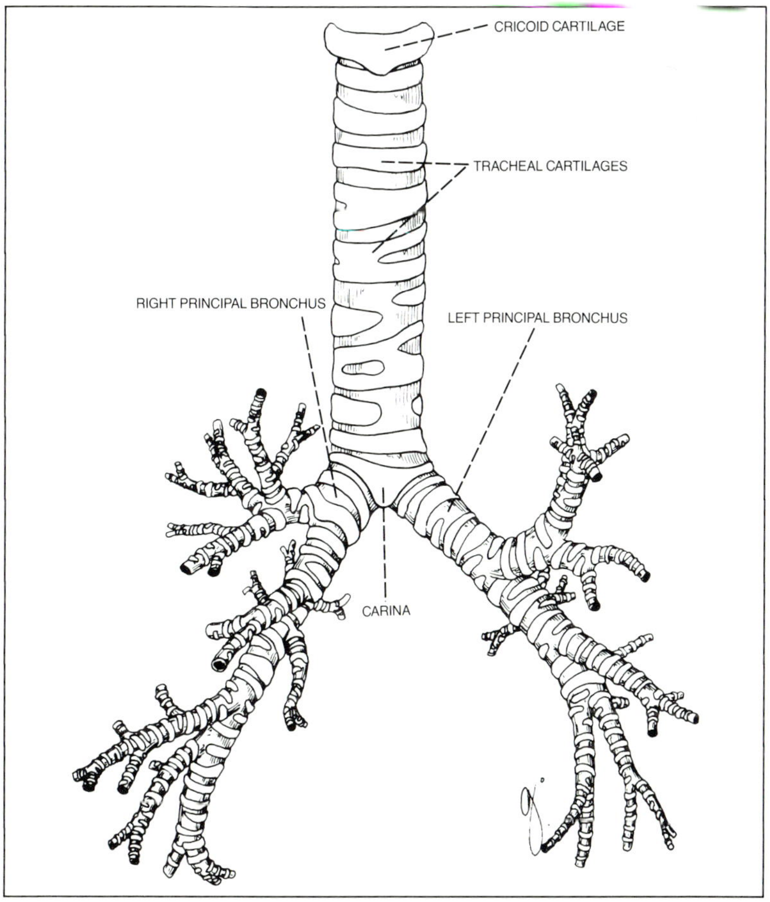

Figure 1–25. The trachea and the bronchi.

Left side. Arch of aorta, left common carotid, left subclavian arteries, left vagus nerve, left phrenic nerve, and pleura.

Nerve Supply. The sensory nerve supply of the upper half of the trachea is from the vagi and the recurrent laryngeal nerves. Sympathetic nerves are given off from the sympathetic trunks and parasympathetic fibers are carried by branches of the vagi (*Fig. 1–26*).

Blood Supply. The upper two-thirds is supplied by the inferior thyroid arteries and the lower third by the bronchial arteries.

Lymphatic Drainage. Into the pretracheal and paratracheal lymph nodes and the deep cervical nodes.

CLINICAL NOTES

Changes in the Tracheal Length with Respiration and Position of the Head and Neck

On deep inspiration the carina may descend by as much as 3 cm. Extension of the head and the neck, as when maintaining an open airway in an anesthetised patient, may stretch the trachea and increase its length by 25 percent.

Tracheotomy. Tracheotomy is a commonly performed procedure in which a tube is inserted into the lumen of the trachea to provide an airway. It may be carried out at two sites in patients with laryngeal obstruction. **High tracheotomy** is performed in the inter-

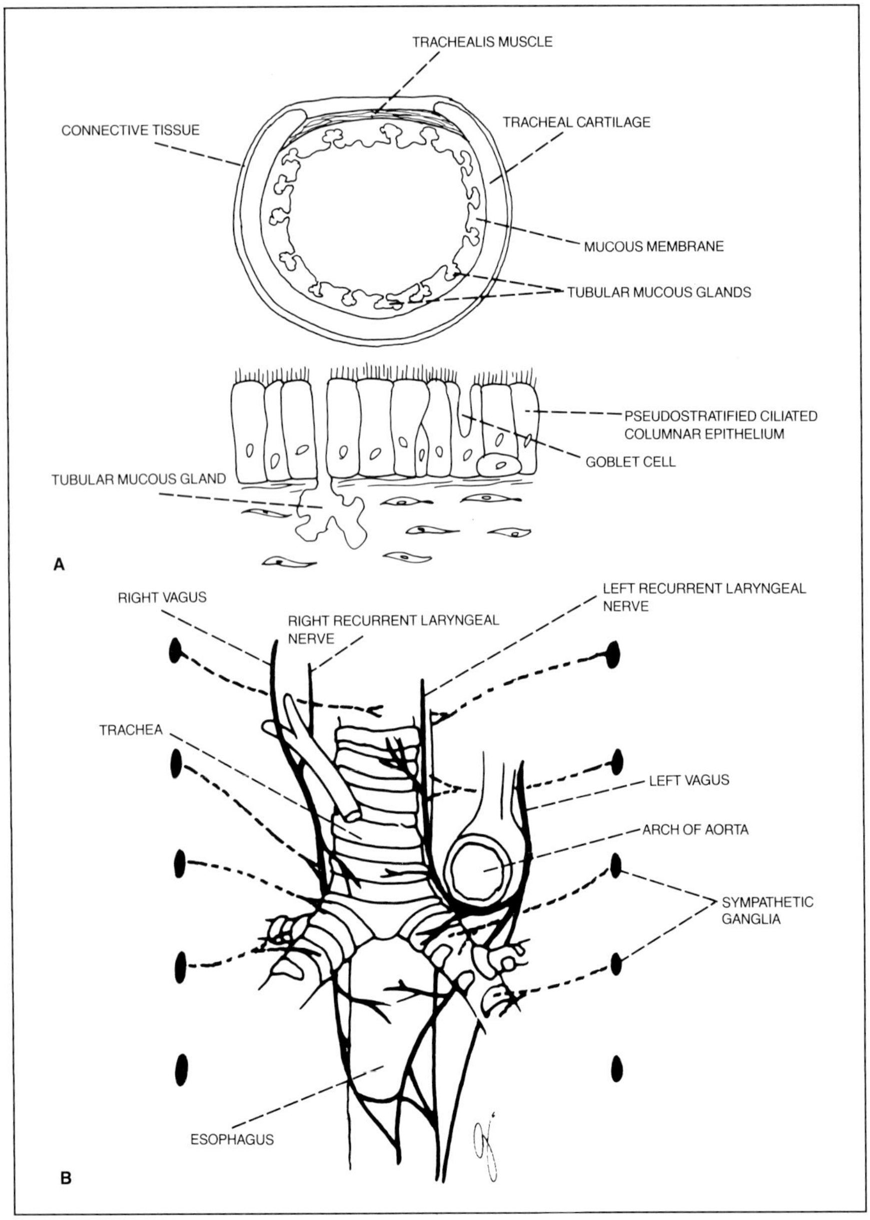

Figure 1–26. A. Cross section of the trachea showing its general structure. **B.** The innervation of the trachea from the vagi and their branches and from branches of both sympathetic trunks.

val between the thyroid cartilage and the cricoid cartilage (laryngotomy) or in the interval between the cricoid cartilage and the isthmus of the thyroid gland. The trachea is steadied by extending the neck over a sandbag. A vertical incision is made through the following structures: (1) skin; (2) superficial fascia (beware of the anterior jugular veins that lie on either side of the midline close together); (3) investing layer of deep cervical fascia; (4) pretracheal fascia (separate the sternohyoid muscles and incise the fascia); and (5) the larynx, which is incised through a vertical incision through the cricothyroid ligament or the trachea, which is incised through a vertical incision after retracting the isthmus of the thyroid gland downward.

Low tracheotomy is a more difficult operation and performed in an operating theater. The trachea is opened below the isthmus of the thyroid gland and above the suprasternal notch. A vertical incision is made through the following structures: (1) skin; (2) superficial fascia; (3) investing layer of deep fascia (the suprasternal space is opened and the jugular arch is found and divided between ligatures); (4) connective tissue containing the inferior thyroid veins and possibly the thyroidea ima artery; (5) pretracheal fascia; and (6) trachea. In incising the trachea, the brachiocephalic vessels must be avoided. In children, the thymus may obscure the trachea.

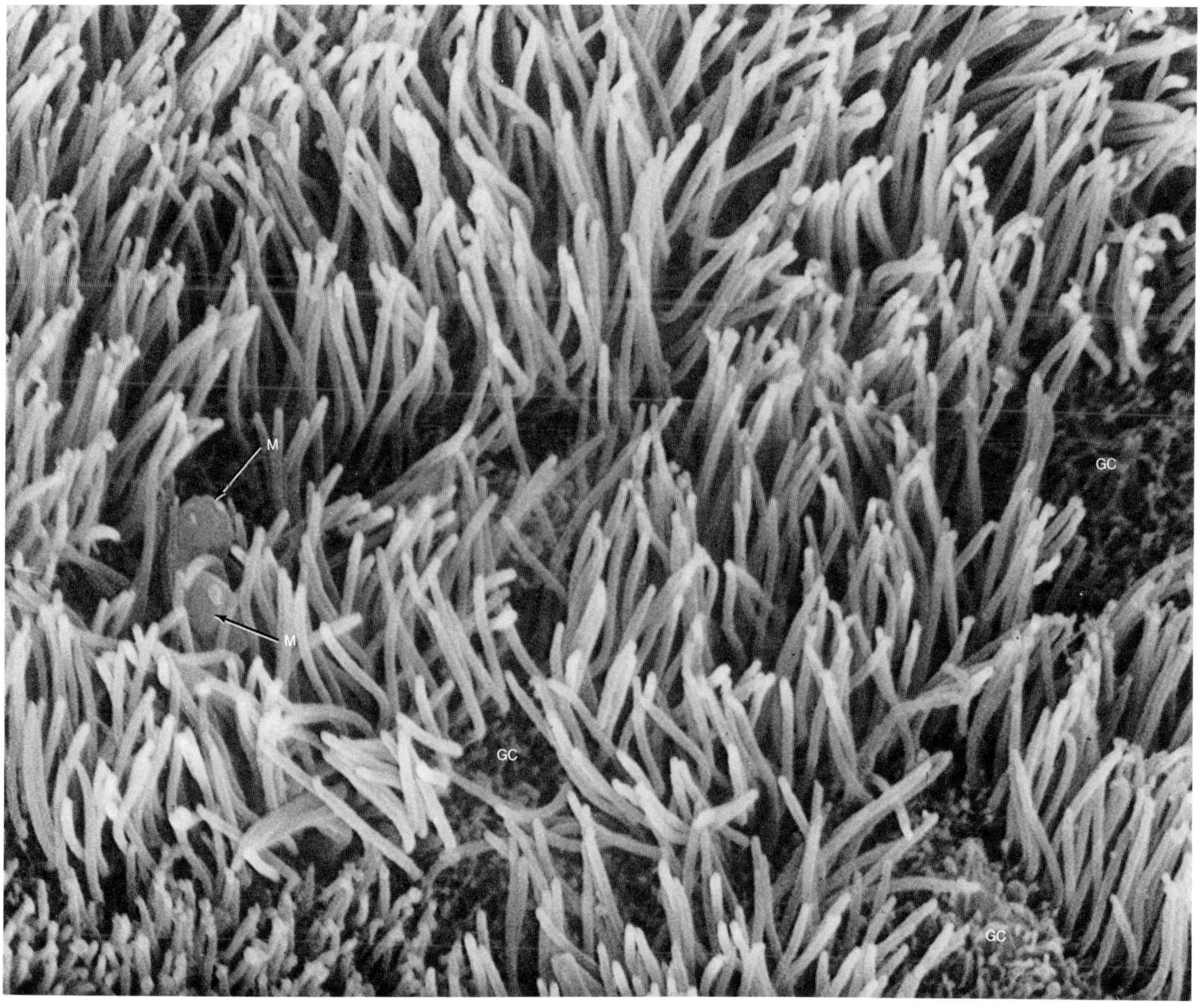

Figure 1–27. Scanning electron micrograph of surface epithelial cells of the mucous membrane of the trachea, showing ciliated columnar cells and goblet cells (GC). M = Droplet of mucus. (×3388) (Courtesy of Dr. M. Koering.)

CLINICAL PROBLEMS

1. A 17-year-old boy was undergoing an appendectomy when the anesthetist noticed a diminution in the movements of the reservoir bag accompanied by heaving efforts of the patient's chest. The anesthetist decided to improve the situation by using a soft nasal tube. On attempting to pass a well-greased tube into the nose, he found it impossible to push it much beyond the nasal vestibule on either side. What are the common anatomic causes of obstruction of the nasal airway? Which is the widest part of the nasal airway?

2. A 45-year-old woman was in the recovery room after a cholecystectomy. She was still unconscious and suddenly started to snore loudly. What is the most likely anatomic cause for the condition? Explain in detail the rationale for treatment.

3. A 12-year-old boy was about to have a simple inguinal herniotomy. Before starting the anesthetic, the anesthetist remarked on the enormous size of the lymphoid tissue seen in the patient's pharynx. List the various lymphoid organs found in the nasal and oral parts of the pharynx. How may the airway be maintained in an individual with large lymphoid organs? What is Waldeyer's ring?

4. Before an operation is scheduled, where it is planned to give an inhalation anesthetic, the anesthetist should perform a complete physical on the upper and lower parts of the respiratory system. Why is it necessary to examine the temporomandibular joints and the movements of the cervical part of the vertebral column?

5. While inducing a general anesthetic in a patient about to undergo a partial gastrectomy, a high pitched crowing noise was noticed. The anesthetist turned to his assistant and said "this patient is developing laryngeal spasm." What is laryngeal spasm? Describe the anatomic reasons for obstruction of the airway in laryngeal spasm.

6. A 5-year-old girl was about to have an operation for an umbilical hernia. The anesthetist decided to give the anesthetic through an endotracheal tube. Having selected what he considered to be the correct size and length, he passed the tube successfully through the glottis, but he could not get it to descend any further. What is the likely anatomic explanation?

7. After a prolonged operation on a 4-year-old boy for intussusception caused by a Meckel's diverticulum, the anesthetist commented on the extreme swelling of the vocal folds that had followed tracheal intubation. Does the structure of the vocal folds in young children differ from those of the adult?

8. Before initiating treatment of a patient with carcinoma of the left lung with metastases in the tracheobronchial nodes, laryngoscopic examination revealed partial paralysis of the left vocal fold. What position does the left vocal fold take up in such circumstances? Explain this abnormal position in anatomic terms. Does the opposite vocal fold change its position to compensate?

ANSWERS TO CLINICAL PROBLEMS

1. The most common cause for difficulty in passing a nasal tube is a deflected nasal septum. This occurs more commonly in the male and is thought to be due to previous trauma to the septum during the period of active growth. Nasal spurs and polyps may also cause difficulty. The widest part of the nasal cavity is near the floor.

2. The most likely cause for blockage of the airway in this patient is the tongue falling back to meet the posterior pharyngeal wall. When unconscious, the tone of the extrinsic muscles of the tongue is lost and the tongue simply obeys the

laws of gravity and falls backward. This may be easily and quickly corrected by pulling the jaw forward. Place two fingers under the symphysis menti and pull forward or press forward the angles of the jaw with the thumbs. The anatomic explanation is that as the mandible is pulled forward, the genioglossus muscles, which run from the mandible to the tongue, pull the tongue forward.

3. The lymphoid tissue around the openings of the mouth and the nasal cavities into the pharynx include (1) the palatine tonsils, (2) the lingual tonsils, (3) the tubal tonsils, and (4) the pharyngeal tonsil. For details of Waldeyer's ring, see page 17. The use of a well-lubricated artificial airway, which may be passed into the pharynx, will provide a clear passage provided the mandible is kept pulled well forward.

4. It may prove difficult to establish a good airway in a patient who has ankylosis of the temporomandibular joint. Moreover, the successful use of a laryngoscope is more difficult if there is difficulty in flexing the cervical part of the vertebral column and extending the atlanto-occipital joints. Alternate methods for intubation, i.e., fiberoptic techniques, nasal intubation, etc., should be prepared for.

5. Laryngeal spasm may occur during the induction of the anesthetic, although it may occur during the administration or the recovery phase. Common initiating factors include direct irritation of the laryngeal mucous membrane (e.g., by anesthetic vapor or instrumentation); hypoxia of the laryngeal muscles (e.g., blocked airway); intravenous barbiturates (enhance respiratory reflexes). Complete or partial laryngeal spasm is caused by contraction of the lateral cricoarytenoid muscles on both sides, which brings about bilateral adduction of the vocal folds and closure of the rima glottidis.

6. The narrowest part of the laryngeal cavity in an adult is the area between the vocal folds. In children under 10 years of age, the narrowest part of the larynx is within the cricoid cartilage.

7. The vocal folds in young children contain a relatively thick submucosa of loose connective tissue compared with that of the adult. It follows that edema due to trauma or an allergic reaction is more likely to cause narrowing of the glottis in a child than in an adult.

8. The left recurrent laryngeal nerve is commonly invaded by bronchogenic carcinoma metastasizing from carcinoma of the left lung. A partial section of the left recurrent laryngeal nerve causes the left vocal fold to assume the adducted midline position. The reason why the abductor rather than the adductor muscles are paralyzed is discussed on page 28. The right vocal fold continues to move normally and its position is normal.

CHAPTER 2

Respiratory System

Lower Airway

BRONCHI

The trachea bifurcates behind the arch of the aorta into the **right** and **left principal** (primary, main) **bronchi** (*Fig. 2–1*). The right bronchus leaves the trachea at an angle of 25 degrees from the vertical, and the left leaves at an angle of 45 degrees from the vertical. In children under 3 years, both bronchi arise from the trachea at equal angles.

The bronchi now divide dichotomously giving rise to several million **terminal bronchioles** that terminate in one or more **respiratory bronchioles** (*Fig. 2–2*). Each respiratory bronchiole gives rise to 2–11 alveolar ducts that enter the alveolar sacs. Arising from the walls of the sacs as diverticula, are the alveoli.

Right Principal Bronchus

The right principal bronchus is wider, shorter, and more vertical than the left and is about 1 inch (2.5 cm) long (*Fig. 2–1*). The azygos vein arches over its superior border. The **superior lobar bronchus** is given off within 2 cm of the commencement and is on a level with the carina. The right principal bronchus now enters the hilum of the right lung where it divides into a **middle and an inferior lobar bronchus.**

Right Superior Lobar Bronchus. The right superior lobar bronchus runs superolaterally to enter the hilum of the lung (*Fig. 2–1*). About 1 cm from its origin it divides into three segmental bronchi, the apical segmental bronchus, the posterior segmental bronchus, and the anterior segmental bronchus. The **apical segmental bronchus** runs superolaterally toward the apex of the lung and separates into an apical and an anterior branch. The **posterior segmental bronchus** runs posterolaterally and divides into a lateral and a posterior branch. The **anterior segmental bronchus** runs anteroinferiorly forming a lateral and an anterior branch.

Right Middle Lobar Bronchus. The right middle lobar bronchus arises from the anterior aspect of the right principal bronchus about 3 cm from its origin (*Fig. 2–1*). It descends anterolaterally and divides into a **lateral segmental bronchus** and a **medial segmental bronchus.**

Right Inferior Lobar Bronchus. The right inferior lobar bronchus is a continuation of the right principal bronchus (*Fig. 2–1*). From the posterior surface of the right inferior bronchus the large **superior (apical) segmental bronchus** arises, which runs posteriorly and divides into medial, superior, and lateral branches. The right inferior lobar bronchus descends posterolaterally and gives off the **medial basal segmental bronchus.** The inferior lobar bronchus continues inferiorly and gives rise to the **anterior basal segmental bronchus,** the **lateral basal segmental bronchus,** and the **posterior basal segmental bronchus.**

In many individuals the right inferior lobar bronchus gives off an additional branch called the **subsuperior (subapical) segmental bronchus,** which runs to a region of the lung between the superior and posterior basal segments.

Left Principal Bronchus

The left principal bronchus is narrower, longer, and more horizontal than the right and is about 2 inches (5 cm) long (*Fig. 2–1*). It passes to the left below the arch of the aorta and in front of the esophagus. On entering the hilum of the left lung it divides into a **superior and an inferior lobar bronchus.**

Left Superior Lobar Bronchus. The left superior lobar bronchus curves laterally and divides after about 1.5 cm into upper and lower branches (*Fig. 2–1*). The upper division gives off the **anterior segmental bronchus** and finally separates into the **apical segmental bronchus** and the **posterior segmental bronchus.** The lower division

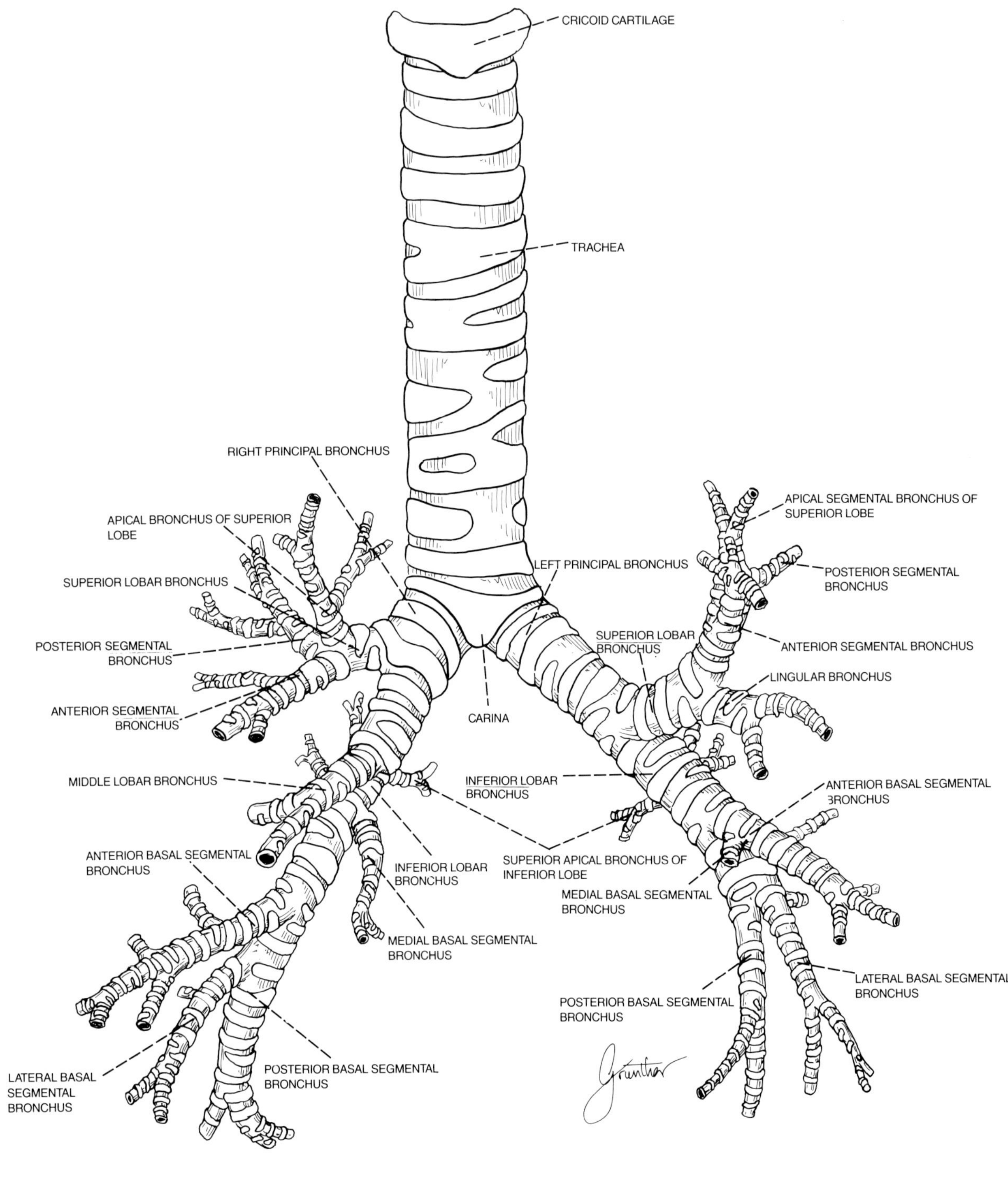

Figure 2–1. The trachea and the principal bronchi and their branches.

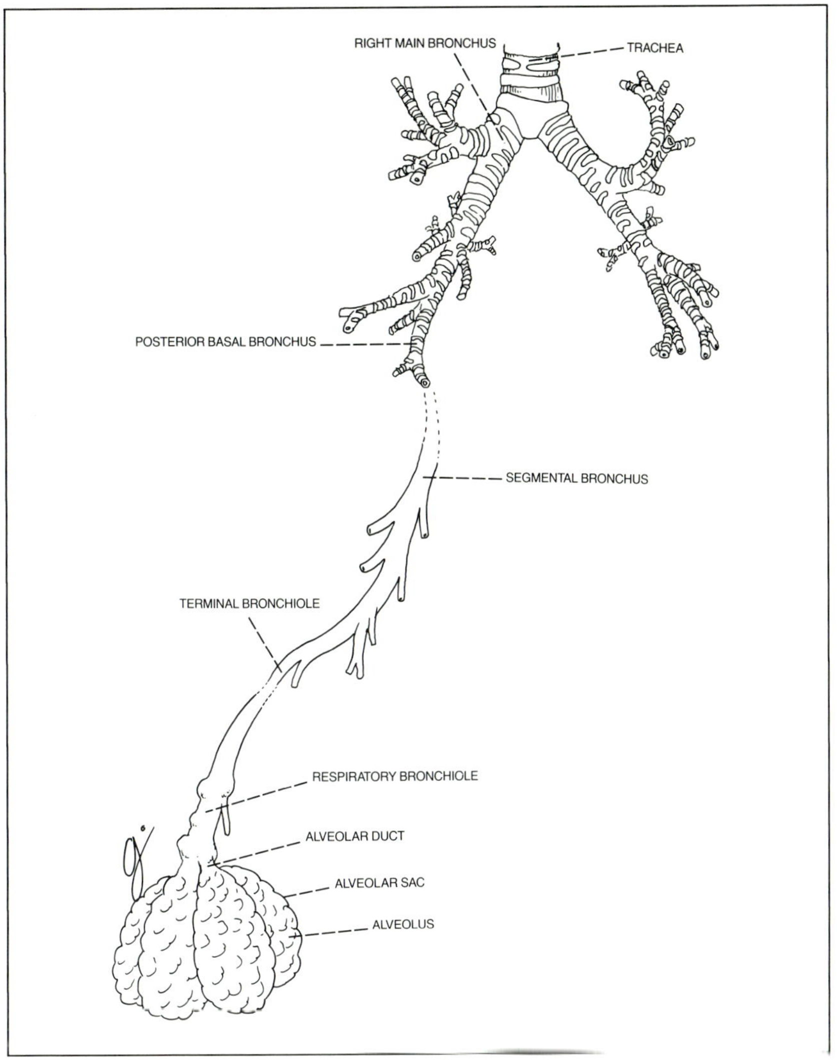

Figure 2–2. Diagram showing the continuity of the trachea, bronchi, bronchioles, alveolar ducts, alveolar sacs, and alveoli.

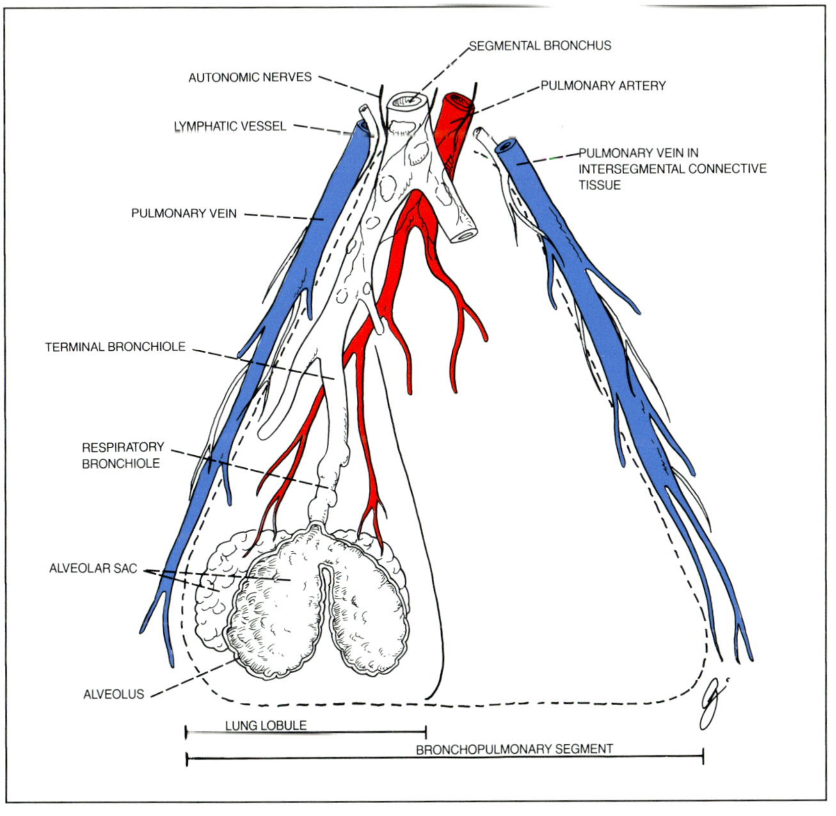

Figure 2–3. A bronchopulmonary segment and a lung lobule. Note that the pulmonary veins lie within the connective tissue septa that separate adjacent segments.

(or lingular bronchus) descends anteriolaterally and divides into the **superior lingular segmental bronchus** and the **inferior lingular segmental bronchus.**

Left Inferior Lobar Bronchus. The left inferior lobar bronchus is a continuation of the left principal bronchus and descends posterolaterally (*Fig. 2–1*). About 1 cm from its origin it gives off the **superior (apical) segmental bronchus** from its posterior wall. This quickly divides into medial, superior, and lateral branches. The left inferior lobar bronchus gives off the **medial basal segmental,** the **anterior basal segmental,** the **lateral basal segmental,** and the **posterior basal segmental bronchi.** In some individuals a **subsuperior (subapical) segmental bronchus** arises from the posterior surface of the left inferior lobar bronchus.

Bronchopulmonary Segments

Each segmental bronchus enters a unit of lung tissue called a bronchopulmonary segment, which is surrounded by connective tissue (*Fig. 2–3*). The segmental bronchus is accompanied by a branch of the pulmonary artery but the tributaries of the pulmonary veins run in the connective tissue between adjacent bronchopulmonary segments. Each segment has its own lymphatic vessels and autonomic nerve supply. The main bronchopulmonary segments (*Fig. 2–4*) are as follows:

Right Lung

Superior lobe	(1) Apical, (2) posterior, (3) anterior
Middle lobe	(4) Lateral, (5) medial
Inferior lobe	(6) Superior (apical), (7) medial basal, (8) anterior basal, (9) lateral basal, (10) posterior basal

Left Lung

Superior lobe	(1) Apical, (2) posterior, (3) anterior, (4) superior lingular, (5) inferior lingular
Inferior lobe	(6) Superior (apical), (7) medial basal, (8) anterior basal, (9) lateral basal, (10) posterior basal

On entering a bronchopulmonary segment, each segmental bronchus divides repeatedly (*Fig. 2–2*). As the bronchi become smaller, the U-shaped bars of hyaline cartilage found in the trachea are gradually replaced by irregular plates of cartilage, which become smaller and fewer in number. The mucous membrane is thinner, and the pseudostratified ciliated columnar epithelium is replaced by ciliated columnar cells with many goblet cells. Smooth muscle is present within the submucosa and, in the smaller bronchi, forms a single complete layer of circularly arranged fibers.

The smallest bronchi divide and give rise to **bronchioles,** which are less than 1 mm in diameter (*Fig. 2–2*). It is the bronchioles that enter the small units of the lungs known as the **lung lobules** (*Fig. 2–3*). Bronchioles possess no cartilage in their walls and are lined with ciliated columnar cells. The submucosa possesses a complete layer of circularly arranged smooth muscle fibers. The seromucous glands found in the submucosa of the larger bronchi disappear.

The bronchioles then divide and give rise to **terminal bronchioles** (*Fig. 2–2*). These are lined with nonciliated columnar or cuboidal cells. The terminal bronchioles divide and give rise to **respiratory bronchioles,** which show delicate, air-containing outpouchings from their walls (*Fig. 2–2*). Gaseous exchange between the blood and the air takes place in the walls of these outpouchings; hence, the name respiratory bronchiole. The diameter of a respiratory bronchiole is about 0.5 mm. The wall of a respiratory bronchiole is lined with cuboidal epithelium, which rests on connective tissue containing interlacing smooth muscle and elastic fibers.

The respiratory bronchioles end by branching into **alevolar ducts** that lead into tubular passages with numerous thin-walled outpouchings called **alveolar sacs** (*Figs. 2–2, 2–3, and 2–5*). The wall of the alveolar duct is lined with cuboidal epithelium supported by connective tissue containing elastic fibers and a few smooth muscle fibers.

The alveolar sacs consists of several alveoli opening into a single chamber (*Figs. 2–3 and 2–5*). The walls of the alveoli (interalveolar septa) are composed of very thin flattened cells supported by delicate connective tissue containing elastic fibers (*Fig. 2–6*). Each alveolus is surrounded by a rich network of blood capillaries. Examination of the wall of an alveolus with the electron microscope (*Fig. 2–7*) shows that the alveoli are in fact lined with simple squamous cells (type I), but there is an occasional cuboidal cell (type II) responsible for secreting surfactant into the lumen. The surfactant reduces the surface tension on the lining cells and permits the alveolar walls to separate from one another as air enters during inspiration. Surfactant is a lipoprotein material and appears in the lungs at about the 30th week of intrauterine life.

Another type of cell seen bulging into the alveolus or residing in the space between adjacent alveoli is the **alveolar phagocyte** (*Fig. 2–6*). Such cells are believed to be derived from the monocytes of the blood. They migrate through the capillary wall to enter the tissue spaces and then pass through the epithelial lining of the alveolus to lie free in the lumen of the alveolus. These cells serve as phagocytes and remove debris, such as carbon particles, from the lumen of the alveolus. They migrate up the bronchial tree and are ultimately swallowed with the mucus.

RESPIRATORY MEMBRANE

Electron microscopic study of the alveolar walls (*Figs. 2–6 and 2–7*) shows that the lumen of the alveolus is separated from the lumen of the blood capillary by the

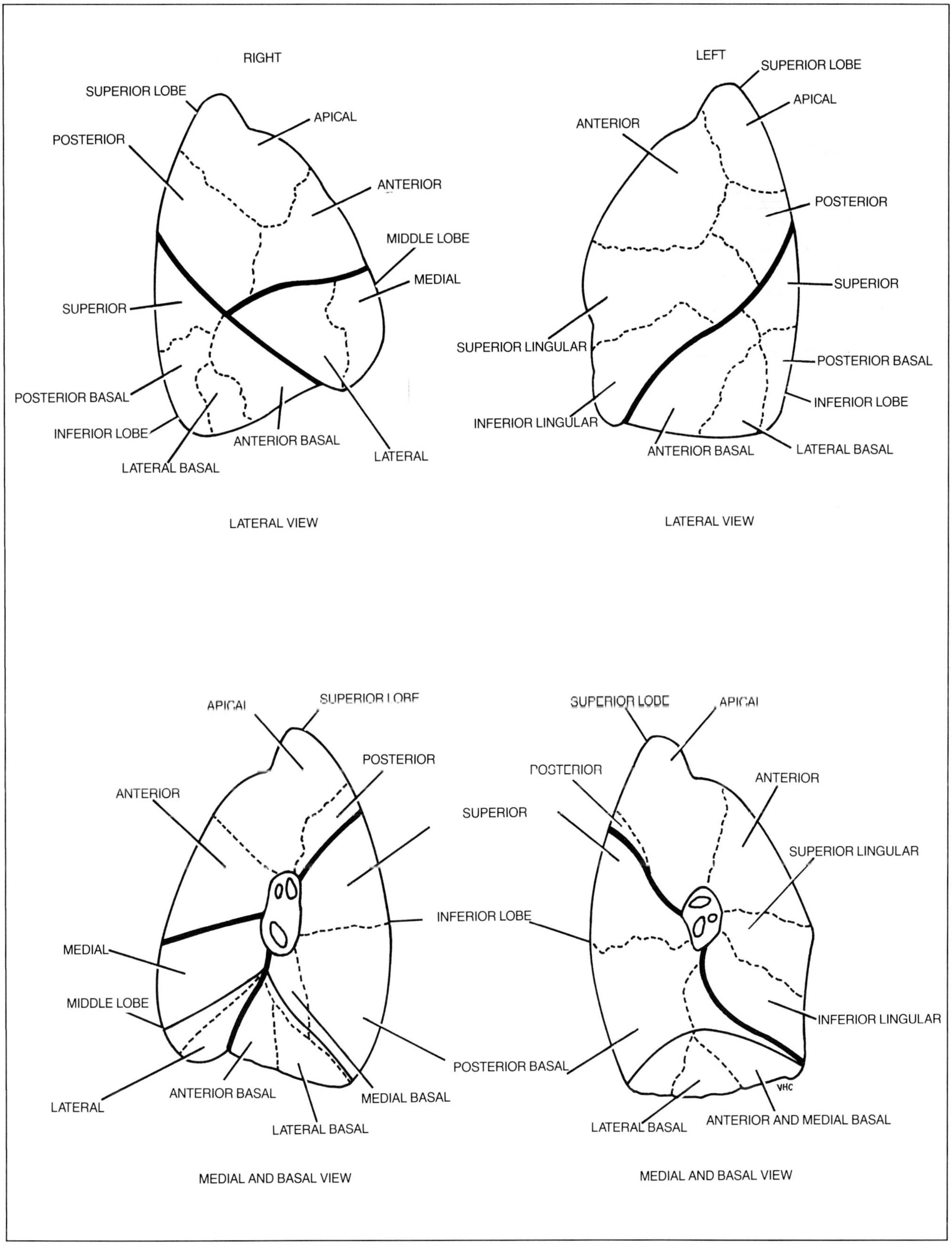

Figure 2–4. Bronchopulmonary segments of each lung.

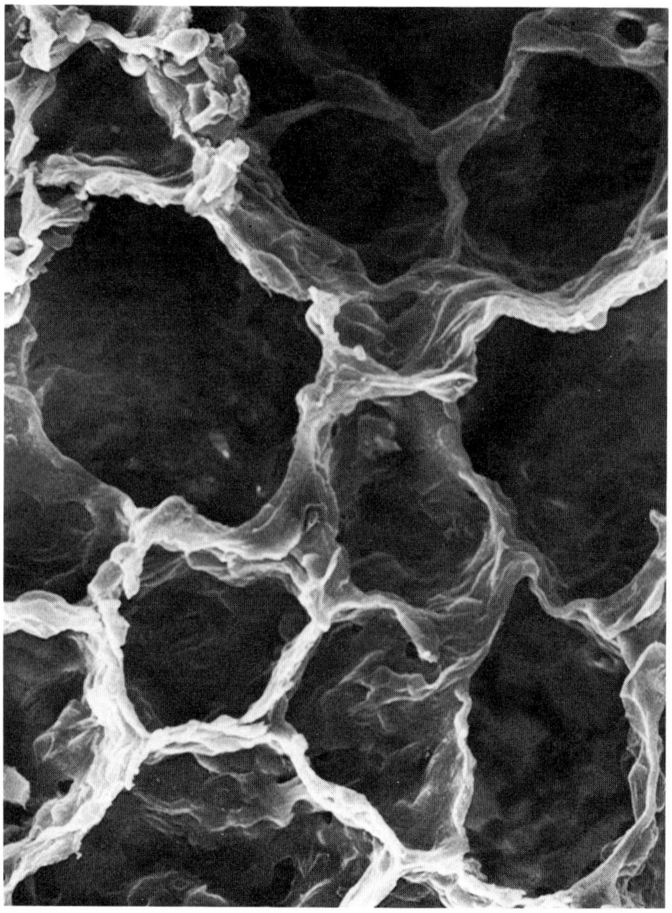

Figure 2–5. Scanning electron micrograph of the lung, showing numerous alveolar sacs. The alveoli are the depressions, or alcoves, along the walls of the alveolar sac. (×520) (Courtesy of Dr. M. Koering.)

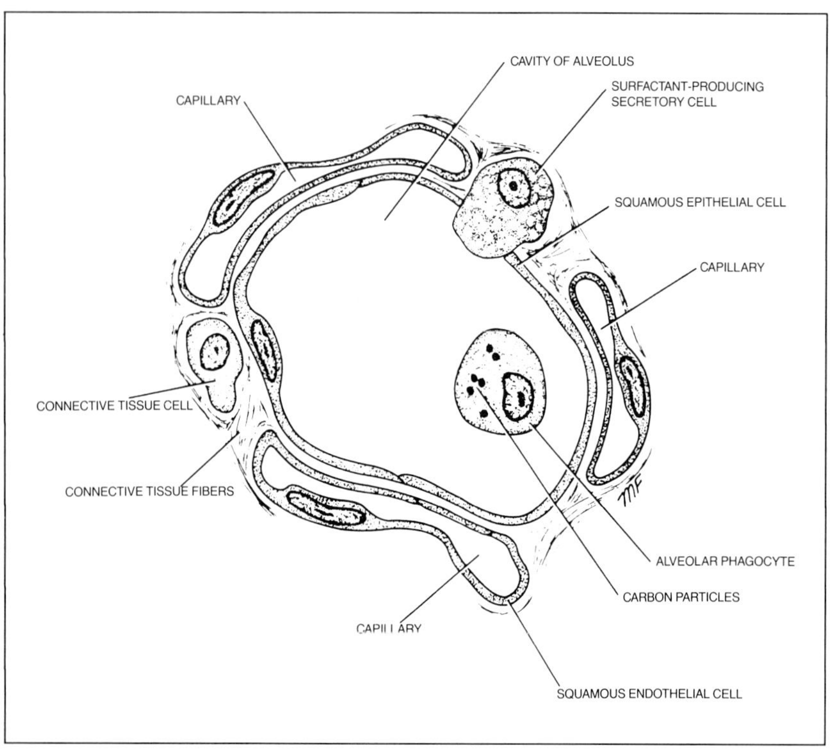

Figure 2–6. Diagram showing the detailed structure of the respiratory membrane.

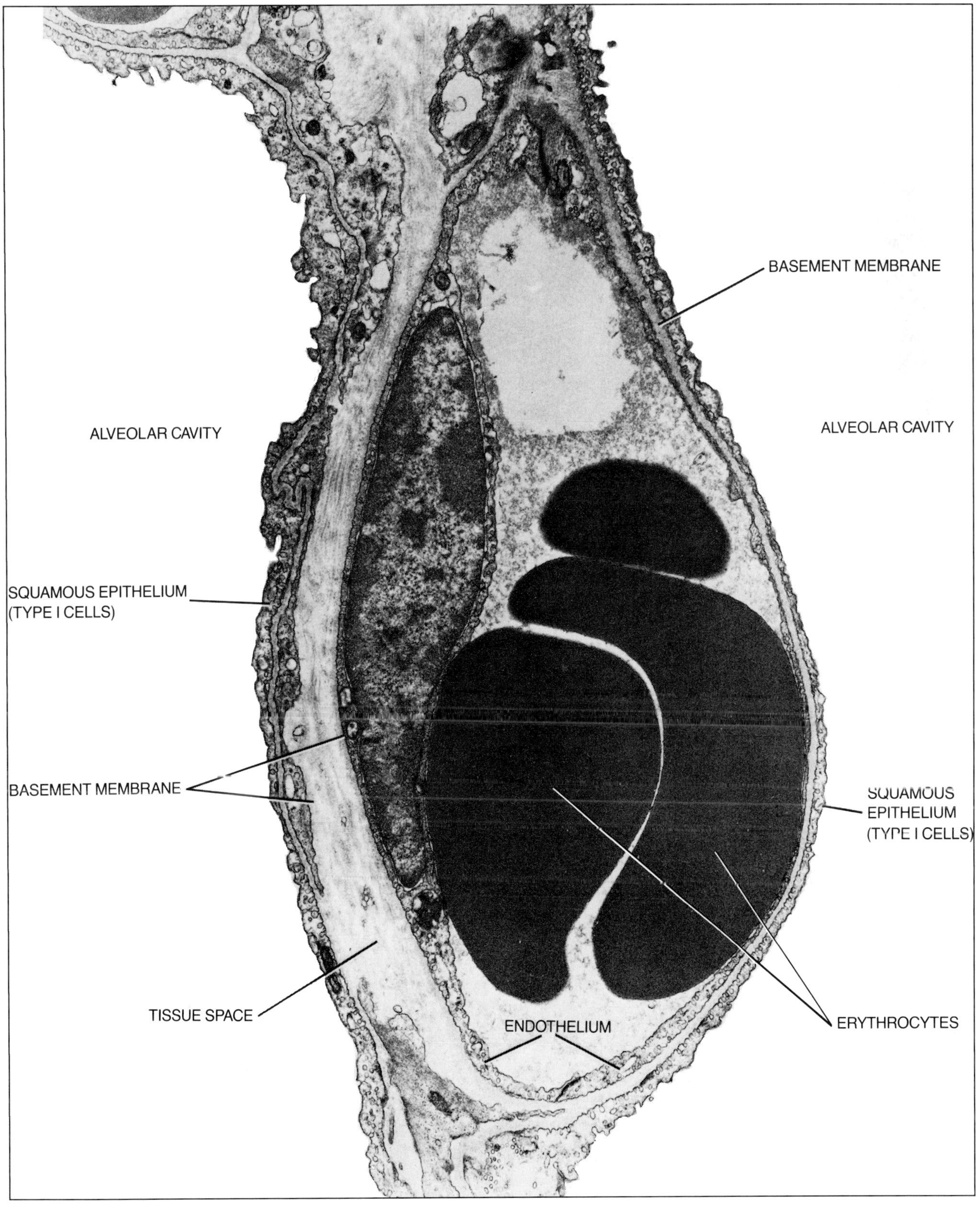

Figure 2–7. Electron micrograph through a capillary in the alveolar wall (interalveolar septum). The alveolar cavity (AC) is lined with squamous epithelium (SE) (type I cells). The capillary containing erythrocytes (RBC) is lined with endothelium (E). The two layers of cells are separated from each other by a basement membrane (BM); the small tissue space (TS) is also visible. (×9400) (Courtesy of Dr. E.R. Weibel.)

following: (1) surfactant-containing fluid produced by the type II alveolar cells, (2) the alveolar squamous epithelium or type I cell, (3) the epithelial basement membrane, (4) a minute tissue space, (5) the blood capillary basement membrane, which fuses with the epithelial basement membrane in many places, and (6) the capillary squamous endothelium.

The combination of these layers is known as the respiratory membrane and averages about 0.5 μ in thickness. It is at the respiratory membrane that gaseous exchange takes place between the air in the alveolar lumen and the blood within the capillary.

The bronchi, the terminal bronchioles, the respiratory bronchioles, the alveolar ducts, the alveolar sacs, the alveoli, and the blood and lymphatic vessels, and the nerves are all held together by connective tissue to form the lung.

It should be remembered that the lungs receive a double blood supply, from the pulmonary and bronchial arteries. The bronchial arteries are able to supply the lungs adequately and keep the lung tissue viable should the pulmonary artery suddenly be blocked, as in pulmonary embolism.

PLEURAE AND LUNGS

The thoracic cavity can be divided into a median partition called the **mediastinum,** that lies between the sternum and the vertebral column, and the laterally placed pleurae and lungs.

Pleurae

The pleurae are two serous sacs surrounding and covering the lungs. Each pleura has two parts: (1) a **parietal layer,** which lines the thoracic wall and covers the thoracic surface of the diaphragm and the lateral surface of the mediastinum; and (2) a **visceral layer,** which covers the outer surfaces of the lungs and extends into the interlobar fissures. The parietal layer of each pleura becomes continuous with the visceral layer at the **hilum** of each lung. Here, they form a cuff that surrounds the structures entering and leaving the lung at the **lung root.** The **pulmonary ligament** is a loose extension of the cuff below the lung root to allow for movement during respiration.

The **pleural cavity** is a slit-like space that separates the parietal and visceral layers of pleura. It normally contains a small amount of **pleural fluid** that lubricates the apposing pleural surfaces. The **costodiaphragmatic recess** is the lowest area of the pleural cavity into which the lung expands on inspiration.

Surface Marking of the Parietal Pleura. The lines, which indicate the limits of the parietal pleura and where they are reflected close to the body surface (*Fig. 2–8*), are as follows:

The pleura extends upward into the neck from the sternoclavicular joint along a curved line about 1 inch (2.5 cm) above the junction of the medial and intermediate thirds of the clavicle. The **anterior border of the right pleura** descends from the sternoclavicular joint to the midline behind the sternal angle. It then descends to the xiphisternal joint. The **anterior border of the left pleura** has a similar course, but at the level of the fourth costal cartilage it turns laterally from the midline to the lateral margin of the sternum, forming the **cardiac notch.** It then turns downward to the xiphisternal joint.

The **lower border of the pleura** on both sides crosses the eighth rib in the midclavicular line and the tenth rib in the midaxillary line, and reaches the 12th rib adjacent to the vertebral column. (See surface marking of lungs on page 46.)

Lungs

The lungs are situated on each side of the mediastinum. They are, therefore, separated from each other by the heart, the great vessels, and other structures in the mediastinum. Each lung is conical in shape and is covered with visceral pleura. It is suspended free in its own pleural cavity, being attached to the mediastinum only by its root.

Each lung has a blunt **apex,** which projects upward for about 1 inch (2.5 cm) above the clavicle; a concave **base,** which overlies the diaphragm; a convex **costal surface,** which corresponds to the chest wall; a concave **mediastinal surface,** which is molded to the pericardium and other mediastinal structures. About the middle of this surface is the **hilum,** a depression where the bronchi, vessels, and nerves enter the lung to form the **root.**

The **anterior border** is thin and overlaps the heart; it is here on the left lung that there is a notch, called the **cardiac notch** (*Fig. 2–8*). The **posterior border** is thick and lies beside the vertebral column.

Lobes and Fissures

The right lung is slightly larger than the left and is divided by the oblique and horizontal fissures into three lobes, the **upper,** the **middle,** and the **lower lobes.** The **oblique fissure** runs from the inferior border upward and backward across the medial and costal surfaces until it cuts the posterior border about 2.5 inches (6.25 cm) below the apex. The **horizontal fissure** runs horizontally across the costal surface at the level of the fourth costal cartilage to meet the oblique fissure in the midaxillary line (*Fig. 2–8*). The middle lobe is thus a small triangular lobe bounded by the horizontal and oblique fissures.

The left lung is divided by a similar oblique fissure into two lobes, the **upper** and the **lower lobes** (*Fig. 2–8*). There is no horizontal fissure in the left lung.

Blood Supply of Lungs. The bronchi, the connective tissue of the lung, and the visceral pleura receive their

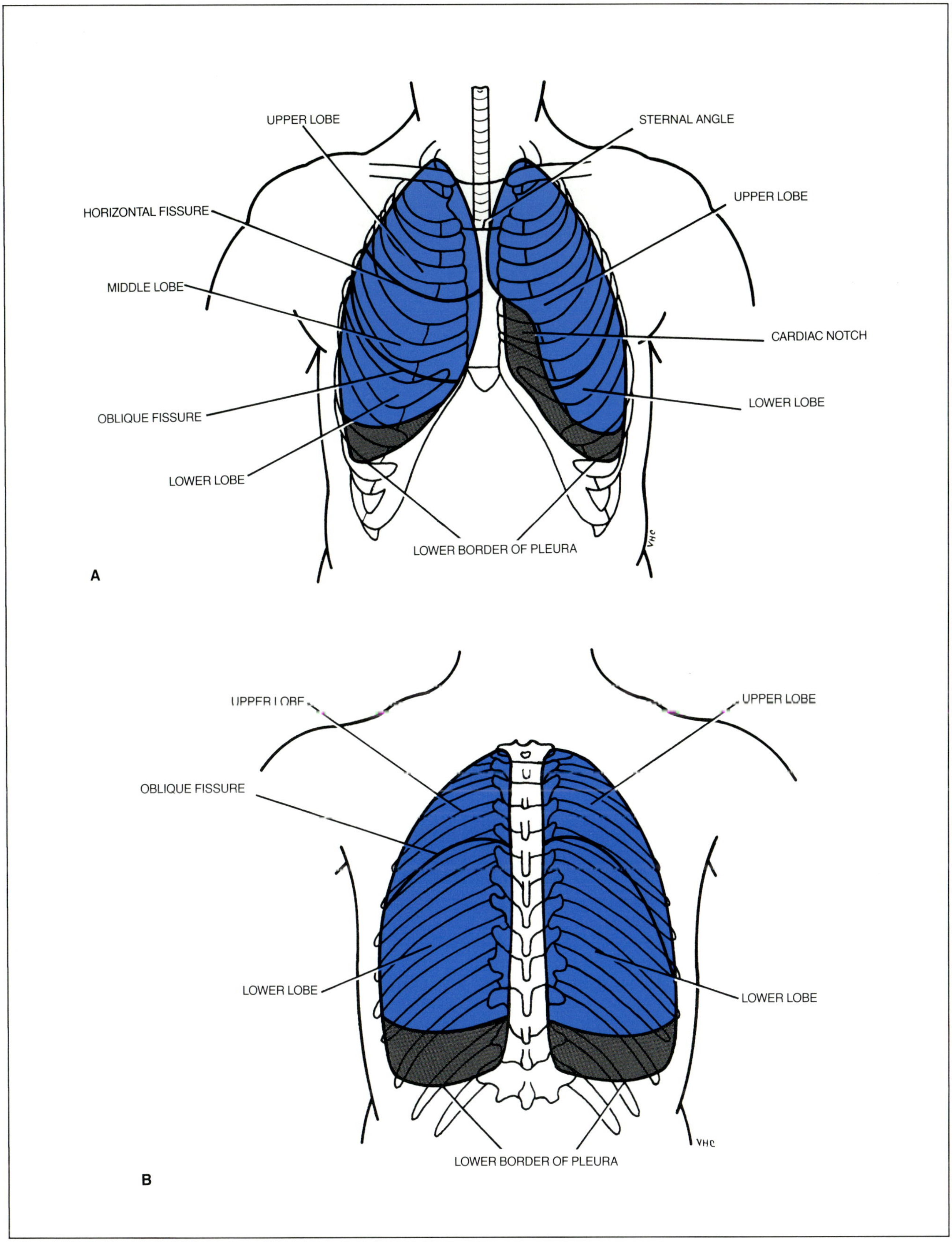

Figure 2–8. Surface markings of lungs and parietal pleura on **(A)** the anterior thoracic wall and **(B)** the posterior thoracic wall.

blood supply from the bronchial arteries, which are branches of the descending aorta. The bronchial veins (which communicate with the pulmonary veins) drain into the azygos and hemiazygos veins.

The alveoli receive deoxygenated blood from the terminal branches of the pulmonary arteries. The tributaries of the pulmonary veins join together and two pulmonary veins leave each lung root.

Lymph Drainage of Lungs. The lymph vessels originate in superficial and deep plexuses. The superficial plexus lies beneath the visceral pleura; the deep plexus travels along the bronchi and toward the root of the lung. The vessels from the deep plexus drain into **pulmonary nodes,** which are located close to the hilum. All lymph from the pulmonary nodes and from the superficial plexus drain into the bronchopulmonary nodes in the hilum. The lymph then drains into the tracheobronchial nodes and from there into the bronchomediastinal lymph trunks.

Nerve Supply of Lungs. Each lung is supplied by the **pulmonary plexus.** The plexus is formed from branches of the sympathetic trunk and receives parasympathetic fibers from the vagus nerve. The sympathetic fibers produce bronchodilation and are vasoconstrictor to the mucosal blood vessels. The parasympathetic fibers produce bronchoconstriction and vasodilation and increase glandular secretion.

Nerve Supply of Pleura

Parietal Pleura. The costal pleura is supplied by intercostal nerves; the mediastinal pleura is supplied by the phrenic nerve; the diaphragmatic pleura is supplied over the domes by the phrenic nerve and around the periphery by the lower intercostal nerves (*Fig. 2–9*).

Visceral Pleura. The visceral pleura receives an autonomic supply from the pulmonary plexus. It is insensitive to common sensations such as pain and touch.

Surface Marking of Lungs. The **apex of each lung** projects up into the neck and extends from the sternoclavicular joint to a point 1 inch (2.5 cm) above the junction of the medial and intermediate thirds of the clavicle (*Fig. 2–8*).

The **anterior border of the right lung** extends from behind the sternoclavicular joint downward and medially to almost reach the midline behind the sternal angle. It then descends to reach the xiphisternal joint (*Fig. 2–8*).

The **lower border of each lung** at the end of quiet inspiration follows a curving line, which crosses the sixth rib in the midclavicular line and the eighth rib in the midaxillary line, and reaches the tenth rib adjacent to the vertebral column (*Fig. 2–8*).

The **posterior border of each lung** extends downward from the spinous process of the seventh cervical vertebra to the level of the tenth thoracic vertebra, and lies about 1.5 inches (4 cm) from the midline (*Fig. 2–8*).

The **oblique fissure** of each lung can be indicated on the surface by a line that extends from the root of the spine of the scapula downward, laterally and anteriorly, following the course of the sixth rib to the costochondral junction. In the left lung the upper lobe lies above and anterior to this line; the lower lobe lies below and posterior to it (*Fig. 2–8*).

In the right lung the **horizontal fissure** may be represented by a line drawn horizontally along the fourth costal cartilage to meet the oblique fissure in the midaxillary line (*Fig. 2–8*). Above the horizontal fissure lies the upper lobe and below it, the middle lobe; below and posterior to the oblique fissure lies the lower lobe.

CLINICAL NOTES

TRACHEOBRONCHIAL ANATOMY AND BRONCHOSCOPY

The mucous membrane of the trachea appears as a glistening pinkish-red lining with spaced whitish rings caused by the underlying cartilages. Just above the carina on the left and anteriorly can be seen the pulsations of the arch of the aorta. The carina appears as a whitish ridge in the midline at the end of the scope (*Fig. 2–10*). The right main bronchus inclines to the right at about 25 degrees and the left at 45 degrees. (The angle of each bronchus is equal, about 55 degrees, in children under 3 years old.) The orifices of the bronchi can then be identified as the distal end of the bronchoscope is advanced down the right and left bronchial trees (*Fig. 2–10*). It is useful to use the figures on the face of a clock and to measure distances from the carina to obtain accurate recognition.

Right Bronchial Tree		
Superior lobar bronchus	3 o'clock	2.5 cm from carina
Middle lobar bronchus	12 o'clock	4.0 cm from carina
Inferior lobar bronchus	6 o'clock	4.5 cm from carina
Left Bronchial Tree		
Superior lobar bronchus	9 o'clock	5 cm from carina
Inferior lobar bronchus	3 o'clock	5.5 cm from carina
Lingular bronchus	Central	5.5 cm from carina

To recognize the more distal bronchi consult Figures 2–1 and 2–10.

Principal Bronchi. The short length of the right principal bronchus makes it difficult to isolate without occluding the superior lobe bronchus. The longer length of the left principal bronchus, however, permits its separation when required for surgery without blocking the lobar bronchi.

Suction catheters and endotracheal tubes are more likely to enter the right more vertical principal

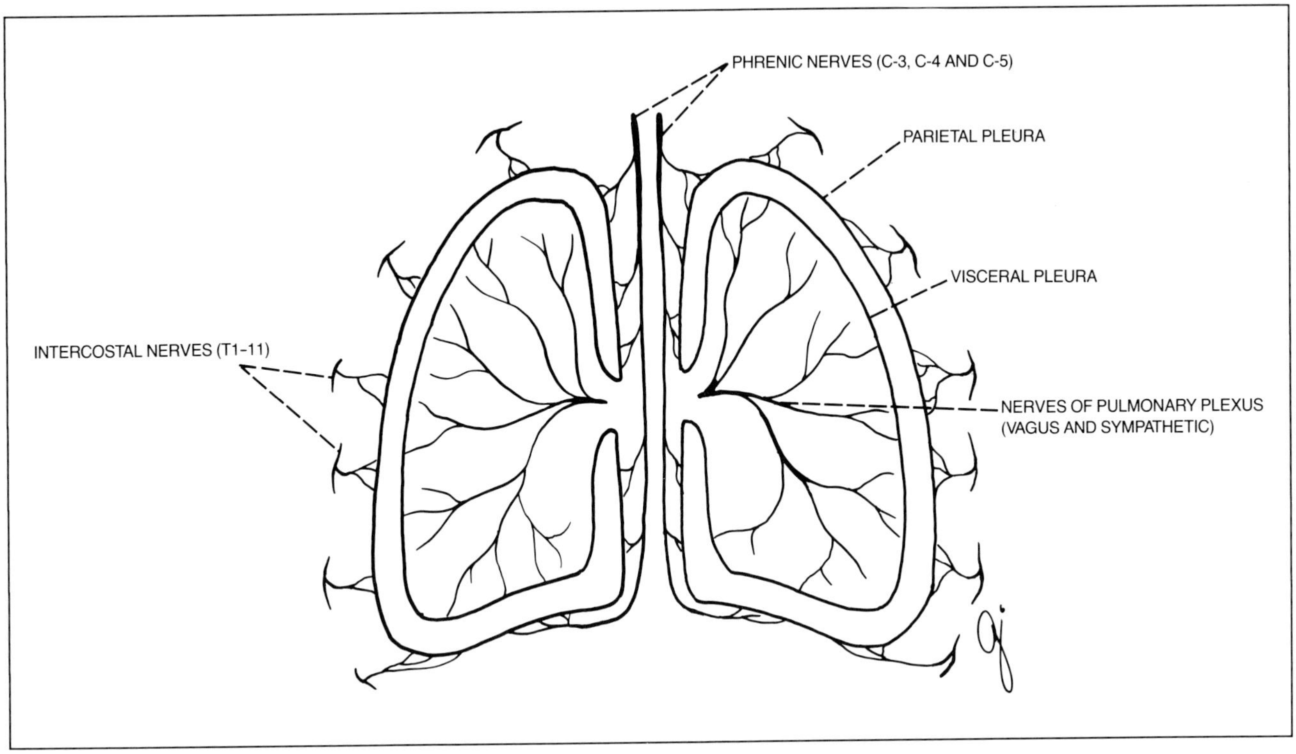

Figure 2–9. Diagram showing the innervation of the parietal and visceral layers of pleura.

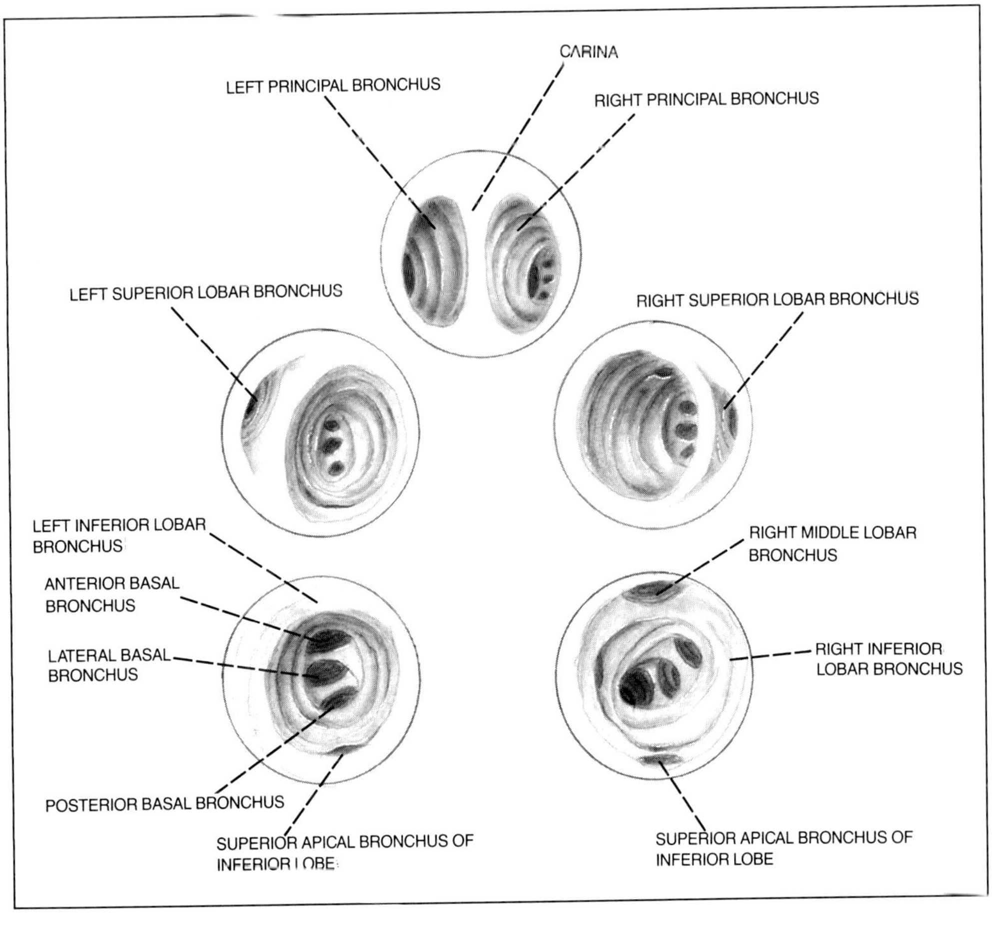

Figure 2–10. Diagrammatic representation of the bronchial tree as seen with a bronchoscope.

bronchus than the obliquely positioned left principal bronchus in adults and older children. In children under the age of 3 years this is not the case.

There is an increase in popularity for the use of double lumen endotracheal tubes for a variety of surgical conditions. These include patients with bronchopleural fistulas, "wet lungs," thoracic aortic aneurysm, and in some institutions, in all unilateral lung procedures. It is, therefore, incumbent on the anesthetist to understand in detail the anatomy. Because both right and left endobronchial intubation is possible with currently available equipment, position of the lobar bronchi in relation to the carina must be kept in mind at all times.

Verification of tube position is done both clinically and by the use of fiberoptic bronchoscopy. Specifics of these techniques are beyond the scope of this book but can be found in any standard anesthesia text.

Postural Accumulation of Secretions in the Unconscious Patient

With the patient lying supine bronchial secretions or inhaled material tends to gravitate to the superior (apical) segment of the inferior lobes. With the patient lying on the side, the inhaled material gravitates to the posterior segments of the superior lobes. When sitting, secretions tend to accumulate in the basal segments of the lower lobes.

THORACIC WALL AND MECHANICS OF RESPIRATION

Thoracic Bones

The thorax has in its walls a bony and cartilaginous cage formed by the sternum, costal cartilages, ribs, and bodies of the thoracic vertebrae (*Fig. 2–11*). The thoracic cage is cone-shaped having a narrow inlet superiorly, and a broad outlet inferiorly. The cage is flattened in its anteroposterior diameter. The function of the cage is to participate in the movements of respiration and protect the underlying thoracic viscera, especially the heart, the lungs, and the upper abdominal viscera, namely, the liver, the spleen, and the stomach.

Sternum

The sternum is a flat narrow bone that lies in the midline of the anterior chest wall and may be divided into three parts: (1) manubrium sterni; (2) body of the sternum; and (3) xiphoid process (*Fig. 2–11*).

The **manubrium sterni** is the triangular upper part of the sternum, and it articulates with the clavicles and the first and upper part of the second costal cartilages on each side. The manubrium has a depression on its superior border called the **suprasternal (jugular) notch.**

The **body of the sternum** articulates above with the manubrium by means of a fibrocartilaginous joint, the **manubriosternal joint.** Below, it articulates with the xiphoid process at the **xiphisternal joint.** On each side are notches for articulation with the lower part of the second costal cartilage and the third to the seventh costal cartilages (*Fig. 2–11*).

The **xiphoid process** is the lowest and smallest part of the sternum. It is a plate of hyaline cartilage that becomes ossified at its proximal end in adult life. The xiphoid process has no ribs or costal cartilages attached to it. However, it does provide attachment for some muscles of the anterior abdominal wall.

The sternal angle (angle of Louis), formed by the articulation of the manubrium with the body of the sternum, can be recognized by the presence of a transverse ridge on the anterior aspect of the sternum. The transverse ridge lies at the level of the second costal cartilage, the point from which all costal cartilages and ribs can be counted. The sternal angle lies opposite the intervertebral disk between the fourth and fifth thoracic vertebrae. The xiphisternal joint lies opposite the body of the ninth thoracic vertebra (*Fig. 2–12*).

Costal Cartilages

Costal cartilages are bars of hyaline cartilage connecting the upper seven ribs to the lateral edge of the sternum, and the eighth, ninth, and tenth ribs to the cartilage immediately above. The cartilages of the 11th and 12th ribs end in the abdominal musculature (*Fig. 2–11*).

Ribs

There are 12 pairs of ribs, the majority of which encircle the trunk from the vertebral column behind to the region of the sternum in front. The ribs are attached posteriorly to the thoracic vertebrae (*Fig. 2–11*). The upper seven pairs are attached anteriorly to the sternum by their costal cartilages. The eighth, ninth, and tenth pairs of ribs are attached anteriorly to each other and to the seventh rib by means of their costal cartilages. The 11th and 12th pairs have no anterior attachment and are referred to as **floating ribs.**

A **typical rib** is a long, twisted flat bone having a rounded, smooth superior border and a sharp, thin inferior border (*Fig. 2–11*). The inferior border overhangs and forms the **costal groove,** which accommodates the intercostal vessels and nerve.

A rib has a **head, neck, tubercle, angle,** and **shaft.** The **head** has two facets for articulation with the numerically corresponding vertebral body and that of the vertebra immediately above. The **neck** is a constricted portion situated between the head and the tubercle. The tubercle has a facet for articulation with the transverse process of the numerically corresponding vertebra. The **angle** is where the rib bends sharply forward. The **shaft** of each rib is attached to the corresponding costal cartilage.

The first rib is **atypical.** It is important because of its close relationship to the lower nerves of the brachial

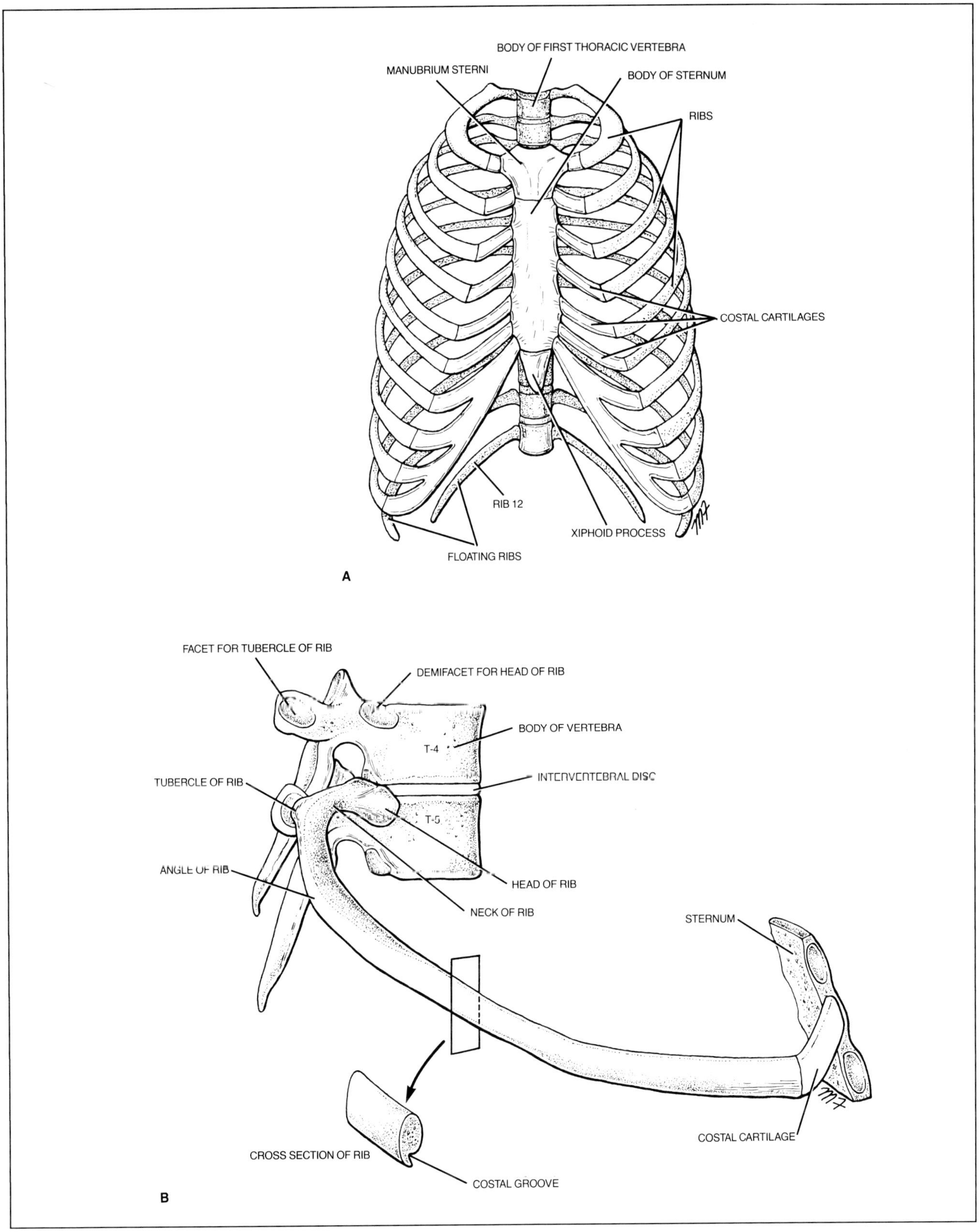

Figure 2–11. A. Anterior view of the bones and cartilages forming the thoracic skeleton. **B.** Fifth right rib as it articulates with the vertebral column posteriorly and with the sternum anteriorly. Note that the rib head articulates with the vertebral body of its own number and that of the vertebra immediately above. Note also the presence of a costal groove along the inferior border of the rib.

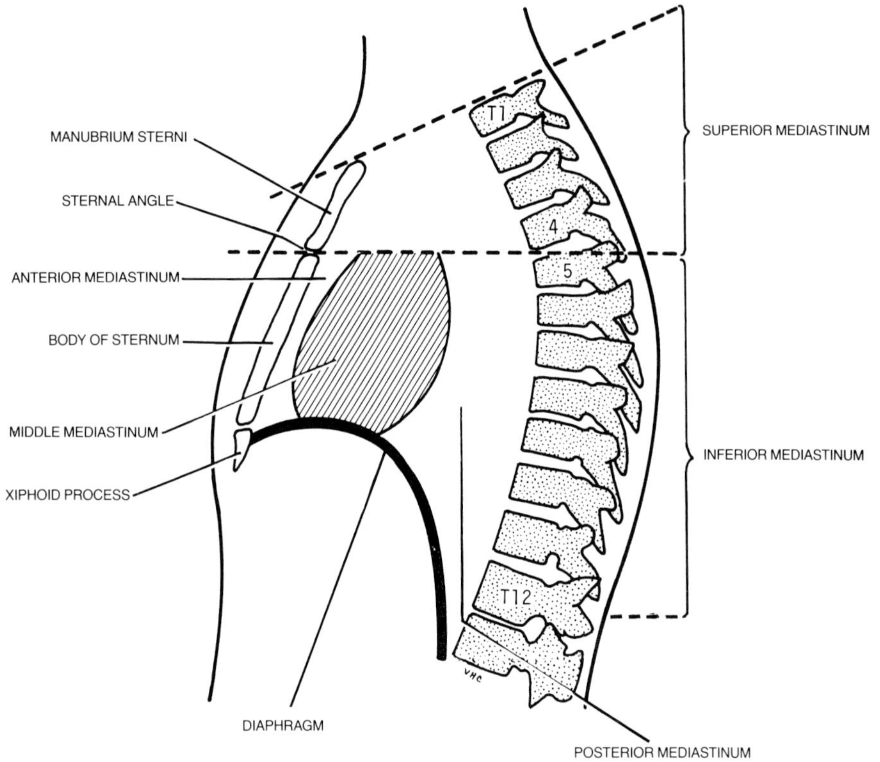

Figure 2–12. Subdivisions of mediastinum.

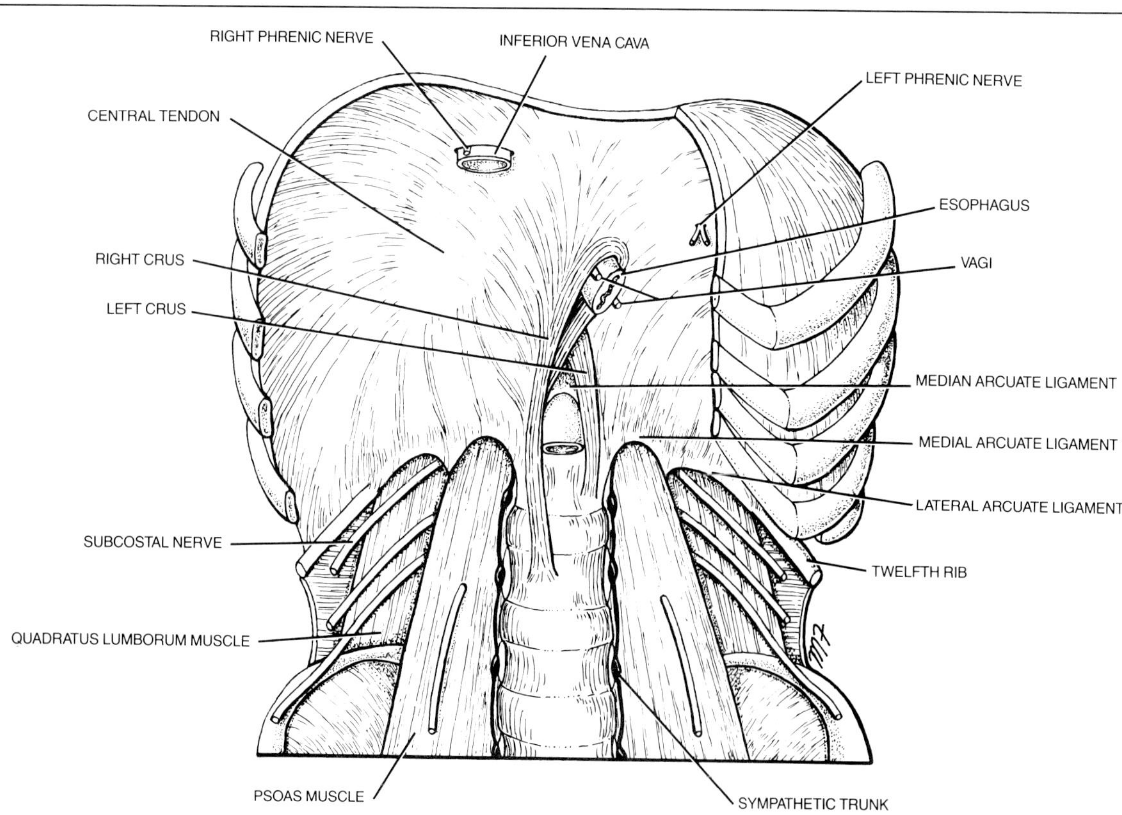

Figure 2–13. Diaphragm as seen from below. The anterior portion of the right side has been removed. Note sternal, costal, and vertebral origins of muscle and important structures that pass through it.

plexus and the main vessels to the arm, namely, the subclavian artery and vein that cross the rib. This rib is flattened from above downward. It has a tubercle on the inner border known as the **scalene tubercle,** for the insertion of the scalenus anterior muscle.

Joints of the Ribs. A typical rib articulates by its head and by its tubercle with the vertebral column posteriorly. Anteriorly, the first seven ribs, with their costal cartilages, articulate with the sternum (*Fig. 2–11*); the eighth, ninth, and tenth ribs and their costal cartilages articulate with the costal cartilages above. The 11th and 12th ribs are free floating ribs anteriorly.

Joints of the Heads of the Ribs. The first and the lower three ribs have a simple synovial joint with its corresponding vertebral body. From the second to the ninth ribs, the head articulates by means of a synovial joint with facets on the corresponding vertebral body and that of the vertebra above. Within these latter joints there is a strong **intra-articular ligament** that connects the head to the intervertebral disc.

Joints of the Tubercles of the Ribs. The medial facet of the tubercle articulates by means of a synovial joint with the transverse process of the vertebra to which it corresponds numerically. (It is absent on the 11th and 12th ribs.)

Costochondral Joints. The costochondral joints are primary cartilaginous joints and no movements are possible.

Joints of the Costal Cartilages with the Sternum. The first costal cartilages articulate with the manubrium by means of primary cartilaginous joints permitting no movement (*Fig. 2–11*). The second costal cartilages articulate with the manubrium and the body of the sternum by a synovial joint. The third to the seventh costal cartilages articulate with the lateral border of the body of the sternum by synovial joints. (The sixth, seventh, eighth, ninth, and tenth costal cartilages articulate with one another along their borders by small synovial joints. The cartilages of the 11th and 12th ribs are embedded in the abdominal musculature.)

Movements of the Ribs and Costal Cartilages. The first ribs and their costal cartilages are fixed to the manubrium and are immobile. The raising and lowering of the ribs during respiration are accompanied by gliding movements in both the joints of the head and the tubercle permitting the neck of each rib to rotate around its own axis.

Joints of the Sternum. The **manubriosternal joint** is a secondary cartilaginous joint between the manubrium and the body of the sternum. The bony surfaces are covered with hyaline cartilage and joined by a disc of fibrocartilage. A small amount of angular movement is possible during respiration. The **xiphisternal joint** is a secondary cartilaginous joint between the xiphoid process (cartilage) and the body of the sternum. The xiphoid process usually fuses with the body of the sternum during middle age.

Thoracic Inlet*

The thorax opens into the root of the neck by a narrow aperture known as the **thoracic inlet** (*Fig. 2–11*). It is bounded by the superior border of the manubrium sterni, by the medial borders of the first ribs, and by the body of the first thoracic vertebra. The thoracic inlet transmits structures that pass between the thorax and the neck (esophagus, trachea, blood vessels, etc.) and for the most part lie close to the midline. On either side of these structures lies the apex of the lung with its visceral and parietal layers of pleura and protected by a deep fascial layer called the **suprapleural membrane.**

Intercostal Spaces

The spaces between the ribs are called intercostal spaces. Each space contains three muscles of respiration: (1) the external intercostal muscle, (2) the internal intercostal muscle, and (3) the transversus thoracis muscle. This latter muscle is related internally to the **endothoracic fascia** and parietal pleura. The intercostal nerves and blood vessels run between the intermediate and deepest layer of muscles (Chap. 6, p. 201). They are arranged in the following order from above downward: intercostal vein, intercostal artery, and intercostal nerve (i.e., VAN).

Thoracic Outlet

Inferiorly, the thorax opens into the abdomen by a wide aperture known as the **thoracic outlet** (*Fig. 2–11*). It is bounded by the xiphisternal joint, the costal margin, and by the body of the 12th thoracic vertebra. It is closed by the diaphragm, which is pierced by the structures that pass between the thorax and the abdomen.

Mediastinum

The mediastinum is a movable partition that lies between the pleurae and the lungs. It extends superiorly to the thoracic inlet and inferiorly to the diaphragm. It extends anteriorly to the sternum and posteriorly to the thoracic vertebrae (*Fig. 2–12*).

For descriptive purposes the mediastinum is divided into superior and inferior mediastina by an imaginary plane passing from the sternal angle anteriorly to the lower border of the body of the fourth thoracic vertebra

---

*The inlet is often loosely referred to as the thoracic outlet as important vessels and nerves emerge from the thorax to enter the neck and upper limbs.

posteriorly. The inferior mediastinum is further subdivided into the middle mediastinum, which consists of the pericardium and the heart; the **anterior mediastinum,** which is a space between the pericardium and the sternum; and the **posterior mediastinum,** which lies between the pericardium and the vertebral column.

Contents of the Superior Mediastinum

Anterior to Posterior. Remains of thymus; brachiocephalic veins, upper part of superior vena cava; branchiocephalic artery, left common carotid artery, left subclavian artery, arch of aorta, both phrenic and vagus nerves, left recurrent laryngeal and cardiac nerves, trachea and lymph nodes, esophagus and thoracic duct, and sympathetic trunks.

Contents of the Anterior Mediastinum. Sternopericardial ligaments, lymph nodes, remains of thymus.

Contents of the Middle Mediastinum. Pericardium, heart and roots of great vessels, phrenic nerves, bifurcation of trachea, lymph nodes.

Contents of the Posterior Mediastinum. Descending thoracic aorta, esophagus, thoracic duct, azygos, and hemiazygos veins, vagus nerves, splanchnic nerves, sympathetic trunks, lymph nodes.

Muscles of the Thorax

Diaphragm. The diaphragm is the most important muscle of respiration (*Fig. 2–13*). It is dome-shaped and consists of a peripheral muscular part, which arises from the margins of the thoracic outlet, and centrally placed tendon. The origin of the diaphragm may be divided into three parts:

1. A **sternal part** consisting of small right and left portions arising from the posterior surface of the xiphoid process.
2. A **costal part** that arises from the deep surfaces of the lower six costal cartilages.
3. A vertebral part arising by means of vertical columns or **crura** and from the arcuate ligaments.

The **right crus** arises from the sides of the bodies of the first three lumbar vertebrae and the intervertebral discs; the **left crus** arises from the sides of the bodies of the first two lumbar vertebrae and the intervertebral disc.

Lateral to the crura, the diaphragm arises from the **medial** and **lateral arcuate ligaments.** The diaphragm is inserted into a **central tendon.** The superior surface of the tendon is partially fused with the inferior surface of the fibrous pericardium. Some of the muscle fibers of the right crus pass up to the left and surround the esophagus in a sling-like loop. These fibers appear to act as a sphincter and assist in preventing regurgitation of the stomach contents into the thoracic part of the esophagus.

Shape of the Diaphragm. As seen in front, the diaphragm curves up into right and left domes. The right dome reaches as high as the upper border of the fifth rib, and the left dome reaches the lower border of the fifth rib. The central tendon lies at the level of the xiphisternal joint. Note that the domes support the right and left lungs whereas the central tendon supports the heart.

Nerve Supply of the Diaphragm. The motor nerve supply is from the phrenic nerves. The sensory nerve supply to the parietal pleura and peritoneum covering the upper and lower surfaces of the central part of the diaphragm is also from the phrenic nerves. However, the sensory nerve supply to the pleura and peritonium covering the upper and lower surfaces of the peripheral part of the diaphragm is from the lower six intercostal nerves.

Openings in the Diaphragm. The diaphragm has three main openings (*Fig. 2–13*).

The **aortic opening** lies anterior to the body of the 12th thoracic vertebra between the crura. It transmits the aorta, the thoracic duct, and the azygos vein.

The **esophageal opening** lies at the level of the tenth thoracic vertebra in a sling of muscle fibers derived from the right crus. It transmits the esophagus, the right and left vagus nerves, the esophageal branches of the left gastric vessels, and the lymphatics from the lower one-third of the esophagus.

The caval opening lies at the level of the eighth thoracic vertebra in the central tendon. It transmits the inferior vena cava and the terminal branches of the right phrenic nerve.

In addition to these structures, the splanchnic nerves pierce the crura, the sympathetic trunk passes posterior to the medial arcuate ligament on each side, and the superior epigastric vessels pass between the sternal and costal origins of the diaphragm on each side. The origins, insertions, nerve supply, and actions of the thoracic muscles are summarized in Table 2–1.

MECHANICS OF RESPIRATION

Respiration consists of two phases, inspiration and expiration. They are accomplished by the alternate increase and decrease of the capacity of the thoracic cavity. The rate varies between 16 and 20 per minute in normal resting subjects, and is faster in children and slower in the elderly.

TABLE 2–1. MUSCLES OF THE THORAX

Name of Muscle	Origin	Insertion	Nerve Supply	Action
Diaphragm	Xiphoid process Lower six costal cartilages First three lumbar vertebrae by crura and medial and lateral arcuate ligaments	Central tendon	Phrenic nerves	Most important muscle of inspiration, increases vertical diameter of thorax by pulling central tendon downward, assists in raising lower ribs
Intercostal Muscles				
External intercostal (fibers pass downward and forward)	Inferior border of rib above	Superior border of rib below	Intercostal nerves	With first rib fixed, they raise ribs during inspiration and thus increase anteroposterior and transverse diameters of thorax. With last rib fixed by abdominal muscles, they lower ribs during expiration
Internal intercostal (fibers pass downward and backward)	Inferior border of rib above	Superior border of rib below	Intercostal nerves	Assists external intercostal muscles
Transversus thoracis (fibers pass transversely; forms incomplete layer of muscle)	Adjacent ribs	Adjacent ribs	Intercostal nerves	Assists external and internal intercostal muscles
Levatores costarum (12 in number)	Transverse processes of 7th cervical to 11th thoracic vertebra	Superior border of ribs	Posterior rami of thoracic spinal nerves	Elevates ribs
Serratus posterior superior	Ligamentum nuchae and upper thoracic spines	Upper ribs	Intercostal nerves	Elevates ribs
Serratus posterior inferior	Lower thoracic and upper lumbar spines	Lower ribs	Intercostal nerves	Lowers ribs

Inspiration

Quiet Inspiration. The capacity of the thoracic cavity can be increased by elongating all its diameters, and this will result in air under atmospheric pressure entering the cavity through the trachea.

Consider now the three diameters of the thoracic cavity and how they may be increased (*Fig. 2–14*).

Vertical Diameter. Theoretically, the roof could be raised and the floor lowered. The roof is formed by the suprapleural membrane and is fixed. On the other hand, the floor is formed by the mobile diaphragm. When the diaphragm contracts, the domes become flattened and the level of the diaphragm is lowered.

Anteroposterior Diameter. If the downward-sloping ribs were raised at their sternal ends, the anteroposterior diameter of the thoracic cavity would be increased and the lower end of the sternum would be thrust forward. This can be brought about by fixing the first rib by the contraction of the scaleni muscles of the neck and contracting the intercostal muscles. By this means all the ribs are drawn together and raised toward the first rib.

Transverse Diameter. The ribs articulate in front with the sternum via their costal cartilages and behind with the vertebral column. Because the ribs curve downward as well as forward around the chest wall, they resemble bucket handles. It follows, therefore, that if the ribs are raised (like bucket handles), the transverse diameter of the thoracic cavity will be increased. As described previously, this can be accomplished by fixing the first rib and raising the other ribs to it by contracting the intercostal muscles.

An additional factor, which must not be overlooked, is the effect of the descent of the diaphragm on the abdominal viscera, and the tone of the muscles of the anterior abdominal wall. As the diaphragm descends on inspiration, the intra-abdominal pressure rises. This rise in pressure is accommodated by the reciprocal relaxation of the abdominal wall musculature. However, a point is reached when no further abdominal relaxation is possible, and the liver and other upper abdominal viscera act as a platform that resists further diaphragmatic descent. On further contraction the diaphragm will now have its central tendon supported from below, and its shortening muscle fibers will assist the intercostal muscles in raising the lower ribs.

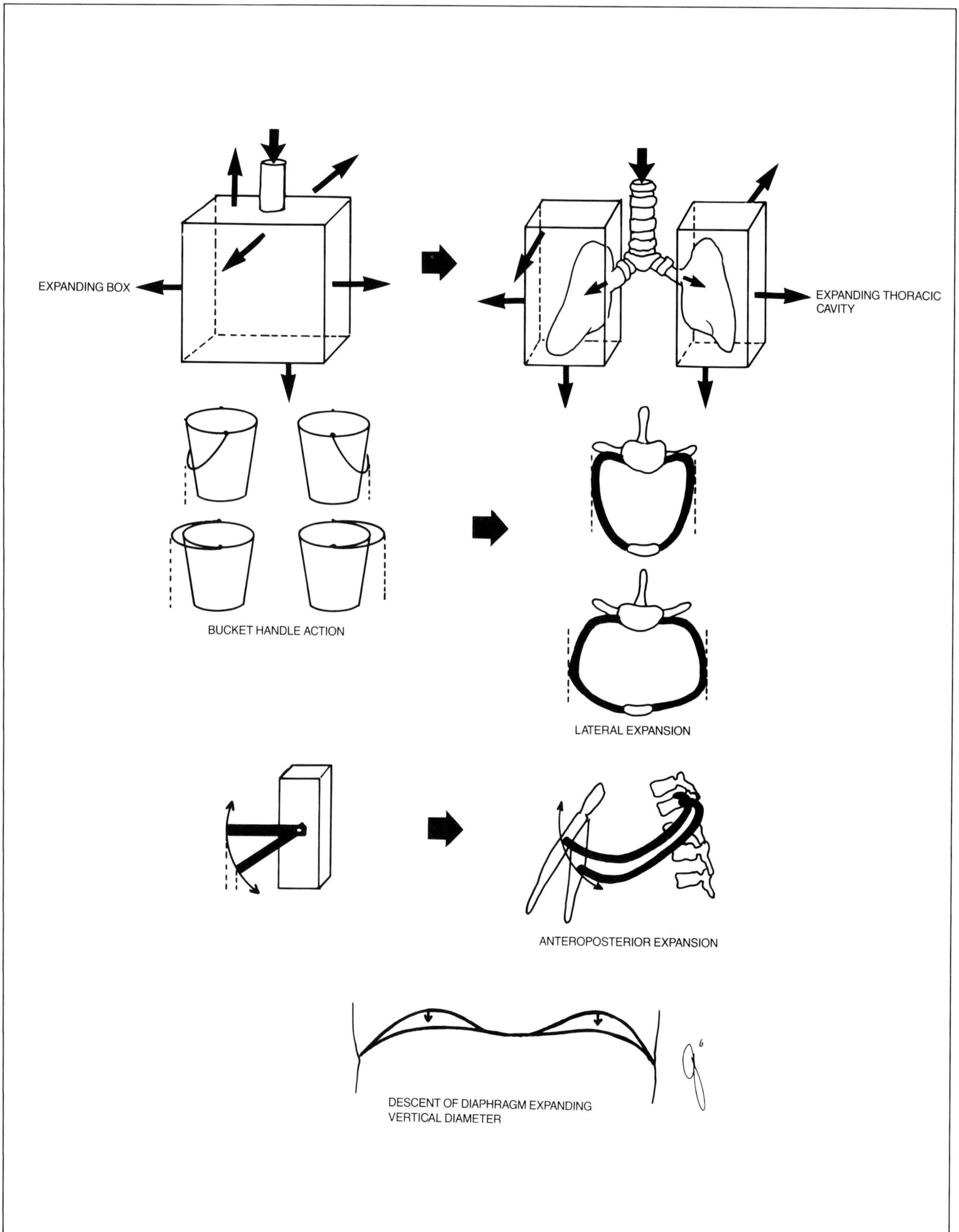

Figure 2–14. Diagrams showing the different ways in which the capacity of the thoracic cavity is increased during inspiration.

Apart from the diaphragm and the intercostals, other less important muscles also contract on inspiration and assist in elevating the ribs, namely, the **levatores costarum muscles** and the **serratus posterior superior muscles.**

Forced Inspiration. In deep forced inspiration there is a maximum increase in the capacity of the thoracic cavity. Every muscle that can raise the ribs is brought into action, including the scalenus anterior and medius and the sternocleidomastoid. In respiratory distress, the action of all the muscles already engaged becomes more violent and the scapulae are fixed by the trapezius, levator scapulae, and rhomboid muscles, enabling the serratus anterior and pectoralis minor to pull up the ribs. If the upper limbs can be supported by grasping a chair back or table, the sternal origin of the pectoralis major muscles can also assist the process.

Lung Changes on Inspiration

In inspiration, the root of the lung descends and the level of the bifurcation of the trachea may be lowered by as much as two vertebrae. The bronchi elongate and dilate and the alveolar capillaries dilate, thus assisting the pulmonary circulation. Air is drawn into the bronchial tree as the result of the positive atmospheric pressure exerted through the upper part of the respiratory tract, and the negative pressure in the intrapleural space brought about by the increased capacity of the thoracic cavity. With the expansion of the lungs, the elastic tissue in the bronchial walls and connective tissue is stretched. As the diaphragm descends, the costodiaphragmatic recess of the pleural cavity opens up, and the expanding sharp lower edges of the lungs descend to a lower level.

Expiration

Quiet Expiration. Quiet expiration is largely a passive phenomenon and is brought about by the elastic recoil of the lungs, the relaxation of the intercostal muscles and diaphragm, and an increase in tone of the muscles of the anterior abdominal wall, which forces the relaxing diaphragm upward. The **serratus posterior inferior muscles** play a minor role in pulling down the lower ribs.

Forced Expiration. Forced expiration is an active process brought about by the active contraction of the musculature of the anterior abdominal wall. The quadratus lumborum muscle also contracts and pulls down the 12th ribs. It is conceivable that under these circumstances some of the intercostal muscles may contract, pull the ribs together, and depress them toward the lowered 12th ribs. The serratus posterior inferior and the latissimus dorsi muscles may also play a minor role.

Lung Changes on Expiration

In expiration, the roots of the lungs ascend along with the bifurcation of the trachea. The bronchi shorten and contract. The elastic tissue of the lungs recoils, and the lungs become reduced in size. With the upward movement of the diaphragm, increasing areas of the diaphragmatic and costal parietal pleura come into apposition, and the costodiaphragmatic recess becomes reduced in size. The lower margins of the lungs shrink and rise to a higher level.

Types of Respiration

In babies and young children the ribs are nearly horizontal. Thus, they have to rely mainly on the descent of the diaphragm to increase their thoracic capacity on inspiration. Because this is accompanied by a marked inward and outward excursion of the anterior abdominal wall, which is easily seen, respiration at this age is referred to as the abdominal type of respiration.

After the second year the ribs become more oblique, and the adult form of respiration is established.

It is interesting to note that in the adult there is a sexual difference in the type of respiratory movements. The female tends to rely mainly on the movements of the ribs rather than the descent of the diaphragm on inspiration. This is referred to as the thoracic type of respiration. The male uses both the thoracic and abdominal forms of respiration, but mainly the abdominal form.

Radiographic Appearances of the Thorax

The more important structures seen in standard posteroanterior and lateral radiographs of the chest are shown in Figures 2–15 and 2–16. Examples of computed tomography scans of the chest are shown in Figure 2–17.

CLINICAL NOTES

Pleurae and Lungs

It is hardly necessary to emphasize the importance of knowing the surface markings of the pleural reflections and the lobes of the lungs. When listening to the breath sounds of the respiratory tract, it should be possible to have a mental image of the structures that lie beneath the stethoscope. It is well to remember that the upper lobes of the lungs are most easily examined from the front of the chest and lower lobes from the back. In the axillae, areas of all lobes may be examined.

The cervical dome of the pleura and the apex of the lungs extend up into the neck so that at their highest point they lie about 1 inch (2.5 cm) above the clavicle. Consequently they are vulnerable to maneuvers that are ordinarily performed by anesthesiologists. Particular care needs be taken when probing for the internal jugular vein for central venous pressure, hyperalimentation, or Swan-Ganz lines.

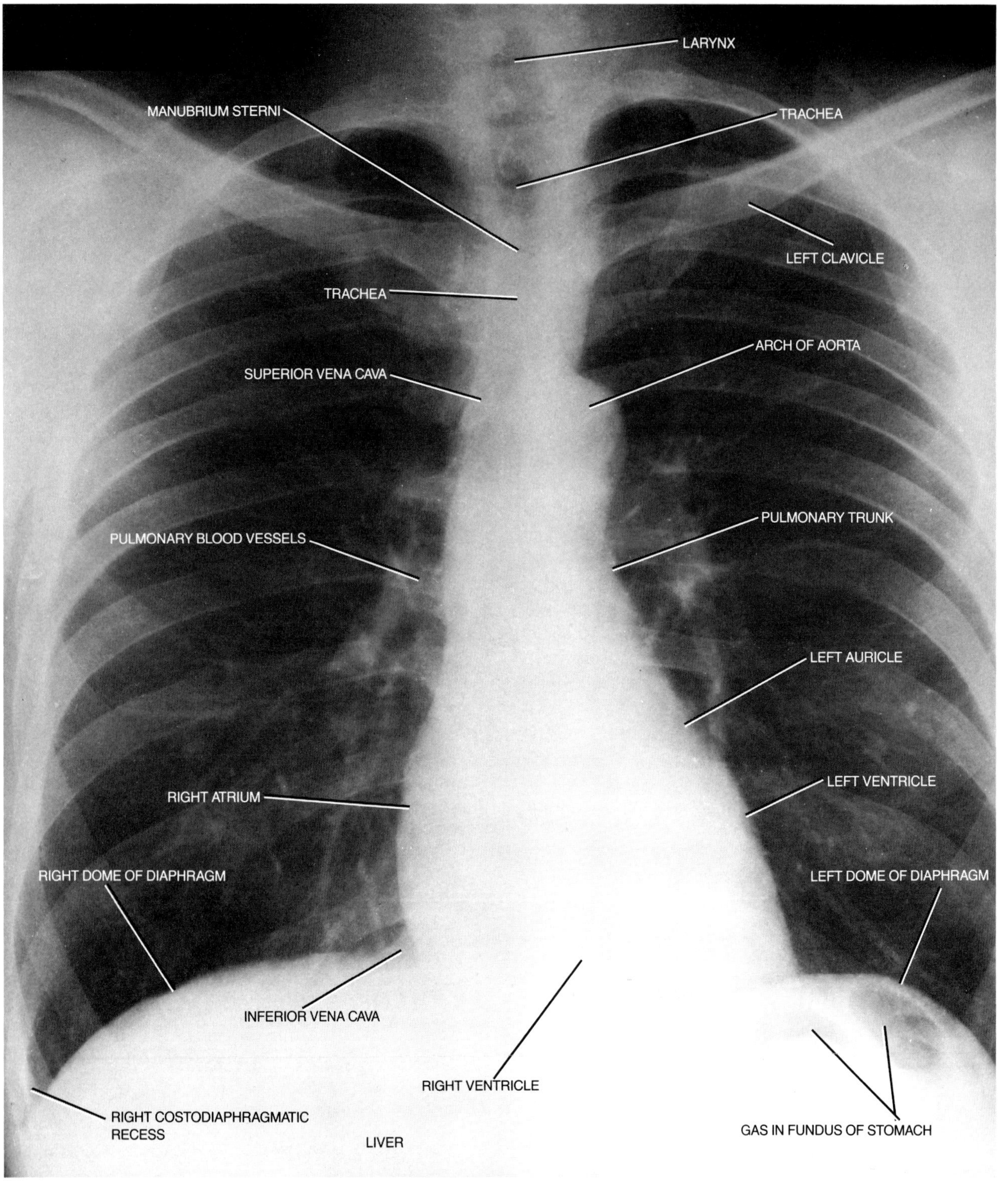

Figure 2–15. Posteroanterior radiograph of chest of a normal adult male.

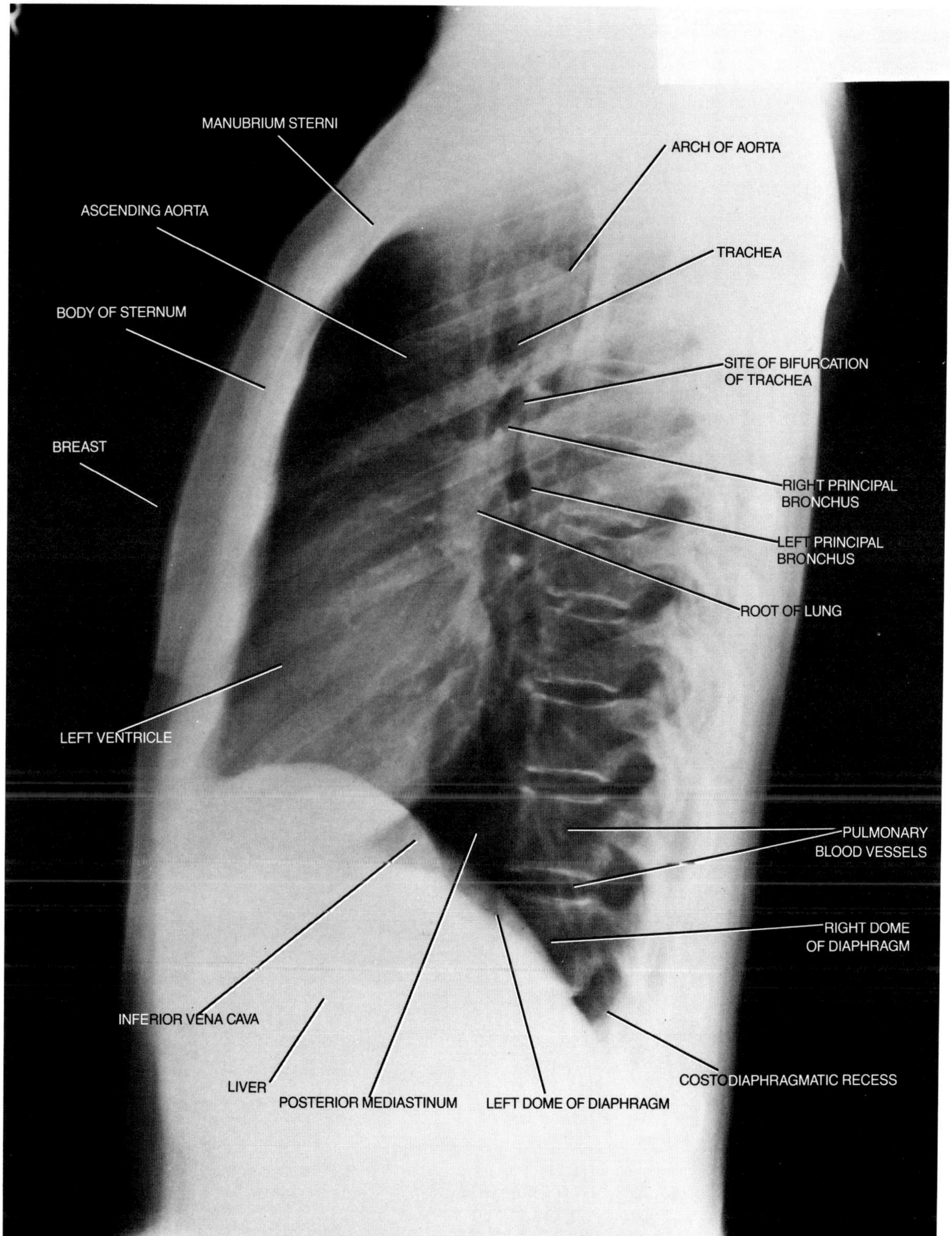

Figure 2–16. Left lateral radiograph of chest of a normal adult female.

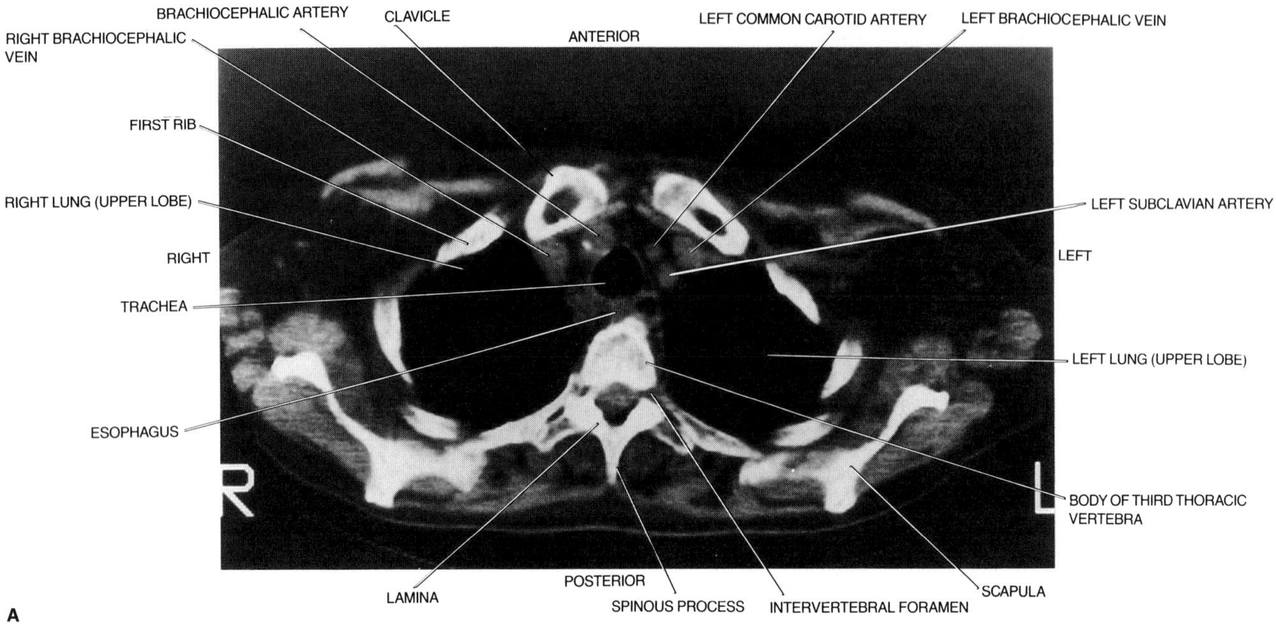

A

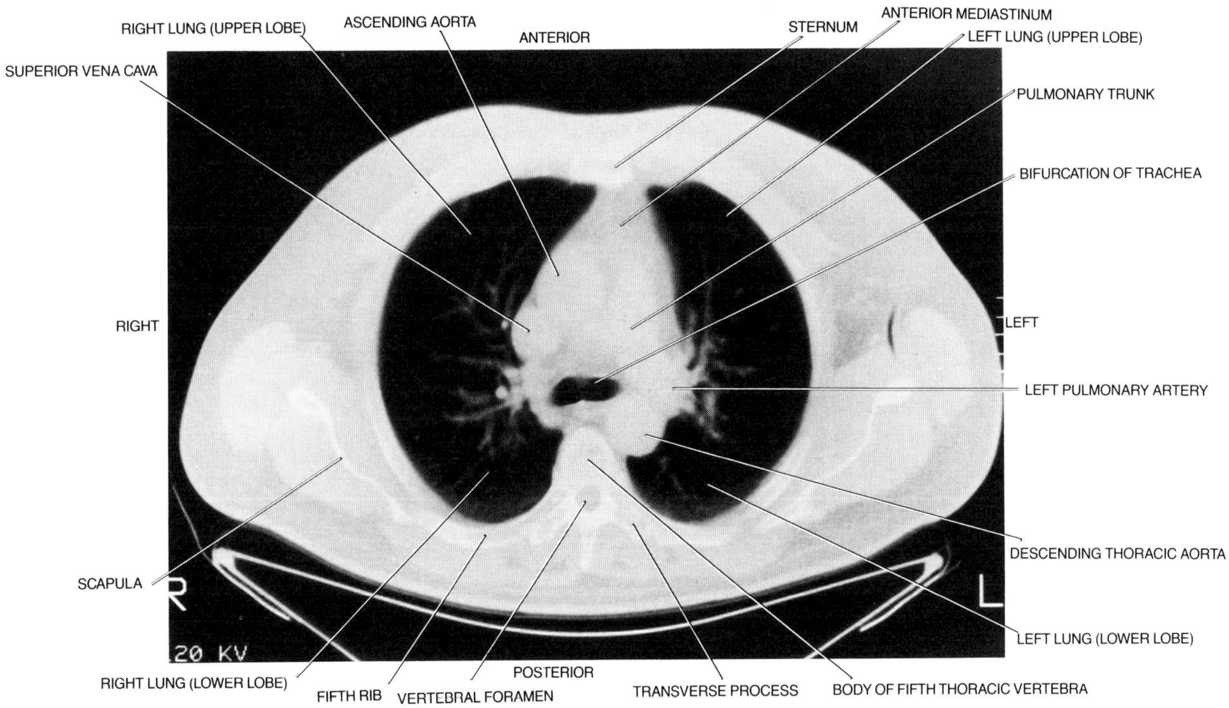

B

Figure 2–17. A. Computed tomography (CT) scan of upper part of thorax at the level of the third thoracic vertebra. The section is viewed from below. **B.** CT scan of middle part of thorax, at level of fifth thoracic vertebra. The section is viewed from below.

In addition, if the supraclavicular approach to the brachial plexus is attempted, a modified technique is recommended. In this technique the needle is inserted higher in the neck (C-6) between the middle and anterior scalene muscles rather than just above the clavicle searching for the plexus as it rides over the first rib. A significant pneumothorax incidence occurs with this latter approach.

Lung tissue and visceral pleura are devoid of pain-sensitive nerve endings, therefore that pain in the chest is always the result of conditions affecting the surrounding structures. Once lung disease crosses the visceral pleura and the pleural cavity to involve the parietal pleura, pain becomes a prominent feature. Lobar pneumonia with pleurisy, for example, produces a severe tearing pain, accentuated by inspiring deeply or coughing. Because the lower part of the costal parietal pleura receives its sensory innervation from the lower five intercostal nerves, which also innervate the skin of the anterior abdominal wall, pleurisy in this area commonly produces pain that is referred to the abdomen. This has sometimes resulted in a mistaken diagnosis of an acute abdominal lesion.

Mediastinum

The mediastinum is a very mobile structure. If air should enter the pleural cavity (pneumothorax), the lung on that side might collapse and the mediastinum would be displaced to the opposite side. On examination, this condition would reveal itself by the trachea and the heart being displaced to the opposite side.

CLINICAL PROBLEMS

1. A child 2 years old inhaled part of a toy, which was later found, after a chest radiograph, to have lodged in a principal bronchus. In a young child into which bronchus does an inhaled foreign body most commonly lodge? In an adult, which is the most likely bronchus to receive a dislodged tooth if it is inhaled?
2. During an emergency operation, an unconscious patient vomited and inhaled part of the stomach contents. After sucking out as much as possible of the material from the bronchial tree, the anesthetist was still very concerned about the possibility that the patient would develop a postoperative lung abscess. If the patient was lying on the left lateral side on the operating table when he vomited, which segments of which lobe of which lung are most likely to become infected?
3. When passing a laryngoscope it is important to get the mouth, the oropharynx, and the larynx into one plane. How do you bring the axes of the oropharynx and the larynx in line? How do you bring the axis of the mouth in line with the other axes? Describe the structures in their correct order that you can view through a laryngoscope from the base of the tongue down to the trachea.
4. Assuming that you have passed a bronchoscope into the left principal bronchus, describe the various orifices that you can see as you slowly lower the bronchoscope into the bronchial tree.
5. Define a bronchopulmonary segment. What is the relationship of the veins, arteries, and nerves to the connective tissue septa that separate adjacent segments? What is the clinical significance of this relationship?
6. As a general rule it can be stated that the further one descends into the bronchial tree there is less cartilage in its walls and more smooth muscle. What is the importance of this fact? Name the first respiratory passage that ceases to have cartilage in its walls in the bronchial tree.
7. Compare the structures of the trachea, a bronchus, a bronchiole, and a respiratory bronchiole.
8. Describe the electron microscopic structure of the respiratory membrane. What is the function of the different types of cells found lining the alveolar walls?

ANSWERS TO CLINICAL PROBLEMS

1. Under the age of 3 years the angulation of the two principal bronchi at the carina is equal on both sides (55 degrees). Both bronchi are, therefore, equally likely to receive an inhaled foreign body. In the adult, however, the right principal bronchus is the most direct continuation of the trachea and, therefore, the right principal bronchus is more likely to receive an inhaled foreign body, such as a tooth.
2. With a patient lying on his left lateral side, inhaled material is likely to accumulate in the posterior segment of the left upper lobe.
3. The axis of the oropharnyx and the larynx are brought into direct line by flexing the cervical part of the vertebral column. The axis of the mouth is brought into line with the oropharnx by extending the atlanto-occipital joints.

   The following structures may be viewed: (1) base of tongue; (2) median glossoepiglottic fold and two lateral glossoepiglottic folds and the valleculae on each side of the median fold; (3) the upper edge of the epiglottis, and aditus into the larynx, bounded in front by the epiglottis with its tubercle, and laterally by the aryepiglottic folds. The rounded elevations of the cuneiform and corniculate cartilages in the folds can be recognized; (4) the reddish fixed vestibular folds; (5) the whitish mobile vocal folds with the rima glottidis; (6) below the glottis can be seen the interior of the trachea with the upper two or three rings.

4. First, the left superior lobar bronchus at 9 o'clock on the lateral wall about 5 cm from the carina. Second, the lingular bronchus is central at 5.5 cm from the carina. On entering the inferior lobar bronchus, the superior (apical) segmental bronchus arises from the posterior surface and the orifices of anterior, lateral, and posterior basal bronchi can be seen straight ahead.

5. A bronchopulmonary segment is a subdivision of a lobe of a lung. It is pyramid shaped, with its apex pointing toward the root of the lung and its base facing the lung surface. Each segment has its own segmental bronchus (tertiary bronchus), a branch of the pulmonary artery, a tributary of the pulmonary vein, lymphatic vessels, and autonomic nerve fibers derived from the pulmonary plexus. All these structures, with the exception of the pulmonary veins, lie within the segment. Tributaries of the pulmonary veins tend to run in the connective tissue septa between adjacent segments. This fact has to be remembered during surgical resection of a segment of the lung.

6. On tracing the respiratory tract downward, one finds that the hyaline cartilaginous tracheal rings are replaced by large, irregular plates of cartilage. These plates become smaller as the bronchi diminish in diameter until they finally disappear in the bronchioles. As the plates of hyaline cartilage in the bronchial walls diminish in size, the amount of smooth muscle increases. The submucosa of the bronchioles possesses a complete layer of circularly arranged smooth muscle fibers. The presence of cartilage in the walls of the air passages keeps the lumen open. Parasympathetic stimulation of the smooth muscle in the walls of the smaller air passages can greatly reduce the size of the lumina and thus reduce the air flow.

7. The structures of the trachea, a bronchus, a bronchiole, and a respiratory bronchiole are described on pages 31 and 40. Note the differences in the epithelium lining these passages.

8. The respiratory membrane measures about 0.5 μ in thickness and consists of the following layers: (1) surfactant-containing fluid lining the alveolar wall produced by the type II alveolar cells, (2) the alveolar squamous epithelium or type I cell, (3) the epithelial basement membrane, (4) a minute tissue space, (5) the blood capillary basement membrane, which fuses with the epithelial basement membrane in many places, and (6) the capillary squamous endothelium.

The squamous epithelial cells (type I) lining the alveolar walls permit gaseous exchange to readily occur between the lumen and the blood. The occasional cuboidal cell (type II) found between the squamous cells, is responsible for the secretion of surfactant.

CHAPTER 3

Cardiovascular System

The cardiovascular system is the transport system of the body and conveys nutrients, gases, hormones, and the products of metabolism. A knowledge of the heart and the principal blood vessels is essential to understand how anesthetic agents are distributed throughout the body. Needless to say, a correctly functioning cardiovascular system is necessary to ensure that the vital organs receive an adequate supply of oxygen.

HEART

The heart is a hollow muscular organ that is somewhat pyramidal in shape and lies within the pericardium in the mediastinum. It is connected at its base to the great blood vessels but otherwise lies free within the pericardium. The heart has three surfaces: a sternocostal surface (anterior), a diaphragmatic surface (inferior); and a base (posterior). It also has an apex that is directed downward, forward, and to the left.

Surface Markings of the Heart (Fig. 3–1)

Superior. Line from the second left costal cartilage to the third right costal cartilage.

Inferior. Line from the sixth right costal cartilage to the apex.

Right Border. Line drawn from a point on the third right costal cartilage 0.5 inches (1.3 cm) from the edge of the sternum downward to a point on the sixth right costal cartilage 0.5 inches (1.3 cm) from the edge of the sternum.

Left Border. Line drawn from the second left costal cartilage 0.5 inches (1.3 cm) from the edge of the sternum to the apex of the heart.

The **apex** of the heart lies at the level of the fifth left intercostal space, 3.5 inches (9 cm) from the midline.

Surface Markings of the Heart Valves. The heart valves may be auscultated in the following areas (*Fig. 3–1*):

- The **tricuspid valve,** over the right half of the lower end of the body of the sternum
- The **mitral valve,** over the apex of the heart
- The **pulmonary valve,** over the medial end of the second left intercostal space
- The **aortic valve,** over the medial end of the second right intercostal space

Structure of the Heart

The heart is divided longitudinally by a vertical septum and transversely by a vertical septum producing four chambers, the right and left atria and the right and left ventricles. The right atrium lies anterior to the left atrium and the right ventricle anterior to the left ventricle (*Fig. 3–2*).

The walls of the heart are composed of cardiac muscle, the **myocardium,** covered externally with serous pericardium, called the **epicardium,** and lined internally with a layer of endothelium, the **endocardium**.

The **skeleton of the heart** (*Fig. 3–5*) consists of fibrous rings that surround the atrioventricular, pulmonary, and aortic orifices and are continuous with the membranous upper part of the ventricular septum.

Microscopic Structure of the Heart

Myocardium. The myocardium forms the main mass of the heart and is composed of cardiac muscle. In the atria, the myocardium is relatively thin, because the pressure within the atria rises by only about 6 mm Hg when the atria contract to force the blood into the ventricles. In the ventricles, the myocardium is thick, and, because the systolic blood pressure in the systemic circulation is about six times higher than in the pulmonary circulation, it is not surprising to find that the myocardium of the

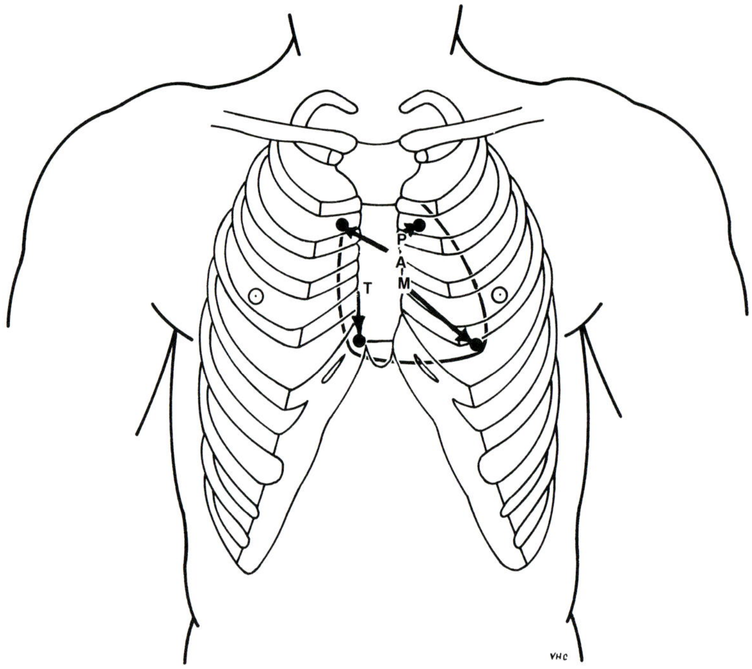

Figure 3–1. Anatomic location of the heart valves. *Arrows* indicate positions where valves may be heard with the least interference. P = pulmonary valve, A = aortic valve, M = mitral valve, T = tricuspid valve.

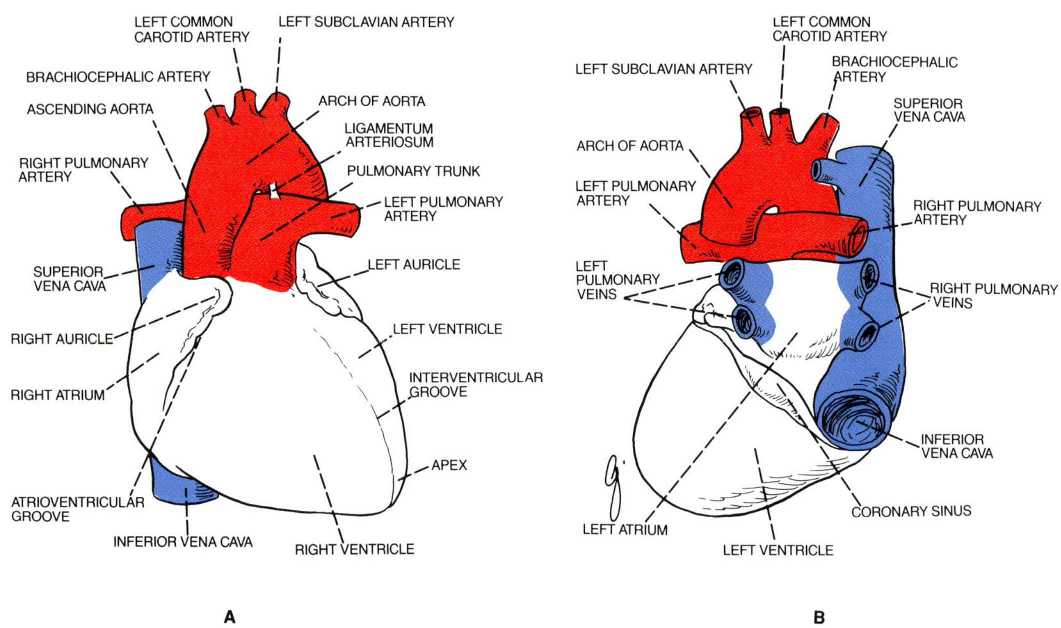

Figure 3–2. A. Anterior surface of the heart and the great blood vessels. **B.** Posterior surface or base of the heart.

left ventricle is about three times as thick as that of the right ventricle.

The cardiac muscle fibers are attached to the fibrous skeleton of the heart and are separated from one another by connective tissue. The arrangement of the muscle fibers is extremely complicated. The superficial fibers are common to the two atria, whereas the deeper fibers are confined to each atrium. In the ventricles, the fibers are arranged in a spiral, and most end in the papillary muscles. The cardiac muscle of the atria and ventricles is completely separated by the fibrous skeleton of the heart, the only connection being the conducting system of the heart.

Cardiac muscle cells are branched and possess centrally placed nuclei (*Fig. 3–3*). Each cell is about 80 μ long and 15 μ in diameter. The sarcoplasm contains myofibrils and shows cross striations similar to those of skeletal muscles. Thus, we find A bands, I bands, H bands, and Z lines. Cardiac muscle cells differ from those of skeletal or smooth muscle in that the cells not only branch but are joined together by a special junctional complex known as the **intercalated disc** (*Fig. 3–4*). The discs are seen as dark lines that run transversely in a stepwise manner across the fibers at the level of the Z lines.

The electron microscopic structure of cardiac muscle is very similar to that of skeletal muscle (*Figs. 3–3 and 3–4*). **Myosin and actin filaments** form the myofibrils, but the latter are not arranged in such discrete groups as in skeletal muscle. The mitochondria are large and numerous, and the tubular invaginations of the sarcolemma that form the T system of cardiac muscle are less frequent, occurring at the Z lines of each sarcomere instead of at the A–I junction as in skeletal muscle. The sarcoplasmic reticulum is not as well developed as in skeletal muscle.

The intercalated discs (*Fig. 3–4*) consist of three regions: (1) the macula adherens (desmosome), (2) the fascia adherens, and (3) the gap junction. The macula adherens provides a spot-like attachment between adjacent muscle fibers. The fascia adherens attaches the actin filaments to the sarcolemma. The gap junctions, which are areas of low electrical resistance, allow the rapid spread of excitation from cell to cell.

Epicardium. The epicardium, the outermost covering of the heart, forms the visceral layer of the **serous pericardium** (*Fig. 3–5*). It consists of an outer layer of flattened mesothelial cells resting on a layer of connective tissue. The connective tissue contains nerves and the coronary arteries and veins. Adipose tissue is sometimes found in large amounts around the coronary blood vessels. A small amount of **pericardial fluid** is produced by the mesothelial cells of the serous pericardium and passes into the pericardial cavity.

Endocardium. The endocardium lines all the internal surfaces of the heart and is continuous with the tunica intima of the blood vessels entering and leaving the heart. It is the layer of the heart wall that lies in contact with the blood. The endocardium consists of an **endothelium** and a **subendothelial layer.** The endothelium consists of squamous cells. The subendothelium is thin and consists of delicate connective tissue that contains numerous elastic fibers and some bundles of smooth muscle. Connecting the endocardium to the myocardium is loose connective tissue containing blood vessels, nerves, and terminal branches of the conducting system.

Skeleton of the Heart

The skeleton of the heart consists of fibrous rings that surround the atrioventricular and arterial orifices (*Fig. 3–5*). It is composed of dense fibrous tissue and provides attachment for the cardiac muscle fibers of the atria and ventricles and cusps of the atrioventricular valves. Its function is to limit the diameter of the atrioventricular and arterial orifices and to provide a solid foundation for the valves of the heart during the different phases of the cardiac cycle.

Borders of the Heart

The right border is formed by the right atrium, the left border by the left auricle and below by the left ventricle (*Fig. 3–2*). The lower border is formed mainly by the right ventricle, but also by the right atrium and the apex by the left ventricle.

Chambers of the Heart

The **right atrium** consists of a main cavity and an auricle (*Fig. 3–5*). Externally at the junction of the two parts is a vertical groove, the **sulcus terminalis,** which on the inside forms a ridge, the **crista terminalis.** The main part of the atrium that lies posterior to the ridge is smooth-walled, whereas the interior of the auricle is trabeculated by bundles of muscle fibers, the **musculi pectinati.**

Openings into the Right Atrium. The **superior vena cava** (*Fig. 3–5*) opens into the upper part of the right atrium; it is devoid of any valve. The **inferior vena cava** (larger than that of the superior vena cava) opens into the lower part of the right atrium; it is guarded by a rudimentary valve. The **coronary sinus** opens into the right atrium between the inferior vena cava and the atrioventricular orifice; it is guarded by a rudimentary valve. The **right atrioventricular orifice** lies anterior to the inferior vena caval opening and is guarded by the tricuspid valve. There are also many small orifices of small veins that drain the wall of the heart and open directly into the right atrium.

Fetal Remnants. In addition to the rudimentary valve of the inferior vena cava, there are the **fossa ovalis** and **anulus ovalis.** These latter structures lie on the atrial septum that separates the right atrium from the left

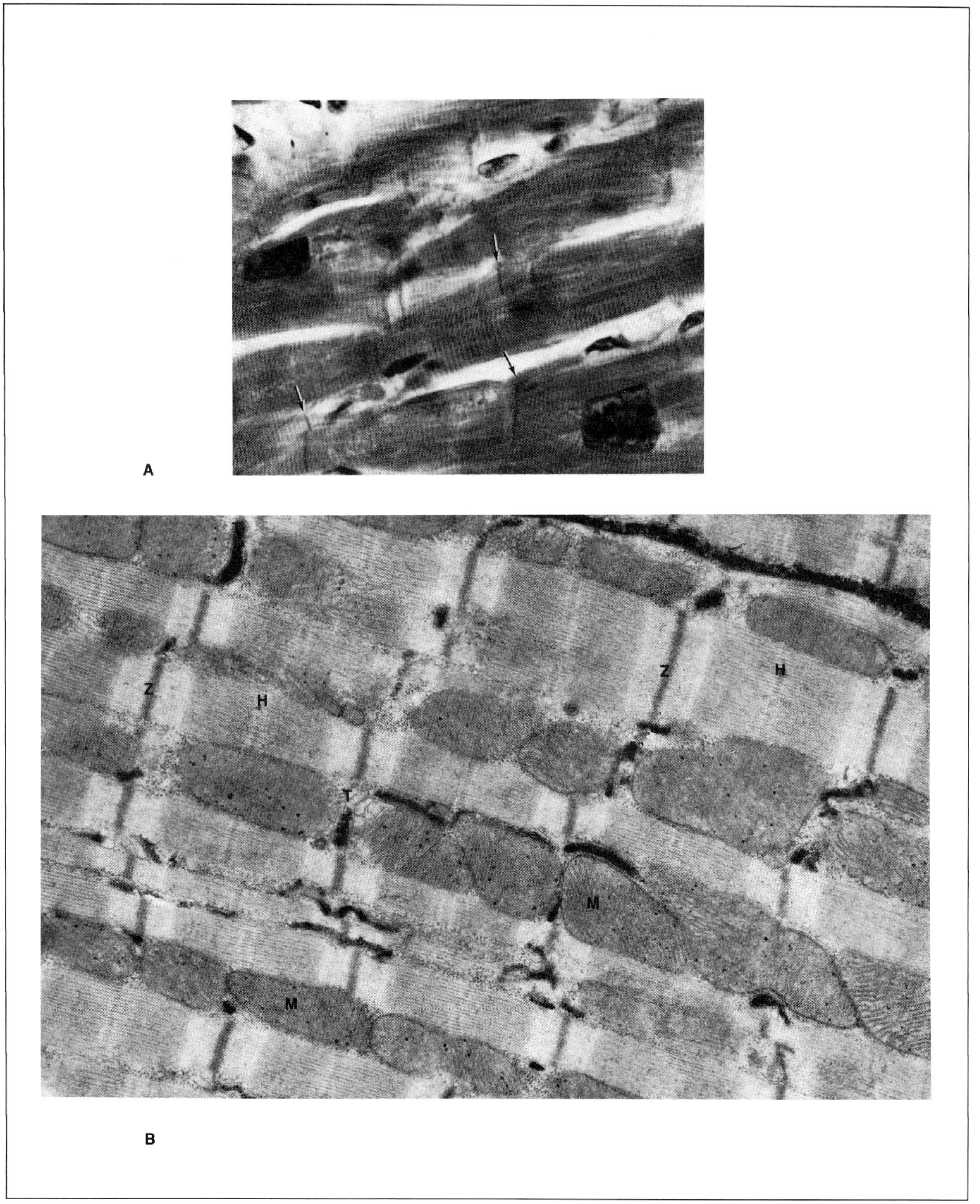

Figure 3–3. A. Photomicrograph of longitudinal section of cardiac muscle, showing branching muscle fibers, crossbanding, centrally placed nuclei, and intercalated discs (*arrows*). (H&E X400) **B.** Electron micrograph of longitudinal section of cardiac muscle, showing numerous mitochondria (M) between myofibrils, and crossbanding. Note that the structure of the crossbands is similar to that in skeletal muscle. Note also the presence of two dark-staining T tubules (T). (Courtesy of Dr. W.G. Forssman.)

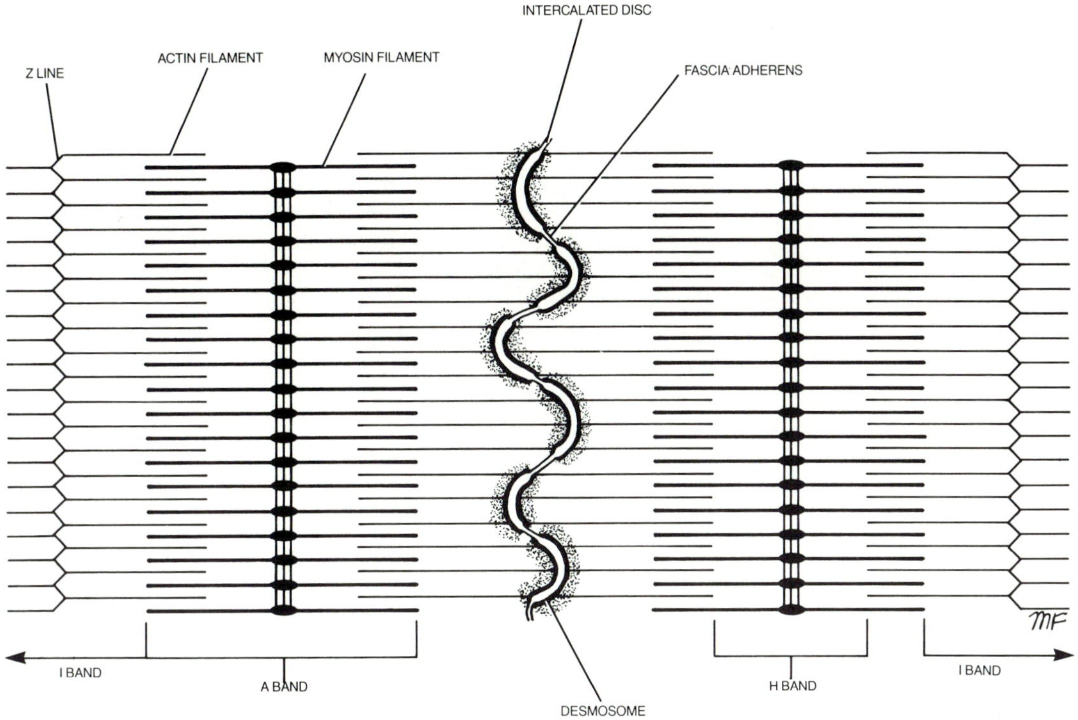

Figure 3–4. Diagram showing the organization of actin and myosin filaments in a myofibril of cardiac muscle. Note the structure of an intercalated disc.

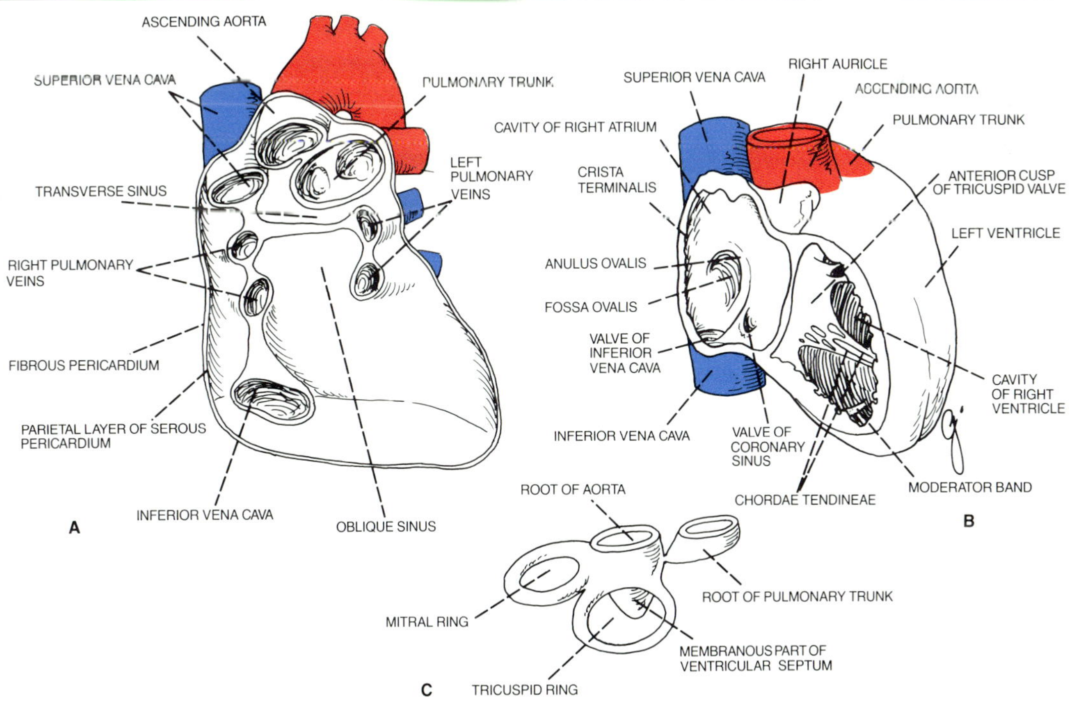

Figure 3–5. A. Interior of the pericardium showing reflection of serous pericardium around the great blood vessels. **B.** Interior of the right atrium and the right ventricle. **C.** Fibrous skeleton of the heart.

atrium. The fossa ovalis is a shallow depression that is the site of the foramen ovale in the fetus (*Fig. 3–5*). The anulus ovalis forms the upper margin of the fossa.

The right ventricle (*Fig. 3–5*) communicates with the right atrium through the atrioventricular orifice, and with the pulmonary trunk through the pulmonary orifice. As the cavity approaches the pulmonary orifice it becomes funnel-shaped and is referred to as the **infundibulum.**

The walls of the right ventricle are much thicker than those of the right atrium and show a number of internal projecting ridges formed of muscle bundles. The projecting ridges are known as **trabeculae carneae** and are of three types. The first type are the **papillary muscles,** which project inward. Their bases are attached to the ventricular wall and their apices are connected by fibrous chords, the **chordae tendineae** (*Fig. 3–5*), to the cusps of the tricuspid valve. The second type are attached at their ends to the ventricular wall, being free in the middle. Of these, one, the **moderator band,** crosses the ventricular cavity from the septal to the anterior wall. It conveys the right branch of the atrioventricular bundle, part of the conducting system of the heart. The third type are simply prominent ridges.

The **tricuspid valve** guards the atrioventricular orifice (*Fig. 3–5*). It consists of three cusps formed by a fold of endocardium with some fibrous tissue enclosed. The bases of the cusps are attached to the fibrous ring of the skeleton of the heart, and to their free edges and ventricular surfaces are attached to the **chordae tendineae.** The chordae tendineae connect the cusps to the papillary muscles. The cusps are anterior, septal, and inferior. The anterior cusp lies anteriorly, the septal cusp lies against the ventricular septum, and the inferior cusp lies inferiorly.

The **pulmonary valve** guards the pulmonary orifice. It consists of three semilunar cusps formed by folds of endocardium with some fibrous tissue enclosed. The curved lower margin of each cusp is attached to the arterial wall. The open mouths of the cusps are directed upward into the pulmonary trunk. At the root of the pulmonary trunk are three dilations called the **sinuses,** and one is situated external to each cusp (see aortic valve). The three semilunar cusps are arranged with one posterior and two anterior.

The **left atrium** (*Fig. 3–2*) consists of a main cavity and an auricle. The interior of the wall of the left auricle has muscular ridges as in the right auricle.

Openings into the Left Atrium. The four **pulmonary veins,** two from each lung, open through the posterior wall and are devoid of valves. The left **atrioventricular orifice** is guarded by the mitral valve.

The **left ventricle** (*Fig. 3–2*) communicates with the left atrium through the atrioventricular orifice, and with the aorta through the aortic orifice. The walls of the left ventricle are three times thicker than those of the right ventricle. There are well-developed trabeculae carneae, two papillary muscles, but there is no moderator band. The part of the ventricle below the aortic orifice is called the **aortic vestibule.**

The **mitral valve** guards the atrioventricular orifice. It consists of two cusps, one anterior and one posterior, which have a similar structure to those of the tricuspid valve. The attachment of the chordae tendineae to the cusps and the papillary muscles is similar to the tricuspid valve.

The **aortic valve** guards the aortic orifice and is precisely similar in structure to the pulmonary valve. One cusp is situated on the anterior wall and two are located on the posterior wall. Behind each cusp the aortic wall bulges to form an **aortic sinus.** The anterior aortic sinus gives origin to the right coronary artery, and the left posterior sinus gives origin to the left coronary artery.

Conducting System of the Heart

The conducting system of the heart consists of (1) the sinoatrial node, (2) the atrioventricular node, (3) the atrioventricular bundle, and (4) the Purkinje plexus.

Sinoatrial Node. The sinoatrial node is located in the wall of the right atrium at the upper part of the sulcus terminalis just to the right of the opening of the superior vena cava (*Fig. 3–6*). The node spontaneously gives origin to rhythmical impulses that spread in all directions through the cardiac muscle of the atria and cause the muscle to contract. The sinoatrial node is sometimes called the **pacemaker,** because it controls the rate of contraction of the entire heart.

The action potential originating in the specialized cardiac muscle fibers of the sinoatrial node spreads to the surrounding muscle fibers and reaches the atrioventricular node in about 0.04 seconds.

Atrioventricular Node. The atrioventricular node is strategically placed in the lower part of the atrial septum (*Fig. 3–6*), just above the fibrous skeleton of the heart, which separates the myocardium of the atria from that of the ventricles. It picks up the excitation wave as it passes through the atrial myocardium.

The speed of conduction of the cardiac impulse through the atrioventricular node (about 0.11 seconds) is such that sufficient time is allowed for the atria to empty their blood into the ventricles before the ventricles start to contract. From the atrioventricular node, the cardiac impulse is conducted to the ventricles by the atrioventricular bundle.

Atrioventricular Bundle. The atrioventricular bundle (bundle of His) is the only pathway of cardiac muscle that connects the myocardium of the atria and the myocardium of the ventricles; it is thus the only route along which the cardiac impulse can travel from the atria to

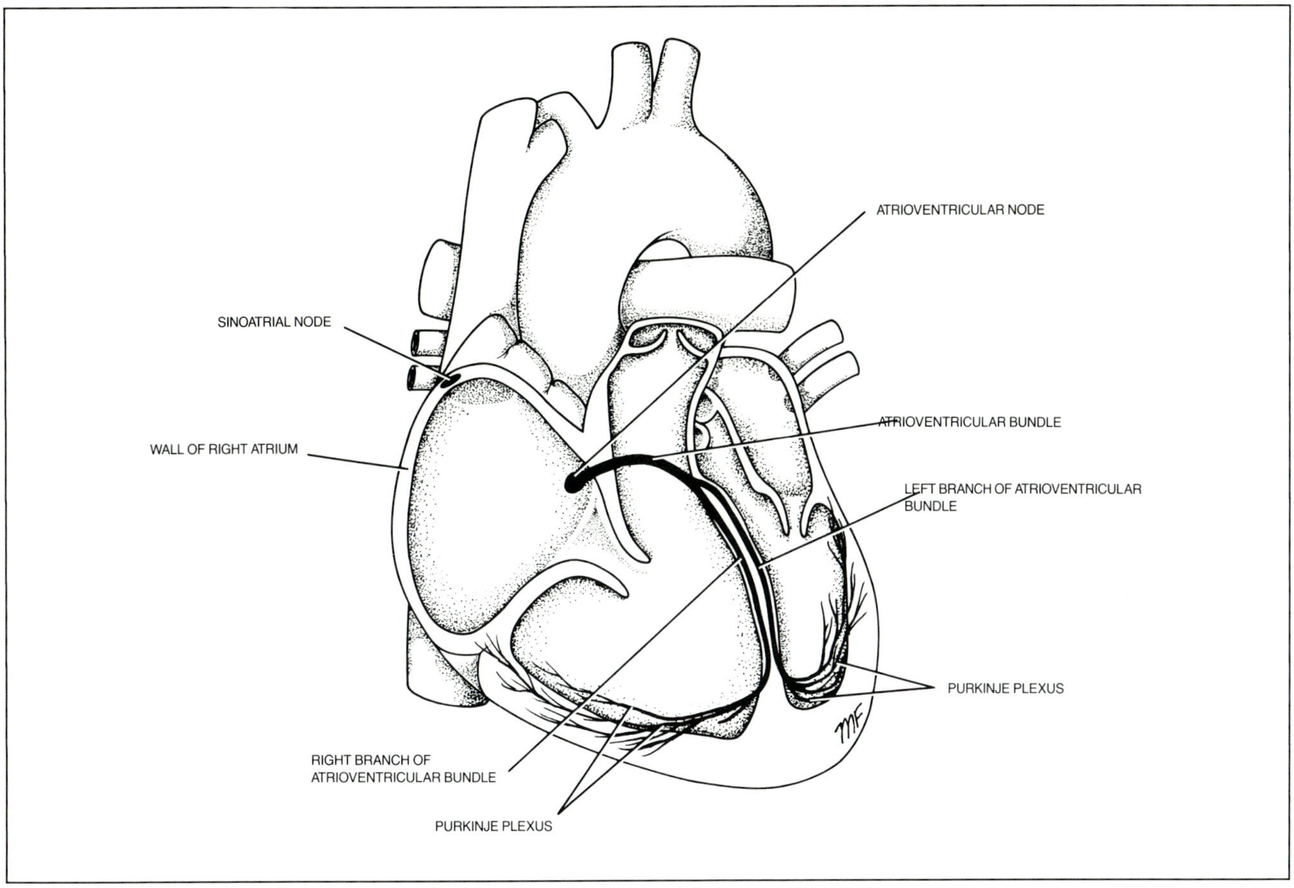

Figure 3–6. The conducting system of the heart.

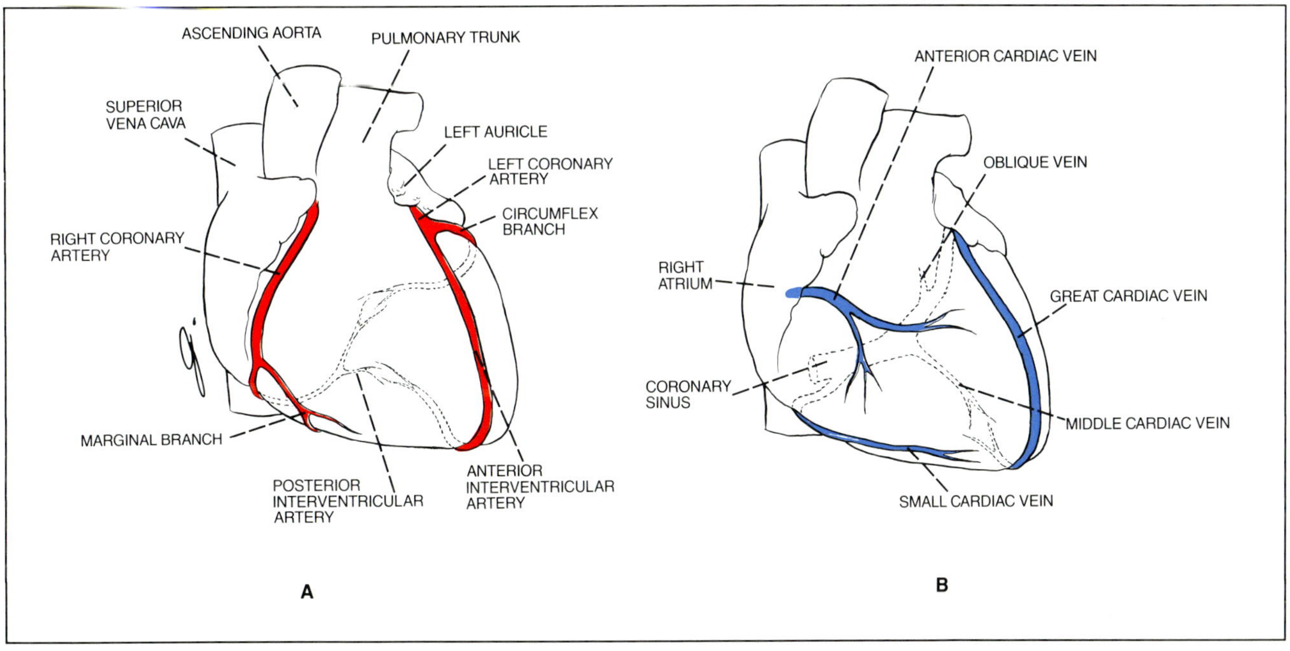

Figure 3–7. **A.** The coronary arteries. **B.** The cardiac veins.

the ventricles (*Fig. 3–6*). The specialized cardiac muscle fibers, which are fast-conducting and make up the atrioventricular bundle, are commonly referred to as **Purkinje fibers.** The atrioventricular bundle divides into right and left terminal branches, which lie on either side of the ventricular septum and extend toward the apex of the heart.

Purkinje Plexus. The right and left terminal branches of the atrioventricular bundle become continuous with small branches that spread around each ventricular chamber and back toward the base of the heart. These fine terminal branches are referred to as the **Purkinje plexus** (*Fig. 3–6*). The cardiac impulse travels from the beginning of the atrioventricular bundle to the terminal branches of the Purkinje plexus in about 0.03 seconds. This rapid spread of the cardiac impulse ensures that the entire myocardium of the ventricles is stimulated to contract at almost exactly the same time.

It can thus be seen that the conducting system of the heart is responsible not only for generating rhythmical cardiac impulses but also for conducting these impulses throughout the myocardium of the heart.

Structure of the Conducting System of the Heart

The sinoatrial and atrioventricular nodes consist of small cardiac muscle fibers embedded in vascular connective tissue. The atrioventricular bundle, its right and left branches, and the subendocardial plexus are composed of cardiac muscle fibers that are usually larger than ordinary cardiac muscle fibers. These specialized fibers are sometimes called **Purkinje fibers.** The fibers have relatively few myofibrils; those that are present tend to be concentrated in the periphery of the cytoplasm. The center of the sarcoplasm is pale staining and contains large amounts of glycogen. Electron microscopic examination reveals many mitochondria, a considerable amount of glycogen, poor development of the sarcoplasmic reticulum, and the absence of a T tubule system. The specialized cells are connected by numerous macula adherens and gap junctions.

The specialized cardiac muscle fibers of the sinoatrial and atrioventricular nodes have a higher rate of intrinsic rhythmical contraction and a slower speed of conduction than do ordinary cardiac muscle fibers. The specialized fibers of the atrioventricular bundle and its terminal branches, however, have a higher speed of conduction.

Blood Supply of the Heart

Arterial Supply (Fig. 3–7). The right and left coronary arteries arise from the ascending aorta immediately above the aortic valve.

The **right coronary artery** arises from the anterior aortic sinus and runs forward between the pulmonary trunk and the right auricle (*Fig. 3–7*). It descends in the right atrioventricular groove, giving branches to the right atrium and right ventricle. At the inferior border of the heart it continues posteriorly along the atrioventricular groove to anastomose with the left coronary artery in the posterior interventricular groove. It gives off a **marginal branch,** which supplies the right ventricle, and a **posterior interventricular branch,** which supplies both ventricles. The posterior interventricular branch anastomoses with the anterior interventricular branch of the left coronary artery.

The **left coronary artery** arises from the left posterior aortic sinus and passes forward between the pulmonary trunk and the left auricle (*Fig. 3–7*). It then enters the atrioventricular groove and divides into an anterior interventricular branch and a circumflex branch. The **anterior interventricular branch** runs downward to the apex of the heart in the anterior interventricular groove. It then passes around the apex of the heart to anastomose with the posterior interventricular branch of the right coronary artery. The anterior interventricular branch supplies the right and left ventricles and the ventricular septum.

The **circumflex branch** follows the atrioventricular groove, winds around the left margin of the heart, and ends by anastomosing with the right coronary artery (*Fig. 3–7*). The circumflex branch supplies the left atrium and the left ventricle.

Variations in the arterial supply of the heart do occur and the most common variations affect the blood supply to the diaphragmatic surface of both ventricles. Here the origin, size, and distribution of the posterior interventricular artery are variable (*Fig. 3–8*). In the case of right dominance the posterior interventricular artery is a large branch of the right coronary artery, whereas in the case of left dominance the posterior interventricular artery is a branch of the left coronary artery.

Anastomoses between the terminal branches of the right and left coronary arteries are not large enough to provide an adequate supply of blood to the cardiac muscle should one of the large branches become blocked by disease. A sudden block of one of the larger branches of either coronary artery will inevitably lead to myocardial infarction.

Arterial Supply of the Conducting System. The sinoatrial node is supplied by the right and left coronary arteries (*Fig. 3–8*), the atrioventricular node and the atrioventricular bundle by the right coronary artery, the right terminal branch of the atrioventricular bundle by the right coronary artery, and the left terminal branch by the right and left coronary arteries.

Venous Drainage of the Heart (Fig. 3–7). Most of the venous blood from the heart wall drains into the right atrium through the coronary sinus. The remainder

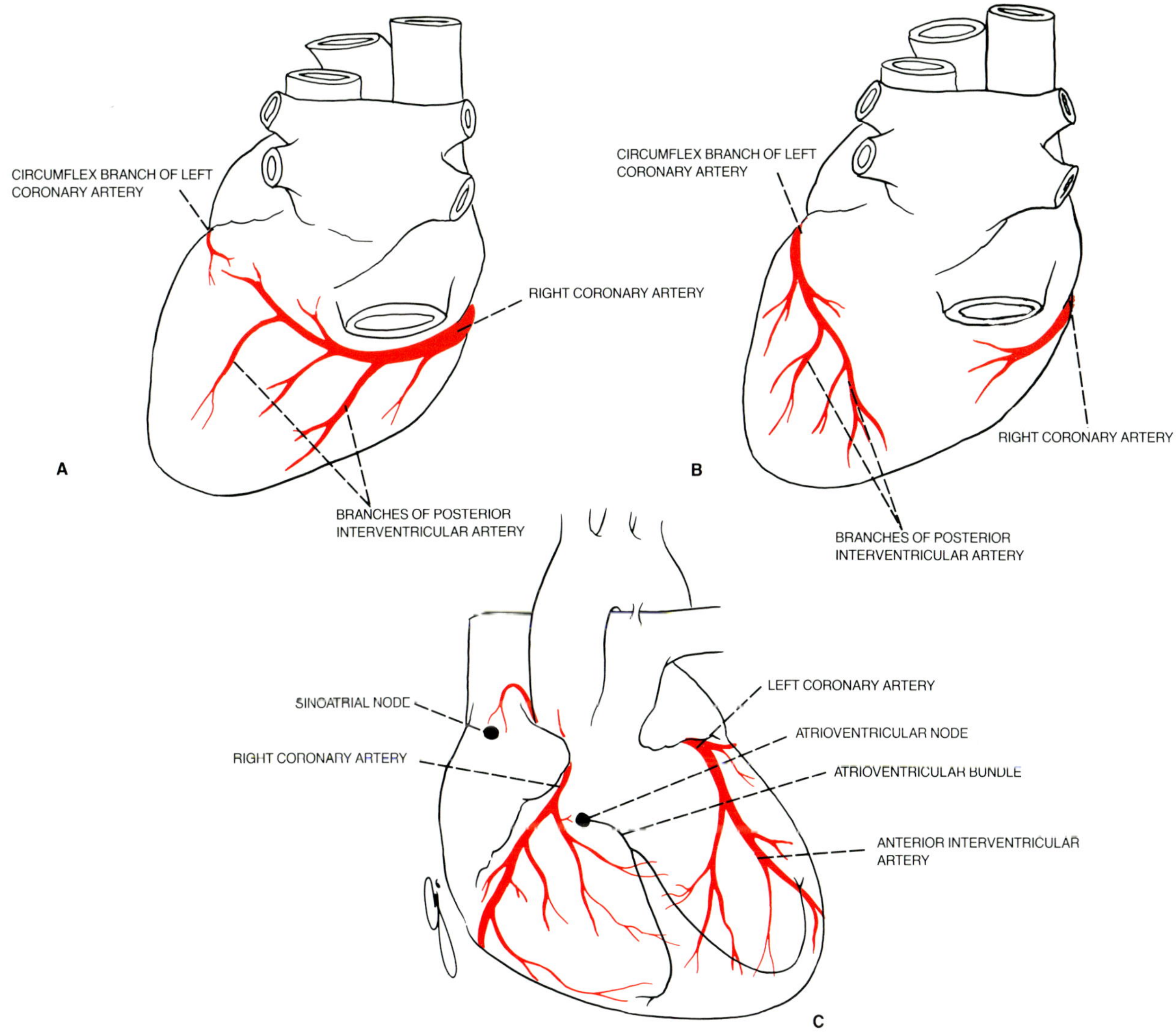

Figure 3–8. **A.** Posterior view of the heart showing the origin and distribution of the posterior interventricular artery in right dominance. **B.** Posterior view of the heart showing the origin and distribution of the posterior interventricular artery in left dominance. **C.** Anterior view of the heart showing the relationship of the blood supply to the conducting system.

goes directly into the right atrium through the **anterior cardiac vein** and small veins, the **venae cordis minimae.**

The **coronary sinus** lies in the posterior part of the atrioventricular groove and is a continuation of the great cardiac vein (*Fig. 3–7*).

The **great cardiac vein** ascends from the apex of the heart in the anterior interventricular groove. It then enters the atrioventricular groove curving to the left side and back of the heart. The **middle cardiac vein** runs from the apex of the heart in the posterior interventricular groove. The **small cardiac vein** accompanies the marginal artery along the inferior border of the heart. All empty into the coronary sinus. The **anterior cardiac vein** drains the anterior surface of the right atrium and the right ventricle and empties directly into the right atrium.

Coronary Circulation. The coronary blood flow in the healthy, resting individual is about 225 ml per minute and is continuous throughout the cardiac cycle, although about 75 percent occurs in diastole. The reason for this is the compression of the small branches of the coronary arteries by the cardiac muscle during systole. Stimulation of the sympathetic nervous system causes slight vasoconstriction of the coronary arteries, whereas parasympathetic stimulation causes minimal vasodilatation. Increased coronary flow is mainly brought about by the increased work of the heart muscle and the local affects of the products of metabolism causing vasodilatation.

Nerve Supply of the Heart

The heart is innervated by sympathetic and parasympathetic fibers of the autonomic nervous system via the cardiac plexuses. The sympathetic supply arises from the cervical and upper thoracic portions of the sympathetic trunks, and the parasympathetic supply comes from the vagus nerves.

Efferent postganglionic fibers pass to the sinoatrial and atrioventricular nodes and are also distributed to the remainder of the heart as nerve plexuses around the coronary arteries.

Afferent fibers running with the vagus nerves take part in cardiovascular reflexes. Afferent fibers running with the sympathetic nerves carry nervous impulses that normally do not reach consciousness. Should the blood supply to the myocardium become impaired, however, pain impulses reach consciousness via this pathway.

The detailed innervation of the heart is described on page 274.

Pericardium

The pericardium is a sac that encloses the heart and the roots of the great vessels (*Fig. 3–5*). It lies in the middle mediastinum.

The **fibrous portion of the pericardium** is attached anteriorly to the sternum by the **sternopericardial ligaments** and above to the walls of the great blood vessels, namely the aorta, pulmonary trunk, superior vena cava, and pulmonary veins. Below it is firmly attached to the central tendon of the diaphragm. The fibrous pericardium limits unnecessary movement of the heart.

The **serous pericardium** lines the fibrous pericardium and coats the heart (*Fig. 3–5*). It is divided into parietal and visceral layers. The parietal layer lines the fibrous pericardium and is reflected around the roots of the great vessels to become continuous with the visceral layer that closely covers the heart (*epicardium*). The slit-like space between the parietal and visceral layers is called the **pericardial cavity.** The cavity contains a small amount of tissue fluid (about 50 ml), which acts as a lubricant to facilitate movements of the heart.

The **transverse sinus** (*Fig. 3–5*) is a passage on the posterior surface of the heart that lies between the reflection of serous pericardium around the aorta and the pulmonary trunk and the reflection around the great veins.

The **oblique sinus** (*Fig. 3–5*) is a recess formed by the reflection of the serous pericardium around the venae cavae and the four pulmonary veins.

Nerve Supply. The fibrous pericardium and the parietal layer of serous pericardium are supplied by the phrenic nerves.

Principal Relations of the Pericardium and the Heart

Anterior. Body of sternum, three to six costal cartilages with intercostal spaces between them, anterior borders of right and left lungs.

Posterior. Fifth to eighth thoracic vertebrae, esophagus, descending thoracic aorta, and bronchi.

Lateral. Mediastinal parietal pleura, phrenic nerve, and lung.

Cardiac Cycle

The series of changes that take place within the heart as it fills with blood and empties is referred to as the **cardiac cycle.** The normal heart beats about 70 times per minute in the resting adult and about 130 times per minute in the newborn child.

Blood is continuously returning to the heart, and during ventricular systole, when the atrioventricular valves are closed, the blood is temporarily accommodated in the large veins and atria. Once ventricular diastole occurs, the atrioventricular valves open, and blood passively flows from the atria to the ventricles. When the ventricles are nearly full, atrial systole occurs and forces the remainder of the blood in the atria into the ventricles. The sinoatrial node initiates the wave of contraction in the atria, which commences around the openings of the large veins and ***milks*** the blood toward the ventricles. By this means there is no reflux of blood into the veins.

The cardiac impulse, having reached the atrioventricular node, is conducted to the papillary muscles by the atrioventricular bundle and its branches (*Fig. 3–6*). The papillary muscles now begin to contract and take up the slack of the chordae tendineae. Meanwhile, the ventricles start contracting and the atrioventricular valves close. The spread of the cardiac impulse along the atrioventricular bundle (*Fig. 3–6*) and its terminal branches, including the Purkinje fibers, ensures that myocardial contraction occurs at almost the same time throughout the ventricles.

Once the intraventricular blood pressure exceeds that present in the large arteries (aorta and pulmonary trunk), the semilunar valve cusps are pushed aside, and the blood is ejected from the heart. On the conclusion of ventricular systole, blood begins to move back toward the ventricles and immediately fills the pockets of the semilunar valves. The cusps now float into apposition and completely close the aortic and pulmonary orifices.

Development of the Heart

The primitive heart is a single tube having an arterial and a venous end. The tube undergoes differential expansion so that several dilatations, separated by grooves, result. From the arterial to the venous end these dilatations are called the **bulbus cordis,** the **ventricle,** the **atrium,** and the right and left horns of the **sinus venosus** (*Fig. 3–9*).

The bulbus cordis and ventricular parts of the tube start to elongate more rapidly than the other parts of the tube, and as the arterial and venous ends are fixed by the pericardium, the tube begins to bend. With a continued rapid increase in length, the bend soon becomes a U-shaped loop with its convexity directed anteriorly and to the right. Later it becomes a compound S-shaped curve with the atrium lying posterior to the ventricle; in addition, the venous and arterial ends are brought close together, as they are in the adult. Meanwhile, the passage between the atrium and the ventricle narrows to form the **atrioventricular canal.** The atrium, ventricle, and bulbus cordis now expand rapidly and undergo some change in position. The atrium expands transversely and appears on either side of the bulbus cordis.

Development of the Atria and Fate of the Sinus Venosus. The primitive atrium divides into two, the right and left atria (*Fig. 3–10*). The atrioventricular canal widens transversely. At the same time, cellular proliferations occur on the ventral and dorsal walls of the canal, producing elevations called the **ventral** and **dorsal atrioventricular cushions** (endocardial cushions). These cushions soon fuse to form a septum called the **septum intermedium,** which divides the atrioventricular canal into right and left halves. Meanwhile another septum, the **septum primum,** develops from the roof of the primitive atrium and grows down to fuse with the septum intermedium. Before fusion occurs, the opening between the lower edge of the septum primum and the septum intermedium is referred to as the **foramen primum.** The atrium now is divided into right and left parts. Before complete obliteration of the foramen primum has taken place, degenerative changes occur in the central portion of the septum primum and a foramen appears, the **foramen secundum,** so that the right and left atrial chambers again communicate. Another thicker septum, the **septum secundum,** grows down from the atrial roof on the right side of the septum primum. The lower edge of the septum secundum overlaps the foramen secundum in the septum primum, but it does not reach the floor of the atrium and does not fuse with the septum intermedium. The space between the free margin of the septum secundum and the septum primum is known as the **foramen ovale** (*Fig. 3–10*).

Before birth, the foramen ovale allows oxygenated blood that has entered the right atrium from the inferior vena cava to pass into the left atrium. The lower part of the septum primum, however, serves as a flap-like valve to prevent blood moving from the left atrium into the right atrium. At birth, owing to raised blood pressure in the left atrium, the septum primum is pressed against the septum secundum and fuses with it, and the foramen ovale is closed. The two atria thus are separated from each other. The lower edge of the septum secundum seen in the right atrium becomes the **anulus ovalis,** and the depression below this is called the **fossa ovalis** (*Fig. 3–11*).

Sinus Venosus. During the formation of the primitive heart tube, the horns of the sinus venosus remain unfused. The sinus venosus at first opens into the primitive atrium. The right horn of the sinus venosus now grows much more rapidly than the left horn. As a consequence of this, the right portion of the sinus venosus and the sinoatrial orifice move to the right and open into that part of the atrium that will become the right atrium. The transversely oriented oval sinoatrial orifice then becomes oriented in a vertical direction, and the margins of the orifice project into the right atrium and form the **right and left venous valves.** The cranial ends of these valves later fuse to form a projection, the **septum spurium** (*Fig. 3–10*). The sinoatrial opening now becomes dilated so that the right horn of the sinus venosus is taken into and becomes part of the right atrium. The septum spurium remains as the **crista terminalis**; in this way the crista represents the line of demarcation between the atrium proper, the interior of which is roughened by the **musculi pectinati,** and the smooth-walled sinus. The left venous valve regresses and the right venous valve becomes the **valve of the inferior vena cava** and the

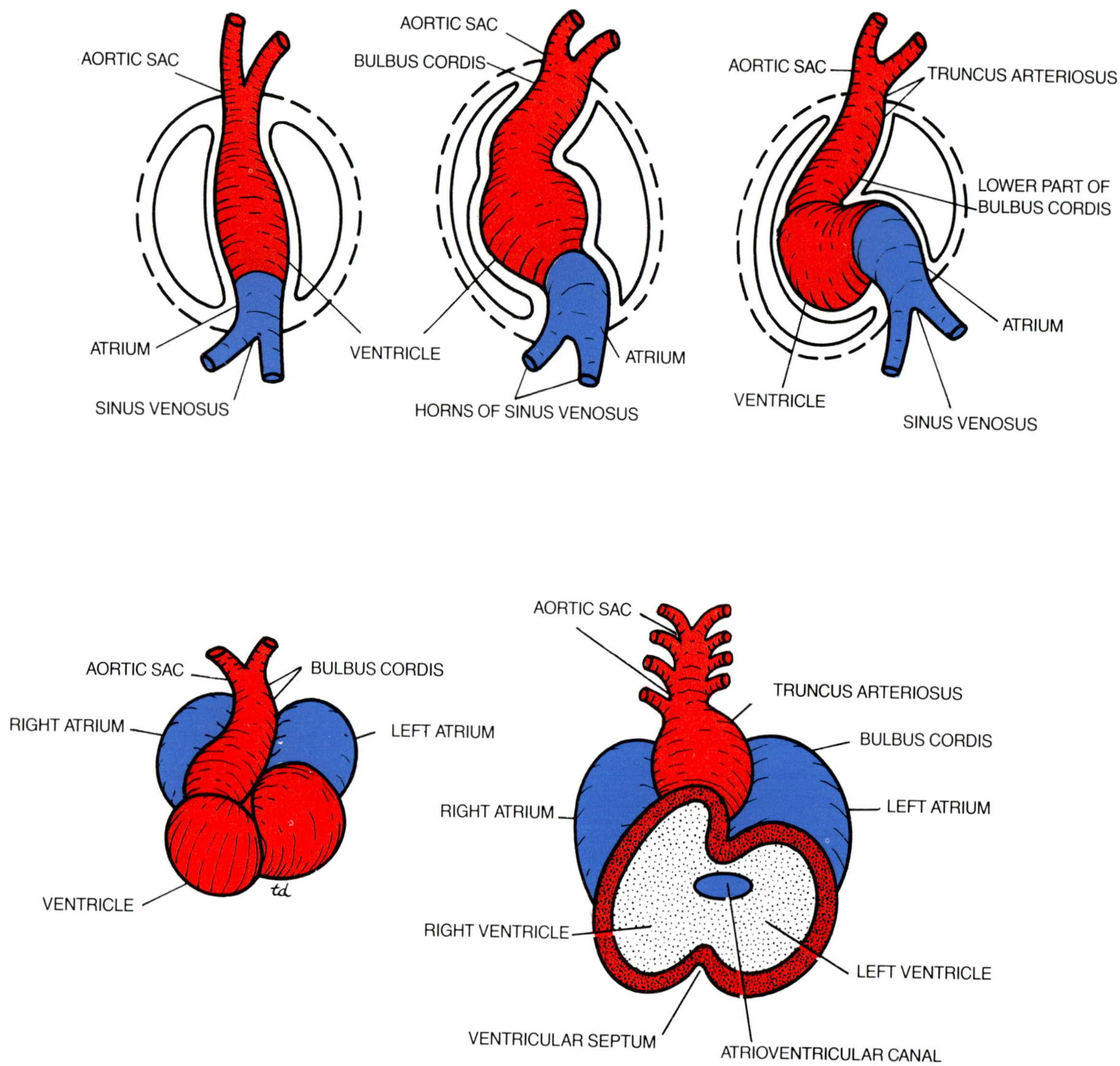

Figure 3–9. The early development of the heart showing the bending of the heart tube within the pericardial cavity. The bottom right diagram shows the interior of the developing ventricles.

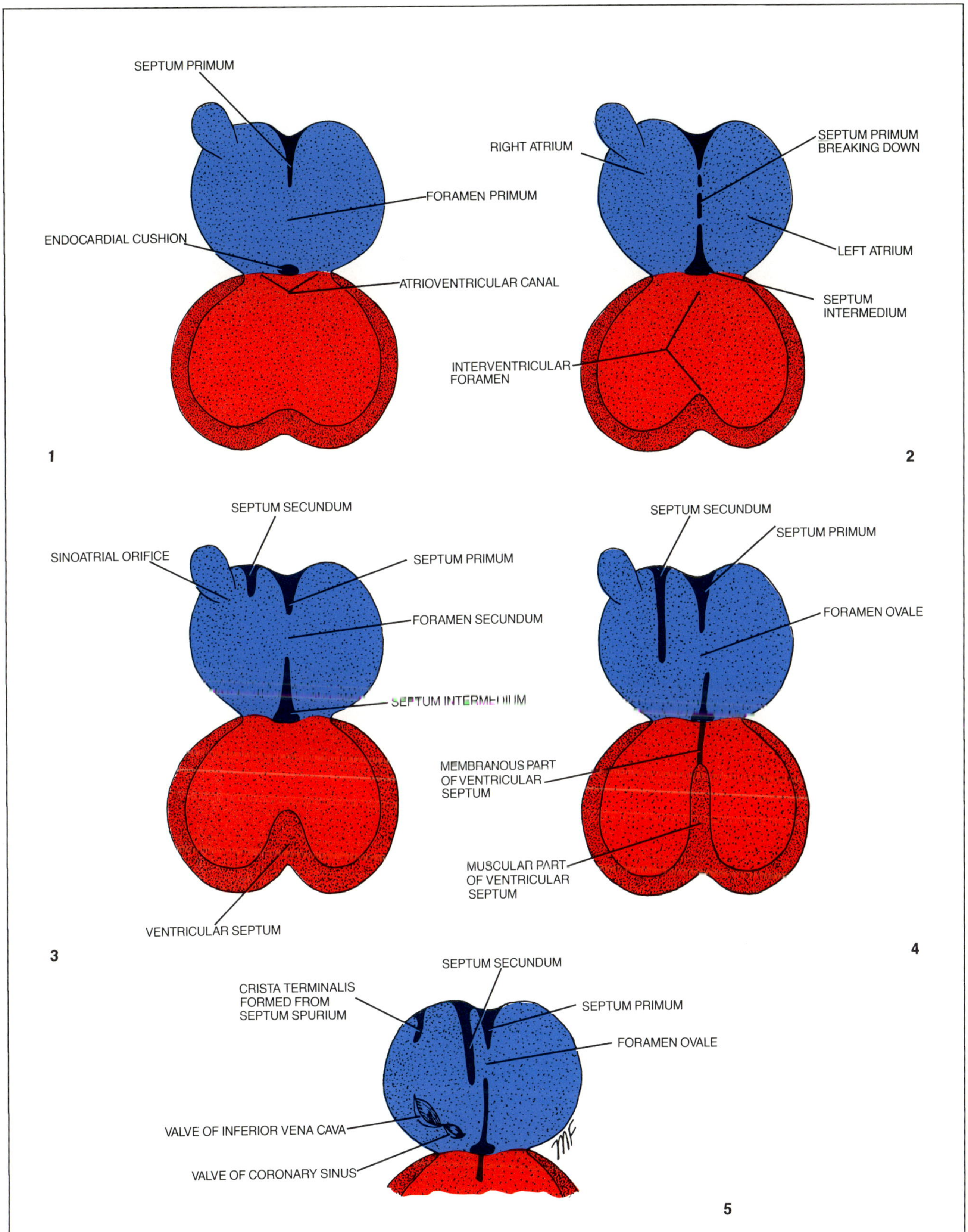

Figure 3–10. The development of the heart showing the division of the primitive atrium into right and left atria by the appearance of the septa. The sinoatrial orifice and the fate of the venous valves also are shown, as is the appearance of the ventricular septum.

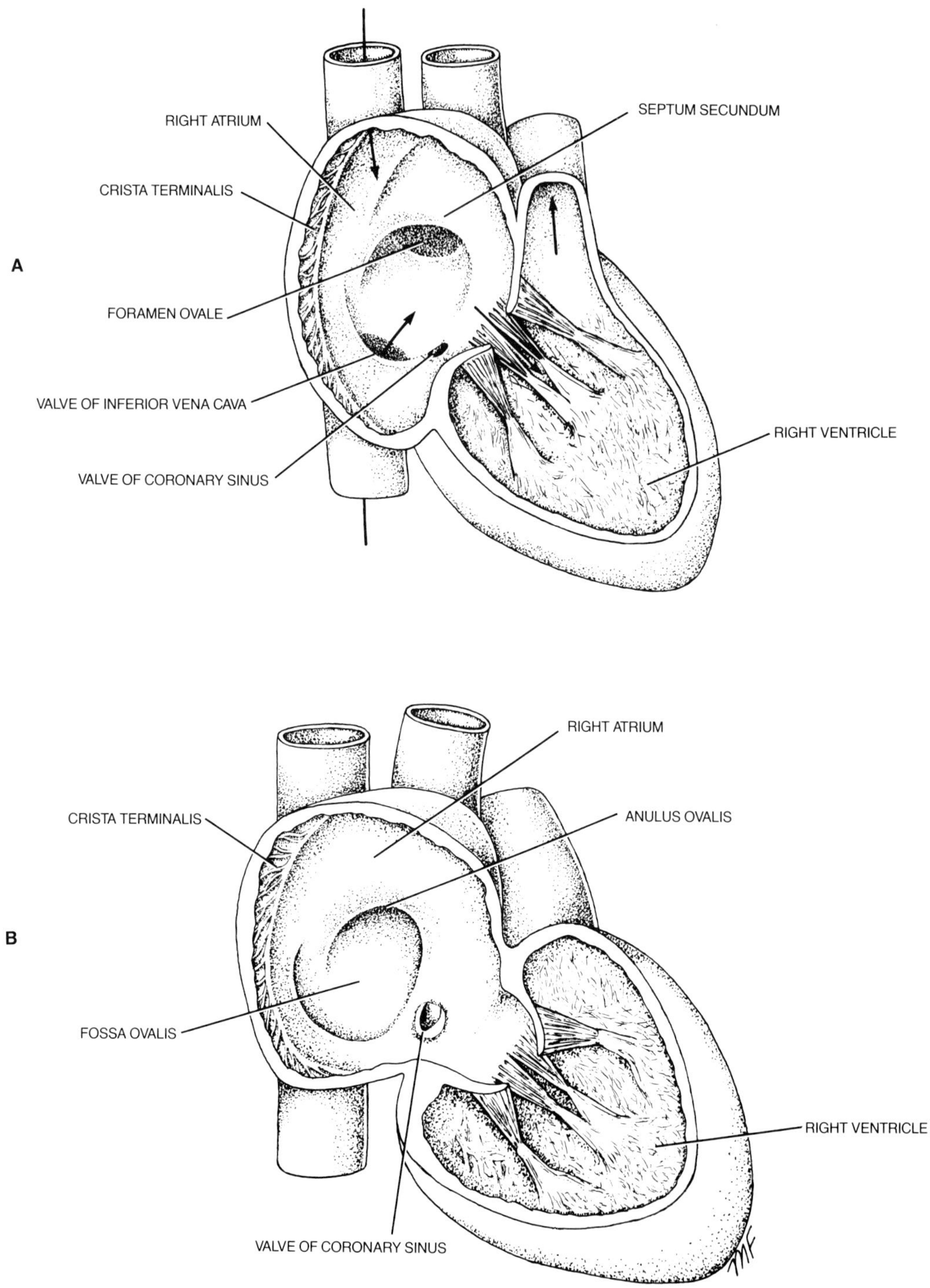

Figure 3–11. A. The interior of the developing heart, showing the formation of the atrial septum; the valve of the inferior vena cava is directing blood to the foramen ovale. **B.** The fully developed heart, showing the region of the foramen ovale that closed at birth. Note the anulus ovalis formed from the lower margin of the septum secundum and the fossa ovalis formed from the septum primum.

valve of the coronary sinus. The left horn of the sinus becomes the **coronary sinus.**

In the left atrium, the root of the pulmonary vein is absorbed into the atrial wall. This process of absorption continues until all four large tributaries of the pulmonary vein open into the left atrium by separate openings. As the result of this absorption, the part of the left atrial wall in the vicinity of the entrance of the pulmonary veins is smooth, whereas the remainder is roughened by the musculi pectinati. The **right and left auricular appendages** later develop as small diverticula from the right and left atria, respectively.

Development of the Ventricles and Fate of the Bulbus Cordis. During the fourth week of gestation, a muscular partition begins to project upward from the floor of the primitive ventricle. This **ventricular septum** first appears as the result of an indentation of the muscular wall that is represented on the external surface as a sulcus (*Fig. 3–10*). Later this muscular septum actively grows and extends upward as a crescentic plate. The space bounded by the crescentic edge of the septum and the endocardial cushions is called the **interventricular foramen.** The primitive ventricle has now been divided into right and left ventricles that communicate with each other through the interventricular foramen and with the cavity of the bulbus cordis.

Meanwhile, spiral subendocardial thickenings, the **bulbar ridges,** appear in the distal part of the bulbus cordis. By the eighth week, the bulbar ridges have grown and fused to form a **spiral aorticopulmonary septum** (*Fig. 3–12*).

The interventricular foramen closes as the result of proliferation of the bulbar ridges and the fused endocardial cushions (septum intermedium). This newly formed tissue grows down and fuses with the crescentic upper edge of the muscular ventricular septum to form the **membranous part of the septum** (*Fig. 3–12*). The enclosure of the interventricular foramen not only shuts off the path of communication between the right and left ventricles, but also ensures that the right ventricular cavity communicates with the pulmonary trunk and the left ventricular cavity communicates with the aorta. In addition, the right atrioventricular opening now connects exclusively with the right ventricular cavity, and the left atrioventricular opening with the left ventricular cavity.

The distal part of the bulbus cordis is known as the **truncus arteriosus.** It is divided by the spiral aorticopulmonary septum to form the roots and proximal portions of the aorta and pulmonary trunk. With the establishment of right and left ventricles, the proximal portion of the bulbus cordis becomes incorporated into the right ventricle as the definitive **infundibulum,** and into the left ventricle as the **aortic vestibule.**

CLINICAL NOTES

SOME IMPORTANT CONGENITAL ANOMALIES OF THE HEART

Atrial Septal Defects

After birth, the raised pressure in the left atrium causes the septum primum to be forced over and to fuse with the septum secundum in such a way that the foramen ovale becomes completely closed (*Fig. 3–11*). In 25 percent of hearts, a small opening persists, but this usually is of such a minor nature that it is of no clinical significance. Occasionally the opening is much larger (*Fig. 3–13*) and results in oxygenated blood from the left atrium passing over into the right atrium, with consequent overworking of the right side of the heart; this may even cause cyanosis. Wide patency of the septum commonly is associated with other important congenital anomalies, such as pulmonary stenosis or transposition of the great arterial trunks. In nearly half the cases of wide patency alone, death results from right-sided congestive heart failure.

Very rarely the foramen primum may persist or the entire atrial septum may be absent. Both of these conditions are more serious than patency of the foramen ovale. When there is only one atrium and two ventricles, the mortality is high.

Ventricular Septal Defects

Localized **ventricular septal defects,** although less frequent than atrial septal defects, are quite common. Normally, the ventricular septum is formed from two sources: the inferior muscular part grows up from the floor of the primitive ventricle and the membranous part is formed as the result of fusion of the lower ends of the bulbar ridges and the septum intermedium (endocardial cushions). The complete ventricular septum is formed when the membranous part fuses with the muscular part. Ventricular septal defects are found almost invariably in the membranous part of the septum. They are oval or circular in shape and measure from 1 to 2 cm in diameter. The blood passes through the defect from left to right, causing enlargement of the right ventricle. Cyanosis is rare, and there may be no symptoms. Small septal defects are relatively benign, but large defects can shorten life if surgery is not performed.

Complete absence of the ventricular septum is rare. In these cases the heart is three-chambered, and if surgery is not possible, the majority of such infants die before reaching adult life.

Defects of the Bulbus Cordis

Normally the bulbus cordis divides into the aorta and pulmonary trunk by the formation of the spiral aorticopulmonary septum. This septum is formed by the fusion of the bulbar ridges. If the bulbar ridges fail to fuse correctly, unequal division of the bulbus cordis may occur, with consequent narrowing of the pulmo-

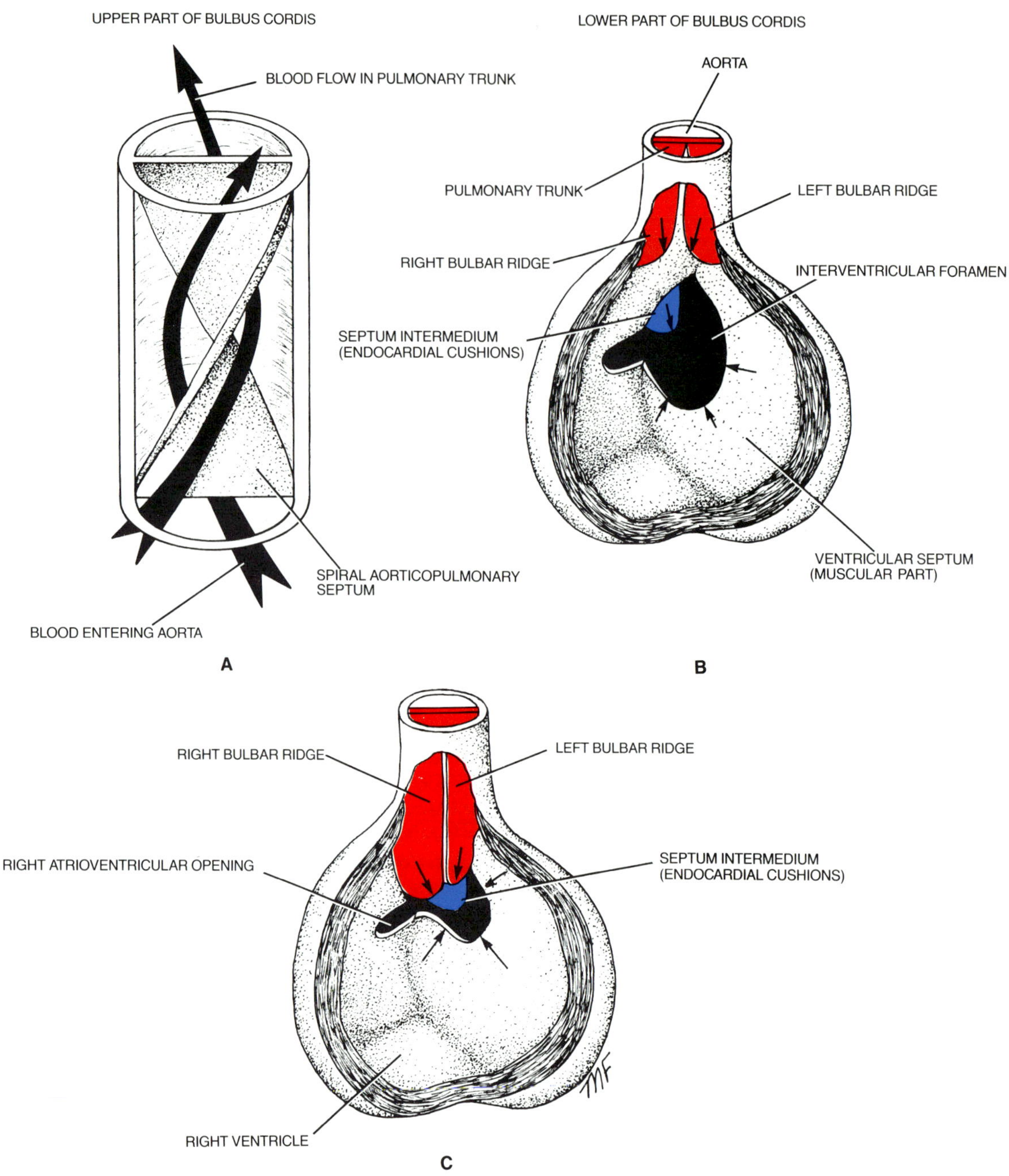

Figure 3–12. The formation of the aorta and pulmonary trunk by the division of the bulbus cordis by the spiral aorticopulmonary septum. **A.** The spiral septum in the truncus arteriosus (upper part of the bulbus cordis). **B.** Lower part of the bulbus cordis. Formation of the spiral septum by fusion of the bulbar ridges. (*red*), which then grow down and join the septum intermedium (*blue*) and the muscular part of the ventricular septum. **C.** The area of the ventricular septum that is formed from the fused bulbar ridges (*red*) and the septum intermedium (*blue*) is called the membranous part of the ventricular septum.

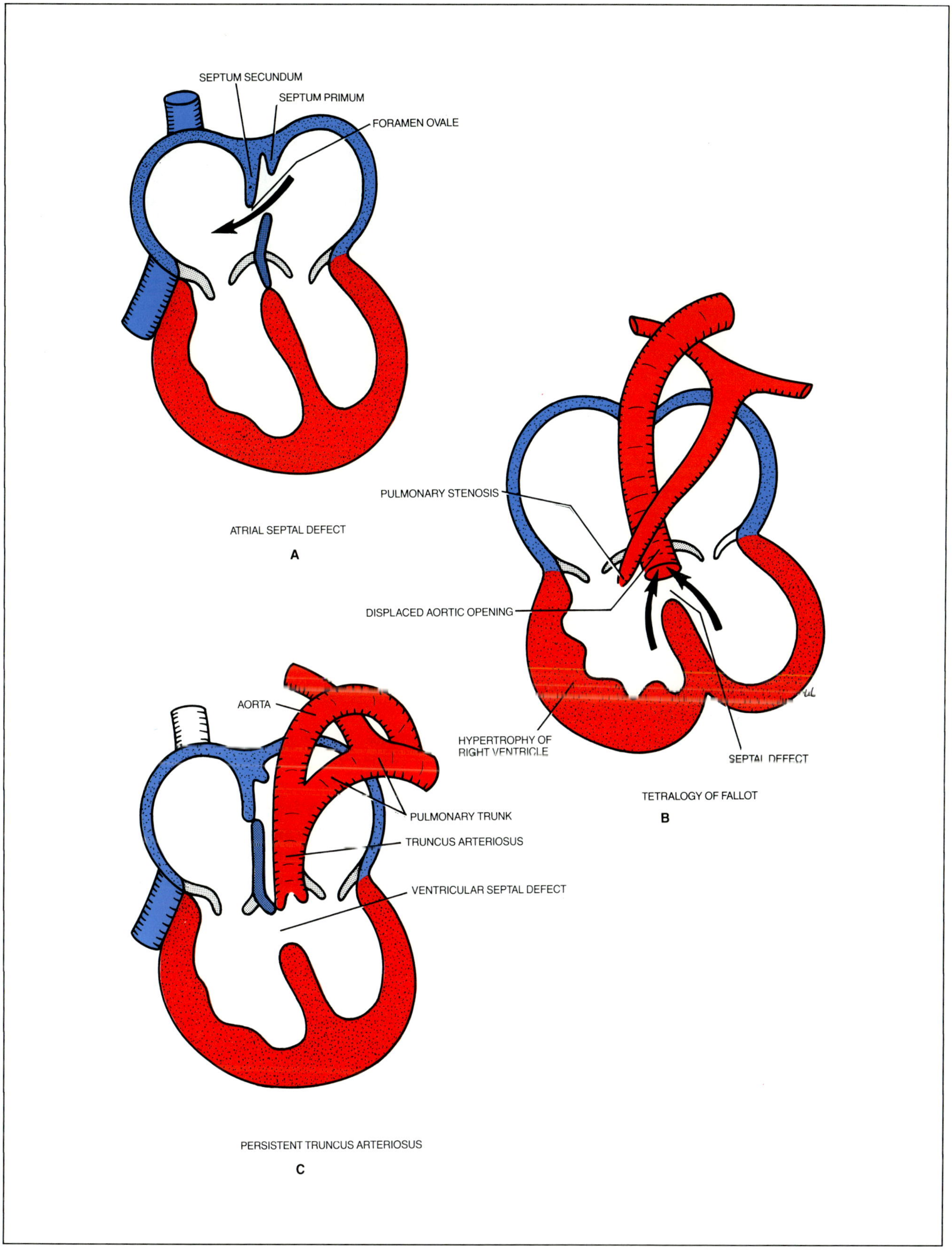

Figure 3–13. Three important congenital anomalies of the heart. **A.** Atrial septal defect. **B.** Tetralogy of Fallot. **C.** Persistent truncus arteriosus.

nary trunk and resulting interference with the right ventricular outflow. The following defects may be present: (1) large ventricular septal defect; (2) stenosis of the pulmonary trunk, which may occur at the infundibulum of the right ventricle or the pulmonary valve, or actually may involve the pulmonary trunk; and (3) exit of the aorta immediately above the ventricular septal defect. The resulting high blood pressure in the right ventricle causes hypertrophy and enlargement of that ventricle. These three defects, together with the enlargement of the right ventricle, are known as the **tetralogy of Fallot** (*Fig. 3–13*). The defects cause congenital cyanosis and considerably limit activity, shortening life to but a few years.

Persistent truncus arteriosus results when the upper part of the bulbus cordis (the truncus arteriosus) does not become divided into the aorta and pulmonary trunk because of failure of the spiral aorticopulmonary septum to develop (*Fig. 3–13*). In this condition, only one artery arises from the heart, and the pulmonary arteries originate from this single trunk. A large ventricular septal defect usually is present.

Myocardial Ischemia

The pain of myocardial ischemia is assumed to be caused by the accumulation of metabolites and by oxygen deficiency, which stimulate the sensory nerve endings in the myocardium. The afferent nerve fibers ascend to the central nervous system through the cardiac branches of the sympathetic trunk and enter the spinal cord via the posterior roots of the upper four thoracic nerves.

It is interesting to note that cardiac pain is not felt in the heart, but is referred to the skin areas supplied by the corresponding spinal nerves. The skin areas supplied by the upper four intercostal nerves and by the intercostobrachial nerve (T_2) are therefore affected. A certain amount of spread of nervous information must occur within the central nervous system, for the pain is sometimes felt in the neck and the jaw.

It must be emphasized that although the coronary arteries anastomose with each other at the arteriolar level, they are essentially **functional end arteries.** A sudden block of one of the large branches of either coronary artery will inevitably lead to myocardial infarction in the vascular area and often to death.

The conducting system of the heart receives its blood supply from the coronary arteries. In the majority of persons, the atrioventricular bundle, for example, is supplied by the posterior interventricular branch of the right coronary artery. Occlusion of the artery supplying the atrioventricular bundle is a serious condition resulting in heart block. The ventricles no longer receive the cardiac impulse from the atria, and they start to beat independently of the atria, at a slower rate.

Pericardium

The pericardium is a fibrous sac lined by a serous membrane. A small amount of fluid (50 ml) is normally present within the pericardial cavity to prevent friction between the visceral and parietal layers of the serous pericardium. The pericardium holds the heart in position, prevents sudden dilatation of its chambers, and aids venous filling of the atria during ventricular systole. The last of these functions occurs as a result of the negative intrapericardial pressure developed by the sudden emptying of the ventricles.

ARTERIES OF THE BODY

Pulmonary Trunk

The pulmonary trunk (*Fig. 3–2*) conveys deoxygenated blood from the right ventricle to the lungs. It leaves the upper part of the right ventricle and runs upward, backward, and to the left. It is about 2 inches (5 cm) long and terminates at the level of the concavity of the aortic arch dividing into right and left pulmonary arteries. The pulmonary trunk is contained within the fibrous pericardium and together with the ascending aorta; it is enclosed in a sheath of serous pericardium.

The **ligamentum arteriosum** is a fibrous band that connects the bifurcation of the pulmonary trunk to the lower surface of the aortic arch (*Fig. 3–2*). It is the remains of the **ductus arteriosus,** which in the fetus conducts blood from the pulmonary trunk to the aorta, thus bypassing the lungs. Following birth, the ductus closes.

Branches. The **right pulmonary artery** runs to the right behind the ascending aorta and the superior vena cava to enter the root of the right lung where it divides into three primary branches, one for each lobe.

The **left pulmonary artery** goes to the left in front of the descending aorta to enter the root of the left lung, where it divides into two primary branches, one for each lobe.

Aorta

The aorta is the main arterial trunk that delivers oxygenated blood from the left ventricle of the heart to the tissues of the body (*Fig. 3–14*). It is divided for purposes of description into the following parts: ascending aorta, arch of aorta, descending thoracic aorta, and abdominal aorta (discussed later).

Ascending Aorta. The ascending aorta (*Fig. 3–14*) begins at the base of the left ventricle and runs upward and forward to lie behind the right half of the sternum at the level of the sternal angle, where it is continuous with the arch of the aorta. It is enclosed with the pulmonary trunk in a sheath of serous pericardium. At its root it possesses three bulges, the **sinuses of the aorta,** one behind each aortic cusp.

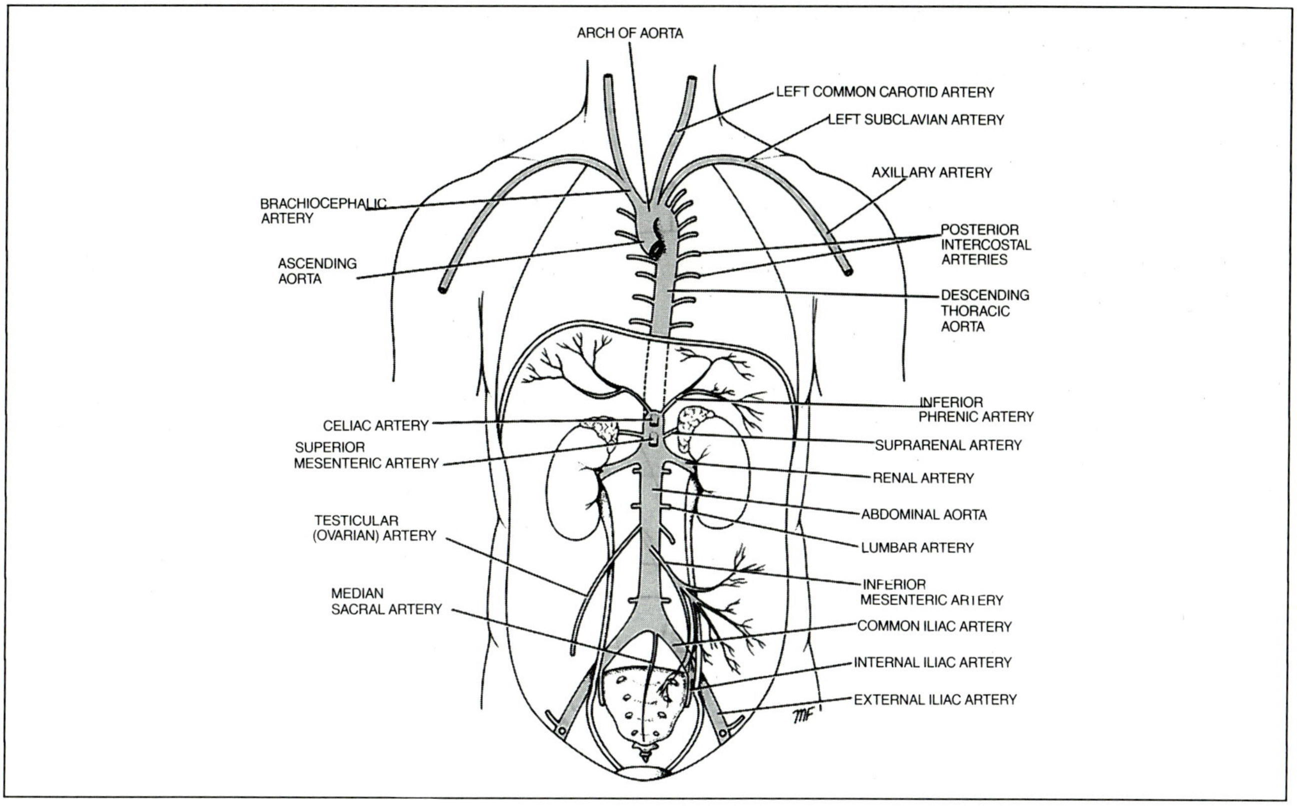

Figure 3–14. Major branches of the aorta.

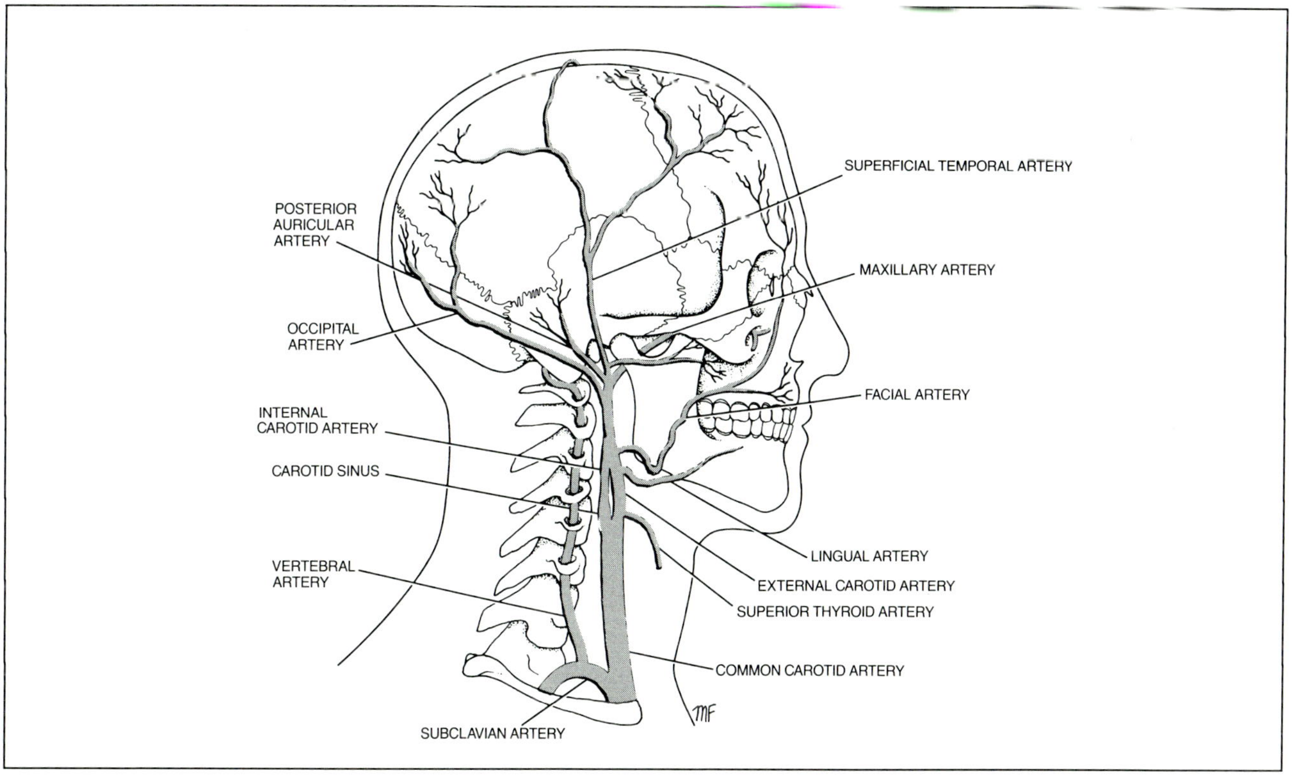

Figure 3–15. Major arteries of the head and the neck. For clarity, the thyrocervical trunk, the costocervical trunk, and the internal thoracic artery, branches of the subclavian artery, are not shown.

Branches. The **right coronary artery** arises from the anterior aortic sinus, and the **left coronary artery** from the left posterior aortic sinus (see page 70).

Arch of Aorta. The arch of the aorta is a continuation of the ascending aorta (*Fig. 3–14*). It lies behind the manubrium sterni and arches upward, backward, and to the left in front of the trachea. (Its main direction is backward.) It then passes downward to the left of the trachea, and at the level of the sternal angle becomes continuous with the descending thoracic aorta.

Branches. The arch of the aorta gives off from its convex surface the **brachiocephalic artery,** the **left common carotid artery,** and the **left subclavian artery** (*Fig. 3–14*).

Descending Thoracic Aorta. The descending thoracic aorta begins as a continuation of the arch of the aorta (*Fig. 3–14*) on the left side of the lower border of the body of the fourth thoracic vertebra (i.e., opposite the sternal angle). It runs downward in the posterior mediastinum inclining forward and medially to reach the anterior surface of the vertebral column. At the level of the 12th thoracic vertebra it passes behind the diaphragm (through the aortic opening) in the midline and becomes continuous with the abdominal aorta.

Branches. The descending thoracic aorta gives off the lower nine **posterior intercostal arteries,** the **subcostal artery,** and **pericardial, esophageal,** and **bronchial arteries.**

Surface Anatomy of Arteries of the Thorax

The **arch of the aorta** and the roots of the **brachiocephalic and left common carotid arteries** lie behind the manubrium sterni.

The **internal thoracic vessels** run vertically downward posterior to the costal cartilages, 0.5 inches (1.13 cm) lateral to the edge of the sternum as far as the sixth intercostal space.

The **intercostal vessels** and nerve (vein, artery, nerve—VAN—is the order from above downward) are situated immediately below their corresponding ribs.

ARTERIES OF THE HEAD AND THE NECK

Common Carotid Artery

The right common carotid artery arises from the brachiocephalic artery behind the right sternoclavicular joint (*Fig. 3–15*). The left artery arises from the arch of the aorta in the superior mediastinum. The common carotid artery runs upward and backward through the neck, under cover of the anterior border of the sternocleidomastoid muscle. At the upper border of the thyroid cartilage it divides into the external and internal carotid arteries.

The common carotid artery at its point of division or the beginning of the internal carotid artery, shows a localized dilatation, called the **carotid sinus,** which contains nerve endings from the glossopharyngeal nerve (*Figs. 3–15 and 3–16*). Here the tunica media is thin and the tunica adventitia is thick. The carotid sinus is a pressoreceptor and assists in the regulation of the blood pressure in the cerebral arteries.

The **carotid body** is a small structure that lies posterior to the common carotid artery at its point of bifurcation (*Fig. 3–16*). It is innervated by the glossopharyngeal nerve and is a chemoreceptor, being sensitive to excess carbon dioxide and reduced oxygen tensions in the blood circulating through the organ. When the chemoreceptors are stimulated the heart and respiratory rates are increased.

The common carotid artery ascends the neck in the **carotid fascial sheath** (*Fig. 3–16*) and is closely related to the internal jugular vein laterally; the vagus nerve lies between these two structures.

Branches. External and *internal carotid arteries* (*Fig. 3–15*).

Circulus Arteriosus

The circulus arteriosus lies in the subarachnoid space in the interpeduncular fossa at the base of the brain. The circle is formed by the anastomosis between the branches of the two internal carotid arteries and the two vertebral arteries. The anterior communicating, the anterior cerebral, the internal carotid, the posterior communicating, the posterior cerebral, and the basilar (formed by the junction of the two vertebral arteries) all contribute to the circle. Cortical and central branches arise from the arterial circle and supply the brain.

The detailed blood supply of the brain and spinal cord are given on pages 84 and 170.

Subclavian Arteries

The **right subclavian artery** arises from the brachiocephalic artery, behind the right sternoclavicular joint (*Fig. 3–15*). It arches upward and laterally over the pleura and comes to lie on the first rib between the scalenus anterior and scalenus medius muscles. At the outer border of the first rib it becomes continuous with the axillary artery.

The **left subclavian artery** arises from the arch of the aorta behind the origin of the left common carotid artery (*Fig. 3–14*). It ascends to the root of the neck and then arches laterally in a manner similar to that of the right subclavian artery.

The scalenus anterior muscle passes anterior to the artery on each side and divides it into three parts: First part, from the origin of the artery to the medial border of the scalenus anterior. Second part is the portion that

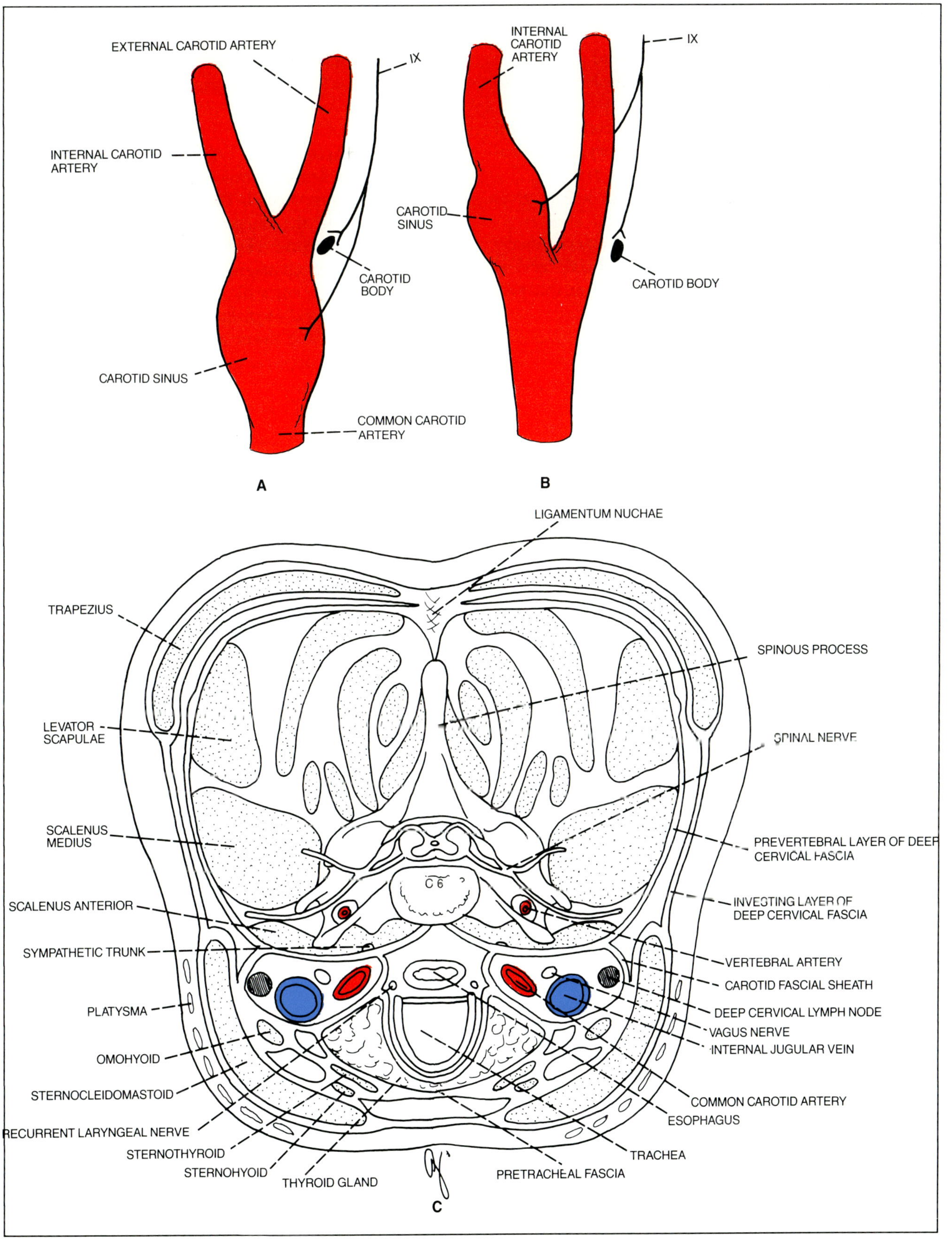

Figure 3–16. The carotid sinus is seen as a dilatation of the common carotid artery just proximal to its division (**A**) or as a dilatation of the commencement of the internal carotid artery (**B**). Note that the carotid sinus, a baroreceptor, and the carotid body, a chemoreceptor, are innervated by branches of the glossopharyngeal nerve. **C.** A cross section of the neck at the level of the sixth cervical vertebra. Note the carotid sheath and its contents.

lies posterior to the scalenus anterior. Third part is from the lateral border of scalenus anterior to the lateral border of the first rib.

The **first part of the subclavian artery** gives off the following branches: the **vertebral artery,** the **thyrocervical trunk,** and the **internal thoracic artery.**

The **second part of the subclavian artery** gives off the **costocervical trunk.**

The **third part of the subclavian artery** usually has no branches. Occasionally the **superficial cervical or suprascapular artery,** or both, arise from this part of the subclavian artery.

ARTERIAL SUPPLY OF THE BRAIN

The brain is supplied by the two internal carotid and the two vertebral arteries. The four arteries anastomose on the inferior surface of the brain and form the **circulus arteriosus** (see page 82).

Internal Carotid Artery

The internal carotid artery ascends the neck within the carotid sheath and perforates the base of the skull by passing through the carotid canal of the temporal bone. The artery then runs horizontally forward through the cavernous sinus and emerges on the medial side of the anterior clinoid process by perforating the dura mater. It enters the subarachnoid space by piercing the arachnoid mater and turns posteriorly to reach the medial end of the lateral cerebral sulcus. Here, it divides into the anterior and middle cerebral arteries.

The **anterior cerebral artery** is the smaller terminal branch of the internal carotid artery (*Fig. 3–17*). It runs forward and medially and enters the longitudinal fissure between the cerebral hemispheres. Here, it is joined by the anterior cerebral artery of the opposite side by the **anterior communicating artery.** It curves backward over the corpus callosum, and anastomoses with the posterior cerebral artery (*Fig. 3–17*). The **cortical branches** supply all the medial surface of the cerebral cortex as far posteriorly as the parieto-occipital sulcus (*Fig. 3–18*). In addition they supply a strip of cortex about 1 inch (2.5 cm) wide on the adjoining lateral surface. A group of **central branches** pierce the anterior perforated substance and help to supply parts of the lentiform and caudate nuclei and the internal capsule.

The **middle cerebral artery,** the largest branch of the internal carotid, passes laterally in the lateral cerebral sulcus (*Fig. 3–17*). **Cortical branches** supply the entire lateral surface of the hemisphere, except for the narrow strip supplied by the anterior cerebral artery, the occipital pole, and the inferolateral surface of the hemisphere, which are supplied by the posterior cerebral artery (*Fig. 3–18*). This artery thus supplies all the motor area except the *leg area* (supplied by the anterior cerebral artery). **Central branches** enter the anterior perforated substance and supply the lentiform and caudate nuclei and the internal capsule.

Vertebral Artery

The vertebral artery, a branch of the first part of the subclavian artery, ascends the neck by passing through the foramina in the transverse processes of the upper six cervical vertebrae (*Fig. 3–15*). It enters the skull through the foramen magnum and pierces the dura mater and arachnoid to enter the subarachnoid space. It then passes forward and medially on the medulla oblongata (*Fig. 3–17*). At the lower border of the pons, it joins the artery of the opposite side to form the **basilar artery.**

Branches of the Cranial Portion of the Vertebral Artery

1. *Meningeal branches* supply the dura in the posterior cranial fossa.
2. *Posterior spinal artery.* This artery arises from the vertebral artery or the posterior inferior cerebellar artery. It descends as two branches close to the posterior roots of the spinal nerves. The branches are reinforced by radicular arteries that enter the vertebral canal through the intervertebral foramina (see page 170).
3. *Anterior spinal artery.* This artery is formed from a branch of each vertebral artery near its termination (*Fig. 3–17*). The single artery descends on the anterior surface of the medulla oblongata and spinal cord in the anterior median fissure (see page 173).
4. *Posterior inferior cerebellar artery.* This artery is the largest branch of the vertebral artery and supplies the inferior surface of the vermis, the central nuclei of the cerebellum, and the under surface of the cerebellar hemisphere (*Fig. 3–17*); it also supplies the medulla oblongata and the choroid plexus of the fourth ventricle.
5. *Medullary arteries* are distributed to the medulla oblongata.

Basilar Artery

The basilar artery, formed by the union of the two vertebral arteries (*Fig. 3–17*), lies in a groove on the anterior surface of the pons. At the upper border of the pons, it divides into the two posterior cerebral arteries.

Branches of the Basilar Artery

1. *Pontine arteries* enter the anterior surface of the pons (*Fig. 3–17*).
2. *Labyrinthine artery* to the internal ear.
3. *Anterior inferior cerebellar artery* supplies the anterior and inferior surfaces of the cerebellum (*Fig. 3–17*) and a few branches pass to the pons and the upper part of the medulla oblongata.
4. *Superior cerebellar artery* supplies the superior surface

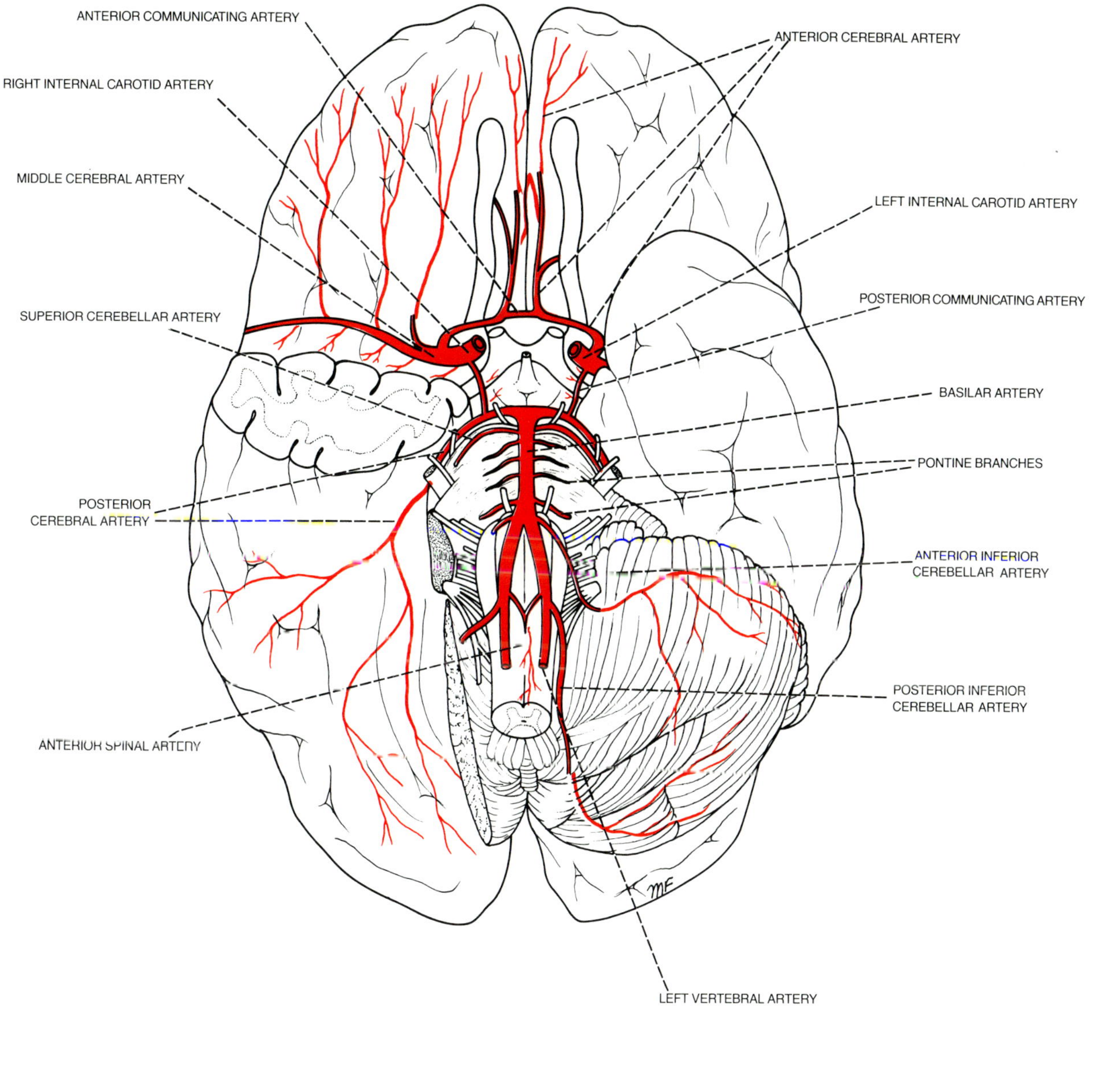

Figure 3–17. The arteries on the inferior surface of the brain showing the formation of the circulus arteriosus.

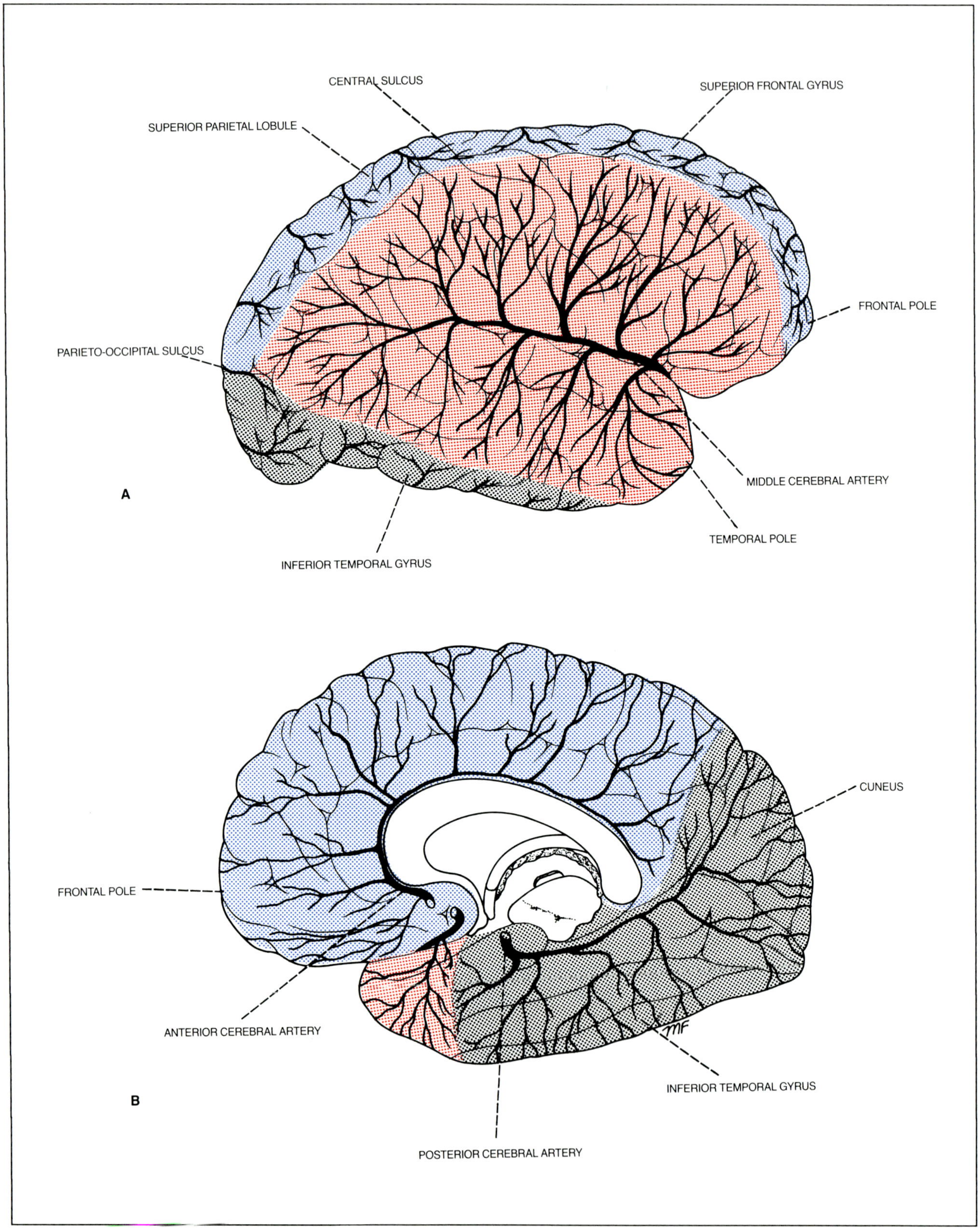

Figure 3–18. Areas of the cerebral cortex supplied by the different cerebral arteries. **A.** The lateral surface of the right cerebral hemisphere. **B.** The medial surface of the right cerebral hemisphere. The area supplied by the anterior cerebral artery is colored *blue,* the area supplied by the middle cerebral artery is *red,* and the one supplied by the posterior cerebral artery is *black.*

of the cerebellum, the pons, the pineal body, and the superior medullary velum.

5. *Posterior cerebral artery*. This artery curves laterally and posteriorly around the midbrain and is joined by the posterior communicating branch of the internal carotid artery (*Fig. 3–17*). **Cortical branches** supply the inferolateral surface of the temporal lobe and the lateral and medial surfaces of the occipital lobe (*Fig. 3–18*). The posterior cerebral artery, therefore, supplies the visual cortex. **Central branches** enter the brain substance and supply parts of the thalamus and the lentiform nucleus, and the midbrain, and pineal, and the medial geniculate bodies. A choroidal branch enters the inferior horn of the lateral ventricle and supplies the choroid plexus; it also supplies the choroid plexus of the third ventricle.

Surface Anatomy of the Arteries of the Head and the Neck

Superficial Temporal Artery. The pulsations of the superficial temporal artery may be felt as it crosses the zygomatic arch, immediately in front of the auricle.

Facial Artery. The pulsations of the facial artery may be felt as it crosses the lower margin of the body of mandible, at the anterior border of the masseter.

Carotid Sheath. The carotid sheath, which contains the carotid arteries, the internal jugular vein, the vagus nerve, and the deep cervical lymph nodes, can be located under a line joining the sternoclavicular joint to a point midway between the tip of the mastoid process and the angle of the mandible. At the level of the upper border of the thyroid cartilage, the common carotid artery bifurcates into the internal and external carotid arteries (*Fig. 3–15*). The pulsations of these arteries can be felt at this level.

Third Part of the Subclavian Artery. This structure occupies the lower anterior angle of the posterior triangle (*Fig. 3–16*). Its course may be indicated by a curved line, which passes upward from the sternoclavicular joint for about 0.5 inches (1.3 cm) and then downward to the middle of the clavicle. It is here, where the artery lies on the upper surface of the first rib, that its pulsations can be felt easily.

ARTERIES OF THE UPPER LIMB

Axillary Artery

The axillary artery begins at the lateral border of the first rib as a continuation of the subclavian artery, and at the lower border of the teres major muscle, it becomes the brachial artery (*Fig. 3–19*). Throughout its course, the artery is closely related to the cords of the brachial plexus and their branches and is enclosed with them in a connective tissue sheath, called the **axillary sheath.** The pectoralis minor muscle crosses in front of the artery dividing it, for purposes of description, into three parts. The first part is from the lateral border of the first rib to the upper border of the pectoralis minor. The second part is the portion that lies posterior to the pectoralis minor muscle. The third part is from the lower border of pectoralis minor muscle to the lower border of the teres major muscle.

Branches of the Axillary Artery

Branch from the First Part. The **highest thoracic artery.**

Branches of the Second Part. The **thoracoacromial artery** and the **lateral thoracic artery.**

Branches of the Third Part. The **subscapular artery** and the **anterior** and **posterior circumflex humeral** arteries.

Brachial Artery

The brachial artery begins at the lower border of the teres major muscle as a direct continuation of the axillary artery (*Fig. 3–19*). It ends at the level of the neck of the radius by dividing into the radial and ulnar arteries.

Branches

1. *Muscular branches*
2. *Nutrient artery* to the humerus
3. *Profunda artery,* large branch that follows the radial nerve into the posterior compartment of the arm
4. *Superior ulnar collateral artery* follows the ulnar nerve
5. *Inferior ulnar collateral artery* takes part in the anastomosis around the elbow joint

Radial Artery

The radial artery, one of the terminal branches of the brachial artery, begins in the cubital fossa at the level of the neck of the radius (*Fig. 3–19*). It descends through the forearm lying superficially throughout most of its course. In the region of the wrist it winds backward, around the lateral side of the carpus to the proximal end of the space between the first and second metacarpal bones. Here, it passes anteriorly into the palm between the two heads of the first dorsal interosseous muscle and joins the deep branch of the ulnar artery forming the **deep palmar arch.**

In the lower part of the forearm the radial artery lies on the anterior surface of the radius and is covered only by skin and fascia. Here, the artery has on its lateral side the tendon of the brachioradialis and on its medial side the tendon of the flexor carpi radialis. It is at this

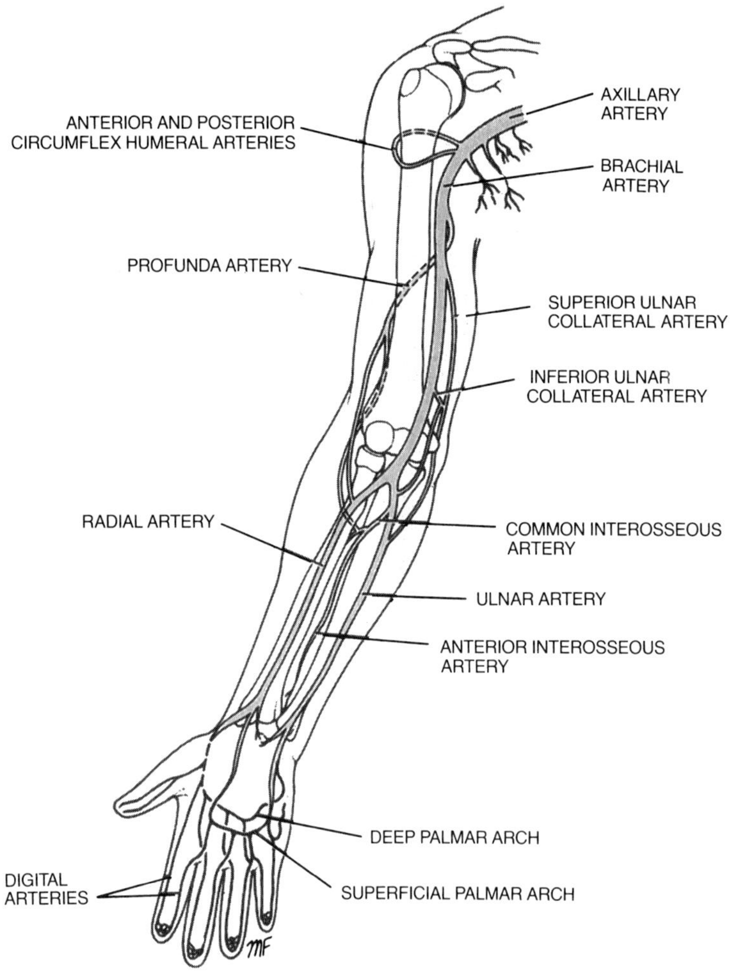

Figure 3–19. Main arteries of the upper limb.

site, just above the wrist, that the radial pulse can be easily palpated.

In the middle third of its course the radial nerve lies on its lateral side.

Branches

1. *Muscular branches*
2. *Recurrent branch,* which takes part in the arterial anastomosis around the elbow joint
3. *Superficial palmar branch,* which arises just above the wrist. It enters the palm and frequently joins the ulnar artery to form the the **superficial palmar arch**
4. *First dorsal metacarpal artery,* which arises just before the radial artery passes between the two heads of the first dorsal interosseous muscle. It supplies the adjacent sides of the thumb and index finger
5. *Arteria princeps pollicis* arises from the radial artery in the palm. It divides into two branches that supply the sides of the thumb
6. *Arteria radialis indicis* arises from the radial artery in the palm. It supplies the lateral side of the index finger

Deep Palmar Arch

The deep palmar arch is deeply placed in the palm and extends from the proximal end of the first interosseous space to the base of the fifth metacarpal bone (*Fig. 3–19*). It is formed by the radial artery and terminates by anastomosing with the deep branch of the ulnar artery.

Branches. Palmar, metacarpal, perforating, and recurrent arteries.

Ulnar Artery

The ulnar artery is the larger of the two terminal branches of the brachial artery (*Fig. 3–19*). It begins in the cubital fossa at the level of the neck of the radius. It descends to the wrist and enters the palm in front of the flexor retinaculum along with the ulnar nerve. It ends by forming the superficial palmar arch, often uniting with the superficial palmar branch of the radial artery.

In its upper course it passes deep to the flexor muscles that arise from the common flexor origin on the medial epicondyle of the humerus. Lower, it becomes superficial and lies between the tendons of flexor carpi ulnaris and the tendons of flexor digitorum superficialis. As the artery passes in front of the flexor retinaculum it is covered only by skin and fasciae and lies just lateral to the pisiform bone. It is here that the pulsations can be palpated.

Branches

1. *Muscular branch*
2. *Recurrent branches, which take part in the arterial anastomosis around the wrist joint*
3. *Common interosseous artery,* which arises from the upper part of the ulnar artery and quickly divides into **anterior and posterior interosseous arteries.** These latter arteries descend on the anterior and posterior surfaces of the interosseous membrane, respectively
4. *Deep palmar branch* arises in front of the flexor reticulum and joins the radial artery to complete the deep palmar arch

Superficial Palmar Arch

This arterial arch lies immediately beneath the palmar aponeurosis and on the long flexor tendons (*Fig. 3–19*). It is a direct continuation of the ulnar artery. The arch is often completed on the lateral side by the superficial palmar branch of the radial artery. The curve of the arch lies across the palm, level with the distal border of the extended thumb. Four digital arteries arise from the arch and supply the medial side of the little finger and the adjacent sides of the little, ring, middle, and index fingers, respectively.

Surface Anatomy of the Arteries of the Upper Limb

The **subclavian artery,** as it crosses the first rib to become the axillary artery, may be palpated in the root of the posterior triangle of the neck.

The **third part of the axillary artery** may be felt in the axilla as it lies in front of the teres major muscle.

The **brachial artery** can be palpated in the arm as it lies on the brachialis and is overlapped from the lateral side by the biceps brachii muscle.

The **radial artery** lies superficially in front of the distal end of the radius, between the tendons of the brachioradialis and flexor carpi radialis. It is here that the radial pulse can be easily felt. If the pulse cannot be detected, try feeling for the radial artery on the other wrist; occasionally a congenitally abnormal radial artery may be difficult to feel. The radial artery may be less easily felt as it crosses the anatomical snuffbox.

The **ulnar artery** may be palpated as it crosses anterior to the flexor retinaculum along with the ulnar nerve. The artery lies lateral to the pisiform bone, separated from it by the ulnar nerve.

ARTERIES OF THE ABDOMEN AND THE PELVIS

Abdominal Aorta

The aorta enters the abdomen through the aortic opening of the diaphragm in front of the 12th thoracic vertebra. It descends on the anterior surfaces of the bodies of the lumbar vertebrae and in front of the fourth lumbar vertebra, it divides into the two common iliac arteries (*Fig. 3–14*). On the right side of the aorta lie the inferior vena cava, the cisterna chyli, and the beginning of the azygos vein. On its left side lies the left sympathetic trunk. On the anterior surface it is related to the stomach, celiac plexus, pancreas, splenic vein, left renal vein, third part of duodenum, coils of small intestine, and peritoneum.

Branches
1. *Three anterior visceral branches*
   a. *Celiac artery*
   b. *Superior mesenteric artery*
   c. *Inferior mesenteric artery*
2. *Three lateral visceral branches*
   a. *Suprarenal artery*
   b. *Renal artery*
   c. *Testicular or ovarian artery*
3. *Five lateral abdominal wall branches*
   a. *Inferior phrenic artery*
   b. *Four lumbar arteries*
4. *Three terminal arteries*
   a. *Two common iliac arteries*
   b. *Median sacral artery*

Celiac Artery
The celiac artery or trunk is very short and arises from the commencement of the abdominal aorta (*Fig. 3–14*). It is surrounded by the celiac plexus and lies behind the lesser sac of peritoneum. It has three terminal branches, the **left gastric,** the **splenic,** and the **hepatic arteries.**

Superior Mesenteric Artery
The superior mesenteric artery arises from the front of the abdominal aorta just below the celiac artery (*Fig. 3–14*). It runs downward and to the right behind the neck of the pancreas and in front of the third part of the duodenum. It continues downward to the right between the layers of the mesentery of the small intestine and ends by anastomosing with the ileal branch of its own ileocolic branch.

The superior mesenteric artery supplies the intestine from halfway along the duodenum to just proximal to the left colic flexure.

Branches
1. *Inferior pancreaticoduodenal artery*
2. *Middle colic artery*
3. *Right colic artery*
4. *Ileocolic artery*
5. *Jejunal and ileal branches*

Inferior Mesenteric Artery
The inferior mesenteric artery arises from the abdominal aorta about 1.5 inches (3.8 cm) above its bifurcation (*Fig. 3–14*). The artery runs downward and to the left and crosses the left common iliac artery. Here, it changes its name and becomes the superior rectal artery.

The inferior mesenteric artery is the artery that supplies the large intestine from the distal one-third of the transverse colon to halfway down the anal canal.

Branches
1. *Superior left colic artery* (left colic artery)
2. *Inferior left colic arteries* (sigmoid arteries)
3. *Superior rectal artery*

The superior rectal artery is a continuation of the inferior mesenteric artery as it crosses the left common iliac artery. The artery supplies the rectum and the upper half of the anal canal and anastomoses with the middle rectal and inferior rectal arteries.

Marginal Artery
The anastomosis of the colic arteries around the concave margin of the large intestine forms a single arterial trunk called the marginal artery. It begins at the ileocecal junction, where it anastomoses with the ileal branches of the superior mesenteric artery, and ends where it anastomoses less freely with the superior rectal artery.

Middle Suprarenal Arteries
The middle suprarenal artery arises on each side of the aorta and runs horizontally and laterally to the suprarenal gland.

Renal Arteries
The renal artery arises on each side of the aorta just below the origin of the superior mesenteric artery (*Fig. 3–14*).

Testicular Arteries
The testicular arteries arise from the front of the aorta just below the origin of the renal arteries (*Fig. 3–14*). Each artery is long and slender and passes obliquely downward and laterally behind the peritoneum. In the female, the **ovarian artery** has a similar abdominal course. Having crossed the external iliac artery at the pelvic inlet, it enters the ovary.

Inferior Phrenic Arteries
The inferior phrenic arteries arise from the aorta just beneath the diaphragm.

Lumbar Arteries
The four pairs of lumbar arteries arise from the back of the aorta and pass round the bodies of the upper four lumbar vertebrae.

Median Sacral Artery
The median sacral artery is a small branch that arises at the bifurcation of the aorta.

Common Iliac Arteries
The right and left common iliac arteries are the terminal branches of the abdominal aorta (*Fig. 3–14*). They arise at the level of the fourth lumbar vertebra and run downward and laterally to end opposite the sacroiliac joint by dividing into external and internal iliac arteries.

Branches. The **external iliac artery** (*Fig. 3–14*) runs along the medial border of the psoas muscle, following the pelvic brim, and gives off the **inferior epigastric** and **deep circumflex iliac branches.** The artery enters the thigh by passing under the inguinal ligament, to become the femoral artery.

The **internal iliac artery** (*Fig. 3–14*) passes down into the pelvis to the upper margin of the greater sciatic foramen where it divides into anterior and posterior divisions. The branches of these divisions supply the pelvic viscera, perineum, buttock, and sacral canal.

Surface Anatomy of the Arteries of the Abdomen

The **aorta** lies in the midline of the abdomen and bifurcates below into the right and left common iliac arteries opposite the fourth lumbar vertebra, i.e., on the intercristal plane (line connecting the highest points on the iliac crests). It may be palpated through the upper part of the anterior abdominal wall just to the left of the midline.

External Iliac Artery. The pulsations of this artery may be felt as it passes under the inguinal ligament to become continuous with the femoral artery. It may be located at a point halfway between the anterior superior iliac spine and the symphysis pubis.

ARTERIES OF THE LOWER LIMB

The main artery of the lower limb is the femoral artery. This artery is supplemented in the gluteal region by the superior and inferior gluteal arteries and in the adductor region by the obturator artery.

Femoral Artery. The femoral artery is a continuation of the external iliac artery (*Fig. 3–20*). It begins behind the inguinal ligament where it lies midway between the anterior superior iliac spine and the symphysis pubis. The artery descends almost vertically toward the adductor tubercle of the femur and ends at the opening of the adductor magnus muscle by entering into the popliteal space as the popliteal artery.

Branches

1. *Superficial circumflex iliac artery* arises just below the inguinal ligament and runs laterally to the region of the anterior superior iliac spine
2. *Superficial epigastric artery* arises just below the inguinal ligament and ascends on to the abdominal wall as high as the umbilicus
3. *Superficial external pudendal artery*
4. *Deep external pudendal artery*. The superficial external pudendal and the deep external pudendal arteries arise just below the inguinal ligament and run medially to supply the skin of the scrotum (or labium majus).
5. *Profunda femoris artery*. This is a large and important branch that arises from the lateral and posterior surface of the femoral artery about 1.5 inches (4 cm) below the inguinal ligament. It supplies structures in the anterior, medial, and posterior fascial compartments of the thigh

Popliteal Artery

The popliteal artery is a continuation of the femoral artery and extends from the opening in the adductor magnus to the lower border of the popliteus muscle, where it divides into anterior and posterior tibial arteries (*Fig. 3–20*). It is deeply placed in the popliteal fossa.

Branches

1. *Muscular branches*
2. *Articular branches to the knee joint*
3. *Terminal branches; anterior and posterior tibial arteries*

Anterior Tibial Artery

The anterior tibial artery arises at the bifurcation of the popliteal artery at the level of the lower border of the popliteus muscle (*Fig. 3–20*). It passes forward between the tibia and the fibula through the upper part of the interosseous membrane to enter the anterior compartment of the leg. It descends with the deep peroneal nerve to the front of the ankle joint where it becomes the dorsalis pedis artery.

Branches

1. *Muscular branches*
2. *Anastomotic branches,* which anastomose with branches of other arteries around the knee and the ankle joints

Dorsalis Pedis Artery

The dorsalis pedis artery begins in front of the ankle joint as a continuation of the anterior tibial artery (*Fig. 3–20*). It ends by passing downward into the sole through the proximal part of the space between the first and second metatarsal bones. Having passed between the two heads of the first dorsal interosseous muscle, it joins the lateral plantar artery and completes the plantar arch.

The artery is superficial in position. On its lateral side lie the tendons of the extensor digitorum longus and on the medial side, the tendon of extensor hallucis longus. Its pulsations can be easily felt.

Branches

1. *Lateral tarsal artery* supplies the dorsum of the foot
2. *Arcuate artery* runs laterally across the bases of the metatarsal bones and gives off metatarsal branches to the toes
3. *First dorsal metatarsal artery* supplies both sides of the big toe

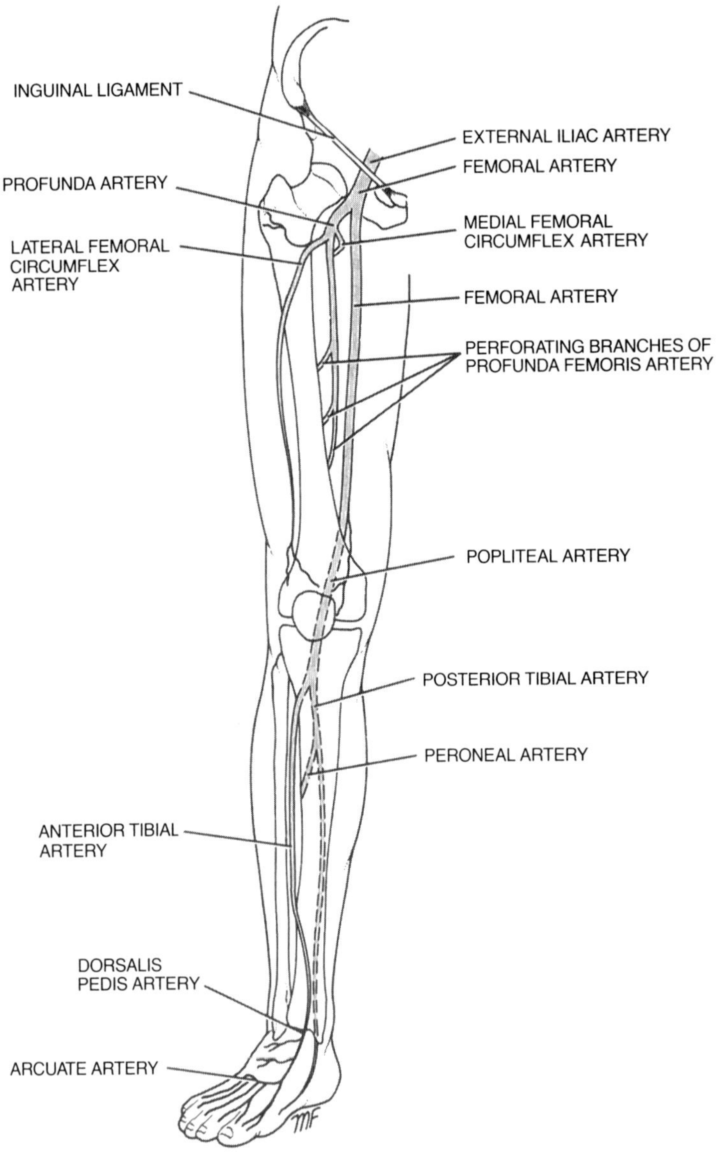

Figure 3–20. Main arteries of the lower limb.

Posterior Tibial Artery

The posterior tibial artery arises at the bifurcation of the popliteal artery at the level of the lower border of the popliteus muscle (*Fig. 3–20*). It descends in the posterior compartment of the leg accompanied by the tibial nerve. The artery terminates behind the medial malleolus by dividing into medial and lateral plantar arteries.

Branches

1. *Peroneal artery,* a large artery that arises close to the origin of the posterior tibial artery
2. *Muscular branches*
3. *Nutrient artery* to the tibia
4. *Anastomotic branches* around the ankle joint
5. *Medial and lateral plantar arteries*

Surface Anatomy of the Arteries of the Lower Limb

The **femoral artery** enters the thigh behind the inguinal ligament, at a point midway between the anterior superior iliac spine and the symphysis pubis. The artery is easily palpated here, as it may be pressed backward against the pectineus and the superior ramus of the pubis.

The **popliteal artery** is deeply placed in the popliteal fossa, and when it is examined with the knee extended, the deep fascia roofing over the fossa is made taut, and the pulse is difficult to detect. With the knee flexed and the patient kneeling or in the prone position, the fascia and neighboring muscles are relaxed, and the pulse is easily felt.

The **dorsalis pedis artery** lies between the tendons of extensor hallucis longus and extensor digitorum longus on the dorsal surface of the foot. It is covered only by skin and fascia and is easily palpated.

The **posterior tibial artery** passes behind the medial malleolus, beneath the flexor retinaculum, and lies between the tendons of flexor digitorum longus and flexor hallucis longus. Gentle palpation, without too great pressure, enables one to detect the artery without much difficulty.

It should be remembered that the dorsalis pedis artery is sometimes absent and is replaced by a large perforating branch of the peroneal artery. In the same manner, the peroneal artery may be larger than normal and replace the posterior tibial artery in the lower part of the leg.

Structure of Arteries

It is customary to classify arteries into elastic, or conducting, and muscular, or distributing. The walls of arteries are thick and strong and consist of three coats: (1) the tunica intima (internal coat), (2) the tunica media (middle coat), and (3) the tunica adventitia (outer coat). These three coats (*Fig. 3–21*) are seen most distinctly in the muscular arteries.

Elastic or Conducting Arteries. Elastic, or conducting, arteries (*Fig. 3–21*) are the large arteries, such as the aorta and pulmonary, brachiocephalic, common carotid, and iliac arteries. These arteries have the following basic structure.

Tunica Intima. This layer consists of a lining of endothelial cells resting on a layer of connective tissue. The tunica intima is bounded externally by a lamina of fenestrated elastic tissue known as the **internal elastic lamina.** This lamina may be difficult to recognize, because it merges with the elastic laminae of the tunica media.

Tunica Media. This layer consists largely of successive layers of fenestrated elastic laminae, between which are bundles of collagen fibers and smooth muscle cells. The outermost elastic lamina corresponds to the **external elastic lamina** found in other kinds of arteries.

The large amount of elastic tissue present in the tunica media of elastic arteries allows the arterial wall to become distended by the blood ejected from the heart during ventricular systole. The elastic recoil of these arteries that occurs when the ventricles are relaxing is largely responsible for maintaining the arterial blood pressure during diastole of the heart.

Tunica Adventitia. This consists of a thin layer of connective tissue that merges with the surrounding connective tissue. It contains nerve fibers and small arteries, the **vasa vasorum.** The arteries supply the outer part of the tunica media.

Muscular or Distributing Arteries. Muscular, or distributing, arteries (*Fig. 3–21*) are the medium-sized arteries, such as the brachial, radial, ulnar, femoral, and tibial arteries. The coronary arteries are also muscular. These arteries have the following basic structure.

Tunica Intima. This layer consists of a lining of endothelial cells resting on a layer of connective tissue. The tunica intima is bounded externally by a distinct lamina of fenestrated elastic tissue known as the **internal elastic lamina.**

Tunica Media. This layer consists largely of concentric layers of smooth muscle cells mixed with a few bundles of collagen fibers and occasional elastic fibers. The tunica media is bounded externally by a distinct layer of elastic tissue known as the **external elastic lamina.** The large number of circularly arranged smooth muscle cells in the tunica media of muscular arteries permits the autonomic nervous system to control the flow of blood through these vessels and thus regulate the blood supply to different organs of the body.

Tunica Adventitia. This layer is composed of a thick layer of connective tissue that merges with the surrounding connective tissues. Numerous nonmyelinated axons, which are postganglionic sympathetic nerve fibers, are found in the tunica adventitia. These nerve fibers stimulate

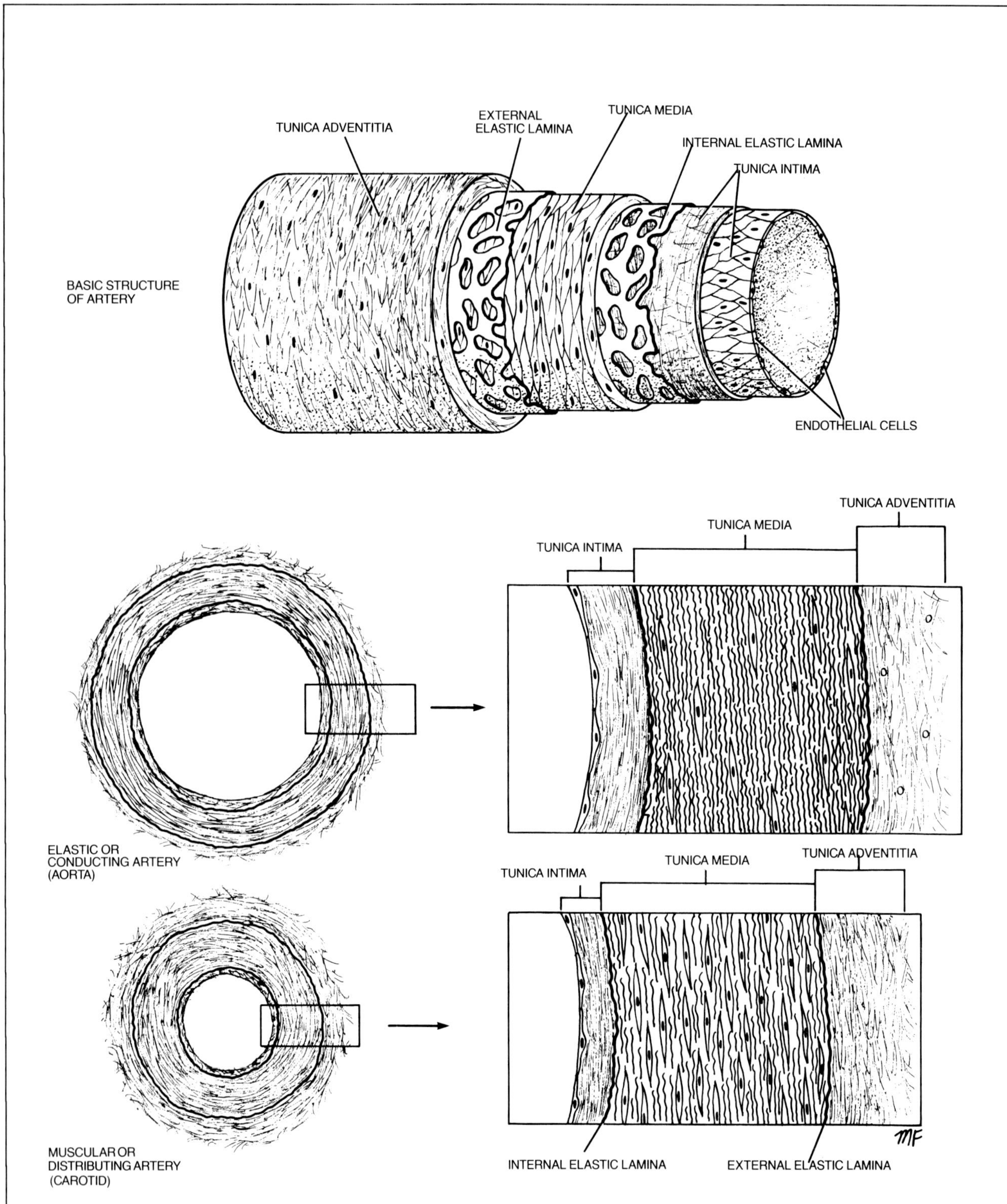

Figure 3–21. Basic structure of an artery and sections of an elastic or conducting artery and a muscular or distributing artery. Note that the tunica media of the aorta contains numerous elastic fibers and only a few smooth muscle fibers, whereas the tunica media of the carotid artery contains many smooth muscle fibers as well as elastic fibers.

the contraction of the smooth muscle cells in the tunica media and thus regulate the diameter of the lumina of these arteries.

Arterioles

An arteriole is an artery having a diameter of less than 0.1 mm; an arteriole characteristically has a very thick wall relative to the size of its lumen (*Fig. 3–22*). Arterioles possess the usual three coats, but the tunica media is made up almost entirely of circularly arranged smooth muscle cells. The tunica adventitia possesses autonomic nerve fibers that bring about contraction of the smooth muscle and thus control the size of the lumen of the vessel.

The terminal branch of many arterioles, sometimes called a **metarteriole,** has a small lumen and only a few smooth muscle cells in its tunica media. At the junction of the metarteriole with the capillary, however, there is a band of smooth muscle called the **precapillary sphincter.** The precapillary sphincter intermittently contracts, opening and closing the arteriolar–capillary junction.

Capillaries

When traced peripherally, arterioles break up into small blood vessels called capillaries (*Fig. 3–22*). The lumen of a typical capillary is about 7 to 9 μ wide, that is, slightly wider than a red blood cell. The wall of a capillary is composed of a single layer of endothelial cells resting on a basement membrane. Under the electron microscope, the endothelial cells can be seen to be held together by tight junctions (maculae occludens). Numerous pinocytotic vesicles are seen; they are involved in transporting materials across the endothelial cells in either direction.

There are two types of capillaries, as demonstrated under the electron microscope: continuous and fenestrated (*Figs. 3–22 and 3–23*). In both types, the endothelial cells form a continuous lining of the capillary, but in the fenestrated capillaries, the cells show tiny holes less than 100 nm in diameter. It is believed that these so-called holes are in fact closed by a diaphragm that is thinner than a cell membrane and that allows dissolved substances to pass through easily.

Occasional cells, called **pericytes,** are found within the basement membrane of the endothelial cells. These cells are noncontracting, and their function is not known.

Venules

Venules receive the blood from the capillaries and are of a larger diameter than capillaries. The structure of a small venule is identical to that of a continuous capillary. Some smooth muscle fibers are situated in the connective tissue surrounding the endothelial cells in large venules.

It is at the capillaries and the small venules that oxygen and nutritive materials pass from the blood into the tissue fluid and the carbon dioxide and the waste products of tissue metabolism pass into the bloodstream. The walls of blood vessels larger than capillaries and venules are impervious to gases and liquids.

NERVE SUPPLY OF THE ARTERIES OF THE UPPER LIMB

The arteries of the upper limb are innervated by sympathetic nerves. The preganglionic fibers originate from cell bodies in the second to eighth thoracic segments of the spinal cord. They ascend in the sympathetic trunk and synapse in the middle cervical, inferior cervical, first thoracic, or stellate ganglia. The postganglionic fibers join the nerves that form the brachial plexus and are distributed to the arteries within the branches of the plexus. For example, the digital arteries of the fingers are supplied by postganglionic sympathetic fibers that run in the digital nerves.

A detailed description of the innervation of the arteries of the upper limb is given on page 279.

NERVE SUPPLY OF THE ARTERIES OF THE LOWER LIMB

Sympathetic innervation of the arteries to the leg is derived from the lower three thoracic and upper two or three lumbar segments of the spinal cord. The preganglionic fibers pass to the lower thoracic and upper lumbar ganglia via white rami. The fibers synapse in the lumbar and sacral ganglia, and the postganglionic fibers reach the blood vessels via branches of the lumbar and sacral plexuses. The femoral artery receives its sympathetic fibers from the femoral and obturator nerves. The more distal arteries receive their postganglionic fibers via the common peroneal and tibial nerves.

A detailed description of the innervation of arteries of the lower limb is given on page 279.

CLINICAL NOTES

ALLEN TEST*

The Allen test is commonly used to determine the patency of the ulnar and radial arteries. With the patient's hands resting in his or her lap, compress the radial arteries against the anterior surface of each radius, and ask the patient to tightly clench the fists. The clenching of the fists closes off the superficial and deep palmar arterial arches. When the patient is asked to open the hands, the skin of the palms is at first white in color, and then normally the blood quickly flows into the arches through the ulnar arteries, causing the palms to promptly turn pink. This establishes that

*The validity of this test has been questioned in the literature.

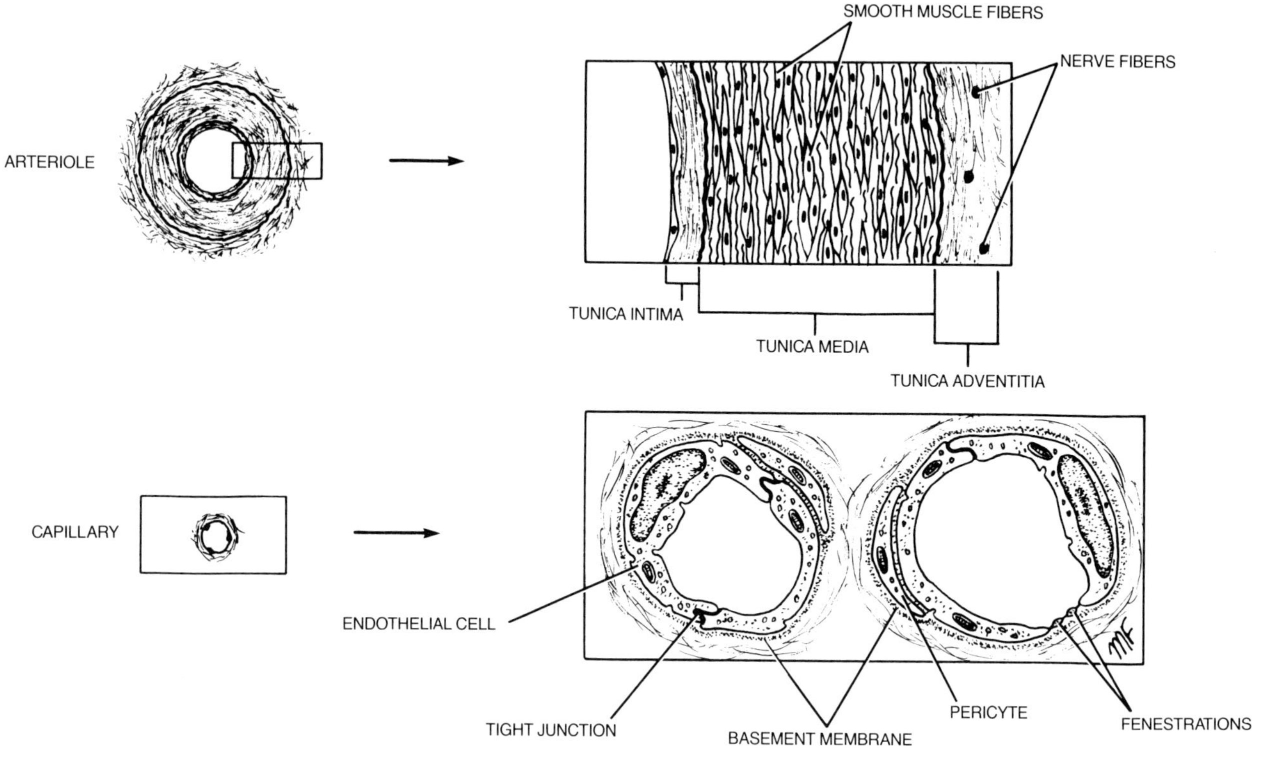

Figure 3–22. Basic structure of an arteriole and a capillary. Note the two types of capillaries, continuous and fenestrated.

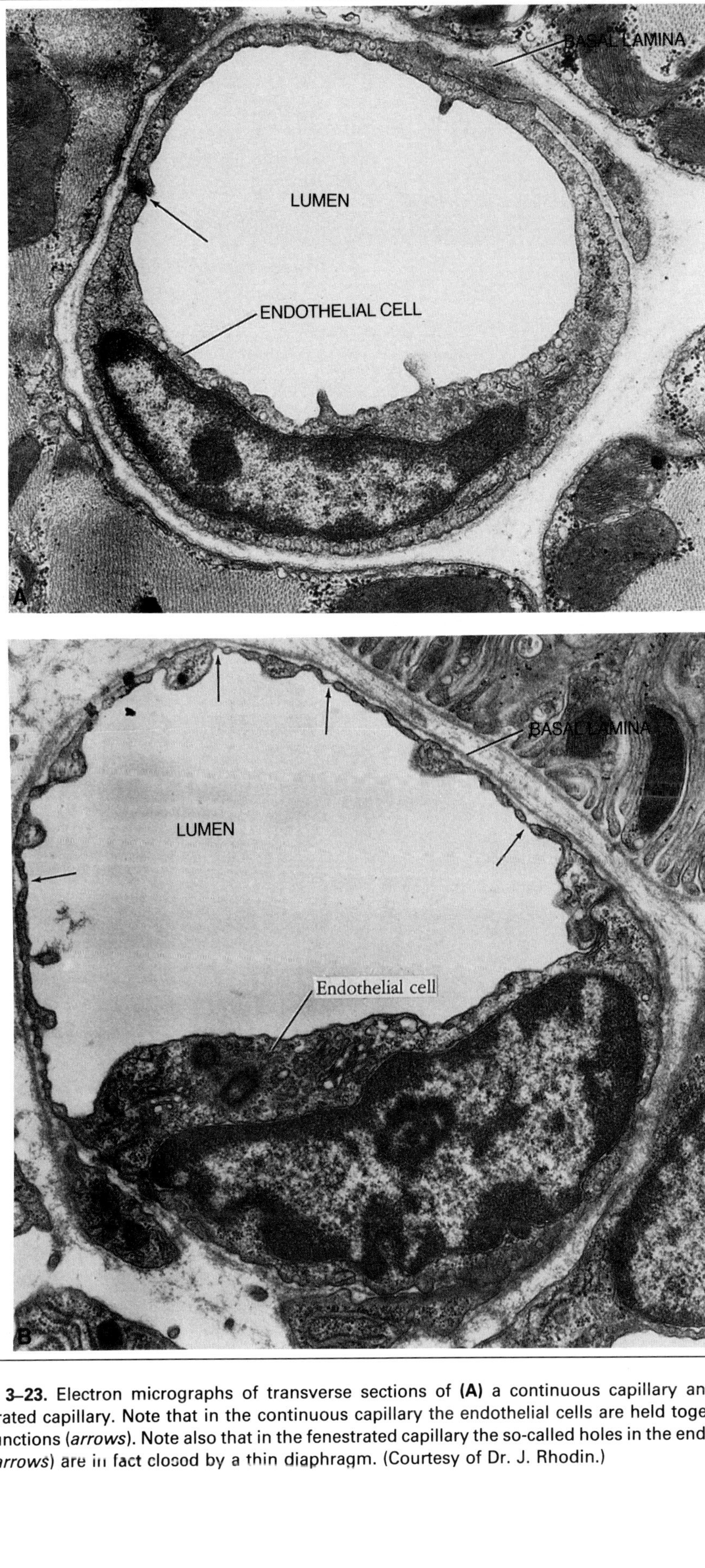

A

B

Figure 3–23. Electron micrographs of transverse sections of **(A)** a continuous capillary and **(B)** a fenestrated capillary. Note that in the continuous capillary the endothelial cells are held together by tight junctions (*arrows*). Note also that in the fenestrated capillary the so-called holes in the endothelial cells (*arrows*) are in fact closed by a thin diaphragm. (Courtesy of Dr. J. Rhodin.)

the ulnar arteries are patent. The patency of the radial arteries can be established by repeating the test, but this time compress the ulnar arteries as they lie lateral to the pisiform bones.

ARTERIAL CANNULATION

Theoretically any palpable artery can be used for cannulation, but in practice the radial, brachial, and femoral arteries are the most satisfactory to use.

Cannulation of the Radial, Brachial, and Femoral Arteries

In the case of the radial artery it is first essential to determine the adequacy of the collateral circulation of the hand by performing the Allen test. This precaution is necessary in case thrombosis occurs during or after cannulation.

The forearm is supinated and the wrist joint is extended to approximately a 50-degree angle (*Fig. 3–24*). The radial artery can then be easily palpated as it lies anterior to the distal end of the radius. After preparing the skin with an antiseptic solution, it is anesthetised with a local anesthetic. The skin over the artery is "nicked" by the tip of a large bore needle. The appropriate gauge needle covering a teflon (or similar) catheter is passed through the arterial wall at an angle of about 25 degrees to the anterior surface of the wrist (*Fig. 3–24*). Once the catheter is within the arterial lumen it can be advanced and the needle can be withdrawn. The catheter is then secured and a suitable dressing applied.

The brachial artery is usually cannulated as it descends into the cubital fossa on the medial border of the biceps brachii muscle. Unfortunately, the brachial artery has been associated with a higher incidence of postcatheterization thrombosis than the radial artery.

The femoral artery can be cannulated as it emerges from behind the inguinal ligament, midway between the anterior superior iliac spine and the symphysis pubis. The femoral pulse can be easily palpated. With one finger steadying the artery, the needle and the catheter can be introduced into the lumen at an oblique angle.

VEINS OF THE BODY

Veins of the Head and the Neck

The veins of the head and the neck may be divided for purposes of description into (1) the veins of the brain, the venous sinuses, the diploic and emissary veins and (2) the veins of the scalp, face, and neck.

Veins of the Brain. The veins of the brain have no valves and no muscle tissue in their walls, and hence are very thin. They include cerebral veins, cerebellar veins, and veins of the brainstem.

Cerebral Veins. These veins may be divided into external veins that lie on the surface of the cerebral hemispheres with the cerebral arteries in the subarachnoid space, and the internal veins that drain the interior of the hemispheres.

The **internal cerebral veins** are formed on each side by the union of the **thalamostriate vein** and the **choroid vein,** that drains the choroid plexus of the lateral ventricle. The internal cerebral veins ran backward in the roof of the third ventricle and below the splenium of the corpus callosum unite to form the **great cerebral vein,** which drains into the straight sinus.

Cerebellar Veins. The superior and inferior cerebellar veins lie on the surface of each cerebellar hemisphere. The former drains into the straight sinus and the latter into the transverse and sigmoid sinuses.

Veins of the Brainstem. These form a superficial plexus that drains into adjacent venous sinuses.

Venous Sinuses. The venous sinuses of the cranial cavity are placed between the layers of the dura mater. Their walls are thick and composed of fibrous tissue; there is no muscular tissue. The sinuses possess no valves. They receive tributaries from the brain, from the diploe of the skull, from the orbit, and from the internal ear.

Superior Sagittal Sinus. This lies in the upper fixed border of the falx cerebri. It begins in front at the foramen cecum, where it occasionally receives a vein from the nasal cavity. It runs backward and at the internal occipital protuberance it usually becomes continuous with the right transverse sinus. On each side the superior sagittal sinus communicates through small openings with cavities called the **venous lacunae.** Numerous arachnoid villi and granulations project into the lacunae.

The superior sagittal sinus receives the superior cerebral veins. At the internal occipital protuberance it is dilated to form the **confluence of the sinuses.** It is here that the sinus becomes continuous with the right transverse sinus and is connected to the left transverse sinus and the occipital sinus.

Inferior Sagittal Sinus. This lies in the lower free margin of the falx cerebri. It runs backward to join the greater cerebral vein to form the straight sinus. It receives cerebral veins from the medial surface of the cerebral hemisphere.

Straight Sinus. This lies at the junction of the falx cerebri with the tentorium cerebelli. It is formed by the union of the inferior sagittal sinus with the great cerebral vein and drains usually into the left transverse sinus. It receives some of the superior cerebellar veins.

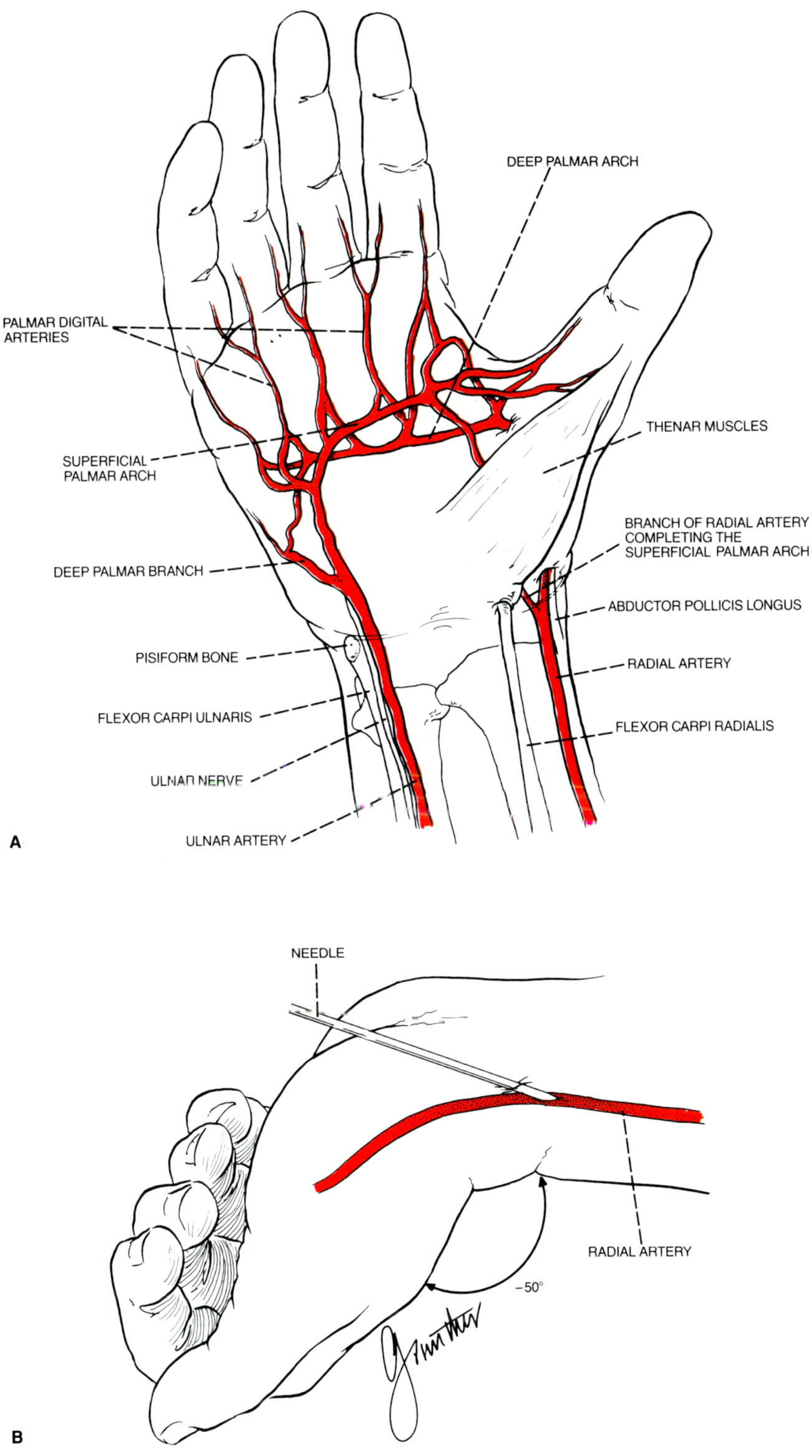

Figure 3–24. A. The positions of the radial and ulnar arteries in front of the wrist. Note that the superficial palmar arch is formed mainly from the ulnar artery and that the deep palmar arch receives its major contribution from the radial artery. **B.** The wrist joint extended during cannulation of the radial artery.

Transverse Sinuses. The right sinus begins at the internal occipital protuberance as a continuation of the superior sagittal sinus; the left sinus is usually a continuation of the straight sinus. Each sinus runs forward and laterally along the attached margin of the tentorium cerebelli. They end on each side by becoming the sigmoid sinus. The transverse sinuses receive the superior petrosal sinuses, and the inferior cerebral and inferior cerebellar veins.

Sigmoid Sinuses. These are a direct continuation of the transverse sinuses. Each sinus curves downward behind the mastoid antrum and leaves the skull through the posterior part of the jugular foramen, where it becomes the internal jugular vein.

Occipital Sinus. This sinus occupies the attached margin of the falx cerebelli. It begins near the foramen magnum where it communicates with the vertebral veins and drains into the confluence of sinuses.

Cavernous Sinuses. Each sinus is situated on the side of the body of the sphenoid bone and extends from the superior orbital fissure in front to the apex of the petrous part of the temporal bone behind. The sinus receives the superior ophthalmic vein (through this vein it communicates with the facial vein), the sphenoparietal sinus, the central vein of the retina, and the cerebral veins. The cavernous sinus drains into the transverse sinus through the superior petrosal sinus.

Intercavernous Sinuses. These sinuses connect the two cavernous sinuses through the sella turcica.

Traveling through the cavernous sinus on each side are the internal carotid artery and the sixth cranial nerve. In the lateral wall lie the third and fourth cranial nerves, and the ophthalmic and maxillary divisions of the fifth cranial nerve.

Superior and Inferior Petrosal Sinuses. The superior petrosal sinus runs along the superior border of the petrous part of the temporal bone. It drains the cavernous sinus into the transverse sinus. The inferior petrosal sinus runs along the inferior border of the petrous part of the temporal bone and drains the cavernous sinus into the internal jugular vein.

Diploic Veins. Diploic veins occupy channels within the bones of the vault of the skull. They communicate with the venous sinuses, the meningeal veins, and the veins of the pericranium.

Emissary Veins. The emissary veins are valveless veins that pass through the foramina in the skull and connect the venous sinuses with the veins outside the skull. They are a route through which infection can travel from outside the skull to the venous sinuses within.

VEINS OF THE SCALP, FACE, AND NECK

Facial Vein

The facial vein is formed at the medial angle of the eye by the union of the **supraorbital** and **supratrochlear veins** (*Fig. 3–25*). It is connected through the superior ophthalmic vein with the cavernous sinus.

Superficial Temporal Vein

The superficial temporal vein is formed as a network on the side of the scalp (*Fig. 3–25*). It enters the substance of the parotid gland and joins the maxillary vein to form the retromandibular vein.

Maxillary Vein

The maxillary vein is a short vein that arises in the infratemporal fossa by the union of the veins that form the pterygoid venous plexus. The **pterygoid venous plexus** is made up of veins that correspond to branches of the maxillary artery.

Retromandibular Vein

The retromandibular vein is formed by the union of the superficial temporal and the maxillary veins (*Fig. 3–25*). It descends into the parotid gland and divides into an anterior branch, which joins the facial vein, and a posterior branch, which joins the **posterior auricular vein** to form the external jugular vein.

External Jugular Vein

The external jugular vein arises just behind the angle of the mandible by the union of the posterior auricular vein with the posterior division of the retromandibular vein (*Fig. 3–25*). It descends obliquely across the sternocleidomastoid muscle, beneath the platysma muscle, and drains into the subclavian vein behind the middle of the clavicle.

The external jugular vein receives the **posterior external jugular vein** from the back of the scalp, the **superficial cervical vein,** the **suprascapular vein** and the **anterior jugular vein.**

Internal Jugular Vein

The internal jugular vein is a large vein that drains blood from the brain, from the face, and from the neck (*Fig. 3–25*). It begins at the jugular foramen in the skull as a continuation of the sigmoid sinus. It descends through the neck in the carotid sheath lateral to the vagus nerve, the internal and common carotid arteries (*Fig. 3–16*). It ends by joining the subclavian vein to form the brachiocephalic vein behind the medial end of the clavicle. It

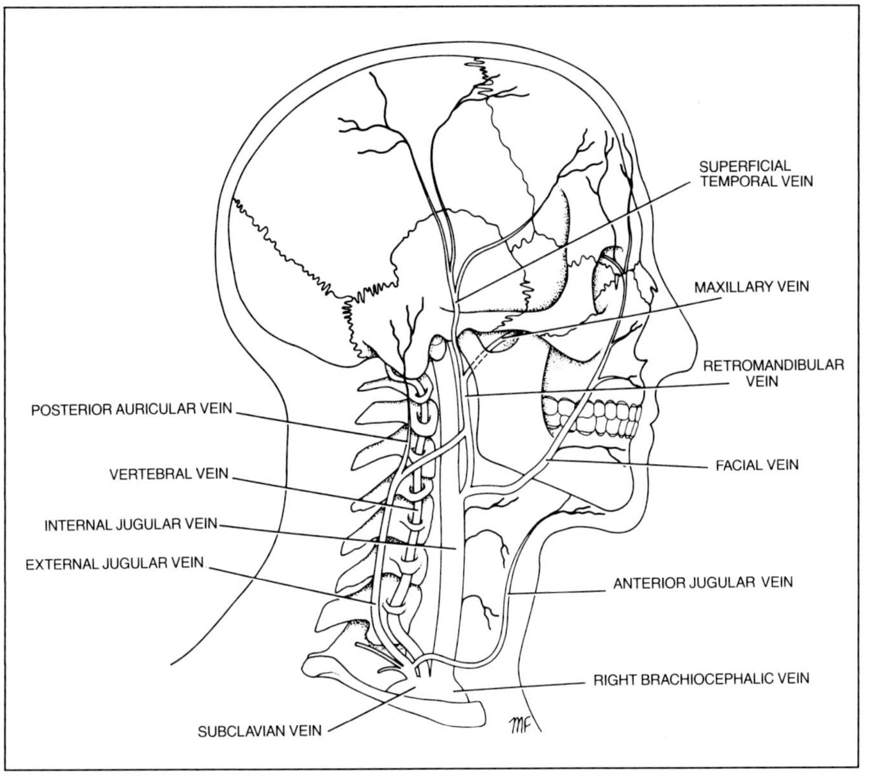

Figure 3–25. Main veins of the head and the neck.

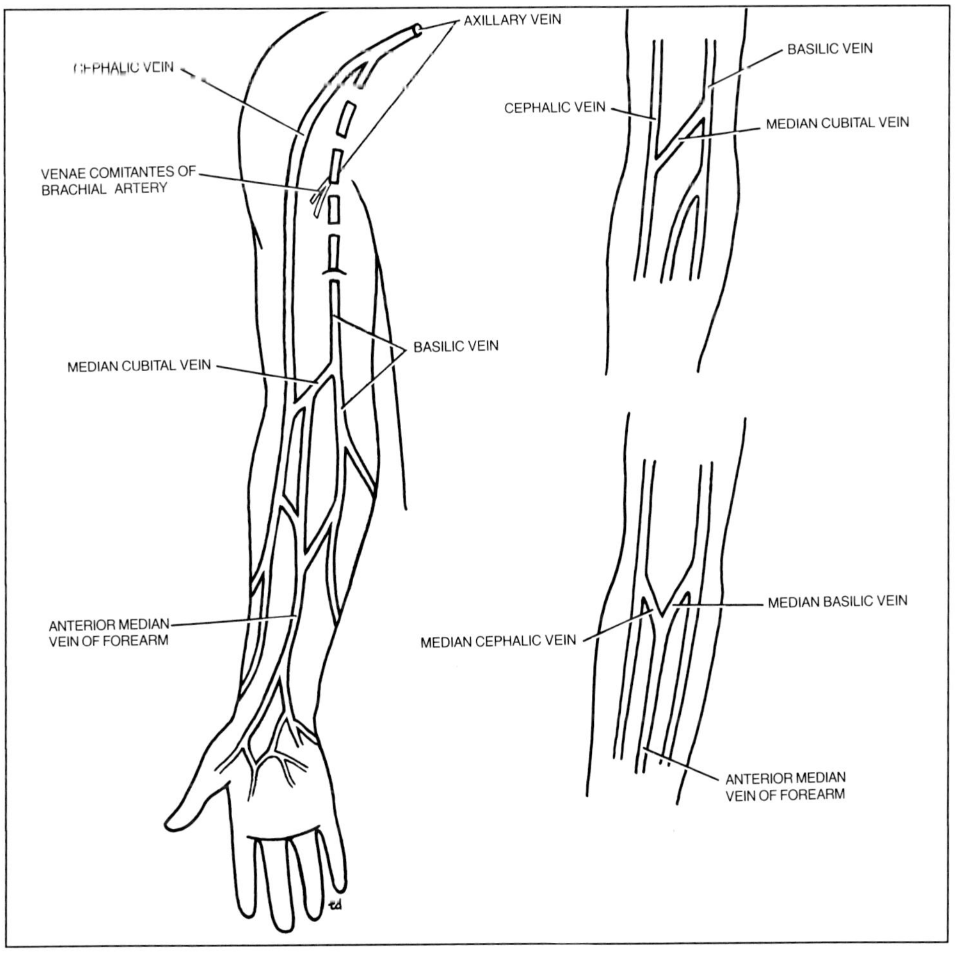

Figure 3–26. Superficial veins of the upper limb. The common variations seen in the region of the elbow are also shown.

is closely related to the **deep cervical lymph nodes** throughout its course.

Tributaries

1. Inferior petrosal sinus
2. Facial vein
3. Pharyngeal veins
4. Lingual veins
5. Superior thyroid vein
6. Middle thyroid vein

Vertebral Vein. The vertebral vein arises from a plexus of veins beneath the skull, which descend through the foramina in the transverse processes of the cervical vertebrae around the vertebral artery (*Fig. 3–25*). On reaching the sixth or seventh cervical transverse process the vertebral vein emerges and descends to open into the brachiocephalic vein.

Subclavian Vein. This vein is a continuation of the axillary vein at the outer border of the first rib (*Fig. 3–25*). It joins the internal jugular vein behind the medial end of the clavicle to form the brachiocephalic vein. It is separated from the subclavian artery by the scalenus anterior muscle. It receives the external jugular vein and it often receives on the left side the **thoracic duct** and on the right side the **right lymphatic duct.**

Surface Anatomy of the Veins of the Head and the Neck

External Jugular Vein. The external jugular vein lies in the superficial fascia deep to the platysma muscle. It passes downward from the region of the angle of the mandible to the middle of the clavicle. It perforates the deep fascia just above the clavicle and drains into the subclavian vein.

Internal Jugular Vein. The internal jugular vein lies on the lateral side of the carotid arteries within the carotid sheath. It can be located by taking a point midway between the tip of the mastoid process and the angle of the mandible and joining it to the sternoclavicular joint. It is overlapped laterally by the sternocleidomastoid muscle.

VEINS OF THE UPPER LIMB

The veins of the upper limb may be divided into superficial and deep groups. The superficial veins are of great clinical importance and lie in the superficial fascia. The deep veins accompany the main arteries.

Superficial Veins

Dorsal Venous Arch or Network. This venous network of superficial veins lies on the dorsum of the hand. It receives most of the blood from the hand and the fingers and is drained on the lateral side by the cephalic vein and on the medial side by the basilic vein.

Cephalic Vein. The cephalic vein arises from the lateral side of the dorsal venous network (*Fig. 3–26*), ascends around the lateral border of the forearm, just lateral to the styloid process of the radius, to reach the anterior aspect of the forearm. It then ascends into the arm and runs along the lateral border of the biceps to reach the interval between the deltoid and pectoralis major muscles, terminating by piercing the deep fascia to enter the axillary vein.

Basilic Vein. This vein arises from the medial side of the dorsal venous network and ascends on the posterior surface of the forearm (*Fig. 3–26*). Just below the elbow it inclines forward to reach the anterior aspect of the forearm. The vein then runs upward medially to the biceps and pierces the deep fascia at about the middle of the arm. The **median cubital vein** links the cephalic and basilic veins in the cubital fossa (*Fig. 3–26*). The basilic vein joins the venae comitantes of the brachial artery at the lower border of the teres major muscle to form the axillary vein.

Median Vein of the Forearm. The median vein of the forearm is a small vein that arises in the palm and ascends on the front of the forearm. It drains into the basilic vein, or the median cubital vein, or divides into two branches; one of which joins the basilic vein (**median basilic vein**) and the other joins the cephalic (**median cephalic vein**).

The superficial veins of the upper limbs are used for taking blood samples or for blood transfusion. They are also used for the introduction of cardiac catheters.

Deep Veins

Venae Comitantes. The deep veins accompany the respective arteries as venae comitantes. The two venae comitantes of the brachial artery join the basilic vein at the lower border of the teres major muscle to form the axillary vein.

Axillary Vein. The axillary vein is formed at the lower border of the teres major muscle by the union of the venae comitantes of the brachial artery with the basilic vein (*Fig. 3–26*). It becomes the subclavian vein at the outer border of the first rib. It receives tributaries that correspond to the branches of the axillary artery.

Surface Anatomy of the Veins of the Upper Limb

The **venous network of superficial veins** can be identified on the dorsum of the hand. These drain upward into a lateral cephalic vein and a medial basilic vein. The **cephalic vein** crosses the anatomical snuffbox and winds around onto the anterior aspect of the forearm. It then ascends into the arm and runs along the lateral border of the biceps. It ends by piercing the deep fascia in the deltopectoral triangle and enters the axillary vein.

The **basilic vein** can be traced from the dorsum of the hand around the medial side of the forearm; it reaches the anterior aspect of the forearm just below the elbow (*Fig. 3–26*). It pierces the deep fascia at about the middle of the arm. The **median cubital vein** (or median cephalic and median basilic veins) links the cephalic and basilic veins in the cubital fossa.

VEINS OF THE THORAX

Brachiocephalic Veins

The **right brachiocephalic vein** is formed at the root of the neck by the union of the right subclavian and the right internal jugular veins. The **left brachiocephalic vein** has a similar origin on the left and passes obliquely downward and to the right behind the manubrium sterni and in front of the large branches of the aortic arch. The left vein joins the right brachiocephalic vein to form the superior vena cava (*Fig. 3–27*).

The tributaries of each vein include, the **vertebral vein,** the **internal thoracic vein,** the **inferior thyroid vein,** and the **first posterior intercostal vein.**

Superior Vena Cava

The superior vena cava collects all the venous blood from the head, the neck, and both upper limbs and is formed by the union of the two brachiocephalic veins. It descends vertically to end in the right atrium of the heart. The azygos vein joins the posterior aspect of the superior vena cava just before it enters the pericardium (*Fig. 3–27*).

Azygos Vein

The azygos vein has a variable origin. It is often formed by the union of the **right ascending lumbar vein** and the **right subcostal vein.** The azygos vein has numerous tributaries that include the eight lower right intercostal veins, the right superior intercostal vein, the superior and inferior hemiazygos veins, and numerous mediastinal veins. It ends by joining the posterior aspect of the superior vena cava.

Inferior Vena Cava

The inferior vena cava is formed in the abdomen (see the next column) and perforates the tendinous part of the diaphragm. It passes through the pericardium and opens into the lower and back part of the right atrium of the heart.

Pulmonary Veins

There are four pulmonary veins (two on each side). They carry oxygenated blood from the lungs and empty into the left atrium. There are no valves.

Cardiac Veins

The cardiac veins are described on page 70.

VEINS OF THE ABDOMEN AND THE PELVIS

External Iliac Vein

The external iliac vein begins on each side as a continuation of the femoral vein behind the inguinal ligament. It ascends along the brim of the pelvis to the sacroiliac joint where it unites with the internal iliac vein to form the common iliac vein (*Fig. 3–27*).

Internal Iliac Vein

The internal iliac vein originates on each side in the pelvis by the union of all the veins that correspond to the branches of the internal iliac artery.

Common Iliac Veins

The common iliac veins are formed by the union of the external and internal iliac veins in front of the sacroiliac joint (*Fig. 3–27*).

Inferior Vena Cava

The inferior vena cava is formed by the union of the common iliac veins behind the right common iliac artery at the level of the fifth lumbar vertebra (*Fig. 3–27*). It ascends on the right side of the aorta, pierces the central tendon of the diaphragm at the level of the eighth thoracic vertebra, and drains into the right atrium of the heart. The inferior vena cava has the following tributaries:

1. Two anterior visceral tributaries, the hepatic veins
2. Three lateral visceral tributaries
   a. Right suprarenal vein (the left suprarenal vein drains into the left renal vein)
   b. Renal veins
   c. Right testicular or ovarian vein (the left vein drains into the left renal vein)
3. Five lateral abdominal wall tributaries
   a. Inferior phrenic vein
   b. Four lumbar veins
4. Three veins of origin
   a. Two common iliac veins
   b. Median sacral vein (sometimes enters the left common iliac vein)

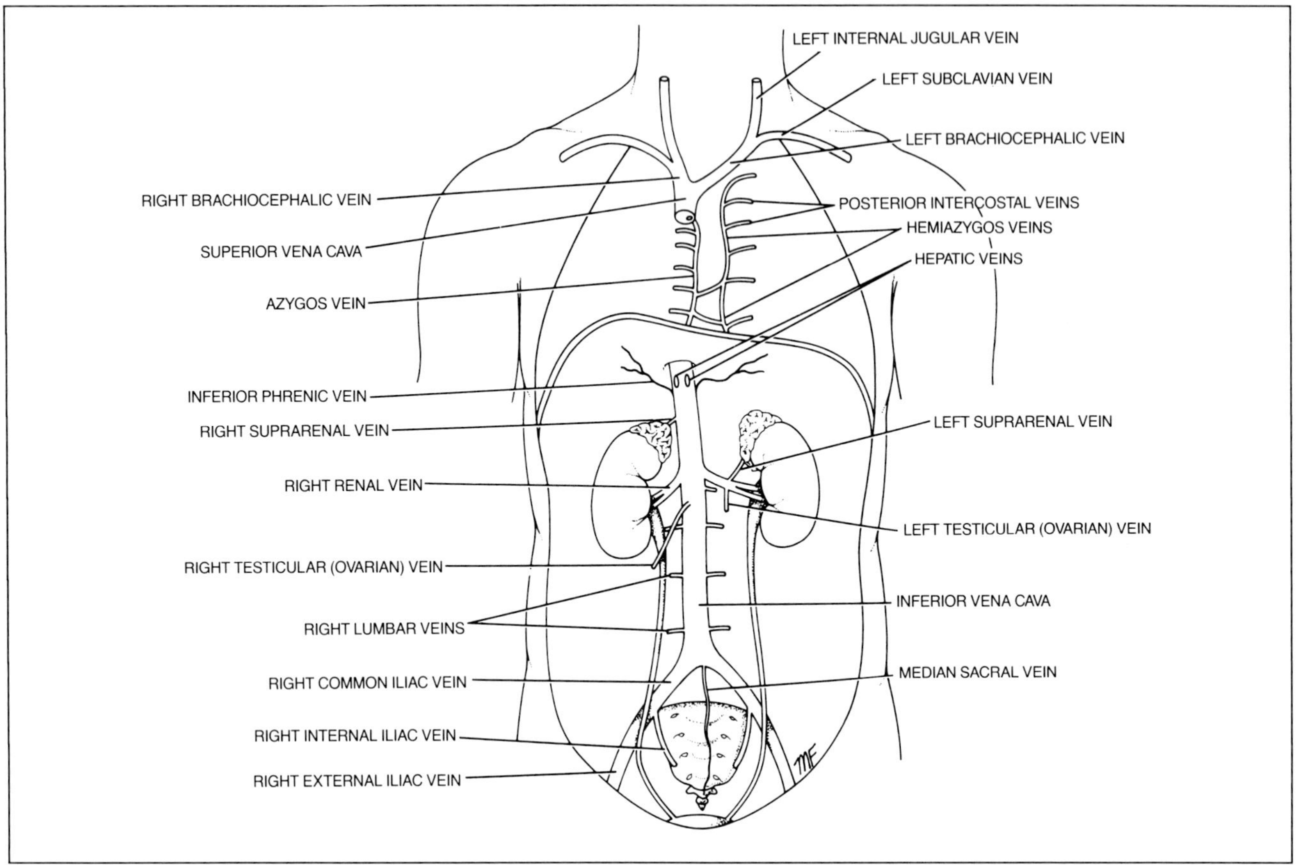

Figure 3–27. Major veins draining into the superior vena cava and the inferior vena cava.

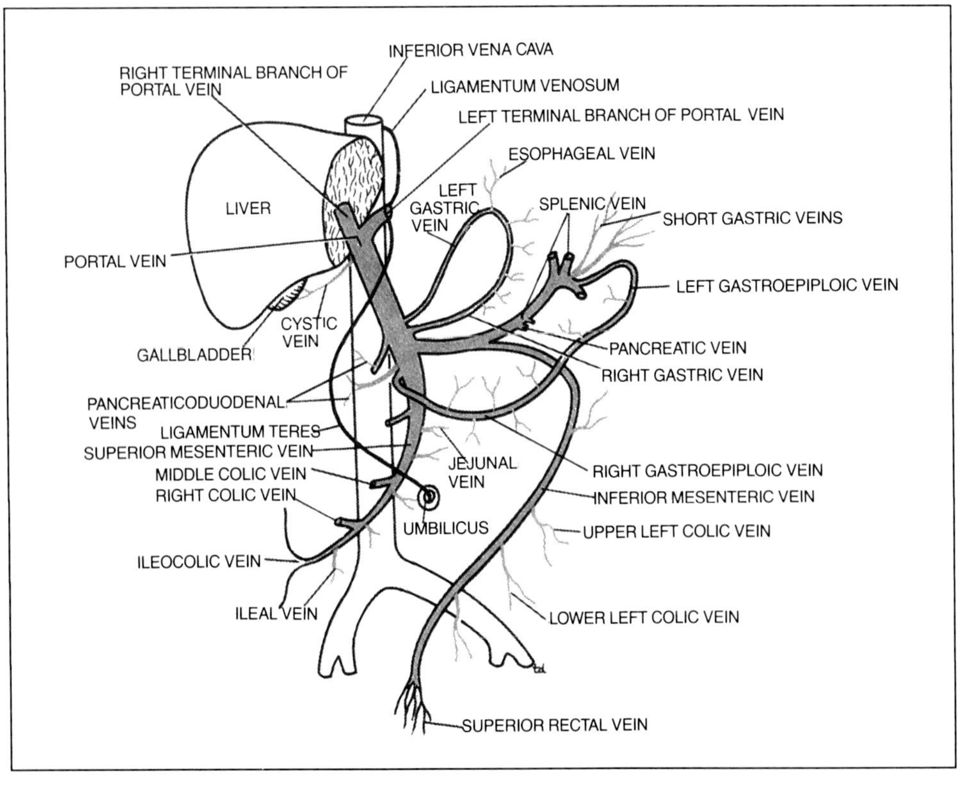

Figure 3–28. Tributaries of the portal vein.

Portal Vein

The portal vein is about 2 inches (5 cm) long, originating from behind the neck of the pancreas by the union of the superior mesenteric and the splenic veins (*Fig. 3–28*). It runs upward and to the right, posterior to the first part of the duodenum, and enters the lesser omentum. Here, the portal vein lies posterior to the bile duct (common bile duct) and hepatic artery. It then passes up in front of the opening into the lesser sac to the porta hepatis, where it divides into right and left terminal branches (*Fig. 3–28*).

The portal venous system begins as a capillary network in the organs it drains and ends in capillary-like vessels, termed sinusoids in the liver. The portal vein drains blood from the alimentary canal from the lower end of the esophagus to half way down the anal canal, the pancreas, the gallbladder, the bile duct, and the spleen.

The tributaries of the portal vein (*Fig. 3–28*) are the **splenic vein,** the **superior mesenteric vein,** the **left gastric vein,** the **right gastric vein,** and the **cystic veins.**

Portal Systemic Anastomoses. These important communications are as follows:

1. At the lower third of the esophagus, the esophageal branches of the left gastric vein (portal tributary) anastomose with the esophageal veins draining the middle third of the esophagus into the azygos veins (systemic tributary)
2. Halfway down the anal canal, the superior rectal veins (portal tributary) draining the upper half of the anal canal anastomose with the middle and inferior rectal veins (systemic tributaries)
3. The **paraumbilical veins** connect the left branch of the portal vein with the superficial veins of the anterior abdominal wall (systemic tributaries). The paraumbilical veins travel in the falciform ligament and accompany the ligamentum teres
4. The veins of the ascending colon, descending colon, duodenum, pancreas, the liver (portal tributaries) anastomose with the renal, the lumbar, and the phrenic veins (systemic tributaries)

VEINS OF THE LOWER LIMB

The veins of the lower limb may be divided into superficial and deep groups. The superficial veins are of great clinical importance and lie in the superficial fascia. The deep veins accompany the main arteries.

Superficial Veins

Dorsal Venous Arch. The dorsal venous arch lies on the dorsum of the foot (*Fig. 3–29*). The greater part of the blood from the whole foot drains into the arch via digital veins, and communicating veins that pass through the interosseous spaces. The dorsal venous arch is drained on the medial side by the great saphenous vein and on the lateral side by the small saphenous vein.

Great Saphenous Vein. The great saphenous vein arises from the medial end of the dorsal venous arch of the foot and ascends directly in front of the medial malleolus (*Fig. 3–29*). It is accompanied by the saphenous nerve and lies in the superficial fascia: It passes up the medial side of the leg, behind the knee and curves forward around the medial side of the thigh. It then runs through the saphenous opening in the deep fascia and joins the femoral vein about 1.5 inches (4 cm) below and lateral to the pubic tubercle. The great saphenous vein possesses numerous valves and is connected to the small saphenous vein by branches that pass behind the knee. A number of **perforating veins** connect the great saphenous vein with the deep veins along the medial side of the calf.

The great saphenous vein receives near its termination the following small named tributaries: (1) the **superficial circumflex iliac vein,** (2) the **superficial epigastric vein,** and (3) the **superficial external pudendal vein.**

Small Saphenous Vein. The small saphenous vein arises from the lateral end of the dorsal venous arch of the foot (*Fig. 3–29*). It ascends behind the lateral malleolus in company with the sural nerve. It then passes up the back of the leg and pierces the deep fascia. It enters the popliteal fossa by passing between the two heads of the gastrocnemius muscle and drains into the popliteal vein. The small saphenous vein communicates with the deep veins and the great saphenous vein.

Deep Veins

Venae Comitantes. The deep veins accompany the respective arteries as venae comitantes. The venae comitantes of the anterior and posterior tibial arteries unite in the popliteal fossa to form the popliteal vein.

Popliteal Vein. The popliteal vein, formed by the union of the venae comitantes of the anterior and posterior tibial arteries, ascends through the popliteal space and at the opening in the adductor magnus muscle, it becomes the femoral vein. It receives numerous tributaries including the small saphenous vein.

Femoral Vein. The femoral vein is a continuation of the popliteal vein at the opening in the adductor magnus muscle. It ascends through the thigh and comes to lie on the medial side of the femoral artery in the intermediate compartment of the femoral sheath. It ends by becoming continuous with the external iliac vein behind the inguinal ligament. The femoral vein receives the great saphenous

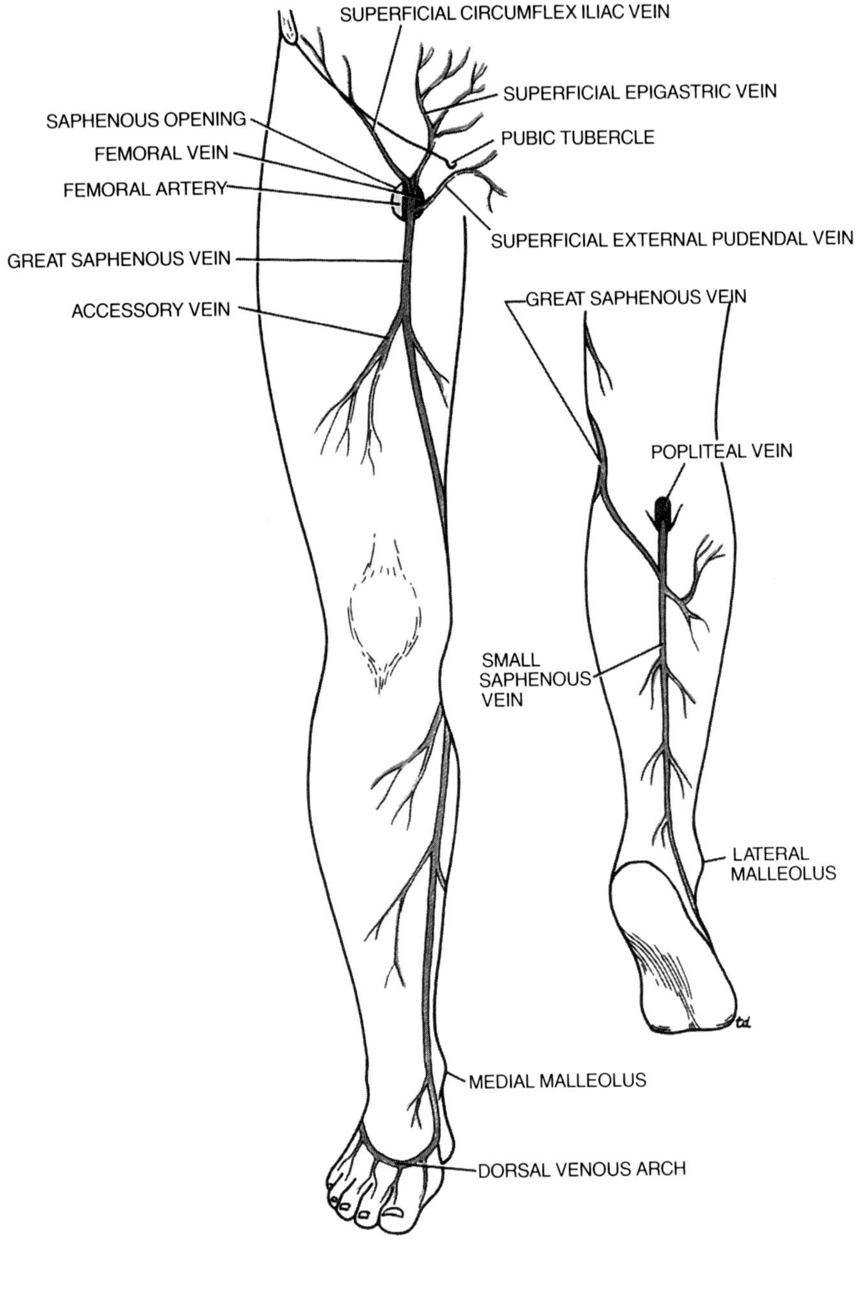

Figure 3–29. Superficial veins of the right lower limb.

vein and veins that correspond to branches of the femoral artery.

Surface Anatomy of the Veins of the Lower Limb

The **dorsal venous arch** can be seen on the dorsal surface of the foot proximal to the toes.

The **great saphenous vein** leaves the medial end of the arch and passes upward **in front** of the medial malleolus (*Fig. 3–29*). It ascends the leg and passes posteromedial to the knee. It then passes up the thigh to join the femoral vein 1.5 inches (4 cm) below and lateral to the pubic tubercle.

The **small saphenous vein** drains the lateral end of the arch and passes up **behind** lateral malleolus. It ascends the leg and joins the popliteal vein in the popliteal fossa.

The **femoral vein** lies in the femoral triangle below the inguinal ligament and on the medial side of the pulsating femoral artery. The surface marking of the artery is the point midway between the anterior superior iliac spine and the symphysis pubis.

Structure of Veins

Small Veins. The smallest veins, the venules, closely resemble capillaries and were discussed previously. The diameters of veins tend to be larger than those of the corresponding arteries. As the venous tributaries unite to form larger veins, their walls become thicker.

The walls of veins, like those of arteries, consist of three coats: (1) the tunica intima, (2) the tunica media, and (3) the tunica adventitia (*Fig. 3–30*).

Medium-sized Veins. These veins include the distal veins of the limbs and the veins of the viscera.

Tunica Intima. This layer consists of a lining of endothelial cells resting on a thin layer of connective tissue.

Tunica Media. This layer is thinner than that found in arteries and consists of circularly arranged smooth muscles fibers mixed with connective tissue.

Tunica Adventitia. This is a thick layer, thicker than the tunica media. It is composed of loose connective tissue containing bundles of collagen fibers and elastic networks. Longitudinally arranged bundles of smooth muscle cells are also frequently found in this layer. Many vasa vasorum are present.

Valves. Many veins are provided with valves that allow the blood to flow toward the heart but not in the opposite direction. Most of the valves have two leaflets (are bicuspid), and each leaflet is composed of a fold of tunica intima. The valves within veins that pass between muscles allow the veins to serve as pumps and aid the return of venous blood to the heart.

Large Veins. These veins include the superior and inferior venae cavae, the portal vein, and the iliac veins; the renal and mesenteric veins have a similar structure.

Tunica Intima. This layer consists of a lining of endothelial cells resting on a thick layer of connective tissue.

Tunica Media. This is a thin layer and consists of circularly arranged smooth muscle fibers mixed with connective tissue.

Tunica Adventitia. This is the thickest layer of the wall. It is composed of loose connective tissue containing bundles of longitudinally running collagen fibers and smooth muscle cells. Elastic networks are also present. Situated within the tunica adventitia are many small blood vessels, the vasa vasorum; lymphatic vessels and nerves are also present.

Nerve Supply of the Veins. Like the arteries, the smooth muscle of the veins is innervated by sympathetic postganglionic fibers that provide vasomotor tone. The origin of these fibers is similar to those of the accompanying arteries (see page 279). A few myelinated afferent fibers are also present.

Other Structures

Sinusoids. Sinusoids are found in the bone marrow, spleen, liver, and some endocrine glands. They are very thin-walled blood vessels having an irregular cross diameter. The sinusoids are lined by endothelium supported by a minimum of connective tissue. In the bone marrow and spleen, there are gaps between the endothelial cells that permit the passage of blood cells. In the liver, the sinusoids are lined by endothelial cells, some of which are phagocytic and are members of the reticuloendothelial system. These are known as **Kupffer cells.** In the suprarenal cortex and the islets of Langerhans in the pancreas, the endothelial cells are fenestrated, as in fenestrated capillaries.

Arteriovenous Anastomoses. In certain areas of the body, notably the skin, the blood is shunted from the arterioles directly into the venules without having to circulate through a capillary network. Such a bypass is referred to as an **arteriovenous anastomosis** or **AV shunt.** The walls of the shunt are thick and muscular and are well supplied with vasomotor nerve fibers.

In the nail beds of the fingers and toes and the auricle of the ear, the arteriovenous anastomoses are branched and tortuous and form a specialized organ called the **glomus.**

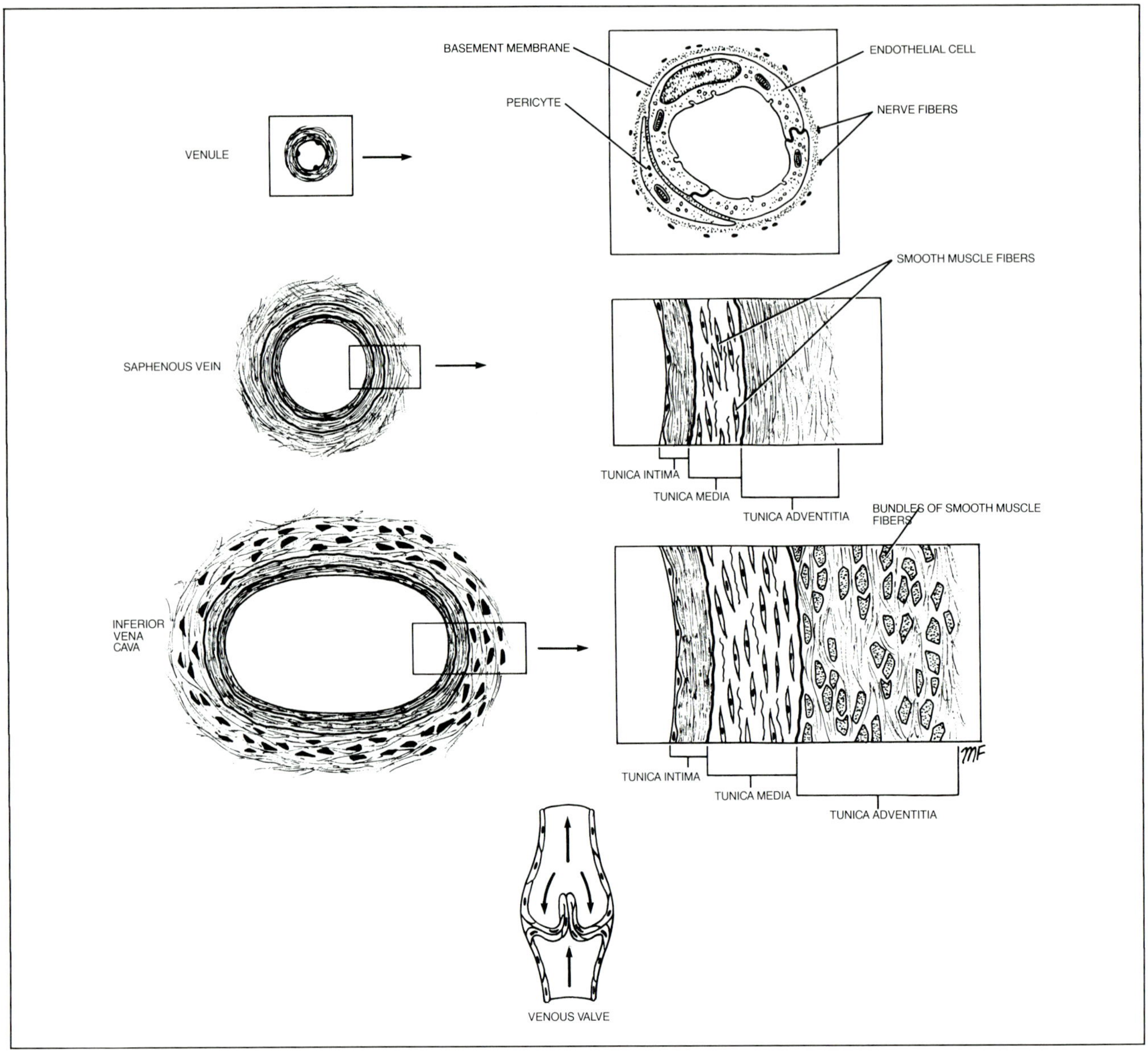

Figure 3–30. Basic structure of a venule, a medium-sized vein, the saphenous vein, and a large vein, the inferior vena cava.

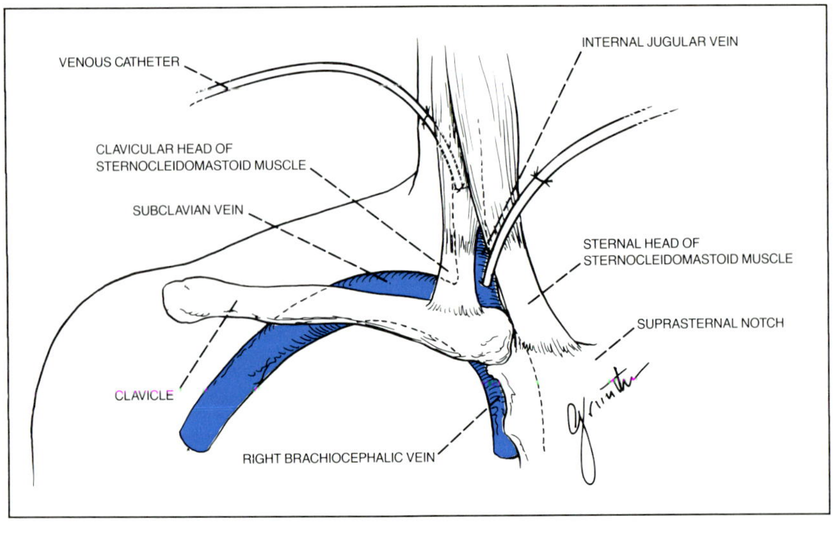

Figure 3–31. Diagram showing two sites at the root of the neck where central venous catheterization can be performed. Note the position of the internal jugular vein beneath the triangle formed by the sternal and clavicular heads of the sternocleidomastoid muscle and the medial end of the clavicle.

Arteriovenous anastomoses control blood flow and assist in temperature regulation of the body. In the skin, for example, the constriction or dilatation of an AV shunt will control the blood flow through the capillaries and thus conserve heat or increase heat loss in the body.

Anatomical End Arteries and Functional End Arteries. Anatomical end arteries are vessels whose terminal branches do not anastomose with branches or arteries supplying adjacent areas. Functional end arteries are vessels whose terminal branches do anastomose with those of adjacent arteries, but the caliber of the anastomosis is insufficient to keep the tissue alive if one of the arteries becomes occluded. Coronary arteries are examples of functional end arteries.

Carotid Sinus. The carotid sinus is a slight dilatation of the terminal part of the common carotid artery or the beginning of the internal carotid artery (*Fig. 3–16*). The carotid arteries are muscular arteries and serve to distribute blood to the head and the neck. In the carotid sinus, the tunica media is thinner than elsewhere and contains many nerve endings of the glossopharyngeal nerve. The thinness of the tunica media permits pressure changes within the lumen of the sinus to be transmitted readily to the sensory endings in the tunica adventitia. For example, a rise in arterial blood pressure will cause an increased number of impulses to ascend the glossopharyngeal nerve, resulting in the slowing of the heart rate and vasodilatation of peripheral blood vessels.

Similar stretch receptors are found in the wall of the aortic arch.

Carotid Body. The carotid body is a highly vascular structure measuring a few millimeters in diameter (*Fig. 3–16*). It is situated close to the bifurcation of the common carotid artery. Similar structures, known as aortic bodies, are located near the arch of the aorta.

The carotid body consists of groups of cells between which lie sinusoid-like capillaries. Numerous nerve fibers that are branches of the glossopharyngeal nerve end on the cells. The cells are chemoreceptors sensitive to excess carbon dioxide and reduced oxygen tensions in the blood circulating through the capillaries of the organ; they are also sensitive to changes in blood pH. When the chemoreceptors are stimulated more afferent nerve impulses ascend the glossopharyngeal nerve and increase the heart and respiratory rates.

CLINICAL NOTES

BASILIC AND CEPHALIC VEIN CATHETERIZATION

The basilic vein is the vein of choice for catheterization because, from the cubital fossa until it reaches the axillary vein, it increases in diameter and is in direct line with the axillary vein. The valves in the axillary vein may be troublesome, but abduction of the shoulder joint may permit the catheter to move past the obstruction.

The cephalic vein does not increase in size as it ascends the arm and frequently divides into small branches as it lies within the deltopectoral groove. One or more of these branches may ascend over the clavicle and join the external jugular vein. In its usual method of termination, the cephalic vein pierces the clavipectoral fascia in the deltopectoral triangle and joins the axillary vein at a right angle.

EXTERNAL JUGULAR VEIN CATHETERIZATION

The external jugular vein may be used for catheterization, but the presence of valves or tortuosity may make the passage of the catheter difficult. Because the right external jugular vein is in the most direct line with the superior vena cava, this is the one most commonly used.

The external jugular vein runs from the region of the angle of the jaw to the midpoint of the clavicle. It is covered by skin, superficial fascia, and the platysma muscle. In the lower part of the posterior triangle of the neck it pierces the investing layer of the deep cervical fascia to drain into the subclavian vein. The external jugular vein has two sets of valves, one pair lies at its entrance into the subclavian vein and the other pair lies about 4 cm superior to the clavicle.

The valves are usually incompetent, therefore if the patient is inclined with the head down at an angle of 30 degrees, the external jugular vein will become distended. The vein is catheterized about halfway between the level of the cricoid cartilage and the clavicle. The passage of the catheter past the valves and around the angle of union of the vein with the subclavian vein is often facilitated by passing a flexible angiographic wire through the catheter lumen to serve as a guide.

INTERNAL JUGULAR VEIN CATHETERIZATION

The internal jugular vein descends through the neck from a point halfway between the tip of the mastoid process and the angle of the jaw to the sternoclavicular joint. The vein runs within the carotid sheath (*Fig. 3–16*) with the carotid arteries on its medial side. It is overlapped above by the anterior border of the sternocleidomastoid muscle and below it is covered laterally by this muscle. Just above the sternoclavicular joint the vein lies deep to a skin depression between the sternal and clavicular heads of the sternocleidomastoid muscle (*Fig. 3–31*). The internal jugular vein has a pair of valves situated just above the junction of the internal jugular vein with the subclavian vein to form the brachiocephalic vein.

The vein can be catheterized from behind, about two fingers' breadths above the clavicle at the posterior border of the sternocleidomastoid (*Fig. 3–31*). The tip of the needle and the catheter is directed toward the suprasternal notch. The skin, the superficial fascia, the platysma, and the deep fascia are pierced, as the needle advances beneath the posterior border of the sternocleidomastoid muscle.

A common site of venipuncture is where the vein lies deep to the triangle formed by the sternal and clavicular heads of the sternocleidomastoid muscle and the medial end of the clavicle (*Fig. 3–31*). With the patient's head turned to the left and the body inclined 30 degrees in the head down position, the anatomical landmarks are usually easily identifiable. The needle and the catheter are inserted into the vein at an angle of 30 degrees in a caudal direction. Once the vein has been entered, the catheter can be advanced along a guide wire.

Complications associated with internal jugular vein catherization include pleural puncture and thoracic duct injury if venipuncture is attempted on the left.

FETAL CIRCULATION

To review the fetal circulation it is suggested that the fetal blood be followed from the placenta through the umbilical vein back to the fetus, then trace the path taken by the blood through the fetus, and finally to follow its return to the placenta by the two umbilical arteries that are branches of the fetal internal iliac arteries (*Fig. 3–32*).

Having circulated through the capillaries of the placental villi, the fetal blood returns through the umbilical vein to the fetus about 80 percent saturated with oxygen and containing many important nutrients, antibodies, and hormones. The oxygenated blood then passes toward the liver, but its greatest volume bypasses the liver, as this organ is not fully functioning, and travels to the inferior vena cava by way of the **ductus venosus;** the remainder is distributed to the liver sinusoids by offshoots of the umbilical vein, and this in turn passes to the inferior vena cava by the hepatic veins. At the same time, the ductus venosus receives poorly oxygenated blood from the gut by way of the left branch of the portal vein. In addition, the inferior vena cava already contains venous blood from the lower part of the trunk and the lower limbs of the fetus. As a consequence of this admixture of blood from these various sources, the inferior vena cava contains blood about 67 percent saturated with oxygen.

Before considering the further passage of the fetal blood, it is necessary to examine the anatomical arrangement of the inferior vena cava and its relationship with the right and left atria in the fetal heart. In the fetus, the opening of the inferior vena cava into the right atrium lies directly opposite the **foramen ovale.** Thus, blood entering the heart through the inferior vena cava is directed through the foramen ovale and enters the left atrium. This process is assisted by the presence of the valve of the inferior vena cava.

The nutritious oxygenated fetal blood, on reaching the foramen ovale, is divided into two streams by the crista dividens, which is the lower margin of the septum secundum. The greater volume of blood enters the left atrium, and the remainder, joined by venous blood from the superior vena cava and coronary sinus, passes from the right atrium into the right ventricle.

The oxygenated blood in the left atrium is joined by a relatively small volume of blood from the nonaerated fetal lungs (*Fig. 3–32*). The left atrial blood then passes into the left ventricle and out into the aorta. The oxygen saturation of this blood now is about 62 percent. This is distributed chiefly to the arteries of the head, the neck, and the upper extremities. Thus, it is apparent that blood that is richer in oxygen and nutrients is transported to the cephalic region rather than the caudal regions of the fetus.

Meanwhile, the right ventricular blood passes into the pulmonary trunk. Only a small portion of this blood passes into the unexpanded lungs, because the vascular resistance is high. Most of the blood bypasses the lungs by being directed through the wide channel, the **ductus arteriosus,** into the descending thoracic aorta.

The now relatively poorly oxygenated blood passes down the descending thoracic and abdominal aortae and supplies the thoracic and abdominal viscera and the lower limbs. The fetal blood, which by this time is loaded with waste products of metabolism and carbon dioxide, returns to the placenta through the right and left umbilical arteries, where the waste products and carbon dioxide are eliminated and oxygen is picked up. The circulatory cycle is then repeated.

Changes That Take Place at Birth

Immediately after birth the **umbilical cord** is tied, thus severing the placental extension of the fetal circulation (*Fig. 3–33*). In cases in which the umbilical cord is not tied, the blood flow may continue for several minutes after delivery, although at a rapidly decreasing rate. A number of factors may contribute to this diminishing flow, which include: (1) the contraction of the uterus and the effect of this on the placental attachment and (2) constriction of the umbilical vessels as the result of mechanical stimulation or the presence in the fetal circulation of catecholamines. The fetal blood volume may be increased by as much as 100 ml if the tying of the cord is delayed. However, generally it is agreed that there is no advantage in delaying the tying beyond a minute after delivery.

The interruption of umbilical flow when the cord is tied results in an immediate fall in blood pressure in

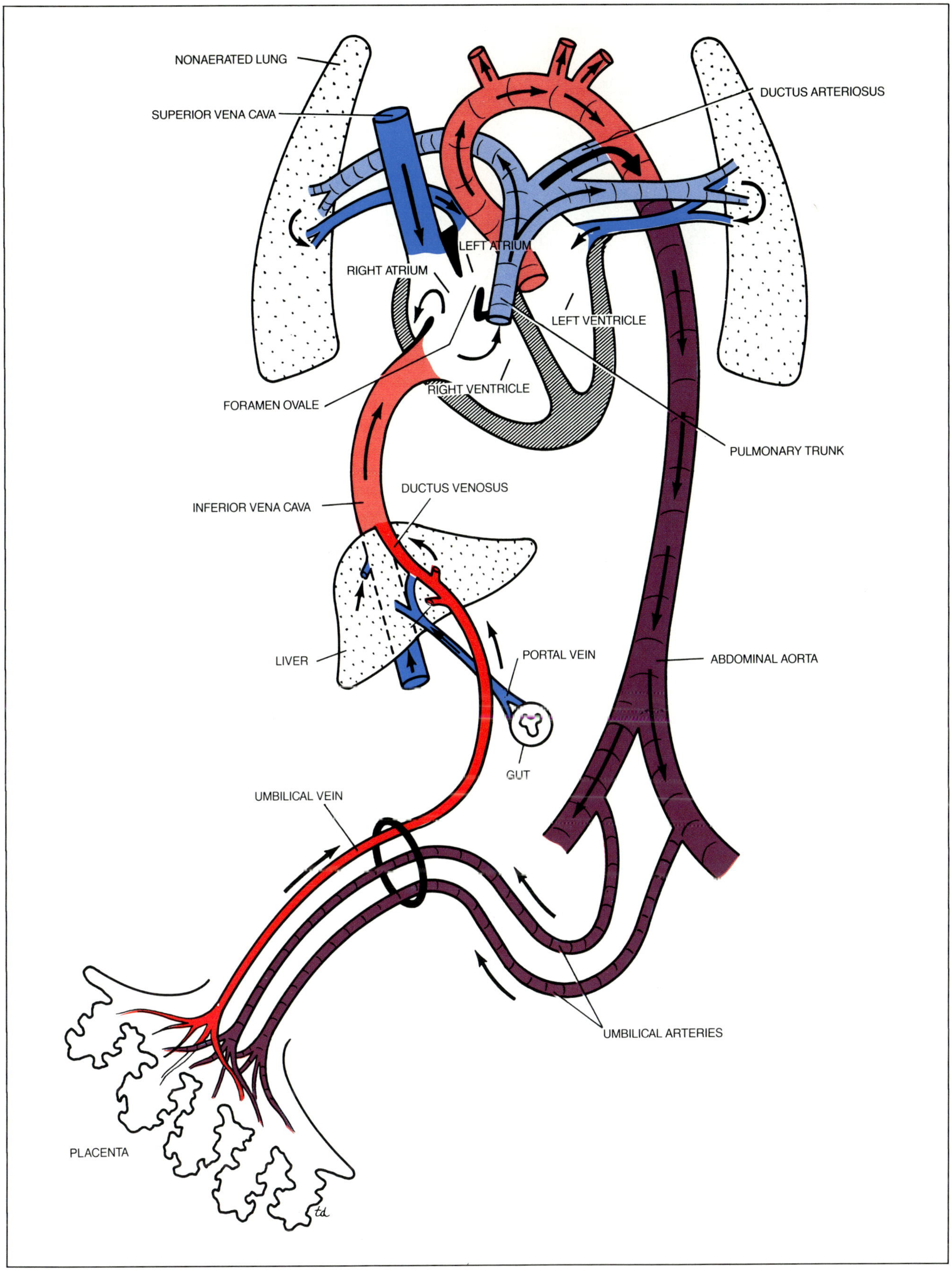

Figure 3–32. Fetal circulation.

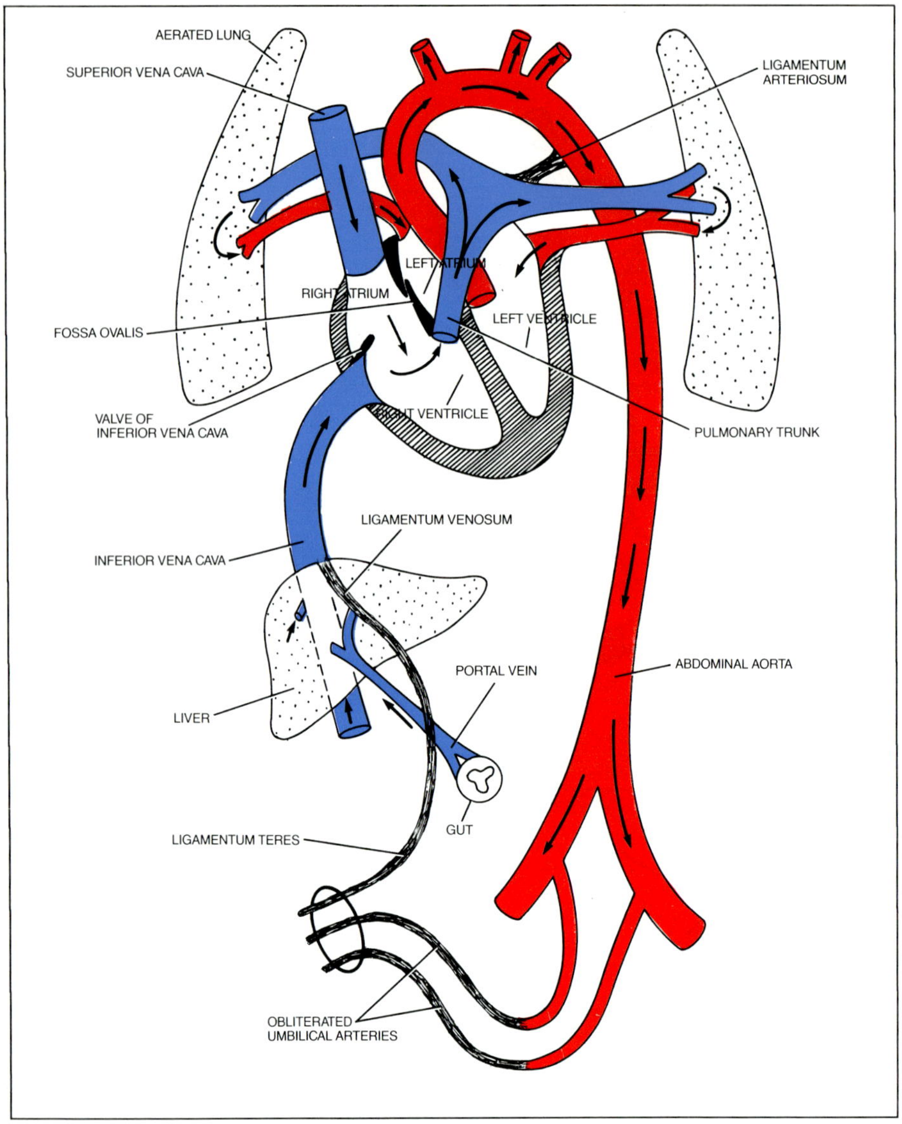

Figure 3–33. The circulatory system after birth. Compare this with the circulation in the fetus (*Fig. 3–32*).

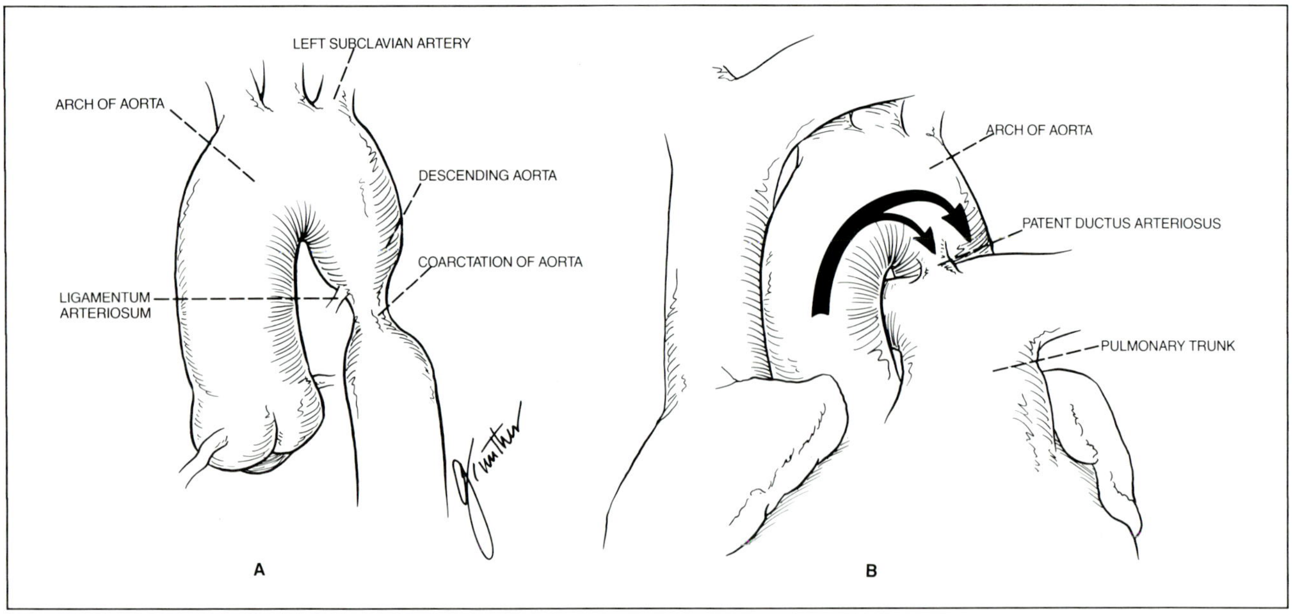

Figure 3–34. Common congenital anomalies. **A.** Coarctation of aorta. **B.** Patent ductus arteriosus.

the inferior vena cava. This fact, coupled with the increased left atrial pressure from the increased pulmonary blood flow, causes the **foramen ovale** to close.

From that moment onward the valve of the foramen ovale is kept closed by the hemodynamic changes, and within a few days of birth the valve becomes attached to the edge of the foramen ovale.

The diminished pulmonary vascular resistance associated with inflation of the lungs causes the direction of flow (from right to left) through the ductus arteriosus to be changed to the neonatal route of left to right. The **ductus arteriosus** constricts as a reaction of its muscle to the raised oxygen tension. It later closes and becomes the **ligamentum arteriosum.** One week after birth its lumen is 2 mm or less in diameter, and by the end of the first month it usually has closed. In addition, the wall of the ductus venosus contracts and the lumen is closed. Later the ductus becomes fibrosed to form the **ligamentum venosum.**

The heart in the newborn is relatively large, higher and more horizontal in the thorax than it is in later life. The pulse rate is rapid with an average of 130 beats per minute. The rate may rise considerably with crying and fall to around 80 during sleep. The peripheral circulation initially is slow in the newborn. The hands and feet may be cold and slightly cyanotic for the first few hours after birth.

CLINICAL NOTES

CONGENITAL ANOMALIES

The Arterial System

Anomalies of the ascending aorta and pulmonary trunk have already been considered under the section on congenital anomalies of the heart.

Patent Ductus Arteriosus

Patent ductus arteriosus is the most frequently seen congenital anomaly of the great vessels (*Fig. 3–34*). The ductus arteriosus represents the distal portion of the sixth left aortic arch and connects the left pulmonary artery to the descending aorta. During fetal life, it has practically the same diameter as the pulmonary trunk and aorta, and blood passes through it from the pulmonary artery to the aorta. After birth it normally constricts, later closes, and becomes the **ligamentum arteriosum.** One week after birth, its lumen is 2 mm or less in diameter. By the end of the first month, the ductus arteriosus has closed in most infants.

It is now believed that a rise in arterial oxygen tension is an important factor in bringing about the initial contraction of the smooth muscle in the wall of the ductus. Congestion and hemorrhage of the vasa vasorum, aided by edema of the tunica intima, may initiate the process of fibrosis. In children born with respiratory distress, the delay in closure of the ductus arteriosus probably is a result of hypoxemia. Persistent patent ductus arteriosus results in aortic blood passing into the pulmonary artery, which raises the pressure in the pulmonary circulation and causes hypertrophy of the right ventricle. This, together with the possibility of bacterial infection of the wall of the pulmonary artery, makes patent ductus arteriosus life-endangering, and thus it should be closed pharmacologically or ligated and divided surgically.

Coarctation of the Aorta

Another congenital anomaly of the arterial system is coarctation of the aorta, a narrowing of the aorta just proximal to, opposite, or distal to the site of attachment of the ligamentum arteriosum (*Fig. 3–34*). It is a condition that arises after birth and is believed to be the result of an unusual quantity of ductus arteriosus muscle tissue incorporated in the wall of the aorta. Because of this excess tissue when the ductus arteriosus contracts, the ductal muscle in the aortic wall also contracts and the aortic lumen becomes narrowed. Later, when fibrosis takes place, the aortic wall also is involved and permanent narrowing occurs.

Clinically, the cardinal sign of aortic coarctation is absent or diminished femoral pulses. In an attempt to compensate for the diminished volume of blood reaching the lower part of the body, an enormous collateral circulation opens that involves the internal thoracic, subclavian, and posterior intercostal arteries. The condition may be treated surgically.

Double Aorta and Right Aortic Arch

Anomalies involving double parts of the arterial system are rare conditions that result from the persistence of aortic arch arteries, which normally disappear. In the case of the right aortic arch, the development of the left arch does not take place. The **subclavian** and **carotid arteries** also may show variations in development caused by the persistence of sections of aortic arch arteries that normally disappear and by the disappearance of other sections that normally remain. The **coronary arteries** occasionally arise by a common trunk from the aorta or very rarely from the pulmonary trunk.

Venous System

In view of the development of complicated patterns of veins from the cardinal veins, it is not surprising that an occasional deviation from the normal is seen in the formation of the large veins. For example, an absent or double inferior vena cava may occur. A left superior vena cava or two superior vena cavae rarely are formed.

CLINICAL PROBLEMS

1. A 7-year-old boy was examined by a pediatrician. The child had a normal color and no symptoms. On physical examination a faint systolic murmur could be heard along the left margin of the sternum. Radiographic examination revealed a slight cardiac enlargement to the right, with enlargement of the pulmonary trunk along the left border of the heart. The lung fields showed a marked increase in pulmonary vasculature. Cardiac catheterization revealed an increased oxygen saturation in the right atrium. A diagnosis of atrial septal defect was made. Explain how an atrial septal defect occurs.

2. A 6-year-old girl was examined by a pediatrician and found to be slightly cyanotic, with clubbing of the fingers and toes, and short in stature for her age. She had a history of repeated attacks of pneumonia and on two occasions she had had right-sided heart failure. The subsequent workup showed that she had a large ventricular septal defect. Explain how a ventricular septal defect occurs. In which part of the septum is the defect commonly found?

3. A 4½-year-old girl with a history of attacks of breathlessness and cyanosis was examined by a pediatrician. The attacks of breathlessness occurred on exertion and during one attack, the child had lost consciousness. It had been noticed that she breathed more easily in the squatting position and she slept easier with her knees drawn up. On physical examination the child was found to be thinner and shorter than normal. Her lips were cyanotic and she had clubbing of the fingers and toes. A loud systolic murmur was present along the left border of the sternum and the heart was considerably enlarged to the right. Angiocardiography made a diagnosis of tetralogy of Fallot certain. Explain this important syndrome. Why is the child breathless on exertion? Why does the child assume the squatting position?

4. A routine physical examination of a 7-year-old boy revealed a continuous machinery-like murmur in the second left intercostal space. The murmur could be heard during systole and diastole. The child was not cyanotic, the heart was not enlarged, and there was no evidence of clubbing of the fingers or toes. A diagnosis of patent ductus arteriosus was made. What is the ductus arteriosus? What factor or factors are responsible for its normal closure?

5. A 13-year-old boy was found to have absent pulses in both femoral arteries. In addition there was a higher blood pressure in both upper limbs than in both lower limbs. An anteroposterior radiograph of the chest showed notching of the necks of the upper ribs on both sides. What is the diagnosis? Why is there notching of the ribs?

6. A 55-year-old patient with a history of angina pectoris is about to undergo a coronary artery bypass operation. Why is the radial artery chosen in preference to the ulnar artery for direct blood pressure monitoring? Why are the upper limb arteries used in preference to the dorsalis pedis artery? What are the important anatomic relations of the radial artery at the site of cannulation? Why do you extend the wrist joint when introducing the cannula?

7. A 4-year-old boy with the tetralogy of Fallot was about to undergo corrective cardiac surgery. An arterial monitoring line had been established through the right femoral artery. What is the surface marking of the femoral artery? What are the anatomical relations of the femoral artery at the site of cannulation?

8. During surgery for cardiac disease, it was decided to monitor the central venous pressure. To measure the central venous pressure where should the tip of the catheter be located? Should the pressure fluctuate with respiration? Explain the fluctuation if it does occur.

9. Why is the basilic vein more often used to establish a central venous pressure line than the cephalic vein?

10. The jugular veins are commonly used to establish a central venous line. Why is it sometimes difficult to pass a catheter from the external jugular vein into the right atrium? What is the surface marking of the internal jugular vein in the neck? What anatomical structures are likely to be damaged when a faulty technique is used in cannulating the internal jugular vein?

ANSWERS TO CLINICAL PROBLEMS

1. Atrial septal defects occur as the result of a failure of the foramen ovale to close or by a persistence of the ostium primum; defects of the ostium primum are less common (see page 77). Small septal defects cause no symptoms and are of no clinical significance. Large atrial septal defects overload the right ventricle and the pulmonary circulation, and should be closed.

Atrial septal defects permit oxygenated blood in the left atrium to flow across into the right atrium, the blood pressure in the left atrium being higher than that in the right atrium. The increased blood flow through the pulmonary valve causes the systolic murmur and the enlargement of the pulmonary trunk. The right-sided cardiac enlargement is produced by the increased amount of blood on the right side of the heart. The increased pulmonary vasculature seen on the chest x-ray was due to the increased blood flow through the lungs.

2. The ventricular septum is formed from below by a muscular part that grows up from the floor of the primitive ventricle, and from above, by the membranous part that is formed from the fusion of the lower ends of the bulbar ridges and the endocardial cushions. The septum is complete when the two parts meet and fuse. Defects occur almost invariably in the membranous part of the septum.

Ventricular septal defects cause overloading of the right side of the heart. Blood under high pressure is shunted through the defect into the low-pressure system within the right ventricle. Small defects are asymptomatic and cyanosis is absent. Large defects, which are less common, cause enlargement of the right ventricle and right sided heart failure.

3. The tetralogy of Fallot is made up of the following: (1) a large ventricular septal defect; (2) stenosis of the pulmonary trunk that may occur at the infundibulum of the right ventricle or at the pulmonary valve; (3) the exit of the aorta lies immediately above the ventricular septal defect and thus communicates with both ventricles; and (4) a right ventricular hypertrophy secondary to the ventricular septal defect and the pulmonary stenosis.

The pulmonary stenosis impairs the pulmonary circulation so that the blood is poorly oxygenated. The lack of oxygen causes impaired growth and development, the cyanosis, and the clubbing of the fingers and toes. The attacks of breathlessness are caused by the failure of the pulmonary circulation to provide sufficient oxygenation of the blood on exertion. The assumption of the squatting position and sleeping with the knees drawn up is beneficial in these patients, as it increases the peripheral resistance in the systemic circulation, decreases the right-to-left shunt through the ventricular septum, and consequently increases slightly the pulmonary blood flow.

4. The ductus arteriosus is the remains of the distal portion of the sixth left aortic arch artery and connects the left pulmonary artery to the descending aorta. In fetal life it serves to bypass blood from the pulmonary trunk to the aorta, at a time when the lungs are nonfunctional. At birth, the ductus arteriosus normally constricts in response to a rise in arterial oxygen tension. Later, fibrosis occurs and the ductus becomes the ligamentum arteriosum.

5. Coarctation of the aorta is a narrowing of the aorta just proximal, opposite, or distal to the site of attachment of the ligamentum arteriosum. It is believed to result from the presence of an unusual quantity of ductus arteriosus muscle tissue incorporated in the wall of the aorta. When the ductus arteriosus contracts after birth, the ductal muscle in the aortic wall also contracts and the aortic lumen becomes narrowed. Later, fibrosis occurs and permanent narrowing occurs. The notching of the lower borders of the ribs is caused by the opening up of the collateral circulation through the subclavian, internal thoracic, and posterior intercostal arteries to carry the blood from above the coarctation to the distal part of the aorta; it is the dilated posterior intercostal arteries that notch the ribs.

6. The radial artery, although usually smaller than the ulnar artery, is the most direct continuation of the brachial artery in the cubital fossa. The dorsalis pedis artery can be easily cannulated and is used extensively during neurosurgery. It has a higher incidence of postoperative arterial thrombosis, however, and sometimes the circulation in the foot is compromised by the inadequate collateral circulation. The radial artery is cannulated 2 or 3 cm proximal to the proximal edge of the flexor retinaculum (i.e., the distal transverse skin crease on the anterior aspect of the wrist). Here, the artery lies anterior to the distal third of the shaft of the radius, medial to the tendon of the brachioradialis, and lateral to the tendon of the flexor carpi radialis. It is covered anteriorly by skin and fascia. The forearm is supinated and the wrist is extended approximately to a 50 degree angle. Extension stretches and stabilizes the artery during the process of introducing the needle and the catheter.

7. The femoral artery emerges from beneath the inguinal ligament at a point halfway between the anterior superior iliac spine and the symphysis pubis. It lies superficially and its pulsations can be easily felt.

   About 3 cm below the inguinal ligament, the femoral artery is related posteriorly to the psoas muscle, which separates it from the hip joint. Medially, it is related to the femoral vein and laterally to the femoral nerve and its branches. Anteriorly, the artery is covered by skin and fascia.

8. Central venous pressure on the right side of the heart can only be obtained if the tip of the catheter is located within the thorax. The common locations used are the superior vena cava and the right atrium. The right brachiocephalic vein is less satisfactory. A correctly placed catheter will enable one to aspirate blood without difficulty and its level fluctuates with respiration. On inspiration, the level falls and on expiration, it rises. The mean right atrial venous pressure in the supine patient at the level of the sternal angle varies between -4 and $+2$ cm of water, assuming that the patient is spontaneously breathing. The venous pressure falls on inspiration because of the decrease in intrathoracic pressure and the venous pressure rises during expiration because of the raised intrathoracic pressure.

9. The basilic vein is used more often than the cephalic vein for the following reasons: (1) The basilic vein increases progressively in diameter from the cubital fossa to the axillary and subclavian veins, whereas the diameter of the cephalic vein increases only slightly as it ascends the upper limb. (2) The basilic vein is in line with the axillary vein, whereas the cephalic vein opens into the axillary vein at a right angle. (3) The basilic vein is directly continuous with the end of the axillary vein, whereas the cephalic vein may bifurcate into several small veins near its termination or may join the external jugular vein.

10. The following difficulties may be encountered on advancing a catheter down the external jugular vein: (1) The catheter tip may catch in the valves. The vein has two pairs of valves, one pair about 4 cm above the clavicle and the

lower pair at its entrance into the subclavian vein. (2) The vein may be small, tortuous, duplicated, and difficult to negotiate. By having the patient in a slight head-down position or applying a positive airway pressure, the vein will become distended. (3) The external jugular vein is not in a direct line with the superior vena cava.

The surface marking of the internal jugular vein is from a point midway between the tip of the mastoid process and the angle of the jaw down to the sternoclavicular joint.

The following antomic structures are closely related to the internal jugular vein and may be damaged by a poorly attempted venous catheterization: (1) internal and common carotid arteries, (2) cervical dome of parietal pleura, (3) brachical plexus, and (4) vagus nerve.

CHAPTER 4

Vertebral Column

The vertebral column is the central bony pillar of the body. It supports the skull, the pectoral girdle, the upper limbs, the thoracic cage, and by way of the pelvic girdle, transmits the body weight to the lower limbs. Within its cavity lies the spinal cord, the roots of the spinal nerves, and the covering of the meninges, to which the vertebral column gives great protection. The anesthetist must have a sound knowledge of the surface anatomy of the vertebral column, the size and architecture of the different vertebrae, the joints that lie between them, and the ligaments and muscles that hold them together and move them. This information is necessary to perform spinal, epidural, and caudal anesthesia, and a variety of nerve blocks (i.e., celiac plexus, paravertebral somatic, and sympathetic).

Composition of the Vertebral Column

The vertebral column (*Fig. 4–1*) is composed of 33 vertebrae: 7 cervical, 12 thoracic, 5 lumbar, 5 sacral (fused), and 4 coccygeal (fused). Because it is segmented and made up of vertebrae, the joints, and pads of fibrocartilage, called intervertebral discs, it is a flexible structure. The intervertebral discs form about one-quarter of the length of the column.

CURVES OF THE VERTEBRAL COLUMN

Curves in the Sagittal Plane

In the fetus the vertebral column has one continuous anterior concavity. As development proceeds, the lumbosacral angle appears. After birth, when the child becomes able to raise his or her head and keep it balanced on the vertebral column, the cervical part of the column becomes concave posteriorly. Toward the end of the first year, when the child begins to stand upright, the lumbar part of the vertebral column becomes concave posteriorly. The development of these secondary curves is largely due to modification in the shape of the intervertebral discs.

In the standing adult (*Fig. 4–1*), the vertebral column exhibits, in the sagittal plane, the following regional curves: cervical—posterior concavity; thoracic—posterior convexity; lumbar—posterior concavity; and sacral—posterior convexity. The degree of curvature varies greatly from one individual to another. During the later months of pregnancy, with the increase in size and weight of the fetus, women tend to increase the posterior lumbar concavity in an attempt to preserve their center of gravity. In the elderly, the intervertebral discs atrophy, resulting in a loss of height and a gradual return of the vertebral column to a continuous anterior concavity.

In the supine position, the curves tend to flatten out, thus straightening the vertebral column. The crest of the lumbar curve is at the third lumbar vertebra.

Curves in the Coronal Plane

In late childhood it is quite common to find the development of minor lateral curves in the thoracic region of the vertebral column. This is normal and is usually caused by the predominant use of one of the upper limbs. For example, right-handed persons will often have a slight right-sided thoracic convexity. Slight compensatory curves are always present above and below such a curvature.

GENERAL CHARACTERISTICS OF A VERTEBRA

Although vertebrae show regional differences, they all possess a common pattern (*Fig. 4–2*).

A **typical vertebra** consists of a rounded **body** anteriorly and a **vertebral arch** posteriorly. These enclose a space, called the **vertebral foramen,** through which runs the spinal cord and its coverings. The vertebral arch consists of a pair of cylindrical **pedicles,** which form the sides of the arch, and a pair of flattened **laminae,** which complete the arch posteriorly. The vertebral arch gives

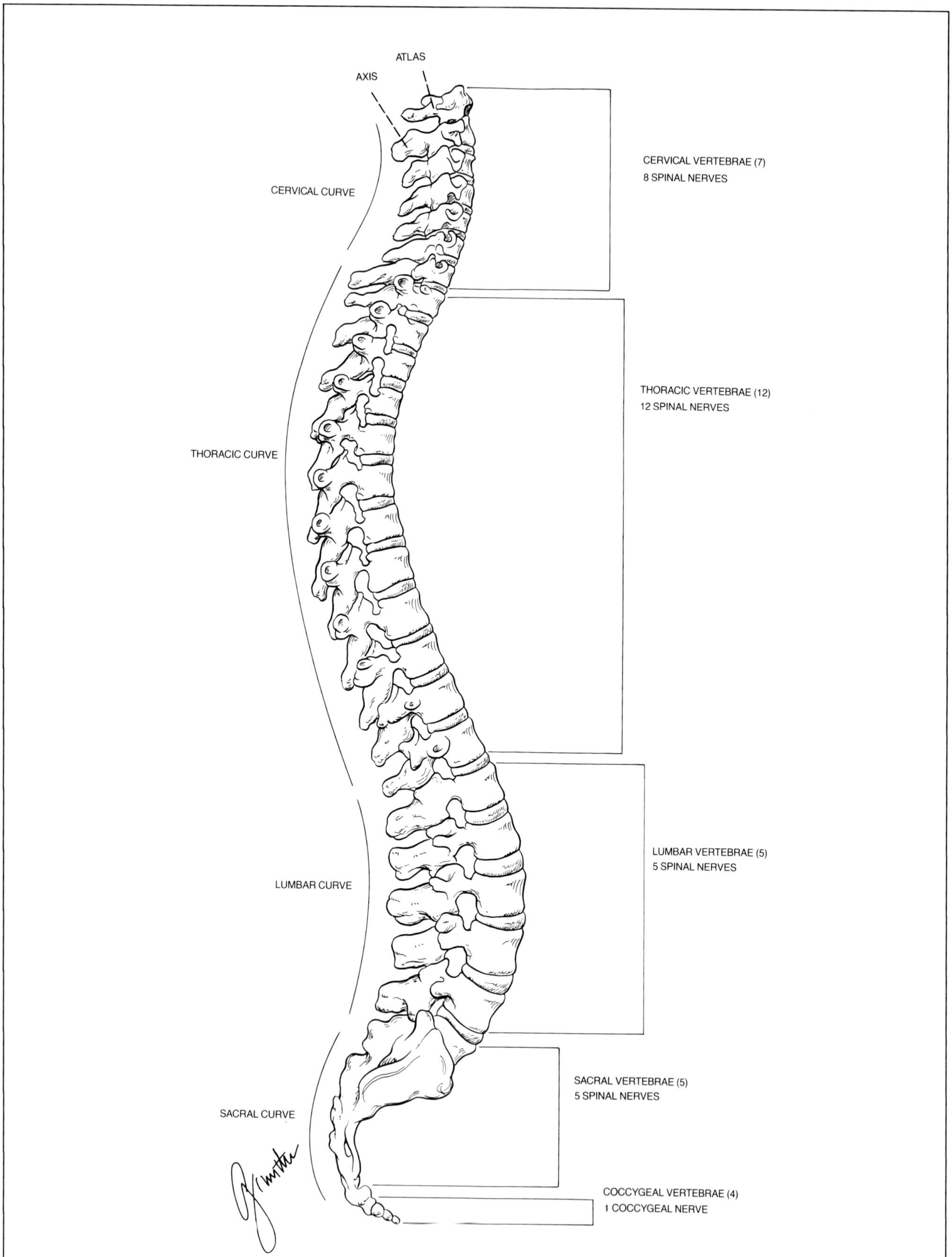

Figure 4–1. Lateral view of vertebral column.

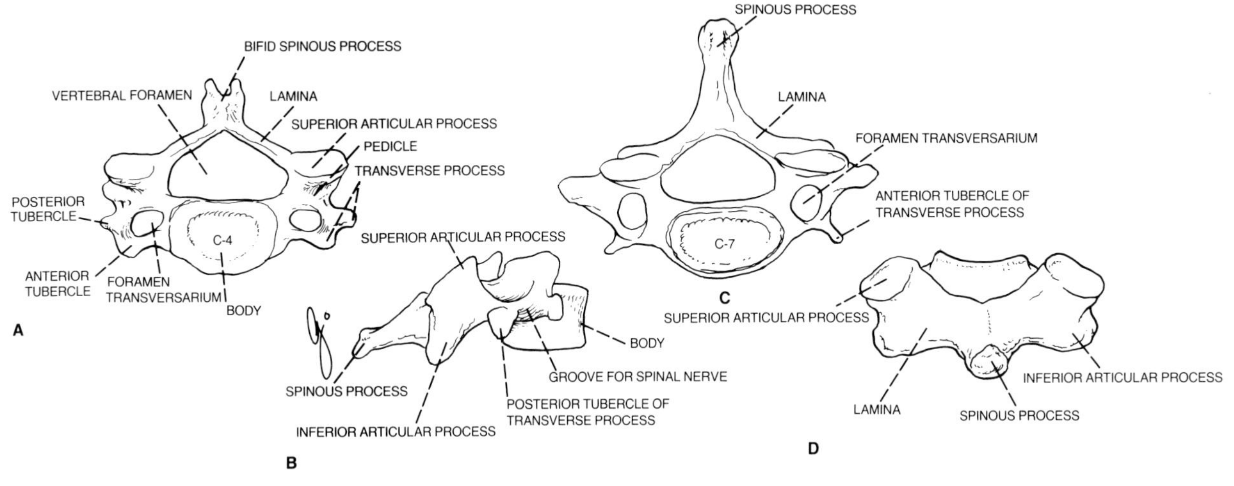

Figure 4–2. Fourth cervical vertebra: **A.** Superior view; **B.** lateral view. Seventh cervical vertebra: **C.** Superior view; **D.** posterior view.

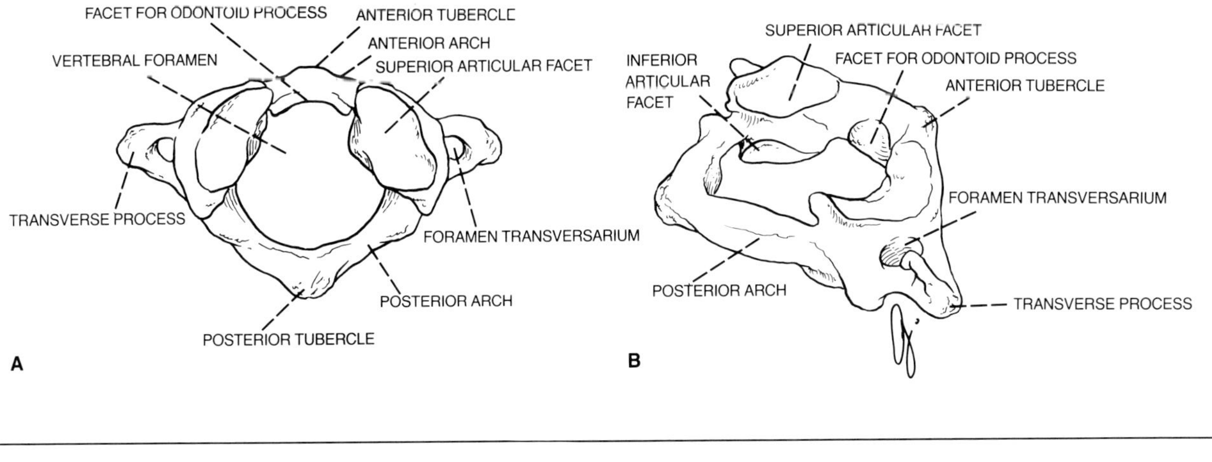

Figure 4–3. Atlas. **A.** Superior view. **B.** Oblique view.

rise to seven processes, one spinous, two transverse, and four articular (*Fig. 4–2*).

The **spinous process,** or **spine** is directed posteriorly from the junction of the two laminae. The **transverse processes** are directed laterally from the junction of the laminae and the pedicles. Both the spinous and transverse processes serve as levers and receive attachments of muscles and ligaments. The **articular processes** are vertically arranged and consist of two superior and two inferior processes. They arise from the junction of the laminae and the pedicles, and their articular surfaces are covered with hyaline cartilage. The two superior articular processes of one vertebral arch articulate with the two inferior articular processes of the arch above, forming two synovial joints (*Fig. 4–9*).

The pedicles are notched on their upper and lower borders, forming the **superior and inferior vertebral notches.** The superior notch of one vertebra and the inferior notch of an adjacent vertebra together form an **intervertebral foramen.** These foramina, in an articulated skeleton, serve to transmit the spinal nerves and the blood vessels. The anterior and posterior nerve roots of a spinal nerve unite within these foramina with their coverings of dura to form the segmental spinal nerves.

Cervical Part of the Vertebral Column

The cervical part of the vertebral column shows a forward convexity and is made up of seven vertebrae.

A **typical cervical vertebra** has the following characteristics (*Fig. 4–2*). Each transverse process possesses a **foramen transversarium** for the passage of the vertebral artery and veins. The spines are small and bifid. The body is small and broader from side to side than from anterior to posterior; there are small synovial joints on each side. The vertebral foramen is large and triangular in shape. The superior articular processes have small, flat articular facets that face backward and upward; the inferior articular processes have facets that face downward and forward.

The first, second, and seventh cervical vertebrae are atypical. The **first cervical vertebra,** or **atlas,** has no body and no spinous process (*Fig. 4–3*). It is merely a ring of bone consisting of anterior and posterior arches and a lateral mass on each side. Each lateral mass has articular surfaces on its upper and lower aspects. The bone articulates above with the occipital condyles, forming the **atlanto-occipital joints,** which allows nodding movements of the head. Below, the bone articulates with the axis, forming the **atlantoaxial joints,** permitting rotation movements of the head.

The **second cervical vertebra,** or **axis,** has a peg-like **odontoid process** that projects from the superior surface of the body and represents the body of the atlas, which has fused with the axis (*Fig. 4–4*).

The **seventh cervical vertebra,** or **vertebra prominens** (*Fig. 4–2*), is so named because it has the longest spinous process. This anatomical feature is used as a landmark for cervical spinal and epidural anesthesias as well as a starting point for counting the lower vertebrae prior to needle insertion. The spinous process is not bifid. The **transverse process** is large, but the **foramen transversarium** is small and does not transmit the vertebral artery.

Thoracic Part of the Vertebral Column

The thoracic part of the vertebral column is concave forward and is made up of 12 vertebrae, together with their intervertebral discs. Thoracic vertebrae have the following characteristics (*Fig. 4–5*): (1) The body is medium-sized and heart-shaped. (2) The vertebral foramen is relatively small and circular. (3) The spines are long and inclined downward. (4) Costal facets are present on the sides of the bodies where the heads of the ribs articulate, and on the transverse processes for articulation with the tubercles of the ribs. (T-11 and T-12 have no facets on the transverse processes.) (5) The superior articular processes bear facets that face backward and laterally, whereas the facets on the inferior articular processes face forward and medially. The inferior articular processes of the 12th vertebra face laterally, as do those of the lumbar vertebrae.

Lumbar Part of the Vertebral Column

The lumbar part of the vertebral column shows a forward convexity and is made up of five vertebrae, together with their intervertebral discs. A typical lumbar vertebra (*Fig. 4–6*) has the following characteristics: (1) The body is large and kidney-shaped. (2) The pedicles are strong and directed backward, laminae are thick, and vertebral foramina are triangular in shape. (3) The transverse processes are long and slender, whereas the spinous process is short, flat, and quadrangular in shape, projecting directly backward. (4) The articular surfaces of the superior articular processes face medially, and those of the inferior articular processes face laterally.

The lumbar vertebrae have no facets for articulation with ribs and no foramina in the transverse processes. The fifth lumbar vertebra articulates with the base of the sacrum at the **lumbosacral joint.**

The intervertebral discs in the lumbar region are thicker than in other regions of the vertebral column. They are wedge-shaped and are responsible for the normal lordosis (forward convexity) found in the lumbar region.

Sacral Part of the Vertebral Column

The sacral part of the vertebral column shows a forward concavity (*Fig. 4–7*) and is made up of five rudimentary vertebrae that are fused together to form a single wedge-shaped bone, the **sacrum.** The upper border, or base, of the bone articulates with the fifth lumbar vertebra and the narrow inferior border with the coccyx. Laterally, the sacrum articulates with the two iliac bones to form

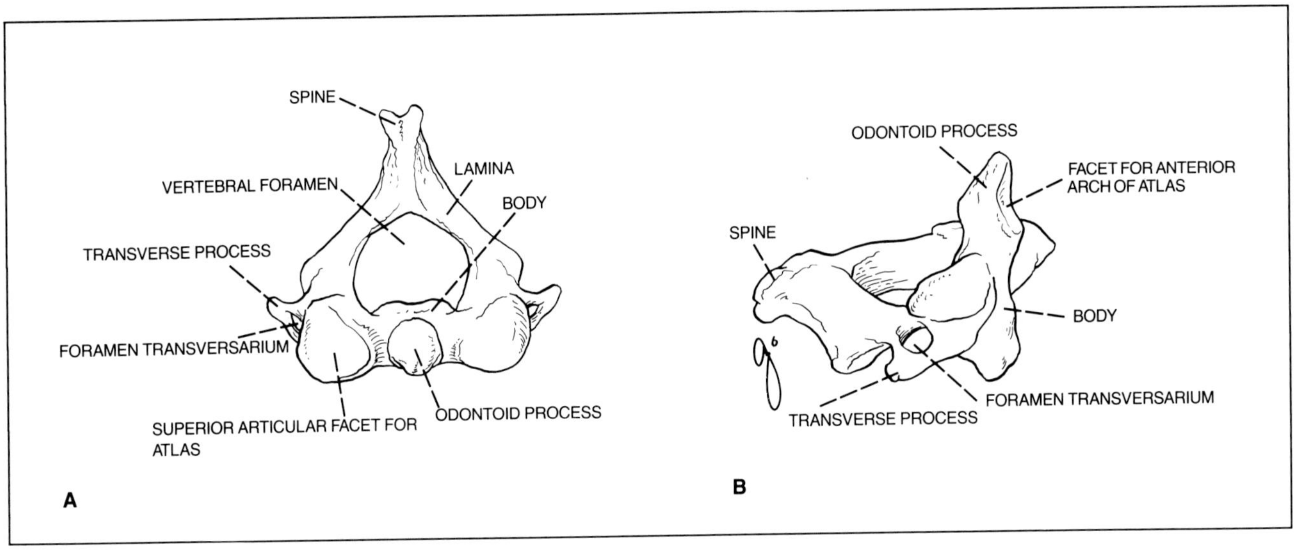

Figure 4–4. Axis. **A.** Superior view. **B.** Oblique lateral view.

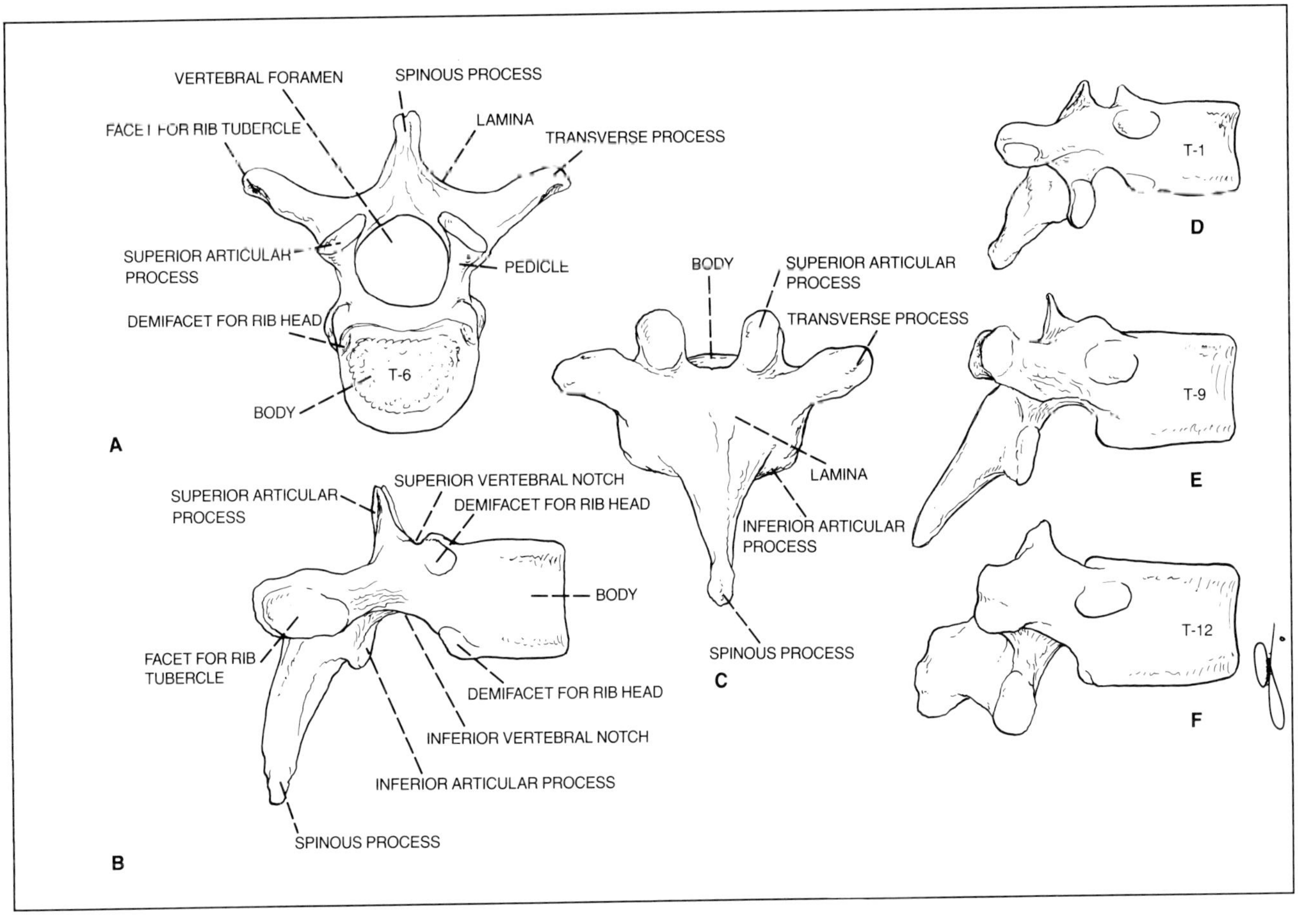

Figure 4–5. Sixth thoracic vertebra. **A.** Superior view, **B.** lateral view, and **C.** posterior view. First thoracic vertebra, lateral view **D.** Ninth thoracic vertebra, lateral view **E.** Twelfth thoracic vertebra, lateral view **F.**

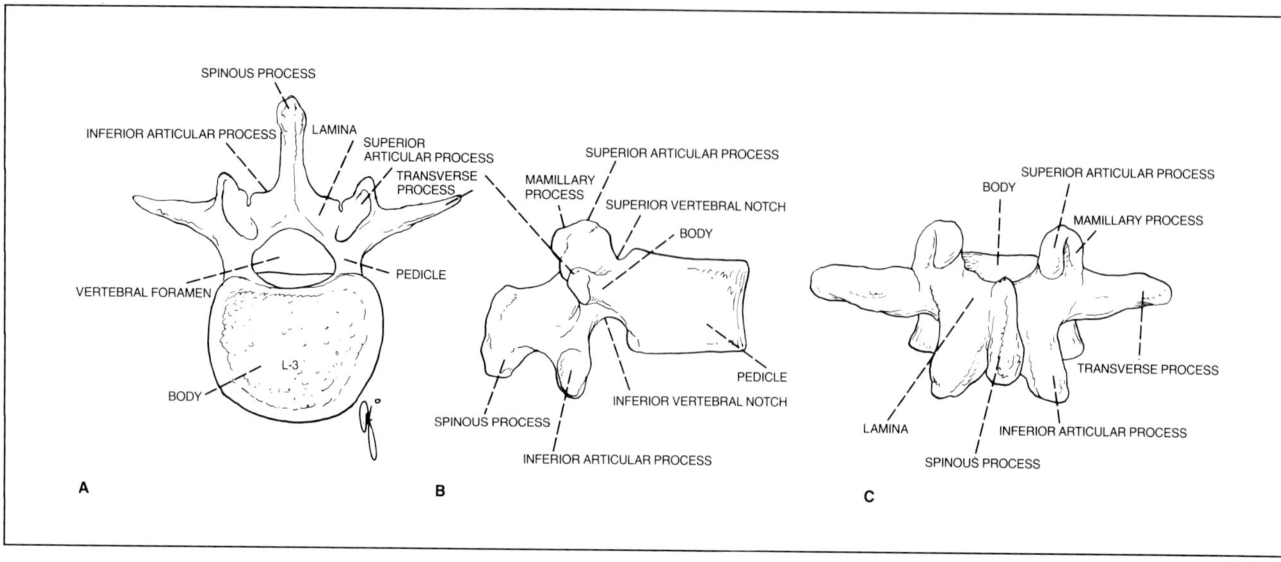

Figure 4–6. Third lumbar vertebra. **A.** Superior view. **B.** Lateral view. **C.** Posterior view.

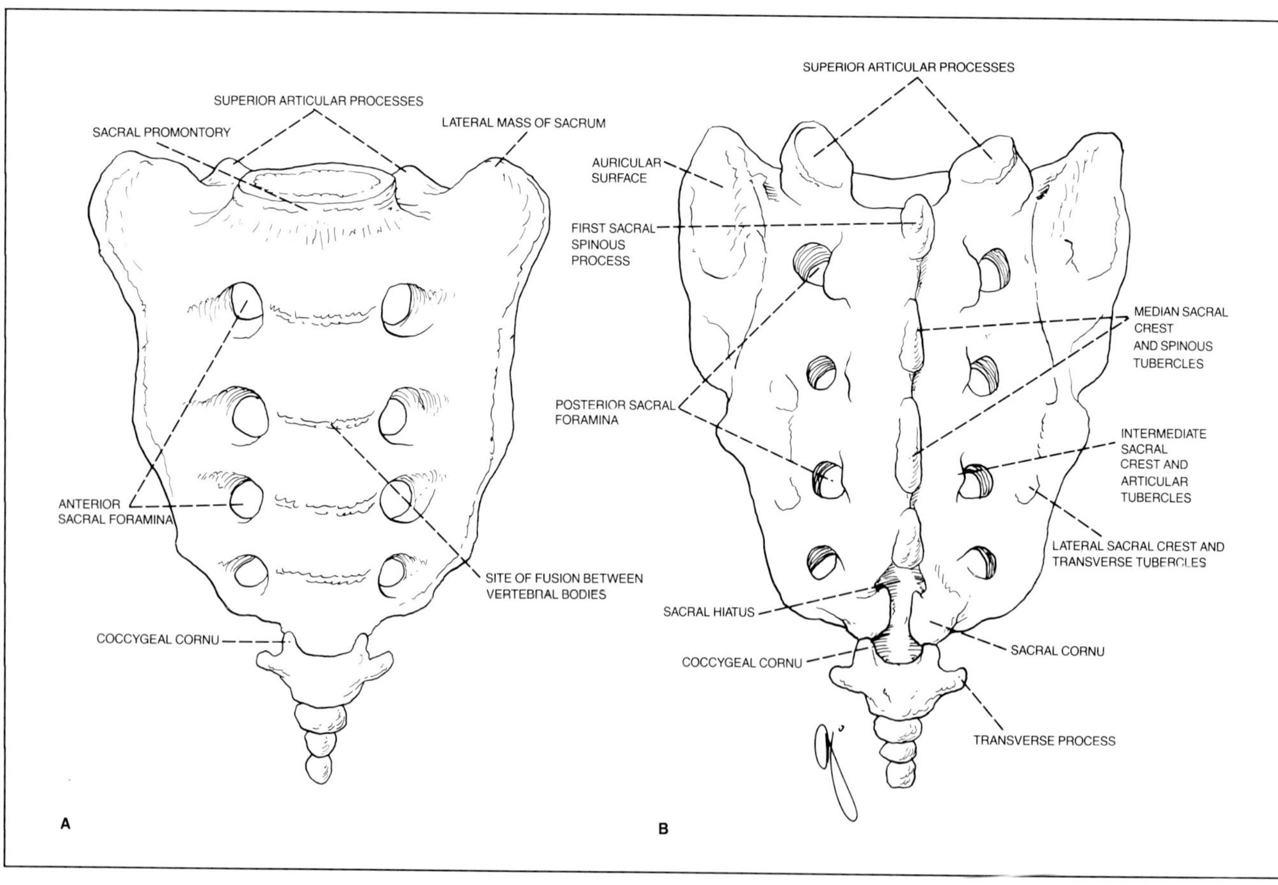

Figure 4–7. Sacrum and coccyx in the adult male. **A.** Anterior view. **B.** Posterior view. Note that the median sacral crest is formed by the fusion of the spinous processes; the tips of the spinous processes form the spinous tubercles. The intermediate sacral crest is formed by the fused articular processes; small projections remain as the articular tubercles. The lateral sacral crest is formed by the fused transverse processes and their remains are known as the transverse tubercles.

the sacroiliac joints. The anterior and upper margin of the first sacral vertebra bulges forward as the posterior margin of the pelvic inlet and is known as the **sacral promontory.** This is an important obstetric landmark used when measuring the size of the pelvis.

The vertebral foramina are present and together form the **sacral canal.** The laminae of the fifth sacral vertebra, and sometimes those of the fourth, fail to meet in the midline, forming the **sacral hiatus.** The **sacral canal** contains part of the cauda equina, filum terminale, and meninges down as far as the lower border of the second sacral vertebra. The lower part of the sacral canal contains the lower sacral and coccygeal nerve roots, the filum terminale, and fibrofatty material.

The anterior and posterior surfaces of the sacrum each have four foramina bilaterally for the passage of the anterior and posterior rami of the upper four sacral spinal nerves.

Coccyx

The coccyx consists of four vertebrae fused together to form a single, small triangular bone, which articulates at its base with the lower end of the sacrum (*Fig. 4–7*). The first coccygeal vertebra is usually not fused, or is incompletely fused with the second vertebra.

Variations in Vertebrae

The number of cervical vertebrae is constant, but the seventh cervical vertebra may possess a **cervical rib.** The thoracic vertebrae may be increased in number by the addition of the first lumbar vertebra, which may have a rib. The fifth lumbar vertebra may be incorporated into the sacrum; this is usually incomplete and may be limited to one side. The first sacral vertebra may remain partially or completely separate from the sacrum and resemble a sixth lumbar vertebra. A large extent of the posterior wall of the sacral canal may be absent because of failure of development of the laminae and the spines.

The coccyx, which usually consists of four fused vertebrae, may have three or five vertebrae. The first coccygeal vertebra may be separate. In this condition the free vertebra usually projects downward and anteriorly from the apex of the sacrum.

Suppression of the development of half a vertebra or the presence of an additional half vertebra will produce a **hemivertebra** causing severe scoliosis.

The laminae may fail to develop adequately and not meet in the midline, producing **spina bifida.** This is most commonly seen in the lumbosacral region.

JOINTS OF THE VERTEBRAL COLUMN

Atlanto-Occipital Joints

The atlanto-occipital joints (*Fig. 4–8*) are synovial joints formed between the occipital condyles, which are found on either side of the foramen magnum of the skull above, and the facets on the superior surfaces of the lateral masses of the atlas below. They are enclosed by a capsule.

Ligaments. The ligaments are the **anterior atlanto-occipital membrane,** which connects the anterior arch of the atlas to the anterior margin of the foramen magnum, and the **posterior atlanto-occipital membrane,** which connects the posterior arch of the atlas to the posterior margin of the foramen magnum.

Movements. Flexion, extension, and lateral flexion; no rotation.

The joints are strengthened by the ligaments that connect the skull to the axis and by the tone of the muscles that surround the joints.

Atlantoaxial Joints

The atlantoaxial joints (*Fig. 4–8*) are three synovial joints, one of which is between the odontoid process and the anterior arch of the atlas, whereas the other two are between the lateral masses of the bones. The joints are enclosed by capsules.

Ligaments. The **apical ligament** is a median-placed structure connecting the apex of the odontoid process to the anterior margin of the foramen magnum.

The **alar ligaments** lie one on each side of the apical ligament and connect the odontoid process to the medial sides of the occipital condyles.

The **cruciate ligament** consists of transverse and vertical bands. The transverse band is attached on each side to the inner aspect of the lateral mass of the atlas and binds down the odontoid process to the anterior arch of the atlas. The vertical portion runs from the posterior surface of the body of the axis to the anterior margin of the foramen magnum.

The **membrana tectoria** is an upward continuation of the posterior longitudinal ligament. It is attached above to the occipital bone just within the foramen magnum. It covers the posterior surface of the odontoid process and the apical, alar, and cruciate ligaments.

Movements. The atlantoaxial joints permit a wide range of rotation of the atlas and thus of the head on the axis.

Joints of the Vertebral Column below the Axis

With the exception of the first two cervical vertebrae, the remainder of the mobile vertebrae articulate with each other by means of cartilaginous joints between their bodies and by synovial joints between their articular processes (*Fig. 4–9*).

Joints between Two Vertebral Bodies

The upper and lower surfaces of the bodies of adjacent vertebrae are covered by thin plates of hyaline cartilage.

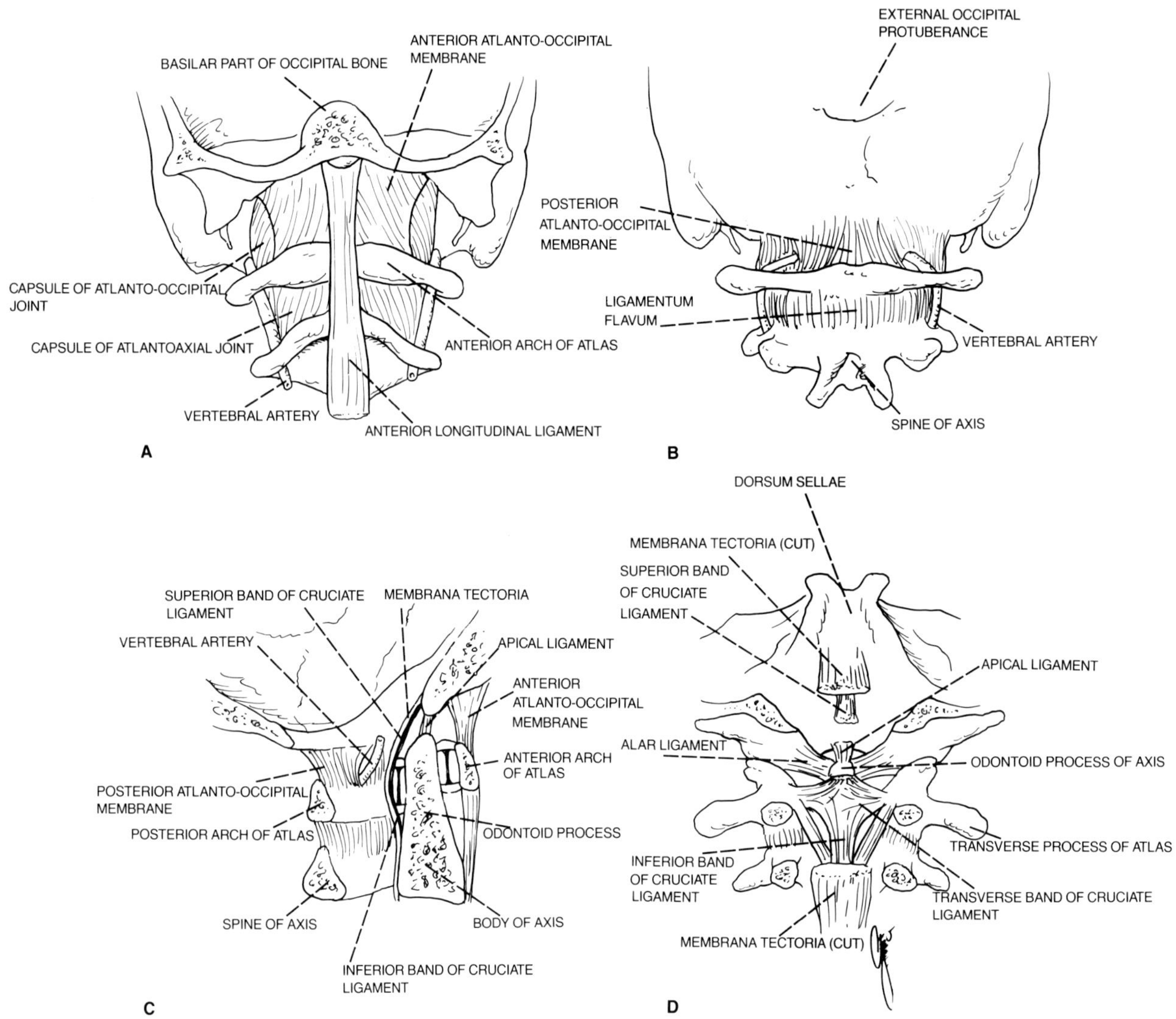

Figure 4–8. Atlanto-occipital joints: **A.** Anterior view; **B.** posterior view. Atlantoaxial joints: **C.** Sagittal section; **D.** posterior view. Note that the posterior arch of the atlas and the laminae and the spine of the axis have been removed.

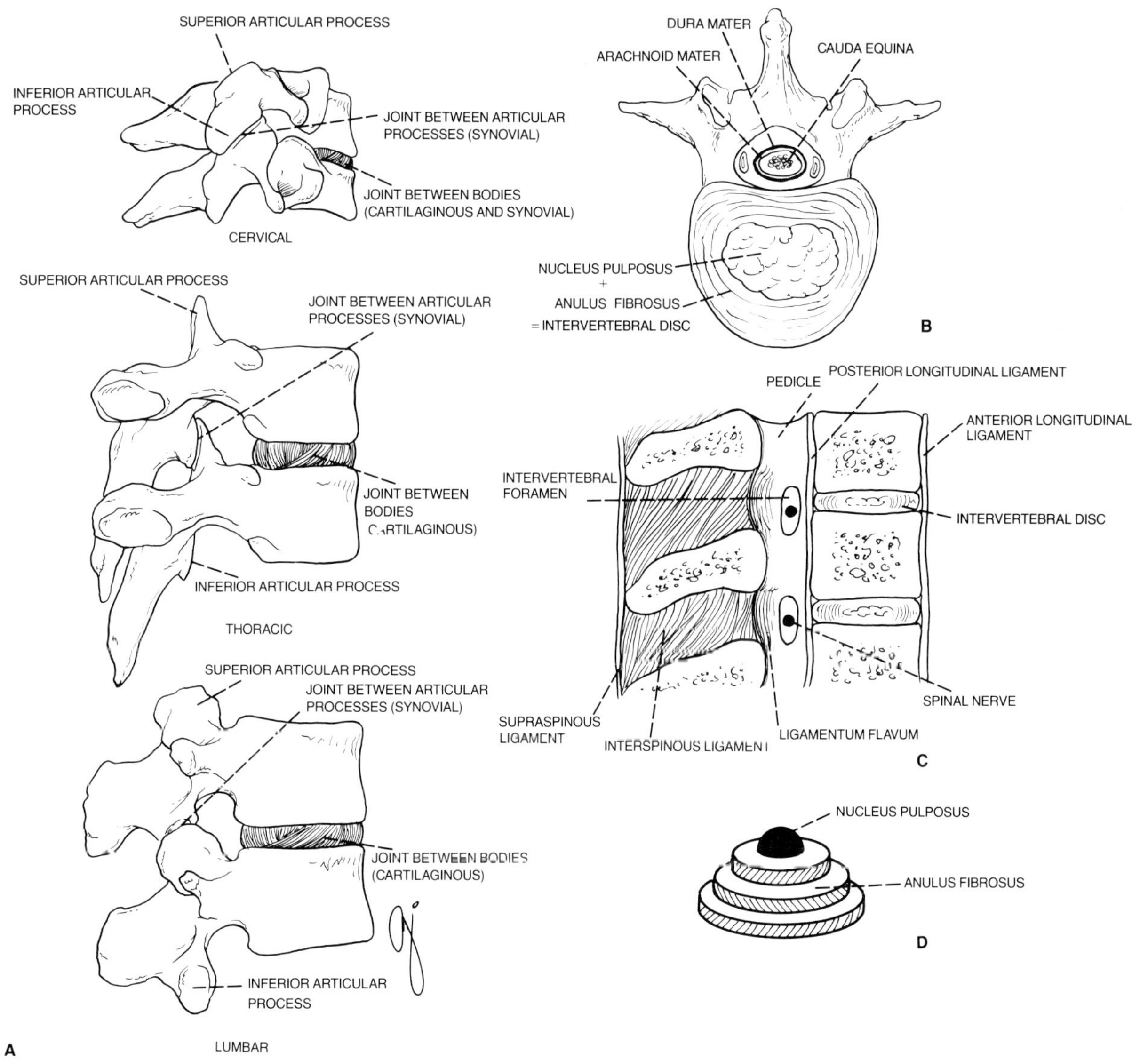

Figure 4–9. A. Joints in cervical, thoracic, and lumbar regions of vertebral column. **B.** Third lumbar vertebra viewed from above, showing the relationship between the intervertebral disc and the cauda equina. **C.** Sagittal section through three lumbar vertebrae, showing ligaments and intervertebral discs. **D.** Structure of an intervertebral disc.

Sandwiched between the plates of hyaline cartilage is an intervertebral disc of fibrocartilage (*Fig. 4–9*). The collagen fibers of the disc strongly unite the bodies of the two vertebrae.

In the lower cervical region, small synovial joints are present at the sides of the intervertebral disc between the upper and lower surfaces of the bodies of the vertebrae.

Ligaments. The **anterior and posterior longitudinal ligaments** run as continuous bands down the anterior and posterior surfaces of the vertebral column from the skull to the sacrum (*Fig. 4–9*). These ligaments hold the vertebrae firmly together but at the same time permit a small amount of movement to take place between them.

Intervertebral Discs. The intervertebral discs are responsible for one-quarter of the length of the vertebral column (*Fig. 4–1*). They are thickest in the cervical and lumbar regions, where the movements of the vertebral column are greatest. Each disc consists of a peripheral part, the anulus fibrosus, and a central part, the nucleus pulposus (*Fig. 4–9*).

The **anulus fibrosus** is composed of fibrocartilage, in which the collagen fibers are arranged in concentric layers or sheets. The collagen bundles pass obliquely between adjacent vertebral bodies, and their inclination is reversed in alternate sheets. The more peripheral fibers are strongly attached to the anterior and posterior longitudinal ligaments of the vertebral column.

The **nucleus pulposus** in the child is an ovoid mass of gelatinous material containing a large amount of water, a small number of collagen fibers, and a few cartilage cells. It is normally under pressure and situated slightly nearer to the posterior than to the anterior margin of the disc.

The upper and lower surfaces of the bodies of adjacent vertebrae that abut onto the disc are covered with thin plates of hyaline cartilage. The semifluid nature of the nucleus pulposus allows it to change shape and permits one vertebra to rock forward or backward on another, as in flexion and extension of the vertebral column. With advancing age the water content of the nucleus pulposus diminishes and is replaced by fibrocartilage. In the elderly, the discs are thin and less elastic, and it is no longer possible to distinguish the nucleus from the anulus.

No discs are found between the first two cervical vertebrae or in the sacrum or the coccyx.

A sudden increase in the compression load on the vertebral column, as when carrying a heavy load or when suddenly flexing the vertebral column, may cause the anulus fibrosus to rupture. This allows the nucleus pulposus to herniate and protrude into the vertebral canal where it may press on the spinal nerve roots, the spinal nerve, or even the spinal cord.

Joints between Two Vertebral Arches

The joints between two vertebral arches consist of synovial joints between the superior and inferior articular processes of adjacent vertebrae. The articular facets are covered with hyaline cartilage, and the joints are surrounded by a capsular ligament.

Ligaments. The **supraspinous ligament** (*Fig. 4–9*) runs between adjacent spines and is thickest and broadest in the lumbar region. It is continuous laterally with the deep fascia. The **interspinous ligament** (*Fig. 4–9*) is thin and connects adjacent spines. It fills in the interval between the supraspinous ligament and the ligamentum flavum. The **intertransverse ligaments,** which run between the transverse processes, are weak bands that are largest in the lumbar region. The **ligamentum flavum** (*Fig. 4–9*) connects the laminae of adjacent vertebrae. The ligament is mainly composed of elastic fibers that are attached to the lower part of the anterior surface of the lamina above and the posterior surface and upper margin of the lamina below. The ligaments are broad, membranous, and thin in the cervical region, thicker in the thoracic region, and thickest in the lumbar region. Their elasticity permits the separation of the laminae in flexion and assists the postvertebral muscles in the movement of extension. In the cervical region, the supraspinous and interspinous ligaments are greatly thickened to form the **ligamentum nuchae.** The latter extends from the spine of the seventh cervical vertebra to the external occipital protuberance of the skull, its anterior border being strongly attached to the cervical spines in between. The supraspinous and interspinous ligaments and the ligamentum flavum are very important structures when performing spinal or epidural anesthesia.

Nerve Supply of Vertebral Joints

The joints between the vertebral bodies are innervated by the meningeal branches of each spinal nerve (*Fig. 4–10*). The meningeal nerves arise from the spinal nerve as it exits from the intervertebral foramen. Each receives branches from an adjacent gray ramus communicans or directly from a thoracic sympathetic ganglion. The meningeal nerves then reenter the vertebral canal through the intervertebral foramen. The sensory and sympathetic fibers are distributed to the meninges, the ligaments, and the intervertebral discs. It is believed that the nerve supply at any particular level is derived from two segmental spinal nerves.

The synovial joints between the articular processes are innervated by small medial branches of the posterior rami of the spinal nerves (*Fig. 4–10*). Here again each joint receives nerve fibers from two adjacent spinal nerves.

MOVEMENTS OF THE VERTEBRAL COLUMN

The extent of movement possible between adjacent vertebrae is relatively small because of the strong ligaments

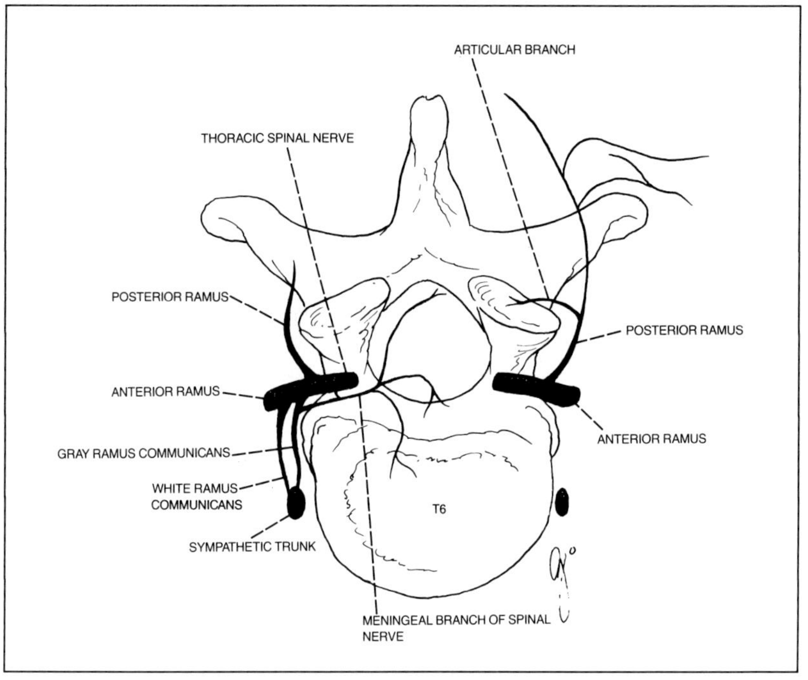

Figure 4–10. Diagram showing the innervation of vertebral joints. Note that at any particular vertebral level the joints receive nerve fibers from two adjacent spinal nerves.

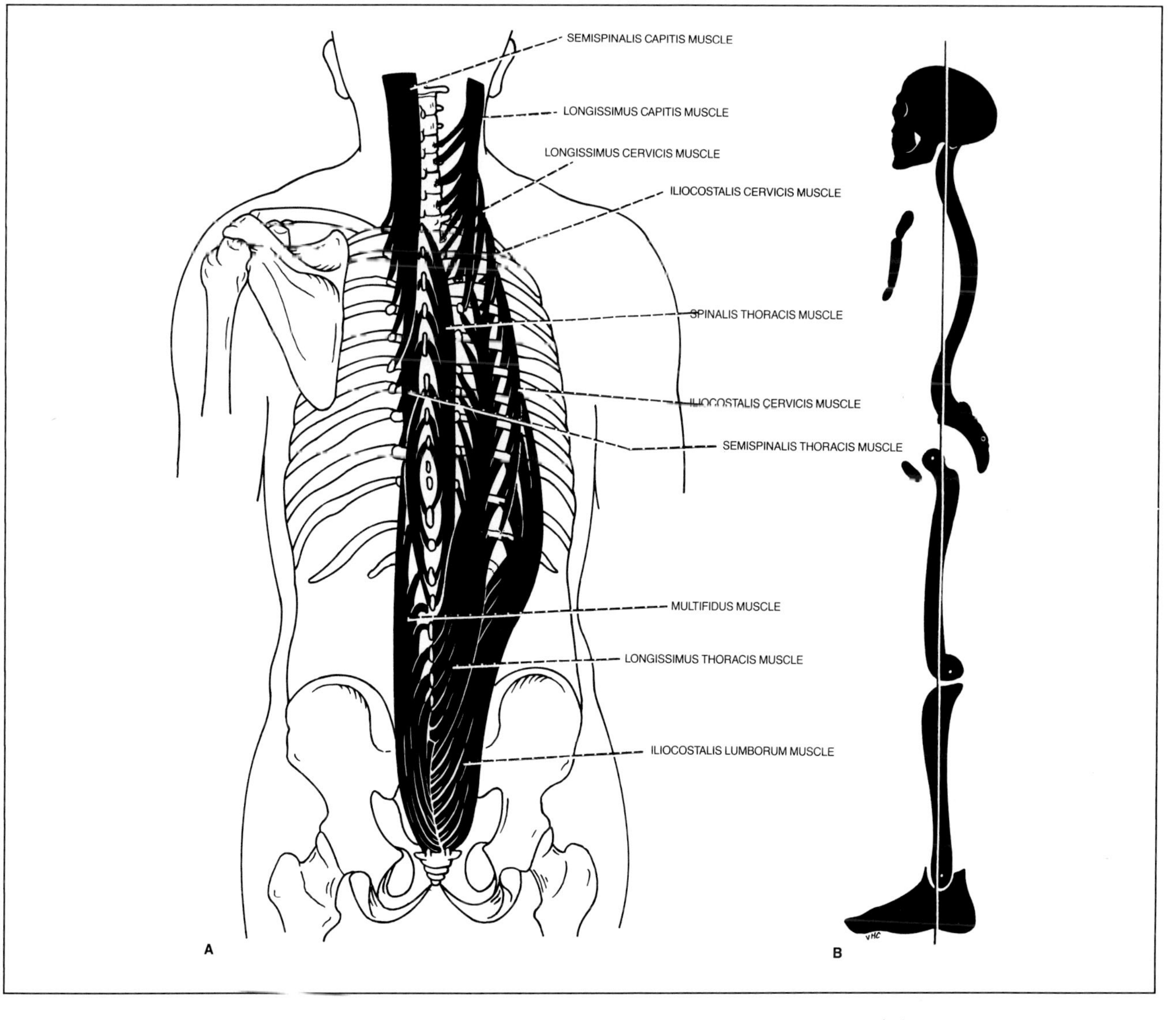

Figure 4–11. A. Arrangement of the deep muscles of the back. **B.** Lateral view of the skeleton showing the line of gravity.

connecting the vertebrae. However, the summation of the movements possible between two vertebrae permits extensive movements to take place of the entire vertebral column.

The following movements are possible: flexion, extension, lateral flexion, rotation, and circumduction.

- *Flexion* is a forward movement, and *extension* is a backward movement. They are both extensive in the cervical and lumbar regions, but restricted in the thoracic region because of the presence of the ribs and costal cartilages that are connected directly or indirectly to the sternum.
- *Lateral flexion* is the bending of the body to one or the other side. It is extensive in the cervical and lumbar regions, but restricted in the thoracic region because of the presence of the ribs.
- *Rotation* is a twisting of the vertebral column. This is most extensive in the lumbar region.
- *Circumduction* is a combination of all these movements.

The type and range of movements possible in each region of the vertebral column is largely dependent on the thickness of the intervertebral discs and the shape and direction of the articular processes. In the thoracic region, as was noted, the ribs, the costal cartilages, and the sternum severely restrict the range of movement. The range of movement is also limited by the tension of the antagonist muscles and the ligaments.

MUSCLES ACTING ON THE VERTEBRAL COLUMN

The vertebral column is moved by numerous muscles, many of which are attached directly to the vertebrae, whereas others, such as the sternocleidomastoid and the abdominal wall muscles, are attached to the skull or to the ribs or fasciae.

Postvertebral Muscles

In the standing position, the line of gravity (*Fig. 4–11*) passes through the odontoid process of the axis, behind the centers of the hip joints, and in front of the knee and ankle joints. It follows that when the body is in this position, the greater part of its weight falls in front of the vertebral column. It is, therefore, not surprising to find that the postvertebral muscles of the back are well developed in man. The postural tone of these muscles is the major factor responsible for the maintenance of the normal curves of the vertebral column.

The deep muscles of the back form a broad, thick column of muscle tissue, which occupies the hollow on each side of the spinous processes (*Fig. 4–11*). They extend from the sacrum to the skull and lie beneath the thoracolumbar fascia. It must be realized that this complicated muscle mass is composed of many separate muscles of varying length. Each individual muscle may be regarded as a string, which, when pulled on, causes one or several vertebrae to be extended or rotated on the vertebra below. Because the origins and insertions of the different groups of muscles overlap, entire regions of the vertebral column can be made to move smoothly.

The spines and the transverse processes of the vertebrae serve as levers that facilitate muscular actions. The muscles of longest length lie superficially and run vertically from the sacrum to the rib angles, the transverse processes, and the upper vertebral spines (*Fig. 4–11*). The muscles of intermediate length run obliquely from transverse processes to the spines. The shortest and deepest muscle fibers run between the spines and between the transverse processes of adjacent vertebrae.

The deep muscles of the back may be classified as follows:

Superficial Vertically Running Muscles

Erector spinae { Iliocostalis, Longissimus, Spinalis }

Intermediate Oblique Running Muscles

Transversospinalis { Semispinalis, Multifidus, Rotatores }

Deepest Muscles

Interspinales
Intertransverse

A summary of the more important muscles acting on the vertebral column, their attachments, and their nerve supply is given in Tables 4–1 through 4–5.

BLOOD SUPPLY OF THE VERTEBRAL COLUMN

Arteries

In the **cervical region,** branches arise from the occipital artery, a branch of the external carotid; the vertebral artery, from the subclavian; the deep cervical artery, from the costocervical trunk; and the ascending cervical artery, from the inferior thyroid artery.

In the **thoracic region,** branches arise from the posterior intercostal arteries and in the **lumbar region,** from the subcostal and lumbar arteries (*Fig. 4–12*). In the **sacral region,** branches arise from the iliolumbar and lateral sacral arteries, branches of the internal iliac artery.

Veins

The veins draining the structures of the back form complicated plexuses extending along the vertebral column from the skull to the coccyx. The veins may be divided into

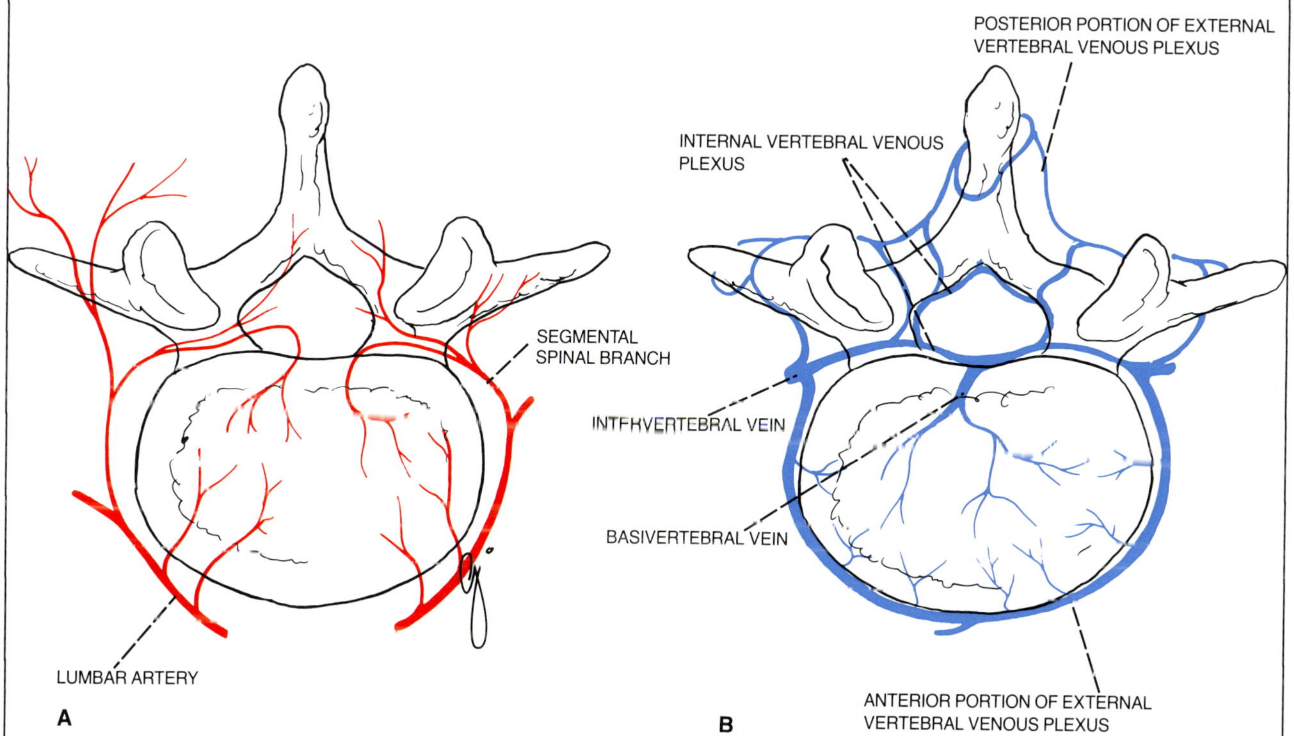

Figure 4–12. **A.** Arterial supply of a lumbar vertebra. **B.** Venous drainage of a lumbar vertebra showing the internal and external vertebral venous plexuses.

TABLE 4–1. ANTERIOR AND LATERAL VERTEBRAL MUSCLES

Name of Muscle	Origin	Insertion	Nerve Supply	Action
Longus colli	Attached to anterior surface of vertebrae between atlas and third thoracic vertebra		Anterior rami of cervical nerves	Flexes cervical part of vertebral column
Longus capitis	Transverse process of lower cervical vertebrae	Occipital bone	Anterior rami of cervical nerves	Flexes head
Rectus capitis anterior	Front of lateral mass of atlas	Occipital bone	Cervical plexus	Flexes head
Rectus capitis lateralis	Transverse process of atlas	Occipital bone	Cervical plexus	Lateral flexion of head
Scalenus anterior	Anterior tubercles of transverse processes of third, fourth, fifth, and sixth cervical vertebrae	Scalene tubercle on inner border of first rib	Cervical spinal nerves	Elevates first rib. Laterally flexes and rotates cervical part of vertebral column
Scalenus medius	Transverse process of axis and posterior tubercles of transverse processes of lower five cervical vertebrae	Upper surface of first rib behind the groove for the subclavian artery	Cervical spinal nerves	Elevates first rib. Laterally flexes and rotates cervical part of vertebral column

(1) those that lie external to the vertebral column and surround it and form the **external vertebral venous plexus** and (2) those that lie within the vertebral canal and form the **internal vertebral venous plexus** (*Fig. 4–12*). There is free communication between these plexuses and the veins in the neck, the thorax, the abdomen, and the pelvis. Above, they communicate through the foramen magnum with the occipital and basilar venous sinuses within the cranial cavity. The internal vertebral plexus lies within the vertebral canal, but outside the dura mater of the spinal cord. It is embedded in areolar tissue and receives tributaries from the vertebrae by way of the **basivertebral veins** (*Fig. 4–12*) and from the meninges and the spinal cord. The internal plexus is drained by the **intervertebral veins,** which pass outward with the spinal nerves through the intervertebral foramina. Here, they are joined by tributaries from the external vertebral plexus and in turn drain into the vertebral, intercostal, lumbar, and lateral sacral veins.

The external and internal vertebral plexuses form a capacious venous network whose walls are thin and whose channels have incompetent valves or are valveless. Free venous blood flow may, therefore, take place between the skull, the neck, the thorax, the abdomen, the pelvis, and the vertebral plexuses, with the direction of flow depending on the pressure differences that exist at any given time between the regions. This fact is of considerable clinical significance.

SURFACE ANATOMY OF THE VERTEBRAL COLUMN AND THE BACK

External Occipital Protuberance

The external occipital protuberance lies at the junction of the head and the neck. If the index finger is placed on the skin in the midline, it can be drawn downward from the protuberance in the **nuchal groove.**

Cervical Vertebrae

The most dominant spinous process that can be felt (*Fig. 4–13*) is that of the **seventh cervical vertebra (vertebra prominens).** Cervical spines 1 through 6 are covered by the **ligamentum nuchae,** a large ligament that runs down the back of the neck connecting the skull to the spinous processes of the cervical vertebrae.

The **transverse processes** are short but easily palpable from the lateral side in a thin neck. The **anterior tubercle of the sixth cervical transverse process (tubercle of Chassaignac)** can be palpated medial to the sternocleidomastoid and against it can be compressed the common carotid artery.

Thoracic and Lumbar Vertebrae

The nuchal groove is continuous below with a furrow that runs down the middle of the back over the tips of the **spines of the thoracic** and **lumbar vertebrae.** The spines of all the thoracic vertebrae and the upper four lumbar vertebrae can be individually palpated. The most prominent spine is that of the first thoracic vertebra; the others may be easily recognized when the trunk is bent forward.

Sacrum. The **spines of the sacrum** are fused with each other in the midline to form the **median sacral crest.** The crest can be felt beneath the skin in the uppermost part of the natal cleft between the buttocks.

The **sacral hiatus** is situated on the posterior aspect of the lower end of the sacrum, and it is here that the **extradural space** (epidural space) terminates. The hiatus lies about 2 inches (5 cm) above the tip of the coccyx and beneath the skin of the natal cleft.

TABLE 4–2. POSTVERTEBRAL MUSCLES

Name of Muscle	Origin	Insertion	Nerve Supply	Action
Splenius capitis	Ligamentum nuchae and first four thoracic spines	Occipital bone; mastoid process of temporal bone	Posterior rami of cervical spinal nerves	Both sides together extend head. Alone, pulls head to one side
Splenius cervicis	Upper thoracic spines	Upper cervical transverse processes	Posterior rami of cervical spinal nerves	Both sides together extend head. One side alone pulls head to one side
***Erector Spinae* (Superficial Vertically Running Muscles)**				
Iliocostalis	Sacrum, iliac crest, and lower ribs	Ribs and transverse processes of cervical vertebrae	Posterior rami of cervical, thoracic, and lumbar spinal nerves	Extension and lateral flexion of vertebral column
Longissimus	Transverse processes of lumbar and thoracic vertebrae	Transverse processes of thoracic and cervical vertebrae, ribs, and mastoid process of skull	Posterior rami of cervical, thoracic, and lumbar spinal nerves	Extension and lateral flexion of vertebral column, extension of head
Spinalis	Spines of lower thoracic and lumbar vertebrae	Spines of upper thoracic vertebrae	Posterior rami of cervical and thoracic spinal nerves	Extension of vertebral column
***Transversospinales* (Intermediate Oblique Running Muscles)**				
Semispinalis Multifidus Rotatores	Transverse processes	Spines	Posterior rami of spinal nerves	Rotation of vertebral column
Deepest Muscles				
Interspinales	Spine below	Spine above	Posterior rami of spinal nerves	Maintains posture
Intertransverse	Transverse process below	Transverse process above		

TABLE 4–3. SUBOCCIPITAL MUSCLES

Name of Muscle	Origin	Insertion	Nerve Supply	Action
Rectus capitis posterior major	Spine of axis	Occipital bone	Posterior ramus of first cervical nerve	Extension of head
Rectus capitis posterior minor	Posterior arch of atlas	Occipital bone	Posterior ramus of first cervical nerve	Extension of head
Obliquus capitis inferior	Spine of axis	Transverse process of atlas	Posterior ramus of first cervical nerve	Rotates face to same side
Obliquus capitis superior	Transverse process of atlas	Occipital bone	Posterior ramus of first cervical nerve	Extends head to same side

Coccyx. The inferior surface and tip of the coccyx can be palpated in the natal cleft about 1 inch (2.5 cm) behind the anus. The anterior surface of the coccyx may be palpated with a gloved finger in the anal canal.

Upper Lateral Part of the Thorax

The upper lateral part of the thorax is covered by the scapula and its associated muscles. The scapula lies posterior to the first to the seventh ribs.

Scapula. The **medial border** of the scapula forms a prominent ridge, which ends above at the superior angle and below at the inferior angle (*Fig. 4–13*).

The **superior angle** can be palpated opposite the first thoracic spine and the inferior angle can be palpated opposite the seventh thoracic spine.

The **crest of the spine of the scapula** can be palpated and traced medially to the medial border of the scapula, which it joins at the level of the third thoracic spine.

The **acromion process of the scapula** forms the lateral extremity of the spine of the scapula. It is subcutaneous and easily located.

Lower Lateral Part of the Back

The lower lateral part of the back is formed by the posterior aspect of the false pelvis and its associated gluteal muscles.

Iliac Crests

The iliac crests are easily palpated along their entire length (*Fig. 4–13*). They lie at the level of the fourth lumbar spine. Each crest ends in front at the **anterior superior iliac spine** and behind at the **posterior superior iliac spine;** the latter lies beneath a skin dimple at the level of the second sacral vertebra and the middle of the sacroiliac joint. The **iliac tubercle** is a prominence felt on the outer surface of the iliac crest about 2 inches (5 cm) posterior to the anterior superior iliac spine at the level of the fifth lumbar spine.

Spinal Cord and Subarachnoid Space

The **spinal cord** in the adult extends down to the level of the lower border of the spine of the first lumbar vertebra (*Fig. 4–14*). In the young child it may extend to the fourth lumbar spine.

The **subarachnoid space** (*Fig. 4–14*), with the **cerebrospinal fluid** it contains, extends down to the lower border of the second sacral vertebra, which lies at the level of the posterior superior iliac spine.

RADIOGRAPHIC APPEARANCES OF THE VERTEBRAL COLUMN

Cervical Region

In the cervical region, x-rays are usually taken in (1) the anteroposterior view and (2) the lateral view.

In the **anteroposterior view,** the atlantoaxial articulation may be demonstrated by asking the patient to keep the mandible in motion while the film is being exposed, or by directing the x-ray tube through the open mouth. By using the latter method, the entire length of the odontoid process may be visualized lying between the lateral masses of the atlas.

Below the level of the third cervical vertebra, the bodies of the vertebrae are shown and the spines are clearly seen. The laminae can be identified. The transverse processes, the foramina transversaria, and the articular processes overlap one another and are difficult to distinguish separately.

In the **lateral view,** the atlanto-occipital joint is difficult to delineate. The anterior and posterior arches of the atlas are well shown (*Fig. 4–15*), and the body of the axis is easily identified. The odontoid process of the axis extends upward, close to the posterior margin of the anterior arch of the atlas. The articular processes are well shown, and the spinous processes can be clearly seen. The transverse processes are difficult to see as they are superimposed on the vertebral bodies. The intervertebral disc spaces between the bodies of adjacent vertebrae are easily defined and are of equal height.

The anterior and posterior surfaces of the vertebral bodies and the posterior wall of the vertebral canal form smooth curved lines that are roughly parallel (*Fig. 4–15*).

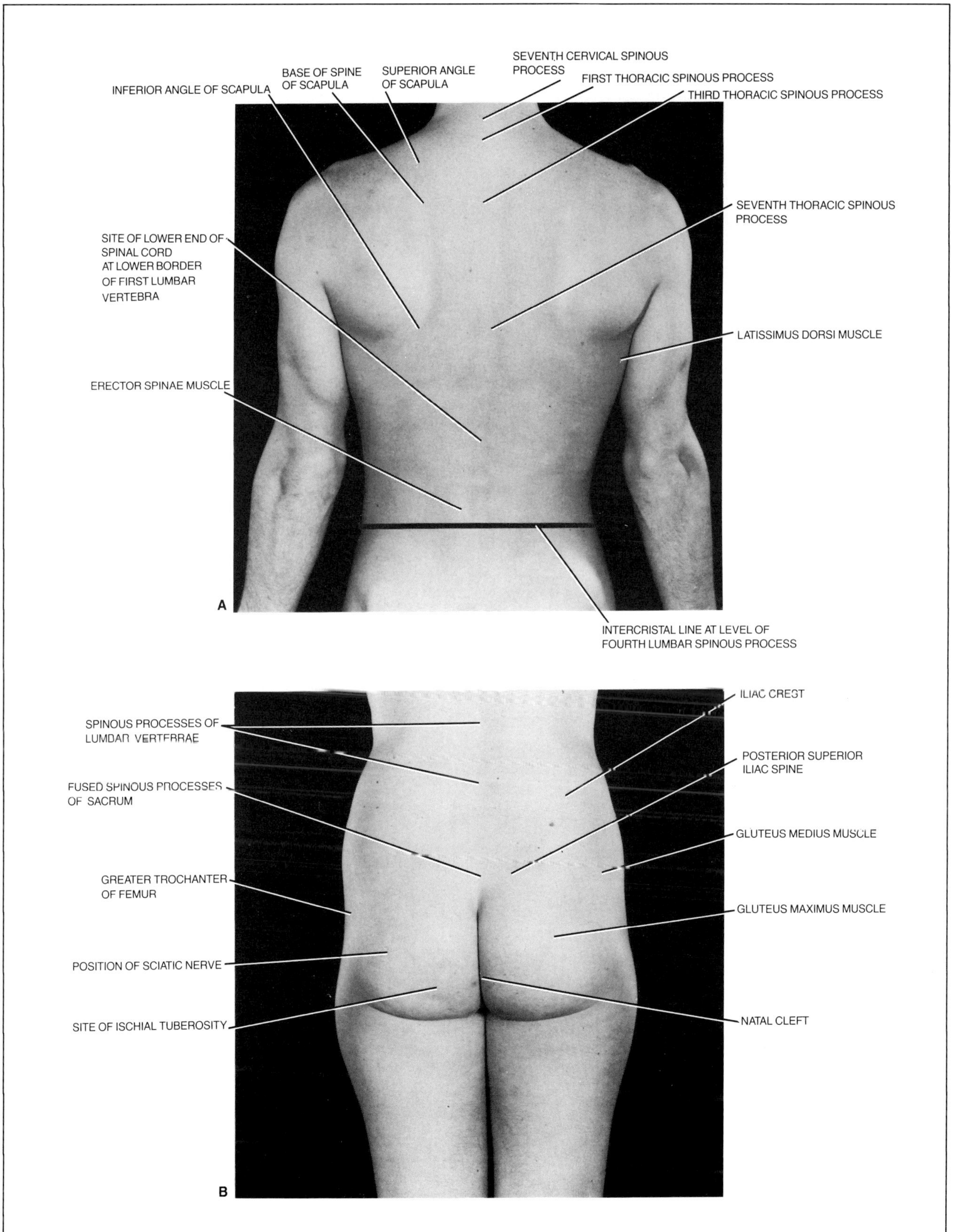

Figure 4–13. A. The back of a 27-year-old male. **B.** The back and gluteal region of a 25-year-old female.

TABLE 4–4. ANTERIOR AND LATERAL ABDOMINAL WALLS MUSCLES ACTING ON VERTEBRAL COLUMN

Name of Muscle	Origin	Insertion	Nerve Supply	Action
External oblique	Lower eight ribs	Xiphoid process, linea alba, pubic crest, pubic tubercle, iliac crest	Lower six thoracic nerves and iliohypogastric and ilioinguinal nerves (L_1)	Assists in flexing and rotation of trunk. Compresses abdominal contents. Pulls down ribs in forced expiration
Internal oblique	Lumbar fascia, iliac crest, lateral two-thirds of inguinal ligament	Lower three ribs and costal cartilages, xiphoid process, linea alba, pubic symphysis. Conjoint tendon with transversus	Lower six thoracic nerves, iliohypogastric and ilioinguinal nerves (L_1)	Assists in flexing and rotation of trunk. Compresses abdominal contents. Pulls down ribs in forced expiration
Rectus abdominis	Symphysis pubis and pubic crest	Fifth, sixth, and seventh costal cartilages and xiphoid process	Lower six thoracic nerves	Flexes vertebral column. Compresses abdominal contents. Accessory muscle of expiration

TABLE 4–5. SUPERFICIAL MUSCLE OF SIDE OF NECK

Name of Muscle	Origin	Insertion	Nerve Supply	Action
Sternocleidomastoid	Manubrium sterni and medial third of clavicle	Mastoid process of temporal bone and occipital bone	Spinal part of accessory nerve and second and third cervical nerves	Muscles of the two sides acting together extend the head and flex the neck. One muscle alone rotates the head to the opposite side

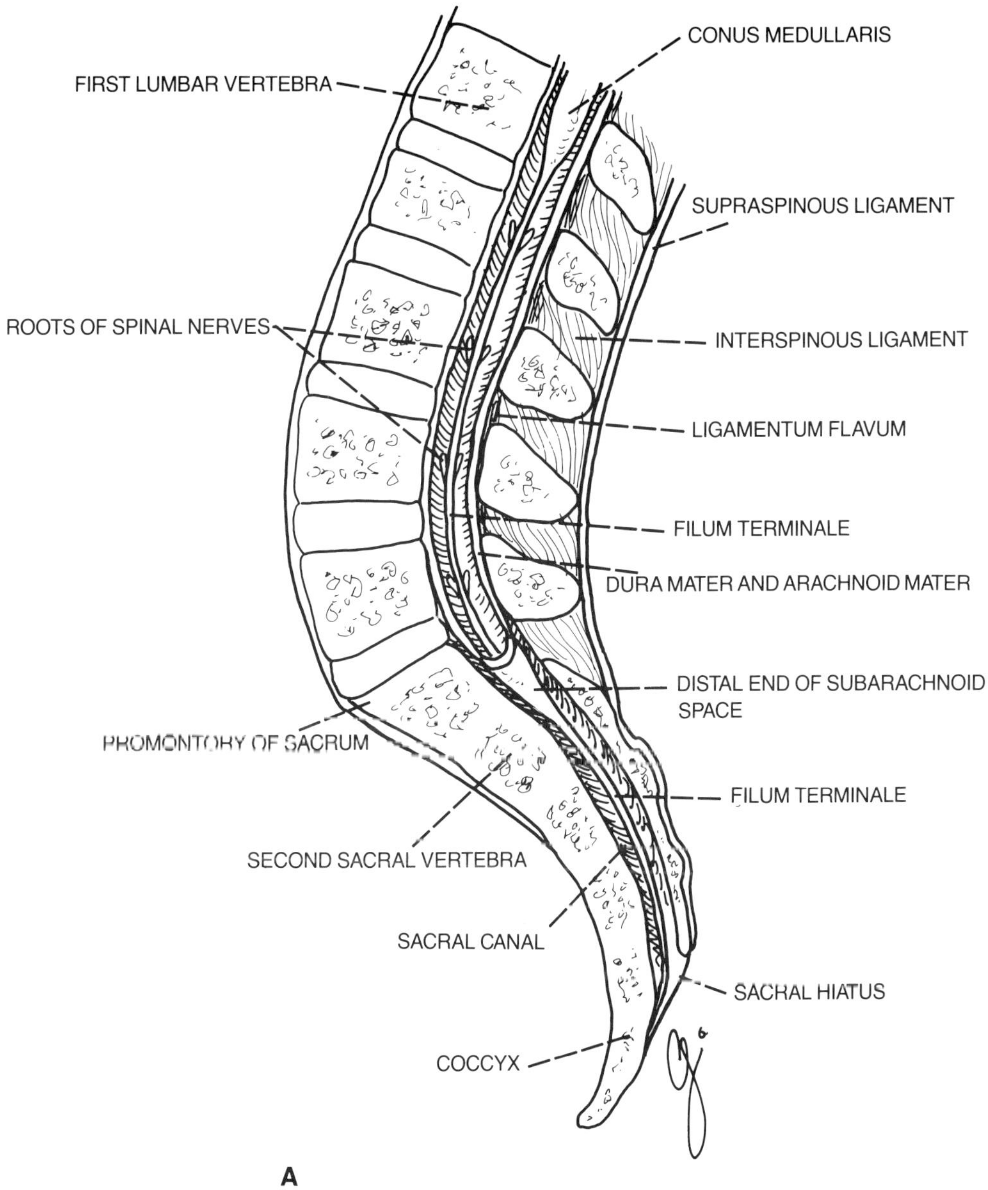

Figure 4–14. A. Median sagittal section through lumbosacral region, showing conus medullaris, filum terminale, and lower limit of subarachnoid space. (*Continued*)

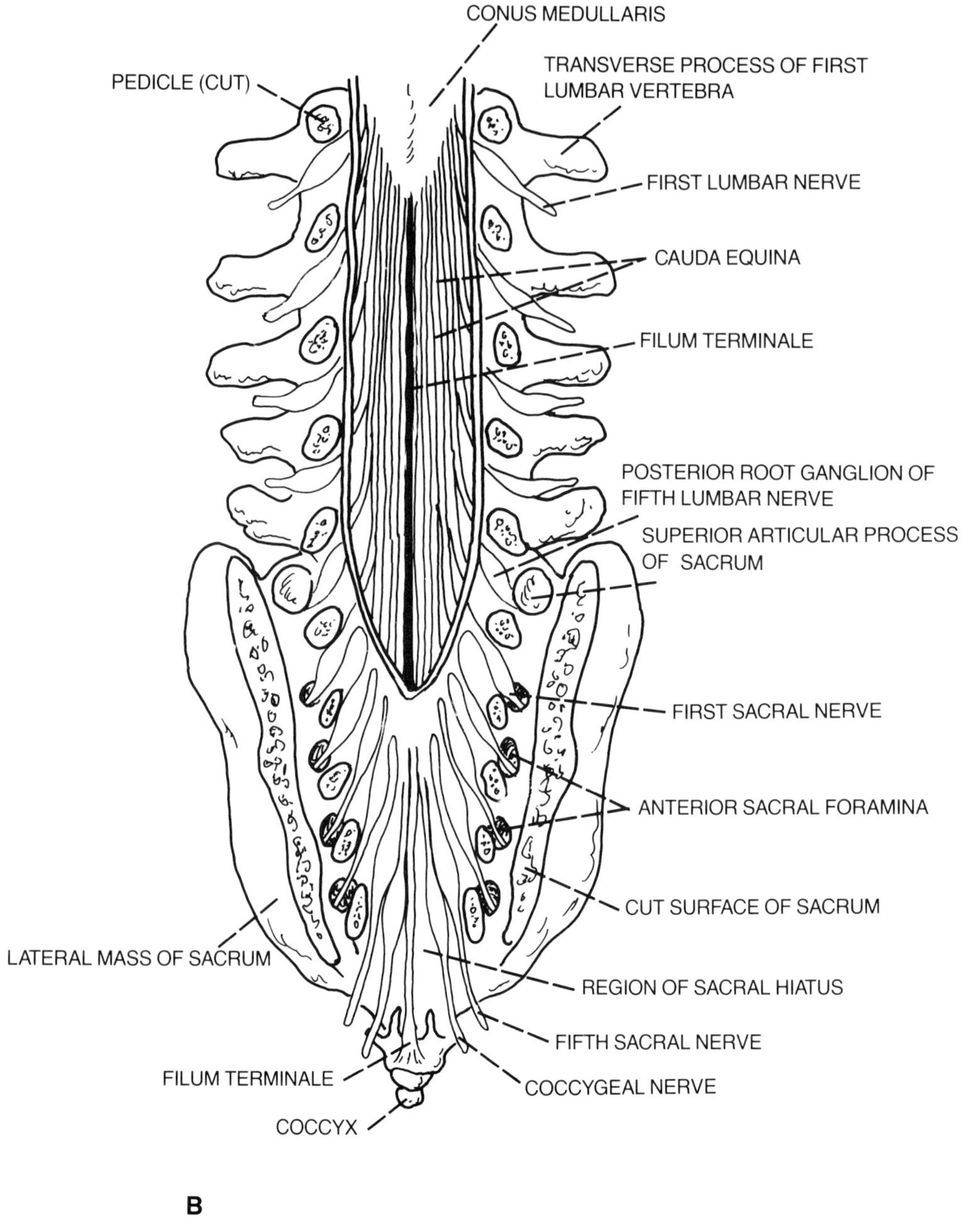

Figure 4–14. (Continued) B. Lumbar vertebrae and sacrum, posterior view. The laminae have been removed to show sacral nerve roots lying within the sacral canal.

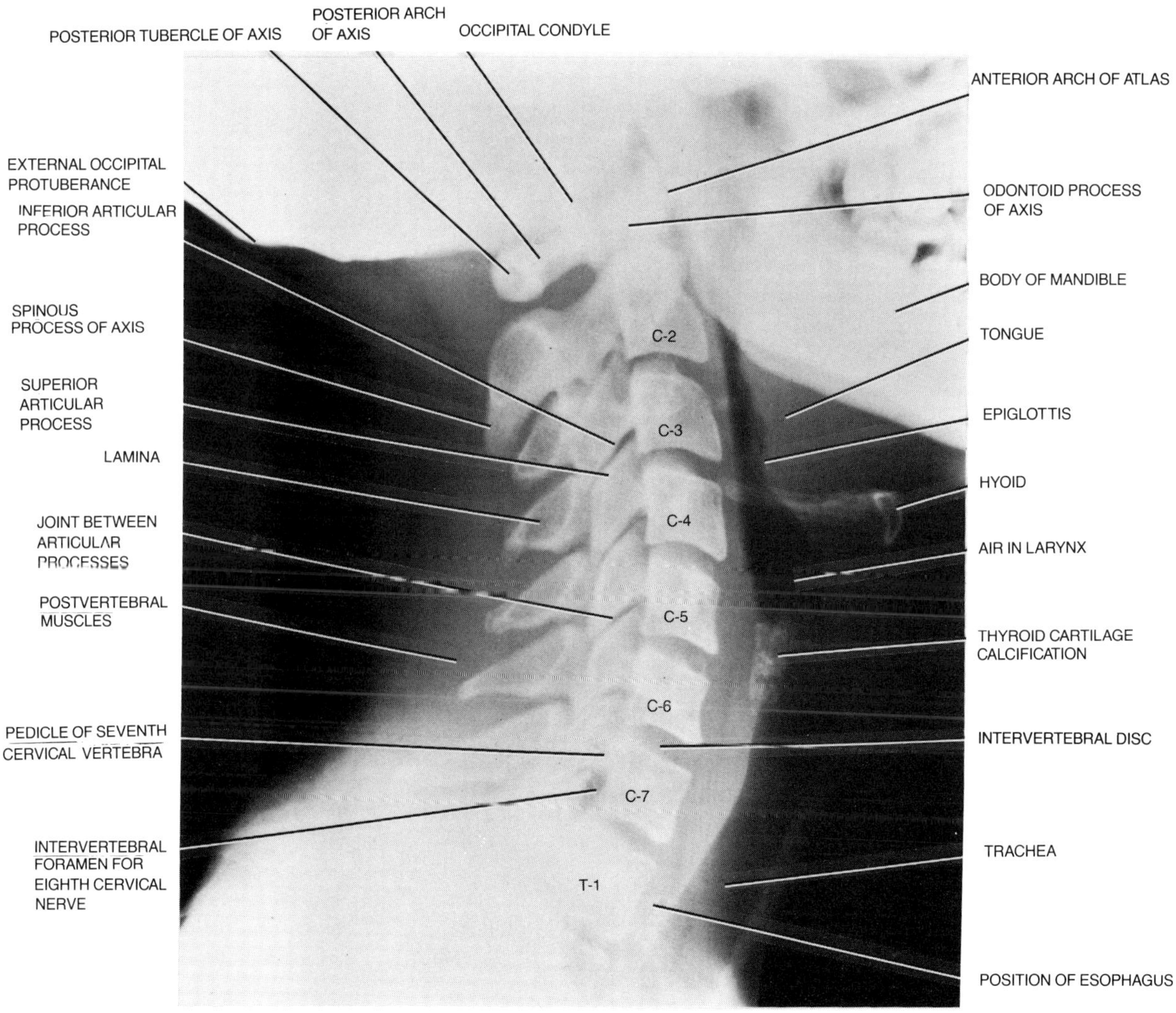

Figure 4–15. Lateral radiograph of the cervical region of vertebral column.

Thoracic Region

In the thoracic region, x-rays are usually taken in (1) the anteroposterior view and (2) the lateral view.

In the **anteroposterior view,** because of the curvature of the thoracic part of the vertebral column, the upper and lower margins of the bodies of adjacent vertebrae overlap (*Fig. 4–16*). The spinous processes and laminae are superimposed on the bodies. The transverse processes can be identified, but they are obscured by the heads and necks of the ribs. The first rib and the 10th, 11th, and 12th ribs on each side articulate only with the bodies of the 1st, 10th, 11th, and 12th thoracic vertebrae, respectively; all the other ribs articulate with two vertebrae. The pedicles are clearly seen as ovoid structures that are superimposed on the lateral parts of the bodies.

The translucent trachea and the heart shadow are superimposed on the thoracic vertebrae.

In the **lateral view,** the rectangular vertebral bodies and the intervertebral disc spaces are clearly seen, even though the ribs and lungs are superimposed on them. The upper four vertebrae are obscured by the shadows of the shoulder girdle.

The pedicles and intervertebral foramina are well demonstrated. The spinous processes, the laminae, the transverse processes, and the ribs are superimposed on one another, however, and their detail is obscured. The vertebral canal is well shown.

Lumbosacral Region

In the lumbosacral region, x-rays are usually taken in (1) the anteroposterior view and (2) the lateral view.

In the **anteroposterior view,** the bodies, the transverse processes, the spinous processes, the laminae, and the intervertebral disc spaces are clearly seen (*Fig. 4–17*). The pedicles produce ovoid shadows, and the articular processes and posterior intervertebral joints can be delineated.

Because of the obliquity of the sacroiliac joint, it is visualized as two lines, the lateral one corresponding to the anterior margin and the medial one to the posterior margin (*Fig. 4–17*). The lower segments of the sacrum and the coccyx are tilted posteriorly and are usually overlapped by the symphysis pubis. In addition, the presence of gas and fecal material in the rectum and sigmoid colon commonly obscures the sacrum. To demonstrate the sacrum in a more direct anteroposterior view, the x-ray tube may be tilted.

In the **lateral view,** the large vertebral bodies, the intervertebral disc spaces, and the intervertebral foramina are clearly seen (*Fig. 4–18*). The pedicles, the articular processes, and the spinous processes are easily visualized. The transverse processes can be identified, but they are superimposed on the sides of the preceding structures. The anterior and posterior surfaces of the vertebral bodies and the posterior wall of the vertebral canal form smooth curved lines that are roughly parallel.

The sacrum on lateral view shows the promontory, the sacral canal, and the fused sacral bodies and the spinous processes (*Fig. 4–18*). Note the localized anterior angulation between the body of the fifth lumbar vertebra and the first sacral vertebra.

Coccyx

The coccyx is not well shown on routine anteroposterior and lateral radiographs. This is caused by its oblique position relative to the film and to the presence of gas and feces in the rectum and the sigmoid colon. These difficulties may be partially overcome by tilting the x-ray tube and evacuating the contents of the rectum and the sigmoid colon.

CLINICAL NOTES

Abnormal Curves of the Vertebral Column

Kyphosis is the term used to describe an exaggeration in the sagittal curvature present in the thoracic part of the vertebral column. It may be due to muscular weakness or structural changes in the vertebral bodies caused by arthritis, fractures, or as the result of surgical procedures. In the elderly, osteoporosis or degeneration of the intervertebral discs, or both, leads to **senile kyphosis,** involving the cervical, thoracic, and lumbar regions of the column.

Lordosis is the term used to describe an exaggeration in the sagittal curvature present in the lumbar region. Lordosis may be due to an increase in the weight of the abdominal contents, as with the gravid uterus or a large ovarian tumor, or it may be due to disease of the vertebral column such as spondylolisthesis. The possibility that it is a postural compensation for a kyphosis in the thoracic region or a disease of the hip joint (congenital dislocation) must not be overlooked.

Scoliosis is the term used to describe a lateral deviation of the vertebral column. This is most commonly found in the thoracic region and may be due to muscular or vertebral defects. Paralysis of muscles due to poliomyelitis may cause severe scoliosis. The presence of a congenital hemivertebra may cause scoliosis. Very often, scoliosis is compensatory and may be due to a short leg or hip disease.

Spinal puncture may be extremely difficult if the curvatures of the vertebral column are drastically changed.

Intervertebral Discs

The physical properties of the intervertebral disc allow one vertebra to rock on another and at the same time serve as a most efficient shock-absorbing mechanism. The resistance of these discs to compression forces is undeniable. Nevertheless, the discs are vulnerable

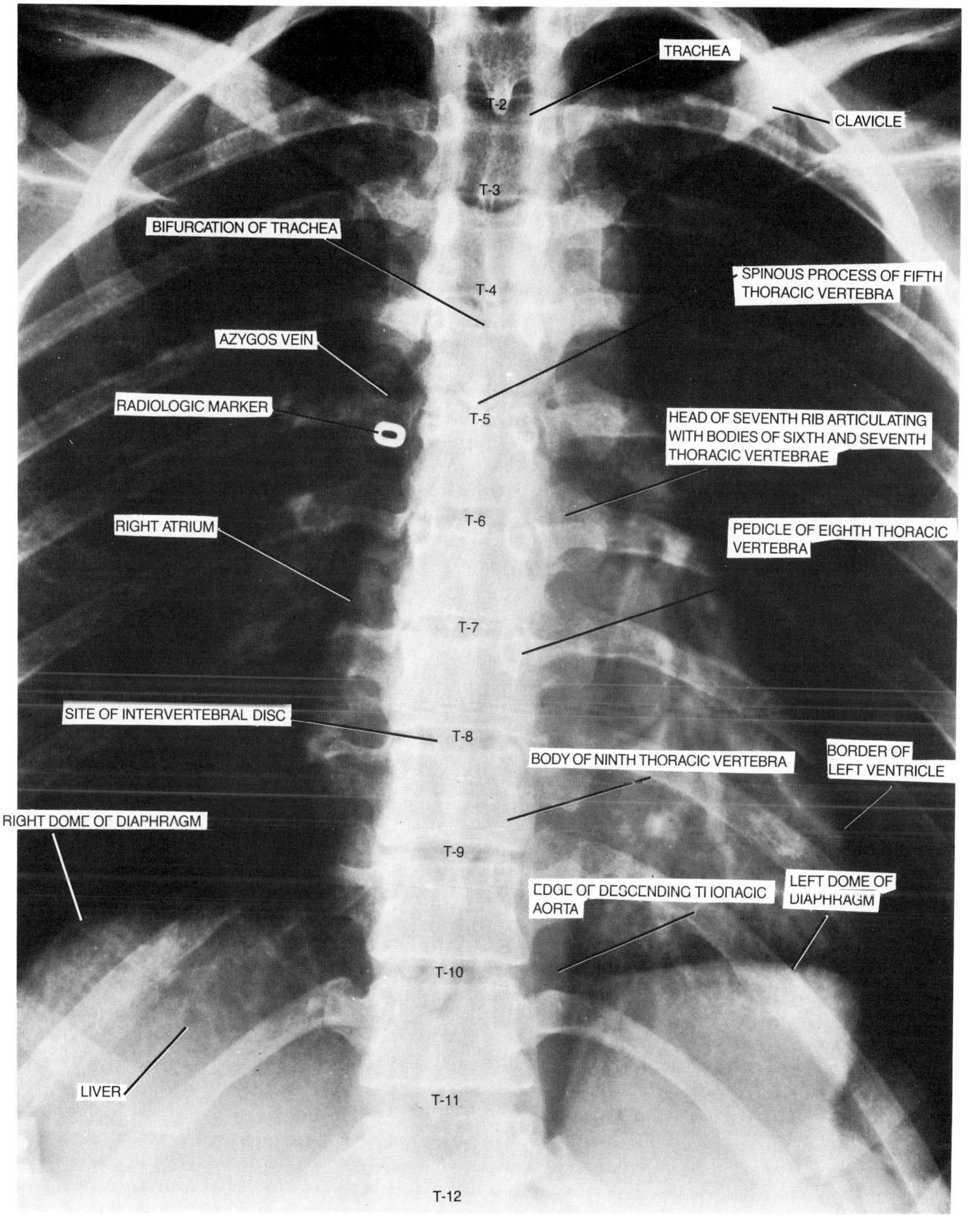

Figure 4–16. Anteroposterior radiograph of the thoracic region of the vertebral column.

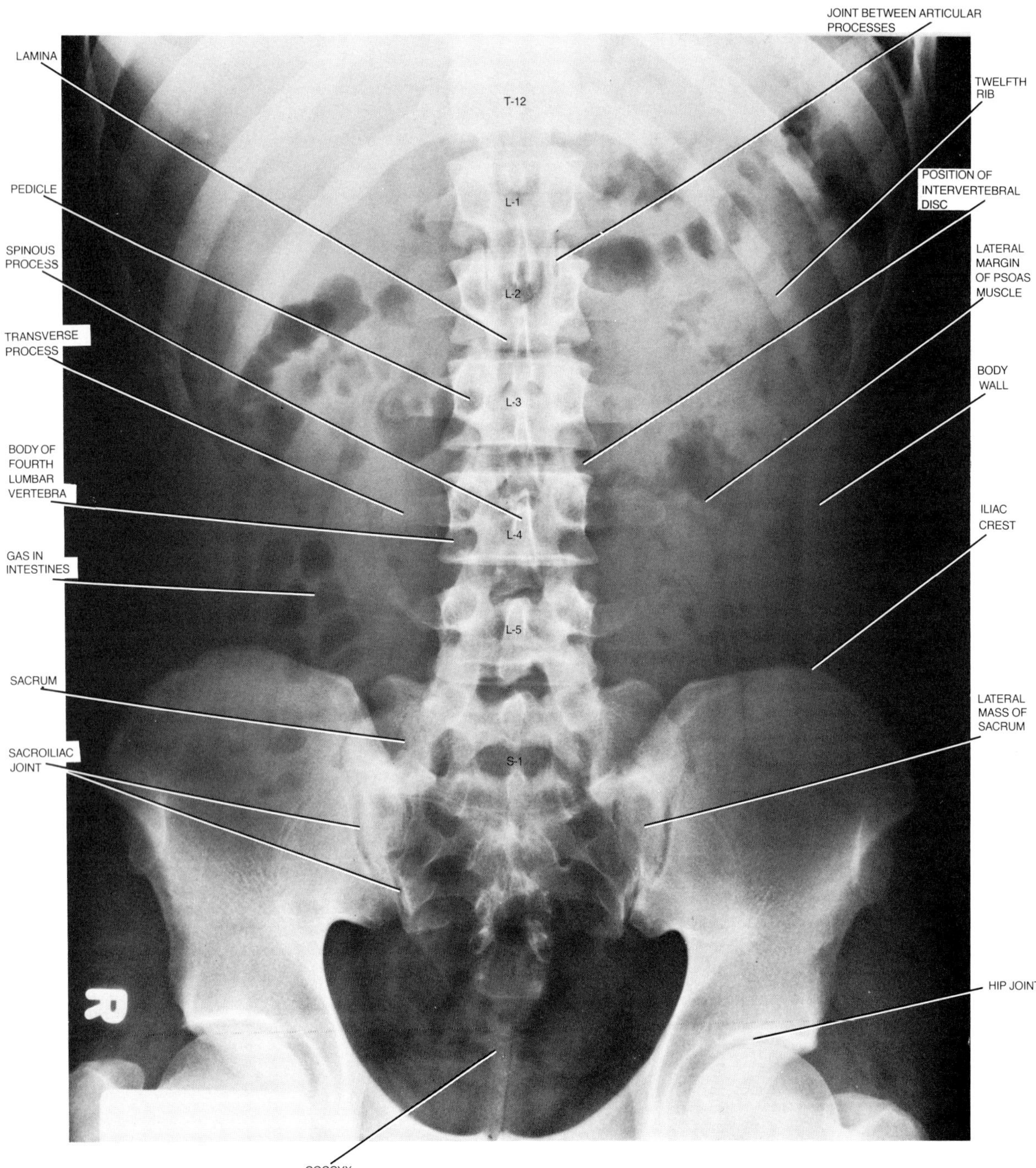

Figure 4–17. Anteroposterior radiograph of lower thoracic, lumbar, and sacral regions of the vertebral column.

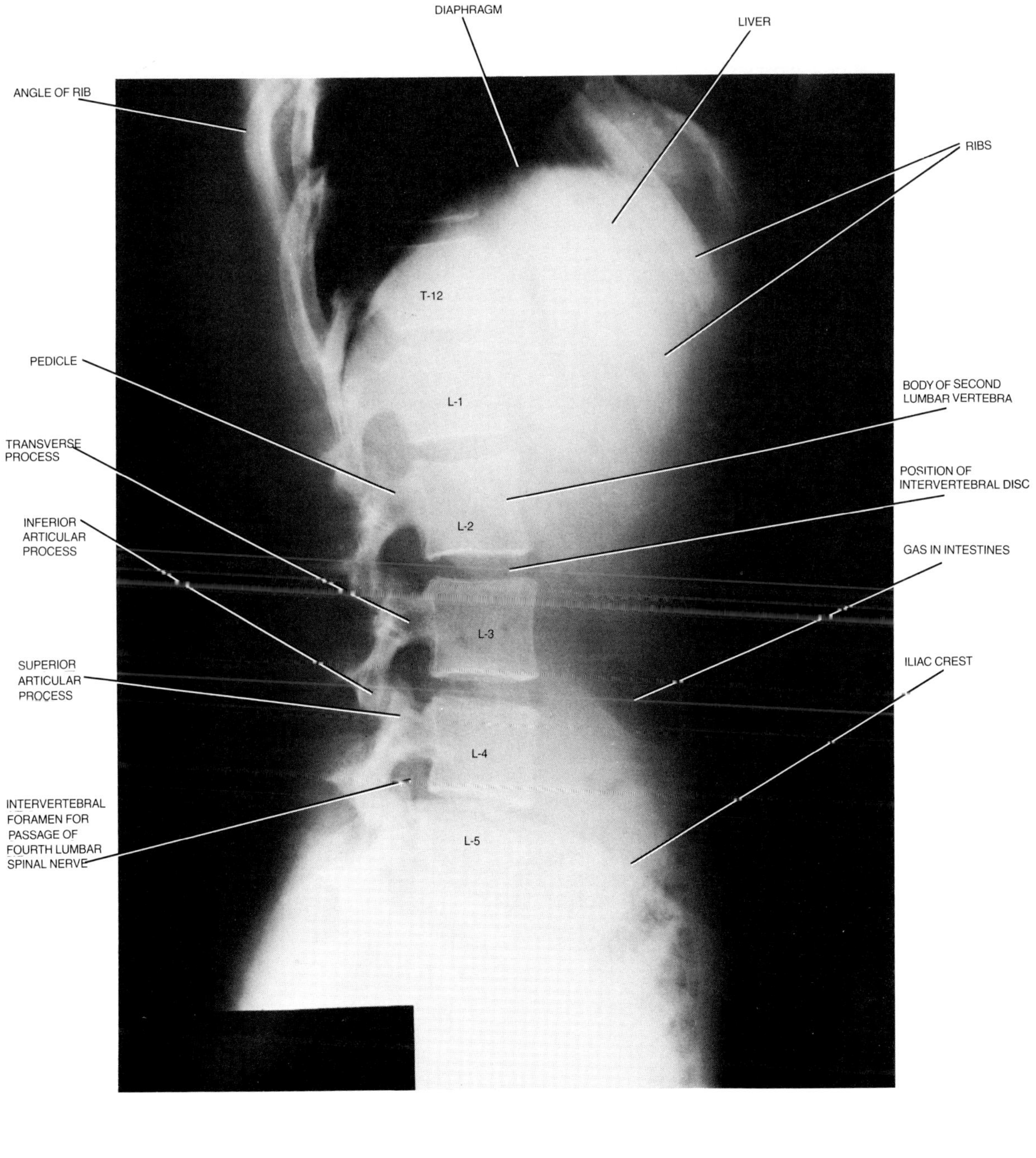

Figure 4–18. Lateral radiograph of lower thoracic, lumbar, and sacral regions of the vertebral column.

to sudden shocks, particularly if the vertebral column is flexed and the disc is undergoing degenerative changes.

The discs most commonly affected are those in areas where a mobile part of the column joins a relatively immobile part, i.e., the cervicothoracic junction and the lumbosacral junction. In these areas the posterior part of the anulus fibrosus ruptures, and the nucleus pulposus is forced posteriorly like toothpaste out of a tube. This is referred to as a **herniation of the nucleus pulposus.** The herniation may result either in a central protrusion in the midline under the posterior longitudinal ligament of the vertebrae or in a lateral protrusion at the side of the posterior ligament close to the intervertebral foramen. The escape of the nucleus pulposus will produce narrowing of the space between the vertebral bodies, which may be visible on radiography. Slackening of the anterior and posterior longitudinal ligaments results in abnormal mobility of the vertebral bodies, producing local pain and subsequent development of osteoarthritis.

Intervertebral Foramina

The intervertebral foramina transmit the spinal nerves and the small segmental arteries and veins, all of which are embedded in areolar tissue. Each foramen is bounded above and below by the pedicles of adjacent vertebrae, in front by the lower part of the vertebral body and by the intervertebral disc, and behind by the articular processes and the joint between them. In this situation the spinal nerve is very vulnerable and may be pressed upon or irritated by disease of the surrounding structures. Herniation of the intervertebral disc, fractures of the vertebral bodies, and osteoarthritis involving the joints of the articular processes or the joints between the vertebral bodies may all result in pressure, stretching, or edema of the emerging spinal nerve.

Dislocations of the Vertebral Column

Dislocations without fracture occur only in the cervical region, as the inclination of the articular processes of the cervical vertebrae permits dislocation to take place without fracture. In the thoracic and lumbar regions, dislocations can occur only if the vertically placed articular processes are fractured.

Fractures of the Vertebral Column

Fractures of the spinous processes, transverse processes, or laminae are caused by direct injury or, in rare cases, by severe muscular activity. Compression fractures of the vertebral bodies are usually caused by an excessive flexion–compression type of injury and take place at the sites of maximum mobility or at the junction of the mobile and fixed regions of the column. It is interesting to note that the body of a vertebra in such a fracture is crushed, whereas the strong posterior longitudinal ligament remains intact. The vertebral arches remain unbroken and the intervertebral ligaments remain intact, therefore, vertebral displacement and spinal cord injury do not occur.

Fracture dislocations are caused by an excessive flexion–compression type of injury and take place at the sites of maximum mobility or the junction of the mobile and fixed regions of the column. Because the articular processes are fractured and the ligaments are torn, the vertebrae involved are unstable, and the spinal cord is usually severely damaged or severed, with accompanying paraplegia.

Spondylolisthesis

In spondylolisthesis, the body of a lower lumbar vertebra, usually the fifth, moves forward on the body of the vertebra below and carries with it the whole of the upper portion of the vertrbral column (*Fig. 4–19*). The essential defect is in the pedicles of the migrating vertebra. It is now generally believed that in this condition the pedicles are abnormally formed, and accessory centers of ossification are present that fail to unite. The spine, the laminae, and the inferior articular processes remain in position, whereas the remainder of the vertebra, having lost the restraining influence of the inferior articular processes, slips forward. Because the laminae are left behind, the vertebral canal is not narrowed, but the nerve roots may be pressed upon, causing low backache and sciatica. In severe cases, the trunk becomes shortened, and the lower ribs contact the iliac crest.

Spinal Cord Injuries

The degree of spinal cord injury at different vertebral levels is largely governed by anatomical factors. In the cervical region, dislocation or fracture dislocation is common, but the large size of the vertebral canal often results in the spinal cord escaping severe injury. The other extreme, when there is considerable displacement, the cord is sectioned and death occurs immediately. Respiration ceases if the lesion occurs above the segmental origin of the phrenic nerves.

In fracture dislocations of the thoracic region, displacement is often considerable, and the small size of the vertebral canal results in severe injury to the spinal cord.

In fracture dislocations of the lumbar region, two anatomical facts aid the patient. First, the spinal cord in the adult extends only down as far as the level of the lower border of the first lumbar vertebra. Second, the large size of the vertebral foramen in this region gives the roots of the cauda equina ample room. Nerve injury may, therefore, be minimal in this region.

ANATOMY OF LUMBAR PUNCTURE

Lumbar puncture may be performed to withdraw a sample of cerebrospinal fluid for examination or to inject drugs to combat infection or induce anesthesia. Fortunately, the spinal cord terminates (*Fig. 4–14*) at the level of the lower border of the first lumbar vertebra in the adult. (In the infant, it may reach as low as the third lumbar vertebra, see also page 153.) The subarachnoid space extends down as far as the lower border of the second sacral vertebra. The lower lumbar

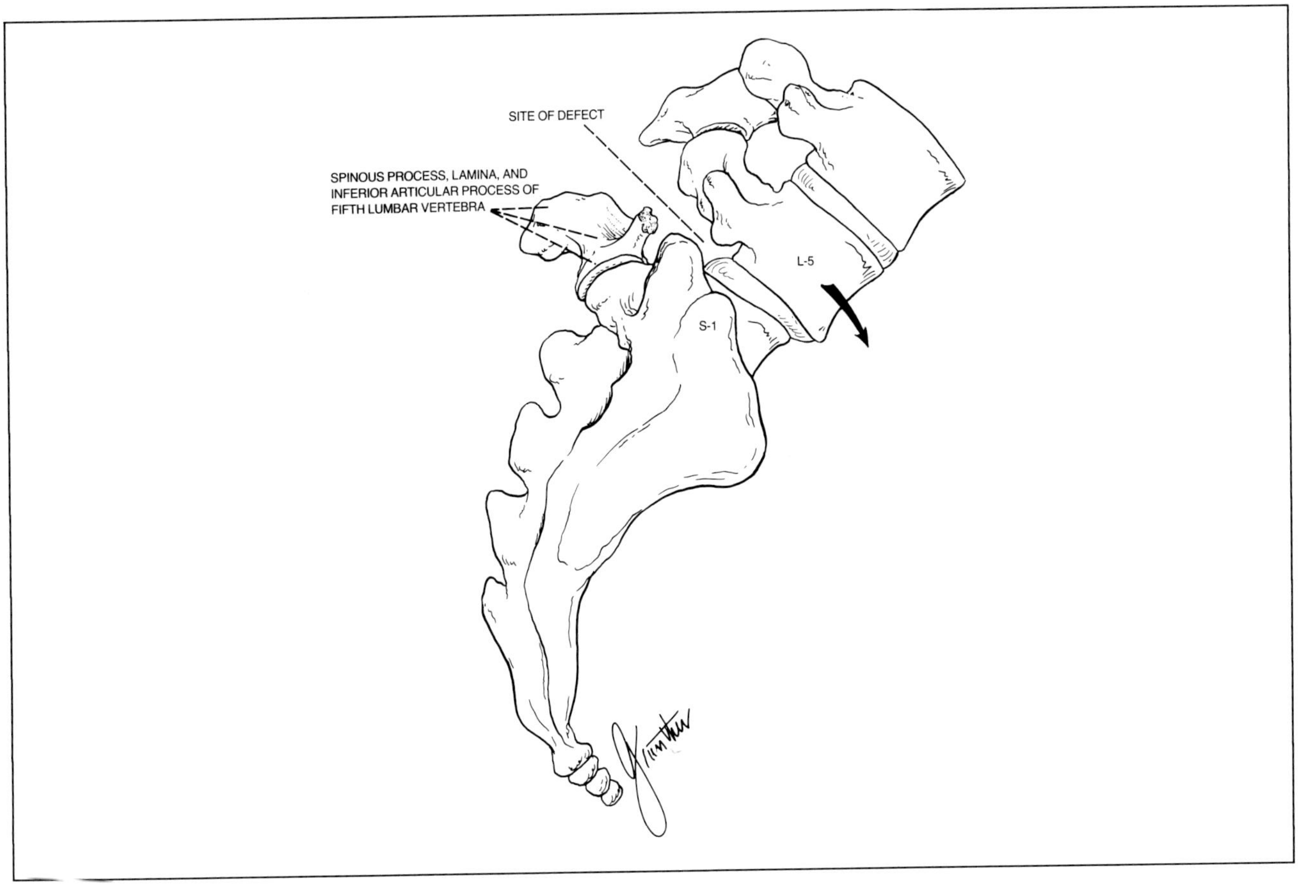

Figure 4–19. Spondylolisthesis. Defect in the pedicles of the fifth lumbar vertebra allow the vertebral body to move anteriorly and inferiorly, carrying with it the whole of the upper portion of the vertebral column.

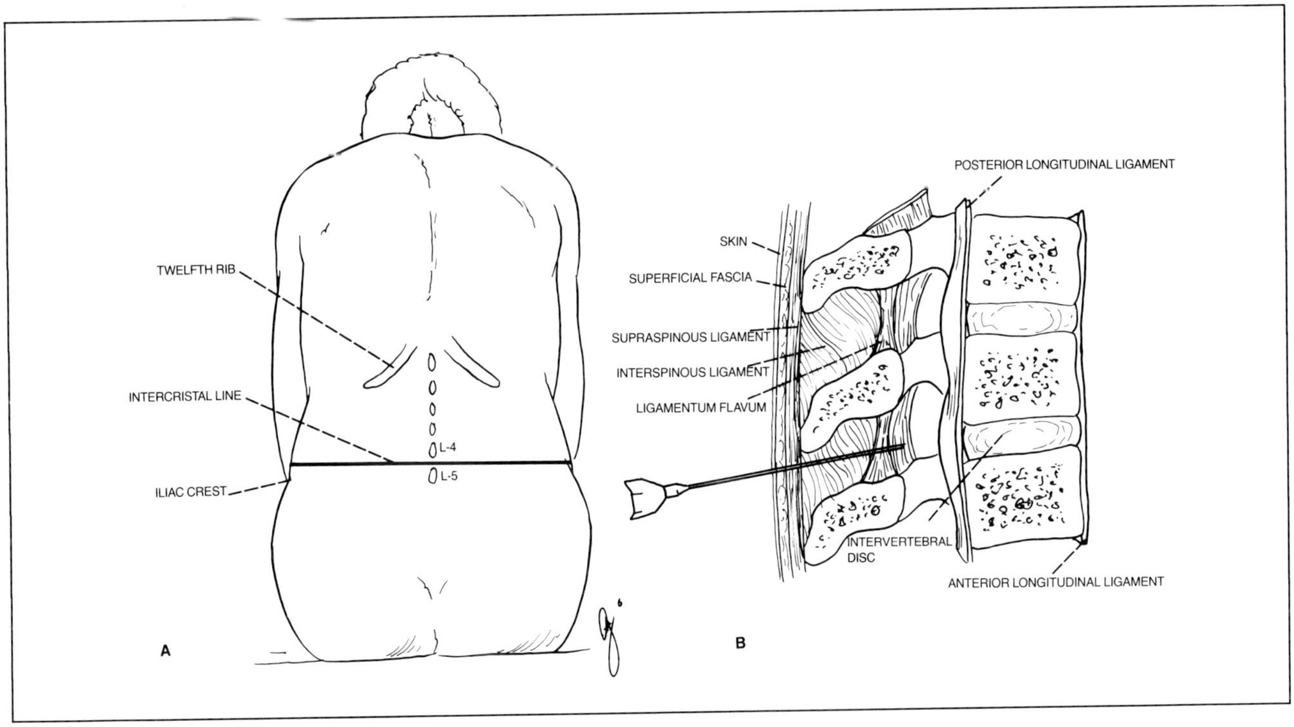

Figure 4–20. A. Important anatomic landmarks when performing a lumbar puncture. **B.** Structures penetrated by lumbar puncture needle before it reaches the dura mater.

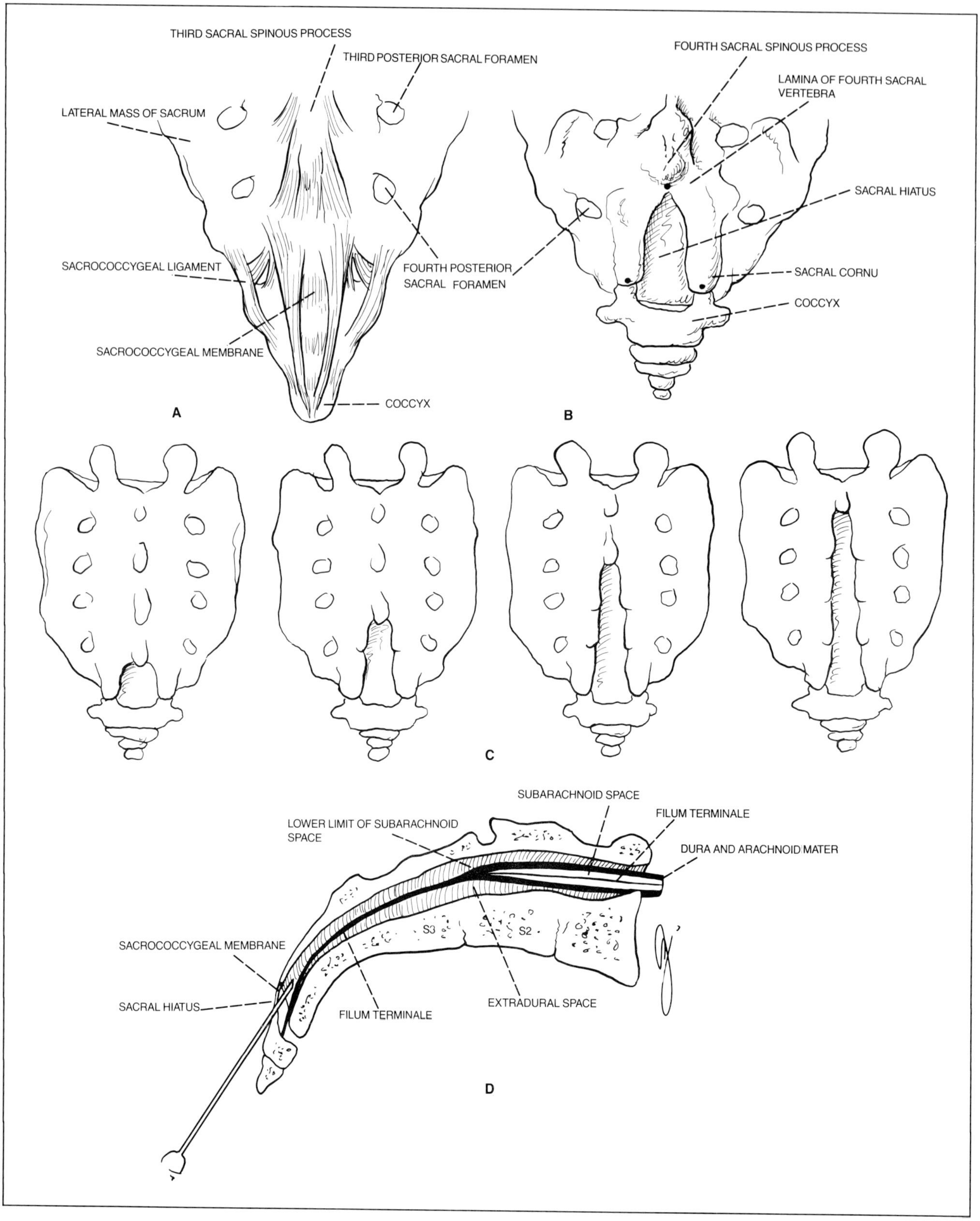

Figure 4–21. **A.** Posterior surface of the lower end of the sacrum and the coccyx showing the ligaments and fibrous membranes covering the sacral hiatus. **B.** The average shape and size of the sacral hiatus. The black dots indicate the position of important bony landmarks. **C.** Variations in the size and shape of the sacral hiatus. **D.** Longitudinal section through the sacrum showing the anatomy of a caudal block.

part of the vertebral canal is thus occupied by the subarachnoid space, which contains the lumbar and sacral nerve roots and the filum terminale (the cauda equina). A needle introduced into the subarachnoid space in this region usually pushes the nerve roots to one side without causing damage.

With the patient in the lateral prone position or in the upright sitting position, with the vertebral column well flexed, the space between adjoining laminae in the lumbar region is opened to a maximum (*Fig. 4–20*). An imaginary line joining the highest points on the iliac crests passes over the fourth lumbar spine (in the female, the line crosses the fourth lumbar spine at a slightly lower level). With a careful aseptic technique and under local anesthesia, the lumber puncture needle, fitted with a stylet, is passed into the vertebral canal above or below the fourth lumber spine (*Fig. 4–20*). The needle will pass through the following anatomical structures before it enters the subarachnoid space: (1) skin, (2) superficial fascia, (3) supraspinous ligament, (4) interspinous ligament, (5) ligamentum flavum, (6) areolar tissue containing the internal vertebral venous plexus in the epidural space, (7) dura mater, (8) arachnoid mater, and (9) cerebrospinal fluid. The depth to which the needle will have to pass will vary from an inch or less in a child to as much as 4 inches (10 cm) in an obese adult.

If the entering needle should stimulate one of the nerve roots of the cauda equina, the patient will experience a fleeting discomfort in one of the dermatomes, or a muscle will twitch, depending on whether a sensory or a motor root was impaled.

The cerebrospinal fluid pressure may be measured by attaching a manometer to the needle. In the recumbent position, the normal pressure is about 120 mm of water.

A block of the subarachnoid space in the vertebral canal, which may be caused by a tumor of the spinal cord or the meninges, may be detected by compressing the internal jugular veins in the neck. This raises the cerebral venous pressure and inhibits the absorption of cerebrospinal fluid in the arachnoid granulations, thus producing a rise in the manometric reading of the cerebrospinal fluid pressure. If this rise fails to occur, the subarachnoid space is blocked and the patient is said to exhibit a positive **Queckenstedt sign.**

It is interesting to note that the cerebrospinal fluid pressure normally fluctuates slightly with the heart beat and with each phase of respiration.

ANATOMY OF CAUDAL ANESTHESIA

Anesthetic solutions injected into the sacral canal through the sacral hiatus pass upward in the loose connective tissue and bathe the spinal nerves as they emerge from the dural sheath.

A successful caudal block depends on an understanding of the normal anatomy of the sacral canal and the anomalies that can occur in this region (*Fig. 4–21*). The sacral canal is curved and follows the general curve of the sacrum. The curve is greater in the adult than in the child. The anterior wall, formed by the fusion of the bodies of the sacral vertebrae, is rough and ridged. The posterior wall, formed by the fusion of the laminae is thin and smooth. At the lower end of the canal is the sacral hiatus, which is triangular or U-shaped and is bounded laterally by the sacral cornua. The size and shape of the sacral hiatus depends on the number of laminae that fail to fuse in the midline posteriorly. The common arrangement is for the hiatus to be formed by the nonfusion of the laminae of the fifth and sometimes the fourth sacral vertebrae. The average distance between the sacral hiatus and the lower end of the subarachnoid space at the second sacral vertebra, is about 47 mm in the adult. The sacral hiatus is closed by the sacrococcygeal membrane, which is composed of dense fibrous tissue.

The sacral canal contains (1) the dural sac (containing the cauda equina), which is tethered to the coccyx by the filum terminale, (2) the sacral and the coccygeal nerves as they emerge from the dural sac surrounded by their dural sheath, and (3) the thin-walled veins of the vertebral venous plexus.

CLINICAL PROBLEMS

1. A 45-year-old man, with a history of right-sided heart failure and pulmonary emphysema, was admitted into hospital for lower abdominal surgery. After consultation with the surgeon, the anesthetist decided to perform an epidural anesthesia between the second and third lumbar vertebrae. What is the extent of the epidural space? Does it extend up into the skull? Where does it end inferiorly? Is the space of uniform size throughout the length of the vertebral canal? If the anesthetic solution does not spread up or down the space, where else may it travel?

2. A 30-year-old man with a history of lumbar disc herniation treated by surgery and vertebral fusion, was to have a spinal anesthetic. Because of the past surgery, it was decided to make a lateral approach with the needle. Name the structures that the spinal needle traversed before entering the subarachnoid space.

3. A 60-year-old man visited his physician complaining of pain in the lower back and down the outer side of his left leg. He gave a history of difficulty in micturition, which had started about 6 weeks previously. A rectal examination revealed an enlarged prostate, which was harder than normal and was more fixed than normal. Radiographic examination of the vertebral column revealed the presence of metastases in the bodies of the fourth and fifth lumbar vertebrae. A diagnosis of carcinoma of the prostate with secondaries in the lumbar part of the vertebral column was made. Can you explain the probable route taken by the carcinoma cells as they spread to the lumbar vertebrae? What is the connection between the prostate and the left-sided sciatica?

4. It was decided to block the right seventh, eighth, and ninth intercostal nerves using a paravertebral approach for a cholecystotomy in a cachectic patient. How does one ascertain the position of the seventh, eighth, and ninth thoracic transverse processes? Is it possible to use the spinous processes of these vertebra as bony landmarks? Do the spinous processes lie opposite the intercostal spaces of the same number?

5. A 55-year-old professor of anesthesiology woke up one morning with a severe pain at the root of the back of the neck and the left shoulder. The previous evening, he had participated in a hard fought tennis tournament at his club. On questioning, it was found that the pain was also referred along the lateral side of the left arm. Movement of the neck caused an increase in the intensity of the pain, which was accentuated by coughing. A lateral radiograph of the neck showed a slight narrowing of the space between the fifth and sixth cervical vertebral bodies. What is the likely cause of the condition? Which nerve root was involved?

6. A 68-year-old man, with a history of osteoarthritis of the lower thoracic and lumbar spine, was admitted to hospital complaining of severe low back pain. On examination, his postvertebral muscles on the right side in the lumbar region felt firmer than normal and were tender to touch. Examination of radiographs of the lumbar vertebrae showed extensive lipping of the vertebral bodies and the presence of large osteophytes encroaching on several intervertebral foramina on the right side. Name the structures that form the boundaries of an intervertebral foramen. What structures normally pass through an intervertebral foramen?

7. A 28-year-old pregnant woman was apprehensive about her future labor. This was her second pregnancy and she had experienced considerable pain during a protracted second stage in her first pregnancy. It was decided to perform a caudal block to anesthetize the afferent fibers from the uterus, the pelvic floor, and the perineum. Having raised an intradermal wheal of local anesthetic in the gluteal (natal) cleft in the midline over the sacral hiatus, the needle was introduced into the sacral canal. Name the structures that the spinal needle

traversed before it entered the sacral canal. What are the bony landmarks used by the anesthetist to identify the sacral hiatus? What types of sacral hiatus exist? What is the explanation for the variability in the shape and size of the sacral hiatus? Is the sacral hiatus ever occluded?

8. A 50-year-old woman, with a thyroid adenoma with symptoms of thyrotoxicosis and cardiac arrythmias, was admitted to hospital for a partial thyroidectomy. Because of the cardiac condition it was decided to perform the operation under heavy sedation combined with a bilateral cervical plexus nerve block. The lateral approach for the nerve block was to be used, and to be performed accurately, the exact position of the transverse process of the sixth cervical vertebra and the tip of the mastoid process had to be determined. How may the transverse process of the sixth cervical vertebra be recognized? What anatomical reasons exist for making the posterior route for cervical plexus block unsatisfactory?

9. A 35-year-old pregnant woman, with a history of placenta previa necessitating a cesarian section in her first pregnancy, was admitted to hospital for a repeat section to deliver her second child. It was decided to perform the operation under a spinal anesthetic. The spinal needle was introduced in the midline between the spinous processes of the third and fourth lumbar vertebrae. How would you determine the surface marking of these processes? The patient was asked to flex her vertebral column as much as possible. Why? Which anatomical structures normally limit flexion of the vertebral column? Name the structures that the spinal needle traverses before it enters the subarachnoid space.

10. A 23-year-old medical student joined a group of other students, who were attempting to lift a car out of a snow drift. Suddenly, he experienced an acute pain in the back. On examination, the deep muscles of the back in the right lumbar region felt firmer than normal and were tender to touch. He said his pain was accentuated by coughing. Evaluation of the muscles of the lower limbs showed some slight weakness in the right quadriceps femoris and a diminished right knee jerk. A diagnosis of herniated intervertebral disc between the third and fourth lumbar vertebrae was made. Which spinal nerve exits between these two vertebrae?

ANSWERS TO CLINICAL PROBLEMS

1. The epidural space extends from the sacral hiatus below to the foramen magnum of the skull above. At the foramen magnum the dura mater surrounding the spinal cord becomes continuous with the meningeal layer of the dura surrounding the brain. It is here that the meningeal layer of dura fuses with the endosteal layer of dura within the skull closing off the epidural space of the vertebral column. Thus, the space does not extend into the skull.

   The epidural space is narrowest where the vertebral canal is smallest. It measures 3 to 5 mm in the cervical region, 3 mm in the midthoracic region, and 4 to 7 mm in the lumbar region.

   If the anesthetic solution does not spread up or down the epidural space, it could spread outward through the intervertebral foramina or through the sacral foramina.

2. In a lateral approach to spinal anesthesia, the spinal needle traverses the following structures: (1) skin and superficial fascia, (2) deep fascia and postvertebral muscles, (3) ligamentum flavum, (4) loose areolar tissue containing thin walled veins, and (5) dura and arachnoid mater.

3. The carcinoma cells probably entered the prostatic venous plexus and traveled, via the internal iliac veins and communicating veins, to the internal vertebral

venous plexus. The increased intra-abdominal pressure associated with micturition, defecation, and coughing assisted the process and resulted in the carcinoma cells reaching the bone marrow of the lumbar vertebral bodies. The prostatic carcinoma had already spread posterolaterally beyond the prostate to involve the sacral plexus of nerves on the left side, hence the left-sided sciatica.

4. The seventh, eighth, and ninth thoracic transverse processes cannot be palpated. The spinous processes, however, can be palpated. Because of the extreme obliquity of the spinous processes in the thoracic region, the tips of the spines of the fifth, sixth, and seventh thoracic vertebrae lie at the level of the seventh, eighth, and ninth transverse processes, respectively. The spinous processes should be counted down from above remembering that the spine of the seventh cervical vertebra is the first to be felt in the lower part of the nuchal groove. The spinous processes of the thoracic vertebrae do not lie opposite the intercostal spaces of the same number because of the obliquity of the spines.

5. This patient had symptoms suggestive of irritation of the left sixth cervical nerve root. The radiographic evidence of narrowing of the space between the fifth and sixth cervical vertebral bodies would suggest a herniation of the nucleus pulposus of the intervertebral disc at this level.

6. An intervertebral foramen has the following boundaries: superiorly and inferiorly, the pedicles of adjacent vertebrae; anteriorly, the lower part of the vertebral body and the intervertebral disc; posteriorly, the superior and inferior articular processes of adjacent vertebrae and the joint between them. The intervertebral foramina transmit the spinal nerves, the small segmental arteries and veins, all of which are embedded in areolar tissue.

7. In a midline caudal block the spinal needle traverses the following structures: (1) skin and fascia and (2) a fibrous ligament (sacrococcygeal membrane) filling in the sacral hiatus.

   The bony landmarks used to identify the sacral hiatus are (1) the posterior superior iliac spines, (2) the median sacral crest, (3) the tip of the coccyx, (4) the sacrococcygeal joint, and (5) the sacral cornua and the sacral hiatus that lies between them. Different types of sacral hiatus exist, which depend on the degree of fusion of the various laminae of the sacral vertebrae to complete the vertebral arch posteriorly. The most common arrangement is for the sacral hiatus to be formed as the result of the failure of fusion of the fifth and sometimes the fourth sacral vertebrae (*see Fig. 4–21*).

   The variability of the size and shape of the sacral hiatus depends on the degree of fusion of the sacral laminae.

   The sacral hiatus is rarely occluded by calcification of the sacrococcygeal membrane or if the posterior wall of the sacral canal is flattened and in contact at its lower end with the anterior wall.

8. The anterior tubercle of the transverse process of the sixth cervical vertebra can be felt on the side of the neck by deep palpation through the sternocleidomastoid muscle at the level of the cricoid cartilage.

   The posterior approach to cervical nerve block is difficult because the landmarks are not easy to identify because of the presence of the ligamentum nuchae and the thick postvertebral muscles.

9. The spinous processes of the third and fourth lumbar vertebrae are easy to palpate, especially if the patient is flexing the vertebral column. The fourth lumbar spine lies on a line connecting the highest points of the iliac crests.

   Flexion of the vertebral column causes separation of the lumbar spines and opens up the interval between the laminae of adjacent vertebrae. This facilitates the passage of the spinal needle into the vertebral canal and the subarachnoid space.

Flexion of the vertebral column is limited by (1) the supraspinous and interspinous ligaments, the ligamentum flavum, and the posterior longitudinal ligament, (2) the ligaments and capsule associated with the joints between the articular processes of the vertebrae, and (3) the tone of the postvertebral muscles.

In a midline spinal anesthetic the needle traverses the following structures: (1) skin and fascia, (2) supraspinous ligament, (3) interspinous ligament, (4) ligamentum flavum, (5) areolar tissue containing the internal vertebral venous plexus in the epidural space, (6) dura mater, and (7) arachnoid mater.

10. A herniated intervertebral disc between the third and fourth lumbar vertebrae may press on the third lumbar spinal nerve as it exits from between these vertebrae. The roots of the fourth lumbar spinal nerve may also be compressed as they cross the posterior surface of the disc to exit between the fourth and fifth lumbar vertebrae.

CHAPTER 5

Spinal Cord and Meninges

SPINAL CORD

The spinal cord is cylindrical in shape and begins superiorly at the foramen magnum where it is continuous with the medulla oblongata of the brain. In the fetus, the spinal cord extends the full length of the vertebral column. After the third fetal month the developing vertebral column grows more rapidly than the spinal cord, therefore, at birth, the coccygeal end of the spinal cord lies at the level of the third lumbar vertebra (*Fig. 5–1*). In the adult, it is at the lower border of the first lumbar vertebra. There is some variation in this level, however, which may be as high as the 12th thoracic vertebra or as low as the third lumbar vertebra. The level of termination is raised slightly during flexion of the vertebral column.

The spinal cord thus occupies the upper two-thirds of the vertebral canal and is surrounded by three meninges, the **dura mater,** the **arachnoid mater,** and the **pia mater** (*Fig. 5–2*). Further protection is provided by the **cerebrospinal fluid,** which surrounds the spinal cord in the **subarachnoid space.**

Enlargements of Spinal Cord

In the cervical region, where the spinal cord gives origin to the brachial plexus, and in the lower thoracic and lumbar regions, where it gives origin to the lumbosacral plexus, there are fusiform enlargements, called the **cervical and lumbar enlargements** (*Fig. 5–2*). Inferiorly, the lumbar enlargement tapers off into the **conus medullaris,** from the apex of which a prolongation of the pia mater, called the **filum terminale,** descends to be attached to the back of the coccyx.

Fissures of the Spinal Cord

The spinal cord possesses, in the midline anteriorly, a deep longitudinal fissure, the **anterior median fissure,** and, on the posterior surface, a shallow furrow, the **posterior median sulcus.**

Roots of the Spinal Nerves

Along the entire length of the spinal cord is attached 31 pairs of spinal nerves by the **anterior or motor roots** and the **posterior or sensory roots** (*Figs. 5–2*). Each root is attached to the cord by a series of **rootlets,** which extend the whole length of the corresponding segment of the cord. Each posterior nerve root possesses a **posterior root ganglion.**

In the upper cervical region, the spinal nerve roots are short and run almost horizontally, but the roots of the lumbar and sacral nerves below the level of the termination of the cord (lower border of the first lumbar vertebra in the adult) form a vertical leash of nerves around the filum terminale, called the **cauda equina** (*see Fig. 4–14*).

The spinal nerve roots pass laterally and the anterior and posterior roots for each spinal cord segment unite in an intervertebral foramen to form a **spinal nerve** (*Figs. 5–2 and 5–3*). After emerging from the intervertebral foramen, each spinal nerve immediately divides into a large **anterior ramus** and a smaller **posterior ramus,** each containing both motor and sensory fibers.

The cervical roots of the **spinal part of the accessory nerve** emerge from the upper five cervical segments and run upward through the foramen magnum.

Structure of Spinal Cord

The central nervous system is composed of large numbers of nerve cells and their processes, which are supported by specialized tissue called **neuroglia. Neuron** is the name given to the nerve cell and all its processes. The long processes of a nerve cell are called **axons** or **nerve fibers.** The interior of the central nervous system is organized into gray and white matter. **Gray matter** consists of nerve cells and the proximal portions of their processes embedded in neuroglia. **White matter** consists of nerve fibers embedded in neuroglia.

The spinal cord is composed of an inner core of

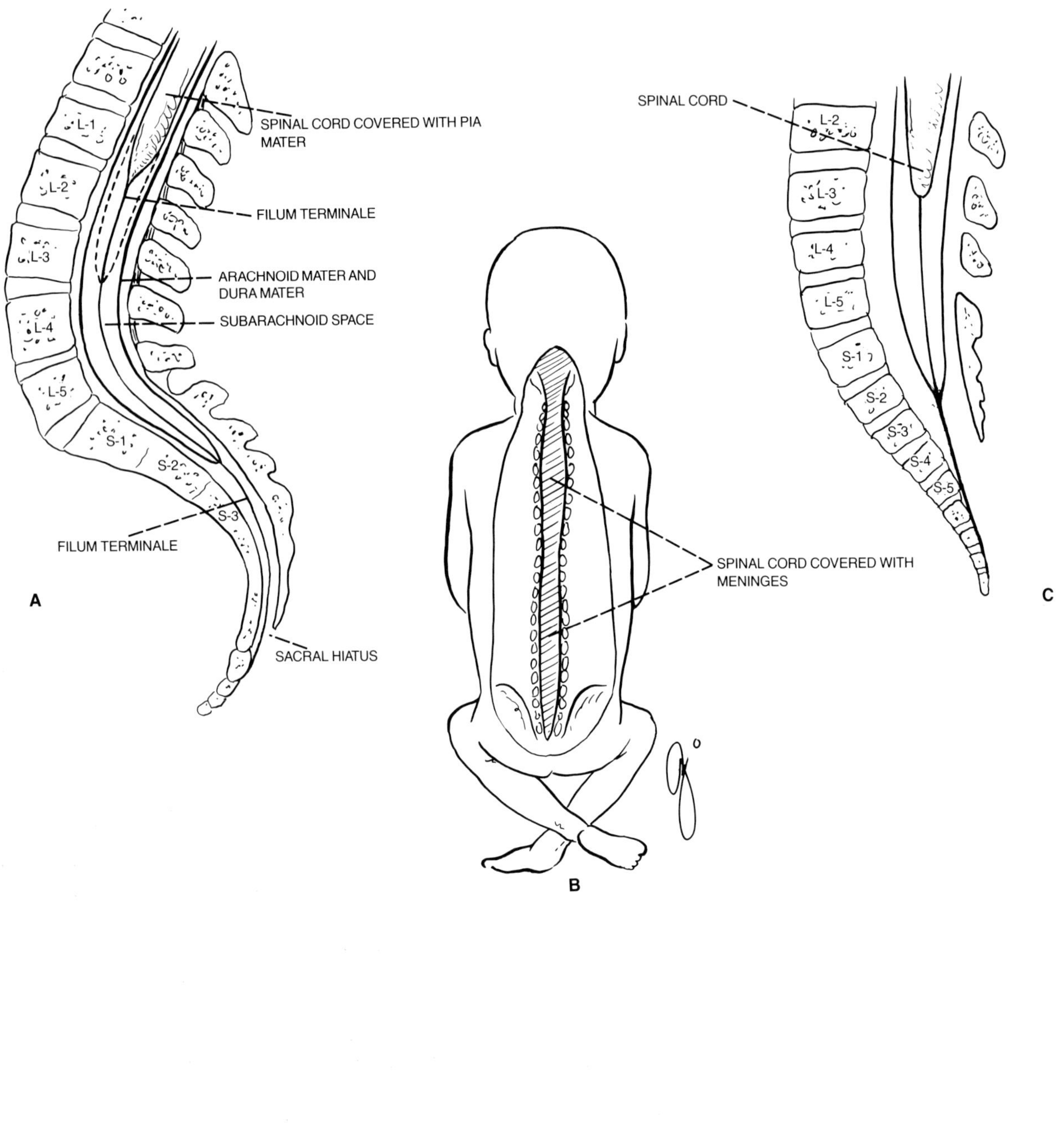

Figure 5–1. **A.** Median sagittal section through the lumbosacral region, showing the conus medullaris, the filum terminale, and the lower limit of the subarachnoid space in the adult. The dotted line indicates the lowest level to which the spinal cord may extend in the adult. **B.** The lowest level of the spinal cord in a 3-month fetus and at birth. **C.** The lowest level of the spinal cord commonly found at birth.

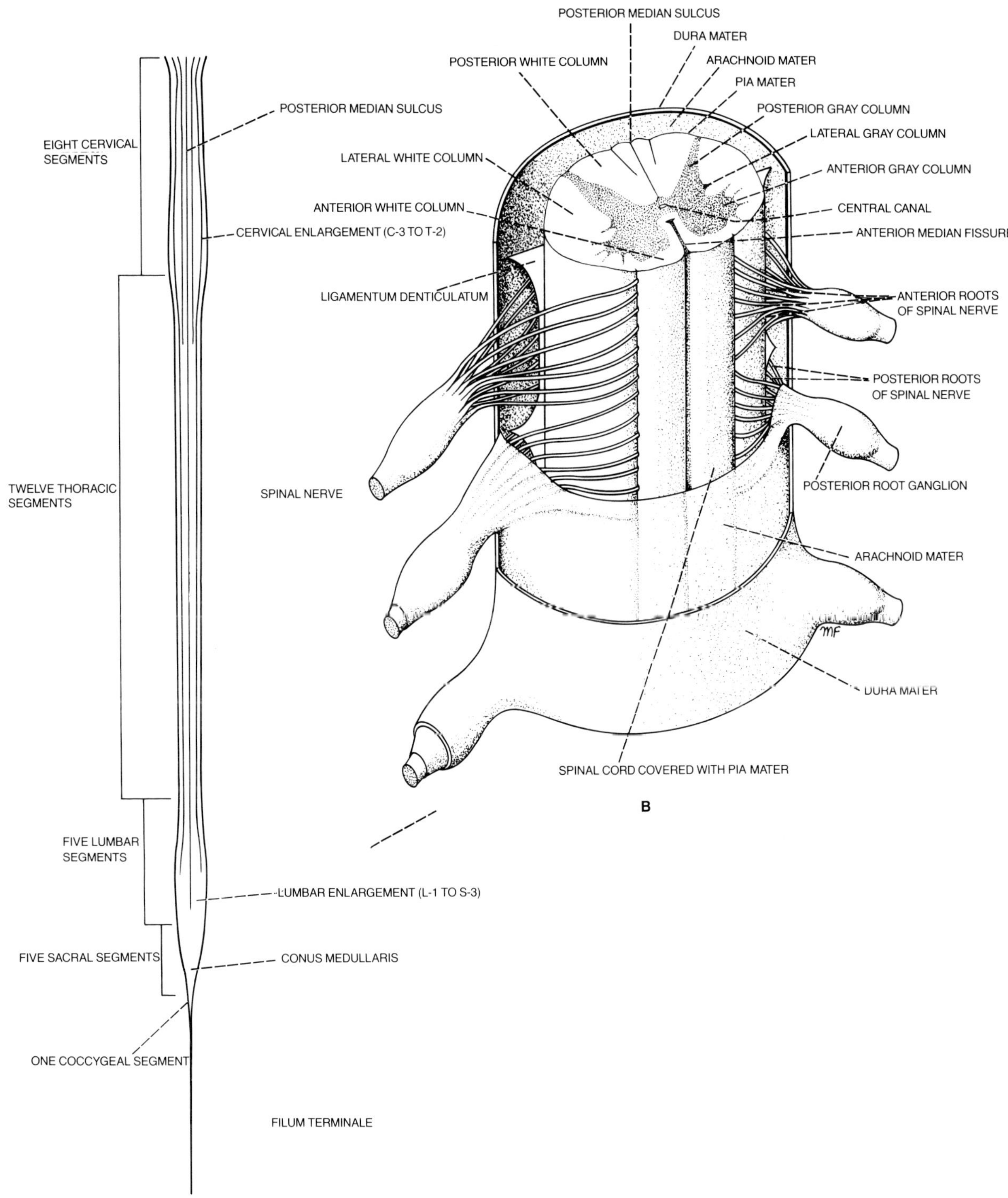

Figure 5–2. Spinal cord. **A.** Posterior view showing cervical and lumbar enlargements. **B.** Three segments of the spinal cord, showing the coverings of dura mater, arachnoid mater, and pia mater.

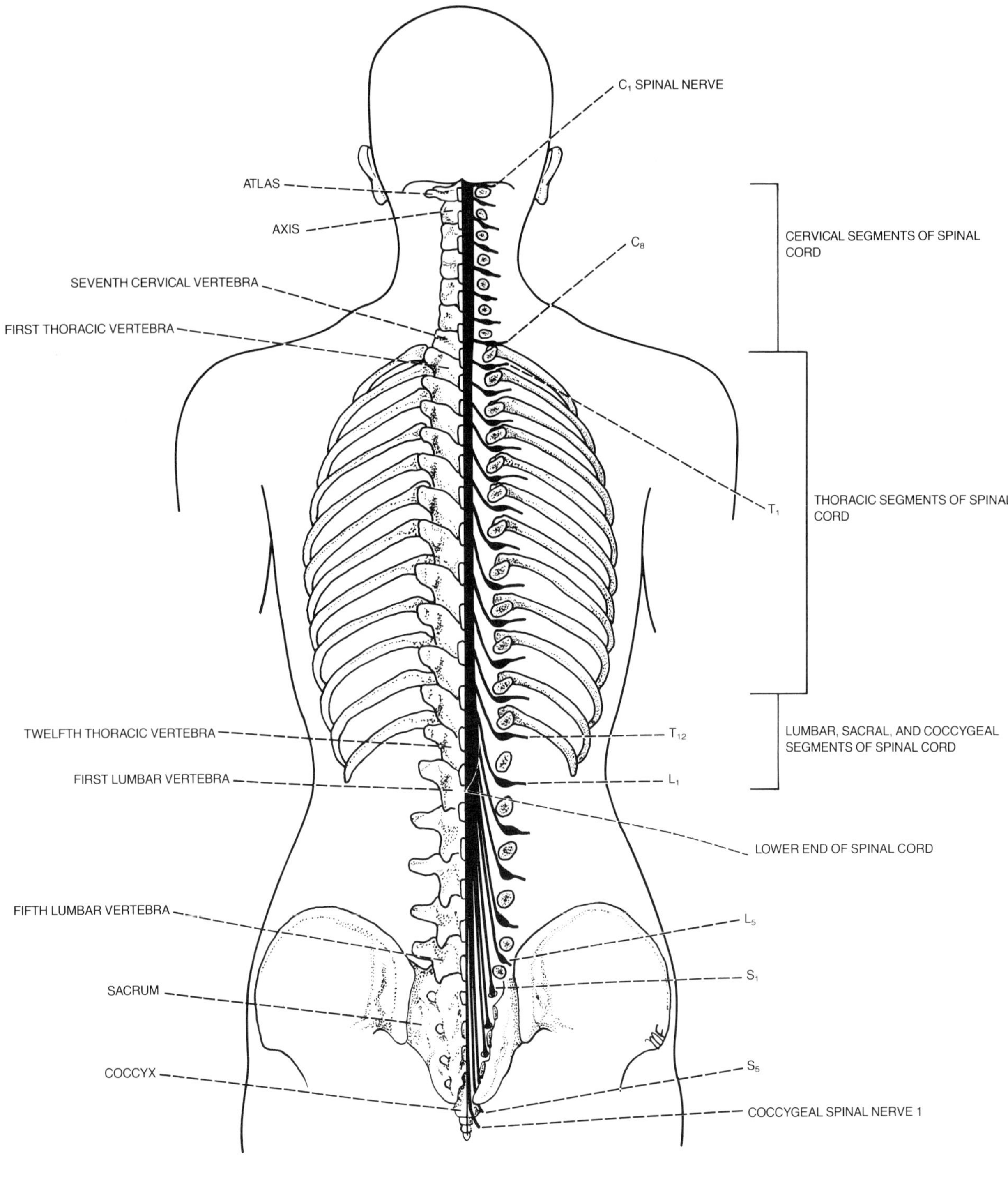

Figure 5–3. Posterior view of spinal cord showing the origins of the roots of the spinal nerves and their relationship to the different vertebrae. On the right, the laminae have been removed to expose the right half of the spinal cord and the nerve roots.

gray matter, which is surrounded by an outer covering of **white matter** (*Fig. 5–2*). The gray matter is seen on cross section as an H-shaped pillar with **anterior and posterior gray columns or horns,** united by a thin **gray commissure** containing the **central canal.**

Organization of the Gray Matter

Nerve Cell Groups in the Anterior Gray Columns. The majority of the nerve cells are large and multipolar in the anterior gray columns, and their axons pass out in the anterior roots of the spinal nerves as **alpha efferents,** which innervate skeletal muscles. The smaller nerve cells are also multipolar and the axons of many of these pass out in the anterior roots of the spinal nerves as **gamma efferents,** which innervate the intrafusal muscle fibers of neuromuscular spindles.

For practical purposes, the nerve cells of the anterior gray column may be divided into three groups or columns, medial, central, and lateral (*Fig 5–4*).

The medial group is present in most segments of the spinal cord and is responsible for innervating the skeletal muscles of the neck and trunk, including the intercostal and abdominal musculature.

The central group is the smallest and is present in some cervical and lumbosacral segments (*Fig. 5–4*). In the cervical part of the cord, some of these nerve cells (segments C-3, C-4, and C-5) specifically innervate the diaphragm and are collectively referred to as the **phrenic nucleus** (*Fig. 5–4*). In the upper five or six cervical segments, some of the nerve cells innervate the sternocleidomastoid and trapezius muscles and are referred to as the **accessory nucleus** (*Fig. 5–4*). The axons of these cells form the spinal part of the accessory nerve. The **lumbosacral nucleus** present in the second lumbar down to the first sacral segments of the cord is made up of nerve cells whose axons have an unknown distribution.

The lateral group is present in the cervical and lumbosacral segments of the cord and is responsible for innervating the skeletal muscles of the limbs (*Fig. 5–4*).

Nerve Cell Groups in the Posterior Gray Columns. There are four nerve cell groups of the posterior gray column, two that extend throughout the length of the cord and two that are restricted to the thoracic and lumbar segments.

The **substantia gelatinosa group** is situated at the apex of the posterior gray column throughout the length of the spinal cord (*Fig. 5–4*). It receives afferent fibers concerned with pain, temperature, and touch from the posterior root.

The **nucleus proprius** is a group of large nerve cells situated anterior to the substantia gelatinosa throughout the spinal cord (*Fig. 5–4*). This nucleus constitutes the main bulk of cells present in the posterior gray column and receives fibers from the posterior white column that are associated with the senses of position and movement (proprioception), two-point discrimination, and vibration.

The **nucleus dorsalis (Clark's column)** is a group of nerve cells situated at the base of the posterior gray column and extending from the eighth cervical segment caudally to the third or fourth lumbar segment (*Fig. 5–4*). The majority of the cells are comparatively large and are associated with proprioceptive endings (neuromuscular spindles and tendon spindles).

The **visceral afferent nucleus** is a group of nerve cells of medium size situated lateral to the nucleus dorsalis; it extends from the first thoracic to the third lumbar segment of the spinal cord. It is believed to be associated with receiving visceral afferent information.

Nerve Cell Groups in the Lateral Gray Columns. The intermediolateral group of cells form the small lateral gray column, which extends from the first thoracic to the second or third lumbar segments of the spinal cord (*Fig. 5–4*). The cells are relatively small and give rise to preganglionic sympathetic fibers.

A similar group of cells found in the second, third, and fourth sacral segments of the spinal cord give rise to preganglionic parasympathetic fibers (*Fig. 5–4*).

Gray Commissure and Central Canal. In transverse sections of the spinal cord, the anterior and posterior gray columns on each side are seen to be connected by a transverse **gray commissure,** so that the gray matter resembles the letter H (*Fig. 5–4*). In the center of the gray commissure is situated the **central canal.**

The central canal is present throughout the spinal cord (*Fig. 5–4*). Superiorly, it is continuous with the central canal of the caudal half of the medulla oblongata and above this it opens into the cavity of the fourth ventricle. Inferiorly, in the conus medullaris, it expands into the fusiform **terminal ventricle** and terminates below within the root of the filum terminale. It is filled with cerebrospinal fluid and is lined with ciliated columnar epithelium, the **ependyma.** Thus, the central canal is closed inferiorly and opens superiorly into the fourth ventricle.

Organization of the White Matter

The white matter, for purposes of description, may be divided into **anterior, lateral,** and **posterior white columns** (*Figs. 5–2 and 5–4*). The anterior column on each side lies between the midline and the point of emergence of the anterior nerve roots; the lateral column lies between the emergence of the anterior nerve roots and the entry of the posterior nerve roots; and the posterior column lies between the entry of the posterior nerve roots and the midline.

Arrangement of Nerve Fiber Tracts. The arrangement of the nerve fiber tracts within the spinal cord has

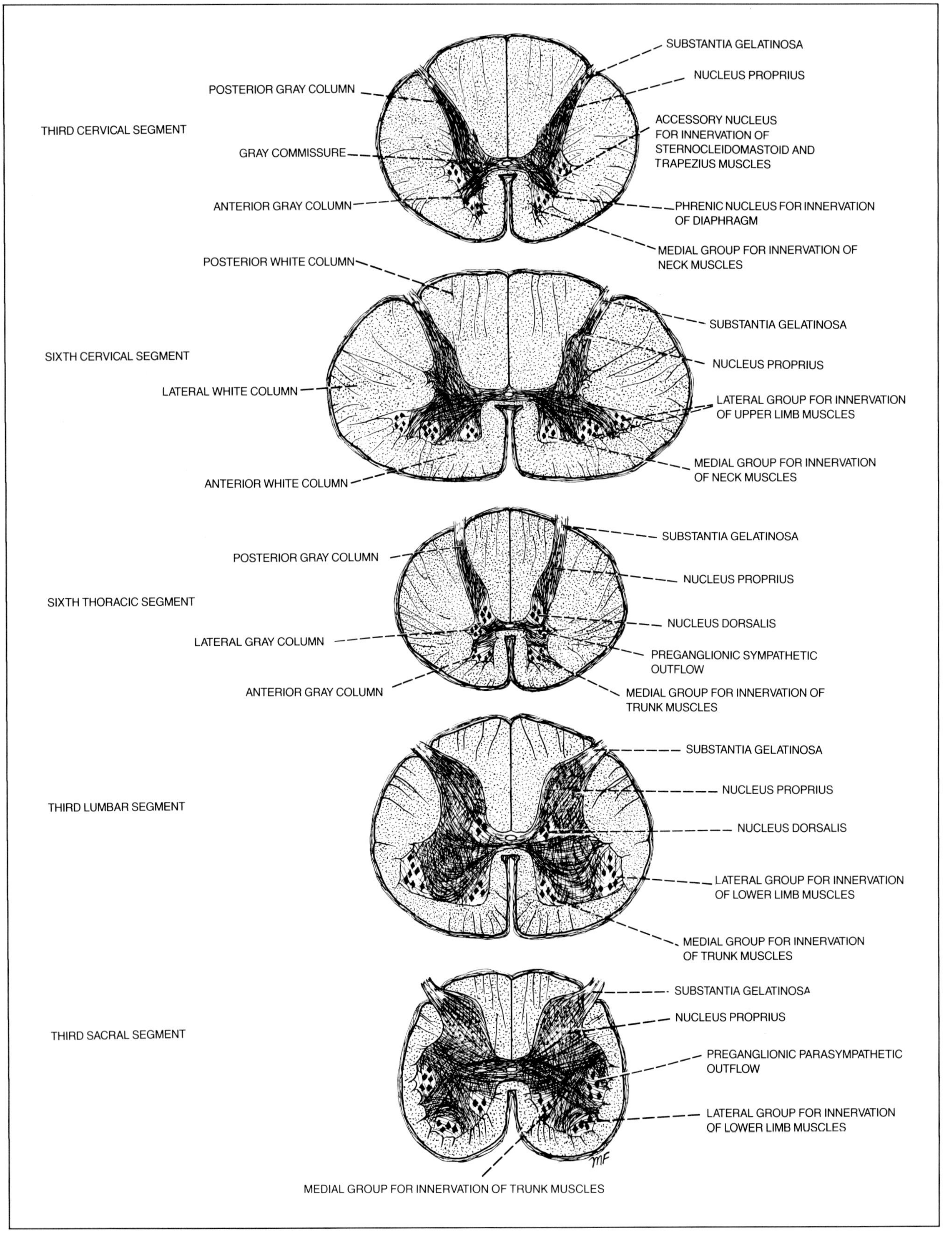

Figure 5–4. Drawings of the transverse sections of the spinal cord at different levels, showing the arrangement of the gray matter and the white matter.

been deduced as the result of animal experimentation and study of the human spinal cord for degenerative nerve fibers resulting from injury or disease. Although some nerve tracts are concentrated in certain areas of the white matter, it is now generally accepted that considerable overlap is present. For purposes of description, the spinal tracts are divided into ascending, descending, and intersegmental tracts and their relative positions in the white matter are described later. A simplified diagram, showing the general arrangement of the major tracts, is shown in Figure 5–5.

Ascending Tracts of the Spinal Cord

Pain and Temperature Pathways

LATERAL SPINOTHALAMIC TRACT. The axons entering the spinal cord from the posterior root ganglion proceed to the tip of the posterior gray column and divide into ascending and descending branches (*Fig. 5–6*). These branches travel a distance of one or two segments in the spinal cord and form the **posterolateral tract of Lissauer** (*Fig. 5–6*). The fibers of the first-order neurons terminate by synapsing with cells in the substantia gelatinosa in the posterior gray column (*Fig. 5–6*). Substance P, a peptide, is thought to be the neurotransmitter at these synapses.

The axons of the second-order neurons now cross obliquely to the opposite side in the anterior gray and white commissures within one spinal segment of the cord and ascend in the contralateral white column as the lateral spinothalamic tract (*Fig. 5–6*). As the lateral spinothalamic tract ascends through the spinal cord, new fibers are added to the anteromedial aspects of the tract, therefore, in the upper cervical segments of the cord, the sacral fibers are posterolateral and the cervical segments are anteromedial. The fibers carrying pain are situated slightly anterior to those conducting temperature.

As the lateral spinothalamic tract ascends through the medulla oblongata, it is accompanied by the anterior spinothalamic tract and the spinotectal tract, which all together form the **spinal lemniscus** (*Fig. 5–6*).

The spinal lemniscus continues to ascend through the posterior part of the pons (*Fig. 5–6*). Many of the fibers of the lateral spinothalamic tract end by synapsing with the third-order neurons in the ventral posterolateral nucleus of the thalamus (*Fig. 5–6*). It is believed that here crude pain and temperature sensations are appreciated and emotional reactions are initiated.

The axons of the third-order neurons in the ventral posterolateral nucleus of the thalamus now pass through the posterior limb of the **internal capsule** and the **corona radiata** to reach the somesthetic area in the postcentral gyrus of the cerebral cortex (*Fig. 5–6*). The contralateral half of the body is represented as inverted, with the hand and mouth situated inferiorly and the leg situated superiorly, and with the foot and anogenital region on the medial surface of the hemisphere. From here, the information is transmitted to other regions of the cerebral cortex to be used by motor areas and the parietal association area. The role of the cerebral cortex is the interpretation of the sensory information at the level of consciousness. (For futher details concerning the pain pathway, see Chapter 9.)

Light Touch and Pressure Pathways

ANTERIOR SPINOTHALAMIC TRACT. Axons enter the spinal cord from the posterior root ganglion and proceed to the tip of the posterior gray column, where they divide into ascending and descending branches (*Fig. 5–7*). These branches travel for a distance of one or two segments of the spinal cord, contributing to the **posterolateral tract of Lissauer.** These fibers of the first-order neurons terminate by synapsing with cells in the substantia gelatinosa (*Fig. 5–7*).

The axons of the second-order neurons now cross very obliquely to the opposite side in the anterior gray and white commissures within several spinal segments, and ascend in the opposite anterolateral white column as the anterior spinothalamic tract (*Fig. 5–7*). As the anterior spinothalamic tract ascends through the spinal cord, new fibers are added to the medial aspects of the tract, therefore, in the upper cervical segments of the cord, the sacral fibers are mostly lateral and the cervical segments are mostly medial.

The anterior spinothalamic tract continues to ascend through the posterior part of the pons and the tegmentum of the midbrain and it terminates by synapsing with the third-order neurons in the ventral posterolateral nucleus of the thalamus (*Fig. 5–7*). Crude awareness of touch and pressure is believed to be appreciated here.

The axons of the third-order neurons in the ventral posterolateral nucleus of the thalamus pass through the posterior limb of the **internal capsule** (*Fig. 5–7*) and the **corona radiata** to reach the somesthetic area in the postcentral gyrus of the cerebral cortex. The contralateral half of the body is represented inverted, with the hand and mouth situated inferiorly, as described previously. The conscious appreciation of touch and pressure depends on the activity of the cerebral cortex.

Discriminative Touch, Vibratory Sense, and Conscious Muscle Joint Sense Pathways

POSTERIOR WHITE COLUMN: FASCICULUS GRACILIS AND FASCICULUS CUNEATUS. Axons enter the spinal cord from the posterior root ganglion and pass directly to the posterior white column on the same side (*Fig. 5–8*). Here, the fibers divide into long ascending and short descending branches. The descending branches pass down a variable number of segments, giving off collateral branches that

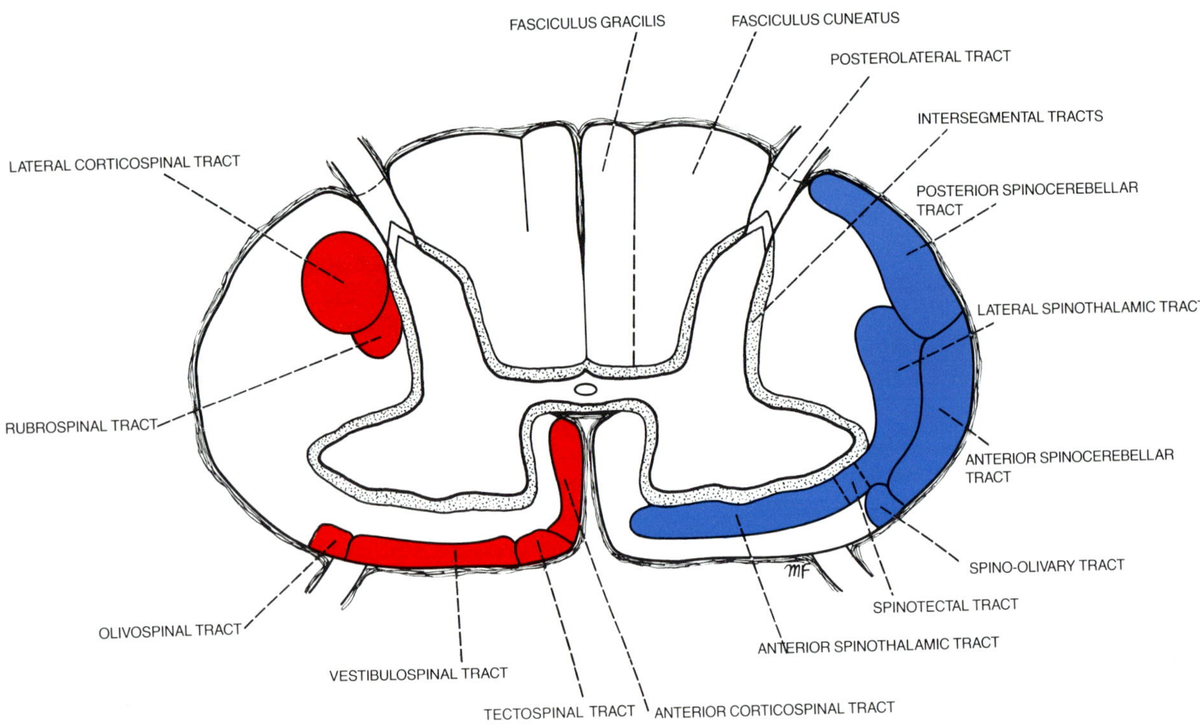

Figure 5–5. Transverse section of spinal cord showing the general arrangement of the ascending tracts on the right side and the descending tracts on the left side.

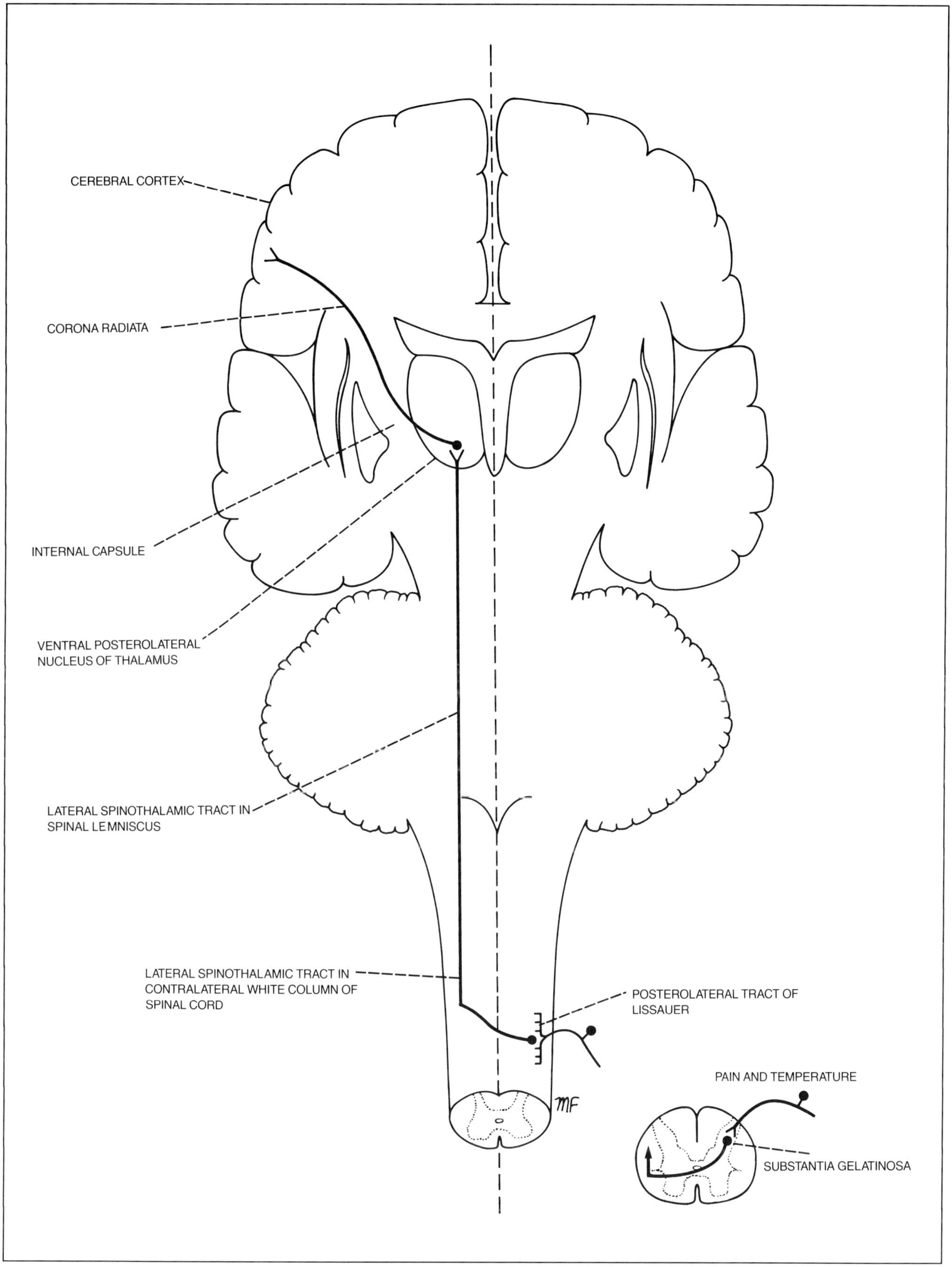

Figure 5–6. Pain and temperature pathways.

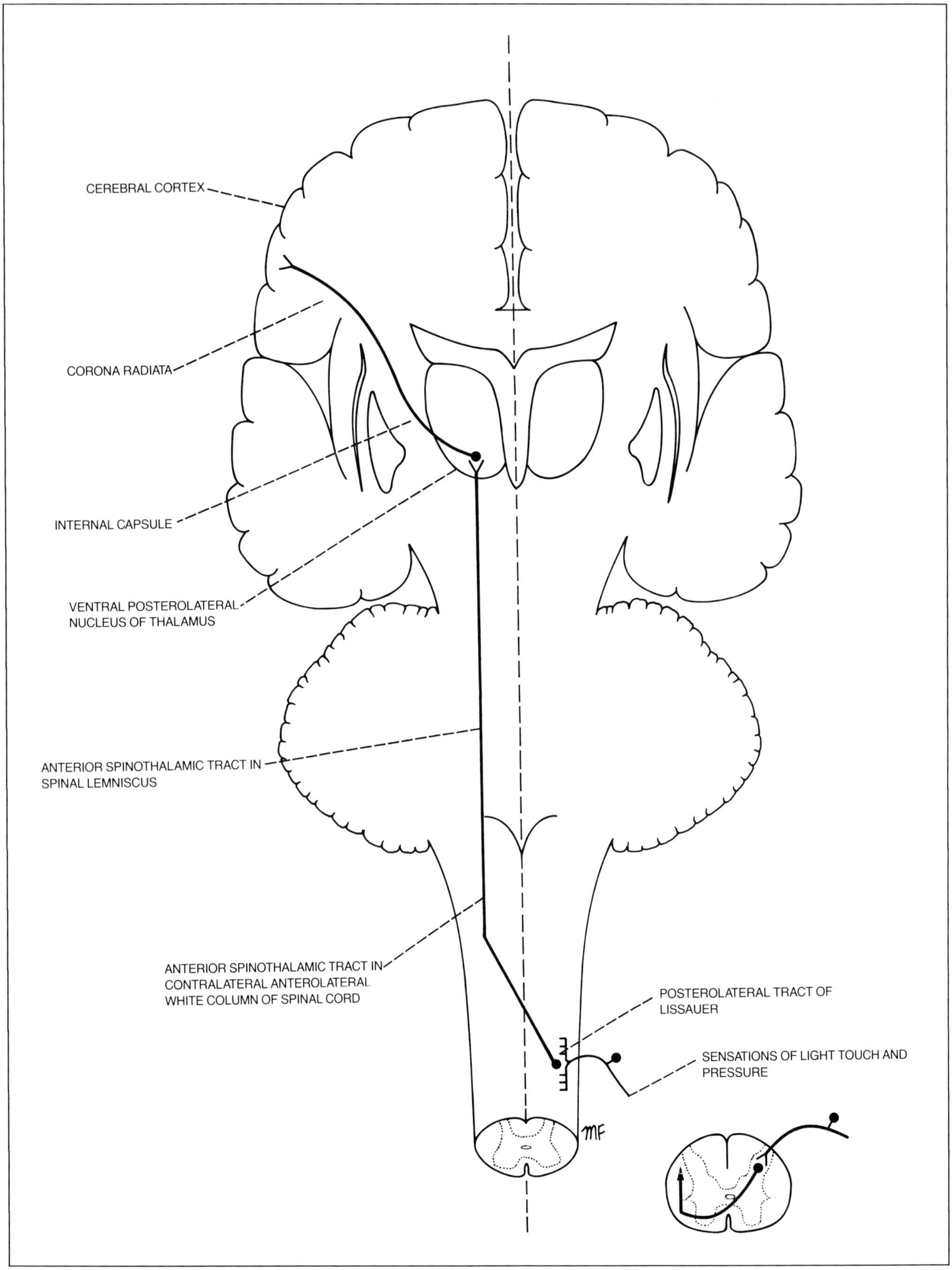

Figure 5–7. Light touch and pressure pathways.

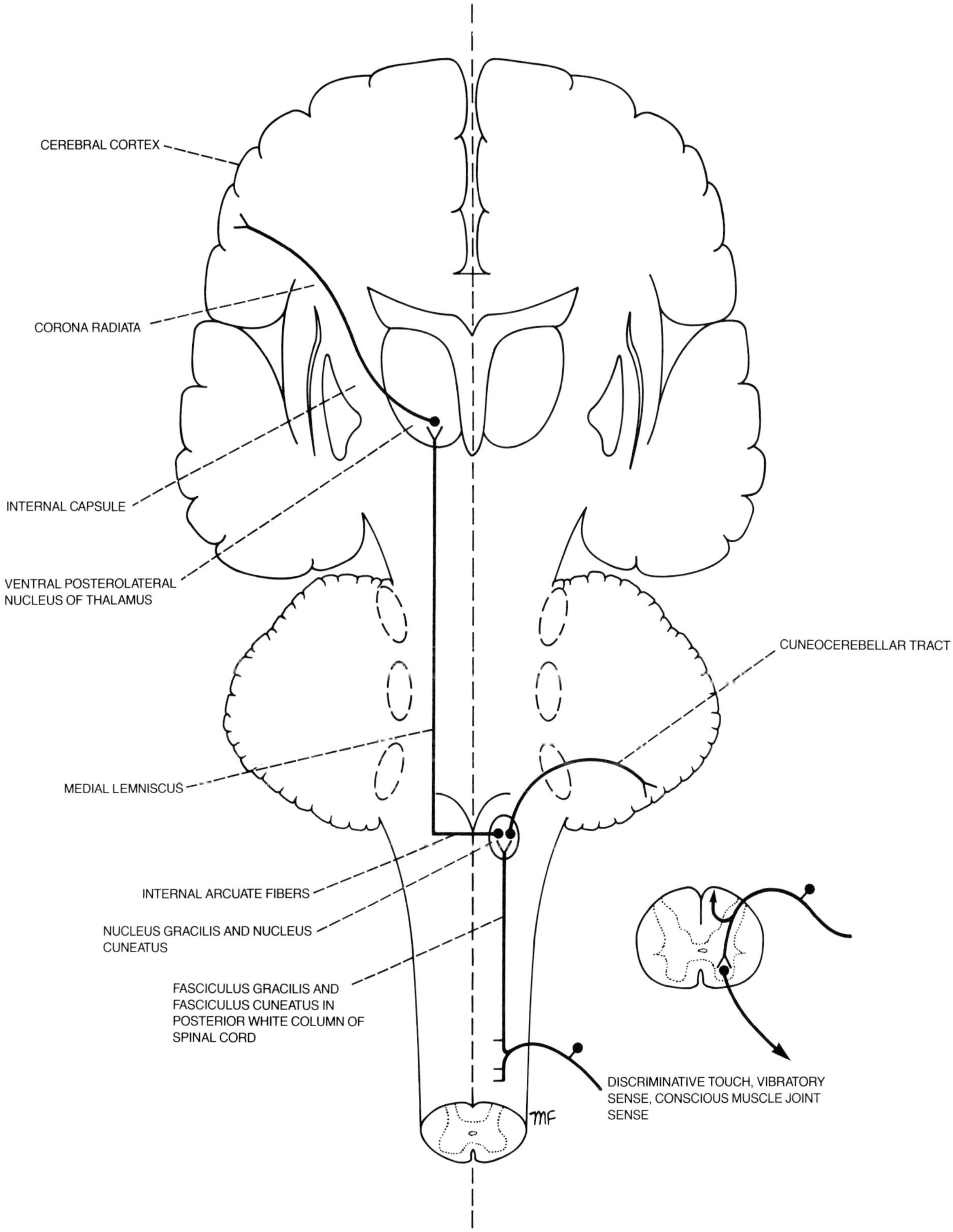

Figure 5–8. Discriminative touch, vibratory sense, and conscious muscle joint sense pathways.

synapse with cells in the posterior gray horn, internuncial neurons (interneurons), and anterior horn cells (*Fig. 5–8*). It is clear that these short descending fibers are involved with intersegmental reflexes.

The long ascending fibers may also end by synapsing with cells in the posterior gray horn, internuncial neurons, and anterior horn cells. This distribution may extend over numerous segments of the spinal cord (*Fig. 5–8*). As in the case of the short descending fibers, they are involved with intersegmental reflexes.

Many of the long ascending fibers travel upward in the posterior white column as the **fasciculus gracilis and fasciculus cuneatus** (*Fig. 5–8*). The fasciculus gracilis is present throughout the length of the spinal cord and contains the long ascending fibers from the sacral, lumbar, and lower six thoracic spinal nerves. The fasciculus cuneatus is situated laterally in the upper thoracic and cervical segments of the spinal cord and is separated from the fasciculus gracilis by a septum. The fasciculus cuneatus contains the long ascending fibers from the upper six thoracic and all the cervical spinal nerves.

The fibers of the fasciculus gracilis and the fasciculus cuneatus ascend ipsilaterally and terminate by synapsing on the second-order neurons in the **nuclei gracilis and cuneatus** of the medulla oblongata (*Fig. 5–8*). The axons of the second-order neurons, called the **internal arcuate fibers,** sweep anteromedially around the central gray matter and cross the median plane decussating with the corresponding fibers of the opposite side in the **sensory decussation** (*Fig. 5–8*). The fibers then ascend as a single compact bundle, the **medial lemniscus,** through the medulla oblongata, the pons, and the midbrain (*Fig. 5–8*). The fibers terminate by synapsing on the third-order neurons in the ventral posterolateral nucleus of the thalamus.

The axons of the third-order neurons leave and pass through the posterior limb of the **internal capsule** and **corona radiata** to reach the somesthetic area in the postcentral gyrus of the cerebral cortex (*Fig. 5–8*). The contralateral half of the body is represented inverted, with the hand and mouth situated inferiorly, as described previously. In this manner, the discriminative touch, vibratory sense, and muscle joint sense are consciously appreciated.

It should be mentioned that a number of fibers in the fasciculus cuneatus from the cervical and upper thoracic segments, having terminated on the second-order neurons of the nucleus cuneatus, are relayed and travel as the axons of the second-order neurons to enter the cerebellum through the inferior cerebellar peduncle of the same side (*Fig. 5–8*). The pathway is referred to as the **cuneocerebellar tract** and the fibers are known as the **posterior external arcuate fibers.** The function of these fibers is to convey information of muscle joint sense to the cerebellum.

Muscle Joint Sense Pathways to the Cerebellum

POSTERIOR SPINOCEREBELLAR TRACT. The axons entering the spinal cord from the posterior root ganglion enter the posterior gray column and terminate by synapsing on the second-order neurons at the base of the posterior gray column (*Fig. 5–9*). These neurons are known collectively as the **nucleus dorsalis (Clark's column).** The axons of the second-order neurons enter the posterolateral part of the lateral white column on the same side and ascend as the posterior spinocerebellar tract to the medulla oblongata. Here, the tract joins the inferior cerebellar peduncle and terminates in the cerebellar cortex (*Fig. 5–9*). Note that it does not ascend to the cerebral cortex.

The posterior spinocerebellar fibers receive muscle joint information, from the muscle spindles, tendon organs, and joint receptors of the trunk and the lower limbs. This information is used by the cerebellum in the coordination of limb movements and in the maintenance of posture.

ANTERIOR SPINOCEREBELLAR TRACT. The axons entering the spinal cord from the posterior root ganglion terminate by synapsing with the second-order neurons in the **nucleus dorsalis** at the base of the anterior gray column (*Fig. 5–9*). The majority of the axons of the second-order neurons cross to the opposite side and ascend as the anterior spinocerebellar tract in the contralateral white column; the rest ascend as the ipsilateral anterior spinocerebellar tract (*Fig. 5–9*). The fibers, having gone through the medulla oblongata and the pons, enter the cerebellum through the superior cerebellar peduncle and terminate in the cerebellar cortex. It is believed that those fibers that crossed over to the opposite side in the spinal cord cross back within the cerebellum (*Fig. 5–9*). The anterior spinocerebellar tract conveys muscle joint information from the muscle spindles, tendon organs, and joint receptors of the trunk and the upper and lower limbs. It is also believed that the cerebellum receives information from the skin and superficial fascia by this tract.

CUNEOCEREBELLAR TRACT. The cuneocerebellar tract fibers have been described previously. They originate in the nucleus cuneatus and enter the cerebellum through the inferior cerebellar peduncle of the same side (*Fig. 5–8*). The fibers are known as the **posterior external arcuate fibers** and their function is to convey information about muscle joint sense to the cerebellum.

Other Ascending Pathways

SPINOTECTAL TRACT. The axons enter the spinal cord from the posterior root ganglion and travel to the gray matter where they synapse on as yet unknown second-order

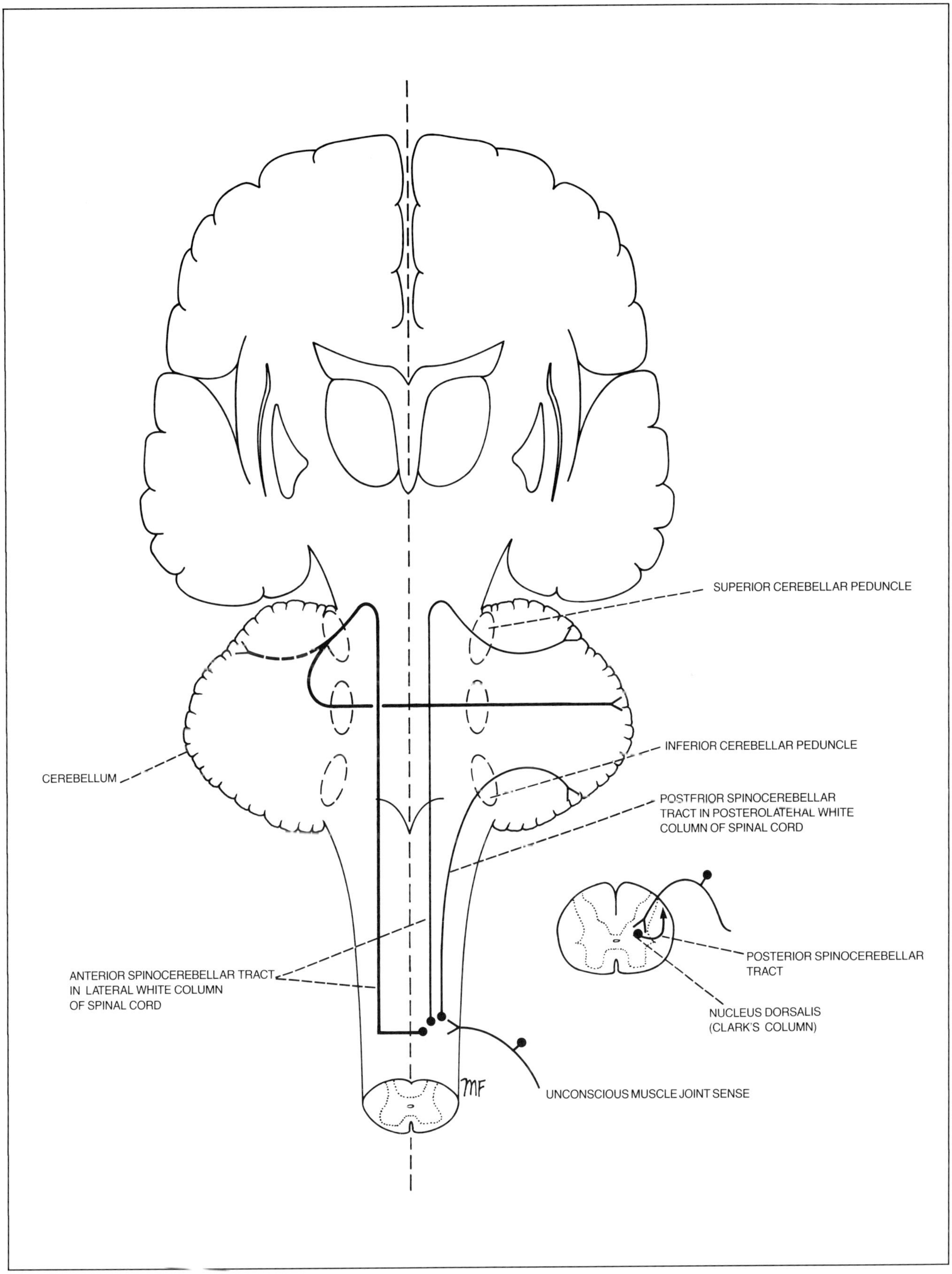

Figure 5–9. Muscle joint sense pathways to the cerebellum.

neurons (*Fig. 5–10*). The axons of the second-order neurons cross the median plane and ascend as the spinotectal tract in the anterolateral white column lying close to the lateral spinothalamic tract. After they pass through the medulla oblongata and the pons, they terminate by synapsing with neurons in the **superior colliculus of the midbrain** (*Fig. 5–10*). This pathway provides afferent information for spinovisual reflexes and brings about movements of the eyes and the head toward the source of the stimulation.

SPINORETICULAR TRACT. The axons enter the spinal cord from the posterior root ganglion and terminate on as yet unknown second-order neurons in the gray matter (*Fig. 5–10*). The axons from these second-order neurons ascend the spinal cord as the spinoreticular tract in the lateral white column mixed with the lateral spinothalamic tract. The majority of the fibers are uncrossed and terminate by synapsing with neurons of the reticular formation in the medulla oblongata, pons, and midbrain (*Fig. 5–10*). The spinoreticular tract provides an afferent pathway for the reticular formation, which plays an important role in influencing levels of consciousness.

SPINO-OLIVARY TRACT. The axons enter the spinal cord from the posterior root ganglion and terminate on as yet unknown second-order neurons in the posterior gray column (*Fig. 5–10*). The axons from the second-order neurons cross the midline and ascend as the spino-olivary tract in the white matter at the junction of the anterior and lateral columns and end by synapsing on third-order neurons in the inferior olivary nuclei in the medulla oblongata (*Fig. 5–10*). The axons of the third-order neurons cross the midline and enter the cerebellum through the inferior cerebellar peduncle. The spino-olivary tract conveys information to the cerebellum from cutaneous and proprioceptive organs.

Visceral Sensory Tract. Sensations that arise in viscera located in the thorax and the abdomen enter the spinal cord through the posterior roots. The cell bodies of the first-order neurons are situated in the posterior root ganglia. The peripheral processes of these cells receive nerve impulses from pain and stretch receptor endings in the viscera. The central processes, having entered the spinal cord, synapse with second-order neurons in the gray matter, probably in the posterior or lateral gray column.

The axons of the second-order neurons are believed to join the spinothalamic tracts and ascend and terminate on the third-order neurons in the ventral posterolateral nucleus of the thalamus. The final destination of the axons of the third-order neurons has not been established, but it may well be in the postcentral gyrus of the cerebral cortex.

It has been reported that sensations from a full rectum and a urinary bladder experienced before defecation and micturition are carried by ascending tracts located in the posterior white columns of the spinal cord.

Many of the visceral afferent fibers that enter the spinal cord branch and participate in reflex activity.

Descending Tracts of the Spinal Cord

The motor neurons situated in the anterior gray columns of the spinal cord send axons to innervate skeletal muscle through the anterior roots of the spinal nerves. These motor neurons are sometimes referred to as the **lower motor neurons** and constitute the final common pathway to the muscles.

The lower motor neurons are constantly bombarded by nervous impulses that descend from the medulla, the pons, the midbrain, and the cerebral cortex, as well as those that enter along sensory fibers from the posterior roots. The nerve fibers that descend in the white matter from different supraspinal nerve centers are segregated into nerve bundles called the **descending tracts.** These supraspinal neurons and their tracts sometimes are referred to as the **upper motor neurons** and they provide numerous separate pathways that can influence motor activity.

Corticospinal Tracts. Fibers of the corticospinal tract arise as axons of pyramidal cells situated in the fifth layer of the cerebral cortex (*Fig. 5–11*). About one-third of the fibers originate from the primary motor cortex (area 4), one-third from the secondary motor cortex (area 6), and one-third* from the parietal lobe (areas 3, 1, and 2); thus, two-thirds of the fibers arise from the precentral gyrus and one-third from the postcentral gyrus. Because electrical stimulation of different parts of the precentral gyrus produces movements of different parts of the opposite side of the body, we can represent the parts of the body in this area of the cortex. Note that the region controlling the face is situated inferiorly, and the one controlling the lower limb is situated superiorly and on the medial surface of the hemisphere. It is interesting to find that the majority of the corticospinal fibers are myelinated and are relatively slow-conducting, small fibers.

The descending fibers converge in the **corona radiata** and then pass through the posterior limb of the **internal capsule** (*Fig. 5–11*). Here, the fibers are organized so that those closest to the genu are related to cervical portions of the body, whereas those situated more posteriorly are related to the lower extremity. The tract then continues through the middle three-fifths of the **basis pedunculi of the midbrain** (*Fig. 5–11*). Here, the fibers concerned with cervical portions of the body are situated medially, whereas those affecting the leg are placed laterally.

* These fibers do not control motor activity but influence sensory input to the nervous system.

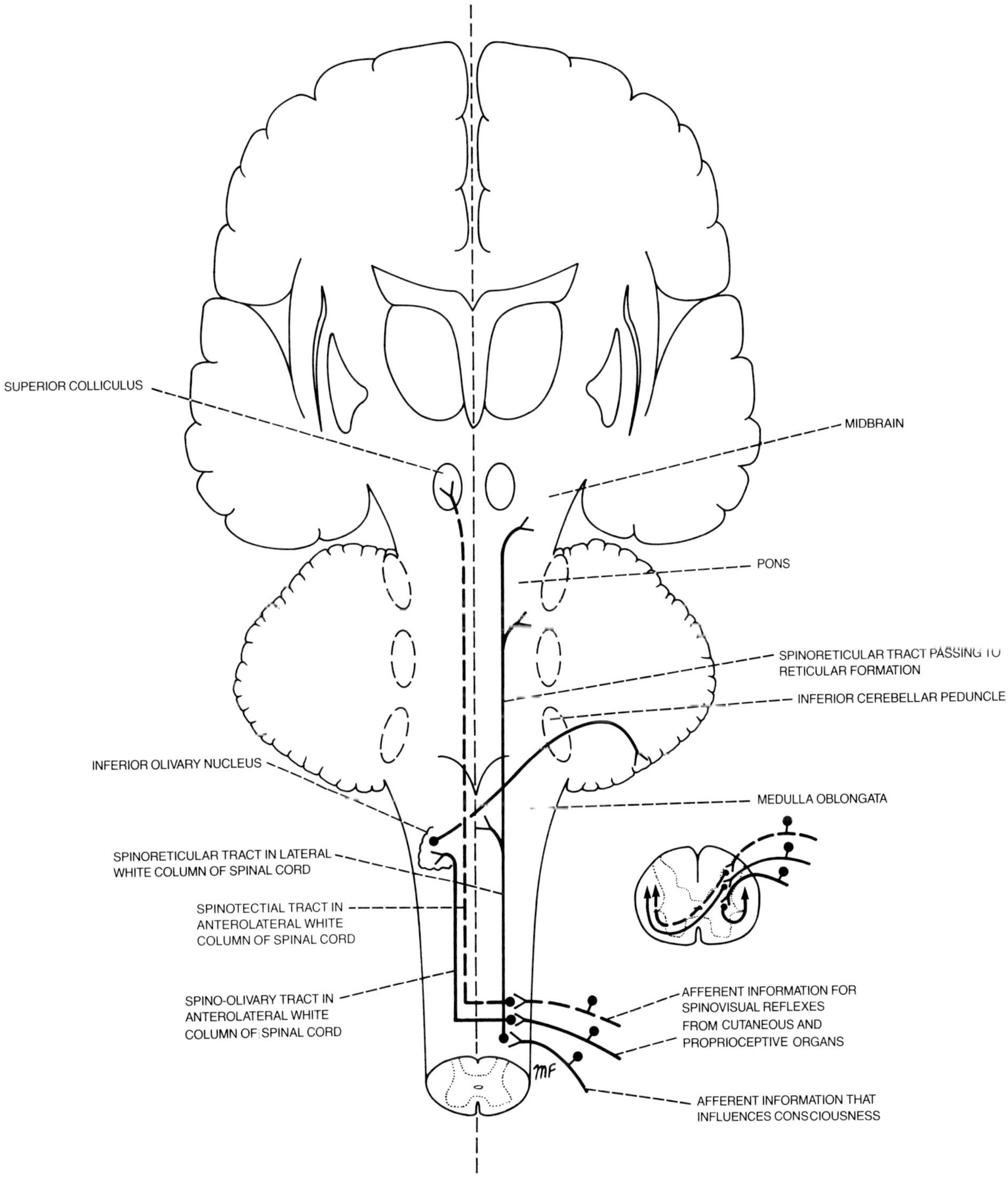

Figure 5–10. Spinotectal, spinoreticular, and spino-olivary tracts.

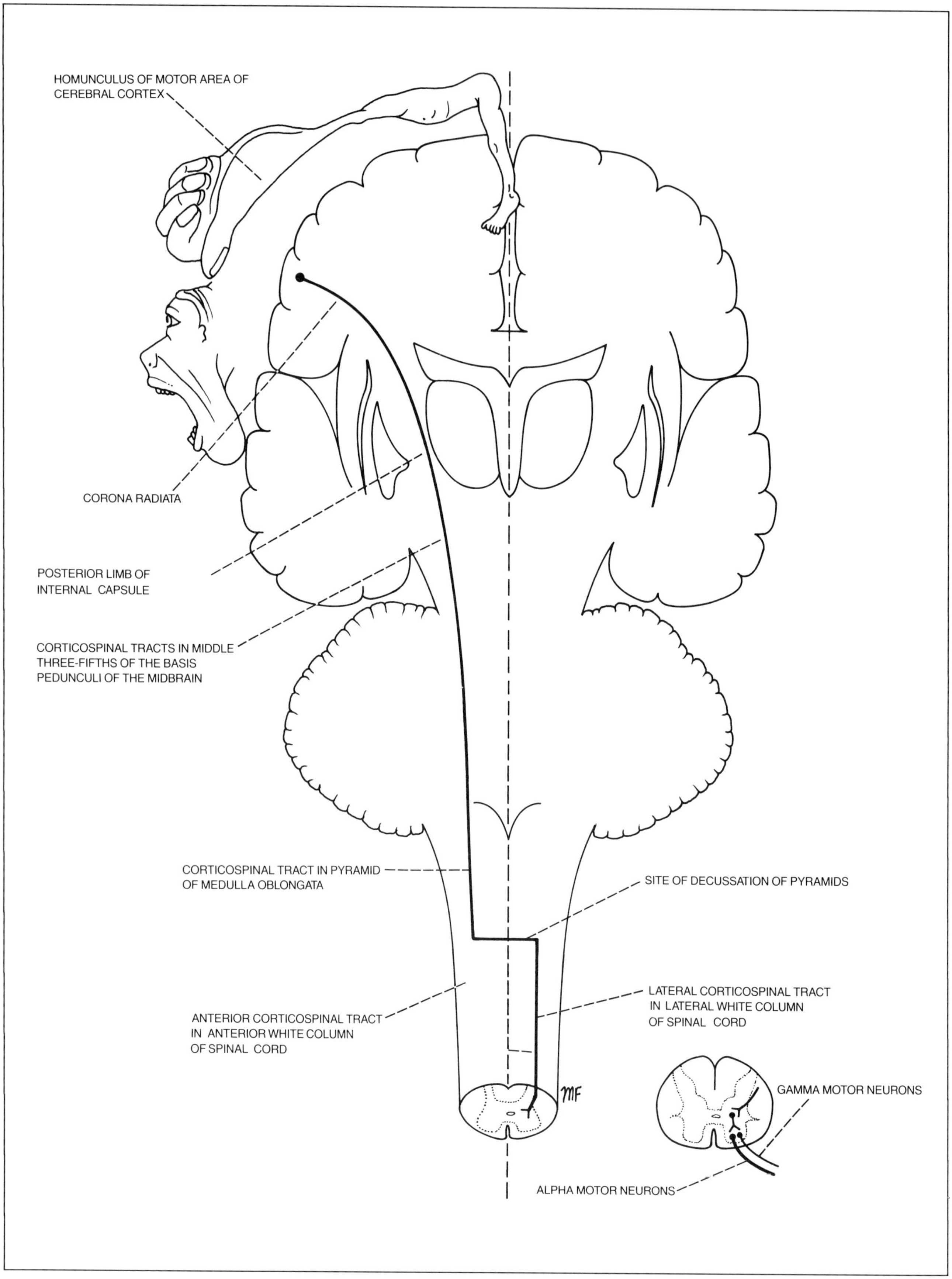

Figure 5–11. Corticospinal tracts.

On entering the pons, the tract is broken into many bundles by the **transverse pontocerebellar fibers.** In the medulla oblongata, the bundles become grouped together along the anterior border to form a swelling known as the **pyramid** (hence the alternative name, **pyramidal tract**). At the junction of the medulla oblongata and the spinal cord, the majority of the fibers cross the midline at the **decussation of the pyramids** (*Fig. 5–11*) and enter the lateral white column of the spinal cord to form the **lateral corticospinal tract** (*Fig. 5–11*). The remaining fibers do not cross in the decussation, but descend in the anterior white column of the spinal cord as the **anterior corticospinal tract** (*Fig. 5–11*). These fibers eventually cross the midline and terminate in the anterior gray column of the spinal cord segments in the cervical and upper thoracic regions.

The lateral corticospinal tract descends the length of the spinal cord and its fibers terminate in the anterior gray column of all the spinal cord segments.

The majority of the corticospinal fibers synapse, via internuncial neurons, with alpha motor neurons and some gamma motor neurons. Only the largest corticospinal fibers synapse directly with the motor neurons.

It is important to understand that the corticospinal tracts are not the sole pathway for serving voluntary movement. Rather, they form the pathway that confers speed and agility to voluntary movements and is thus used in performing rapid skilled movements. Many of the simple, basic voluntary movements are believed to be mediated by other descending tracts.

Reticulospinal Tracts. Throughout the midbrain, the pons, and the medulla oblongata groups of scattered nerve cells and nerve fibers exist that are collectively known as the **reticular formation.** From the pons, these neurons send axons, which are mostly uncrossed, down into the spinal cord and form the **pontine reticulospinal tract.** From the medulla, similar neurons send axons, which are crossed and uncrossed, to the spinal cord and form the **medullary reticulospinal tract.**

The reticulospinal fibers from the pons descend through the anterior white column, whereas those from the medulla oblongata descend in the lateral white column. Both sets of fibers enter the anterior gray columns of the spinal cord and may facilitate or inhibit the activity of the alpha and gamma motor neurons. By this means the reticulospinal tracts influence voluntary movements and reflex activity.

Tectospinal Tract. Fibers of the tectospinal tract arise from nerve cells in the **superior colliculus** of the midbrain. The majority of the fibers cross the midline soon after their origin and descend through the brainstem close to the **medial longitudinal fasciculus.** The tectospinal tract descends through the anterior white column of the spinal cord close to the anterior median fissure. The majority of fibers terminate in the anterior gray column in the upper cervical segments of the spinal cord by synapsing with internuncial neurons. These fibers are believed to be concerned with reflex postural movements in response to visual stimuli.

Rubrospinal Tract. The **red nucleus** is situated in the tegmentum of the midbrain at the level of the superior colliculus. The axons of neurons in this nucleus cross the midline at the level of the red nucleus and descend as the rubrospinal tract through the pons and the medulla oblongata to enter the lateral white column of the spinal cord. The fibers terminate by synapsing with internuncial neurons in the anterior gray column of the cord.

The neurons of the red nucleus receive afferent impulses through connections with the cerebral cortex and the cerebellum. This is believed to be an important indirect pathway by which the cerebral cortex and the cerebellum can influence the activity of the alpha and gamma motor neurons of the spinal cord. The tract facilitates the activity of the flexor muscles and inhibits the activity of the extensor or antigravity muscles.

Vestibulospinal Tract. The **vestibular nuclei** are situated in the pons and the medulla oblongata beneath the floor of the fourth ventricle. The vestibular nuclei receive afferent fibers from the inner ear through the vestibular nerve and from the cerebellum. The neurons of the lateral vestibular nucleus give rise to the axons that form the vestibulospinal tract, which descends uncrossed in the medulla and through the length of the spinal cord in the anterior white column. The fibers terminate by synapsing with internuncial neurons of the anterior gray column of the spinal cord.

The inner ear and the cerebellum, by means of this tract, facilitate the activity of the extensor muscles and inhibit the activity of the flexor muscles to maintain balance.

Descending Autonomic Fibers. The higher centers of the central nervous system associated with the control of autonomic activity are situated in the cerebral cortex, the hypothalamus, the amygdaloid complex, and the reticular formation. Although distinct tracts have not been recognized, it is known from studies of spinal cord lesions, that descending autonomic tracts do exist and probably form part of the reticulospinal tract.

The fibers arise from neurons in the higher centers and cross the midline in the brainstem. They are believed to descend in the lateral white column of the spinal cord and terminate by synapsing on the autonomic motor cells in the lateral gray columns in the thoracic, the upper lumbar (sympathetic outflow), and the midsacral (parasympathetic) levels of the spinal cord.

Intersegmental Tracts. Short ascending and descending tracts, which originate and end within the spinal cord, exist in the anterior, lateral, and posterior white columns. The function of these pathways is to interconnect the neurons of different segmental levels, and they are particularly important in intersegmental spinal reflexes.

Reflex Arc. A **reflex** may be defined as an involuntary response to a stimulus. It depends on the integrity of the reflex arc (*Fig. 5–12*). In its simplest form, a reflex arc consists of the following anatomical structures: (1) a receptor organ, (2) an afferent neuron, (3) an effector neuron, and (4) an effector organ. Such a reflex arc involving only one synapse is referred to as a **monosynaptic reflex arc.** Interruption of the reflex arc at any point along its course would abolish the response.

In the spinal cord, reflex arcs play an important role in maintaining muscle tone, which is the basis of body posture. The receptor organ is situated in the skin, the muscle, or the tendon. The cell body of the afferent neuron is located in the posterior root ganglion, and the central axon of this first-order neuron terminates by synapsing on the effector neuron. Because the afferent fibers are of large diameter and are rapidly conducting, and because of the presence of only one synapse, a very quick response is possible.

Physiological studies of the electrical activity of the effector neuron show that after the very quick monosynaptic discharge, there is a prolonged asynchronous discharge. The reason for this later discharge is that the afferent fibers entering the spinal cord frequently branch and synapse with many internuncial neurons, which ultimately synapse with the effector neuron (*Fig. 5–12*). These additional neuronal circuits prolong the bombardment of the effector neurons after the initial stimulation by the afferent neuron has ceased. The presence of internuncial neurons also results in the spread of the afferent stimulus to neurons at different segmental levels of the spinal cord.

To understand reflex skeletal muscle activity, it is important to know the **law of reciprocal innervation** (*Fig. 5–12*). Simply stated, it means that the flexor and extensor reflexes of the same limb cannot be made to contract simultaneously. Therefore, the afferent nerve fibers responsible for flexor reflex muscle action must have branches that synapse with the extensor motor neurons of the same limb, causing them to be inhibited.

Another interesting property of spinal reflexes should be pointed out. The evocation of a reflex on one side of the body causes opposite effects on the limb of the other side of the body. This **crossed extensor reflex** (*Fig. 5–12*) may be demonstrated as follows. Afferent stimulation of the reflex arc that causes the ipsilateral limb to flex results in the contralateral limb being extended.

Influences of Higher Neuronal Centers on the Activities of Spinal Reflexes. The spinal segmental reflex arc involving motor activity is greatly influenced by higher centers in the brain. These influences are mediated through the corticospinal, reticulospinal, tectospinal, rubrospinal, and vestibulospinal tracts. In the clinical condition known as spinal shock, which follows the sudden removal of these influences by severance of the spinal cord, the segmental spinal reflexes are depressed. When the spinal shock disappears in a few weeks, the segmental spinal reflexes return and the muscle tone is increased. This so-called **decerebrate rigidity** is caused by the overactivity of the gamma efferent nerve fibers to the muscle spindles, which results from the release of these neurons from the higher centers. The next stage may be **paraplegia in extension** with domination of the increased tone of the extensor muscles over the flexor muscles. Some neurologists believe that this condition is caused by incomplete severance of all the descending tracts with persistence of the vestibulospinal tract. Should all the descending tracts be severed, the condition of **paraplegia in flexion** occurs. In this condition, the reflex responses are flexor in nature and the tone of the extensor muscles is diminished.

Renshaw Cells and Lower Motor Neuron Inhibition. Lower motor neuron axons give off collateral branches as they pass through the white matter to reach the anterior roots of the spinal nerve. These collaterals synapse on neurons, described by Renshaw, which in turn synapse on the lower motor neurons (*Fig. 5–12*). These internuncial neurons are believed to provide feedback on the lower motor neurons, inhibiting their activity.

Blood Supply of the Spinal Cord

Arterial Supply. The spinal cord receives its arterial supply from three small longitudinally running arteries that are reinforced by small segmentally arranged arteries that enter the vertebral canal through the intervertebral foramina. These vessels anastomose on the surface of the cord and send branches into the substance of the white and gray matter. Considerable variation exists as to the size and segmental levels at which the reinforcing arteries occur.

Posterior Spinal Arteries. The posterior spinal arteries arise either directly from the vertebral arteries inside the skull or indirectly from the posterior inferior cerebellar arteries. Each artery descends on the posterior surface of the spinal cord close to the posterior nerve roots and gives off branches that enter the substance of the cord (*Fig. 5–13*). The posterior spinal arteries supply the posterior one-third of the spinal cord.

It should be noted that the posterior spinal arteries

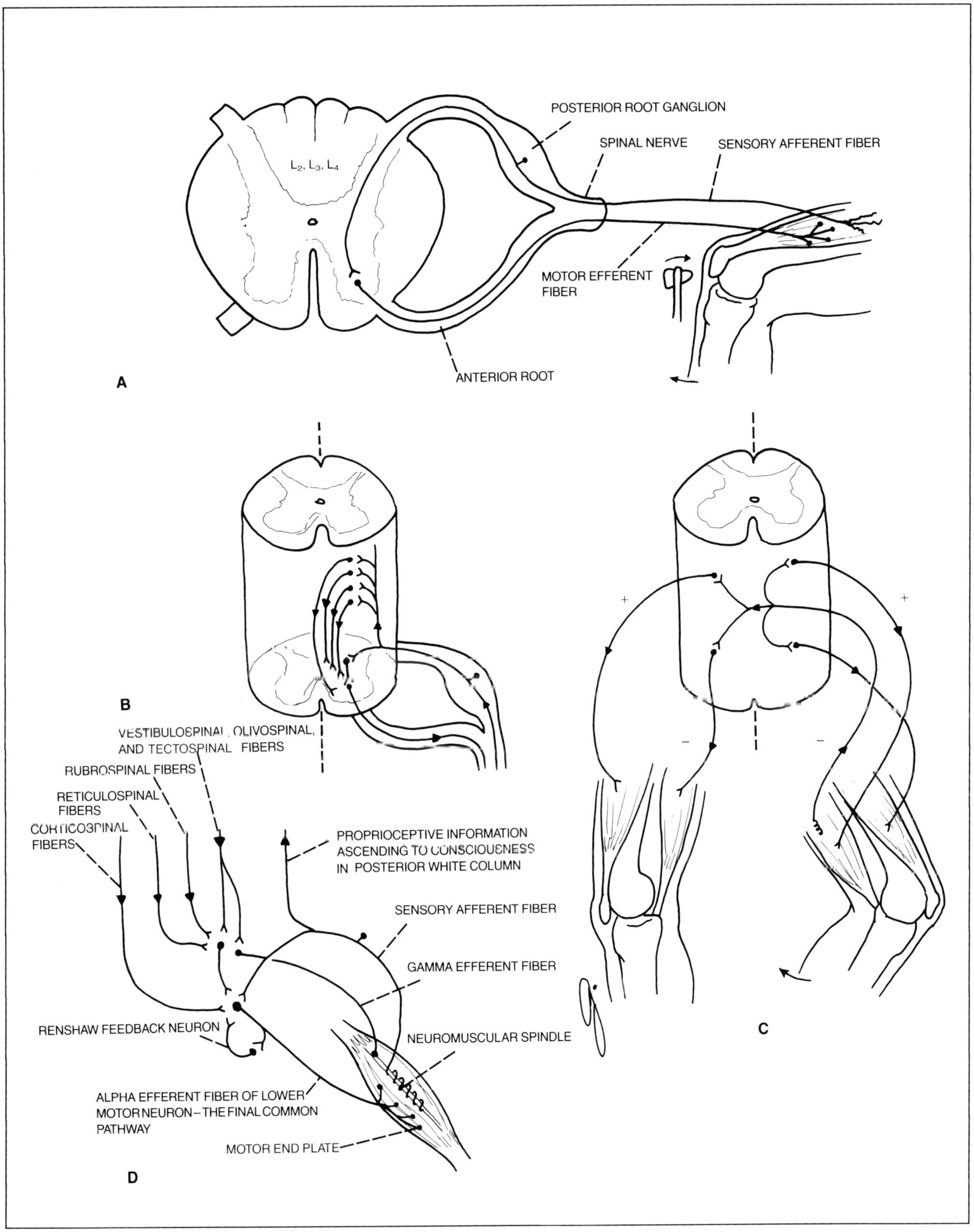

Figure 5–12. A. A simple monosynaptic reflex arc. **B.** Multiple branching of afferent fibers entering the spinal cord and the presence of many internuncial neurons that synapse with the effector neuron. **C.** Reciprocal innervation and the crossed extensor reflex. **D.** Multiple neurons synapsing with the lower motor neuron. A Renshaw feedback neuron is also shown.

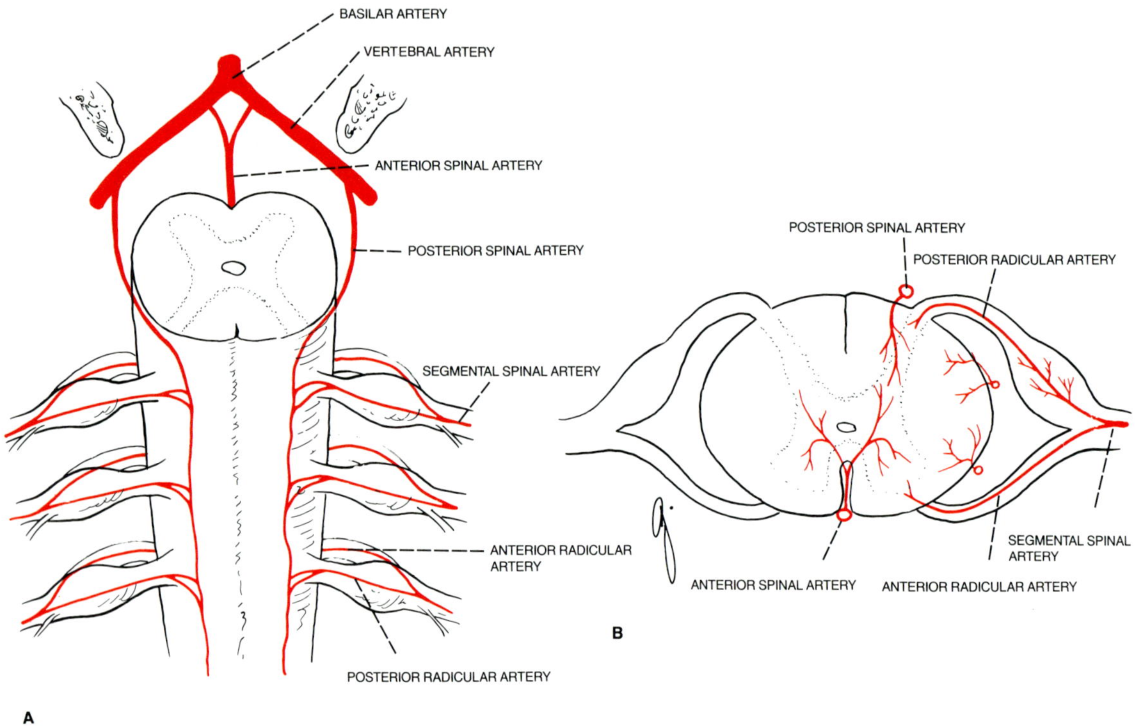

Figure 5–13. A. Arterial supply of spinal cord showing formation of two posterior and one anterior spinal arteries. **B.** Transverse section of the spinal cord showing the segmental spinal arteries and the radicular arteries.

are small in the upper thoracic region and that the first three thoracic segments of the spinal cord are particularly vulnerable to ischemia should the segmental or radicular arteries in this region be occluded.

Anterior Spinal Arteries. The anterior spinal arteries arise from the two vertebral arteries within the skull and unite to form a single artery. The anterior spinal artery then descends within the anterior median sulcus (*Fig. 5–13*). The artery, although larger than the posterior spinal arteries, varies in diameter as it descends, being smallest in the mid-thoracic region. Branches from the anterior spinal artery enter the substance of the cord and supply the anterior two-thirds of the spinal cord.

In the upper and lower thoracic segments of the spinal cord the anterior spinal artery may be extremely small. Should the segmental or radicular arteries be occluded in these regions, the fourth thoracic and the first lumbar segments of the spinal cord would be particularly liable to ischemic necrosis.

Segmental Spinal Arteries. At each intervertebral foramen the longitudinally running posterior and anterior spinal arteries are reinforced by small segmental arteries on both sides (*Fig. 5–13*). In the neck, the arteries are derived from the vertebral arteries, and from the ascending cervical branch of the inferior thyroid artery or from the deep cervical branch of the costocervical trunk.

In the thoracic region, they arise from the posterior intercostal branches of the descending aorta. In the lumbar region, they arise from the lumbar arteries, and in the sacral region, from the lateral sacral or middle sacral arteries. Once inside the vertebral canal, the sacral radicular arteries ascend to the conus medullaris along the roots of the cauda equina.

Having entered the vertebral canal, each segmental spinal artery gives rise to **anterior and posterior radicular arteries** that accompany the anterior and posterior nerve roots to the spinal cord. Additional feeder branches anastomose with the anterior and posterior spinal arteries. The number and size of the feeder arteries vary considerably from one individual to another. One large and important feeder artery, the **great anterior medullary artery of Adamkiewicz** enters the vertebral canal in the lower thoracic or upper lumbar regions. It is unilateral and in the majority of persons enters the spinal cord from the left side. The importance of this artery lies in the fact that it may be the major source of blood to the lower two-thirds of the spinal cord.

Pial Plexus. Superficial branches from the various spinal arteries form an anastomosing plexus in the pia mater. Penetrating branches from the plexus supply the periphery of the spinal cord.

Venous Drainage. The veins of the spinal cord drain into six tortuous longitudinal channels that communicate superiorly within the skull with the veins of the brain and the venous sinuses. They drain mainly into the internal vertebral venous plexus.

The blood supply to the spinal cord is surprisingly meagre considering the importance of this nervous tissue. The longitudinally running arteries are of small and variable diameter and the reinforcing segmental arteries vary in number and in size. Ischemia of the spinal cord can easily follow minor damage to the arterial supply, as the result of regional anesthesia, pain block procedures, aortic surgery, or any operation in which severe hypotension occurs.

MENINGES OF THE SPINAL CORD

The spinal cord, like the brain, is covered by three meninges, the dura mater, the arachnoid mater, and the pia mater (*Fig. 5–14*).

Dura Mater

The dura mater is the most external membrane and is a dense fibrous sheet that encloses the spinal cord and the cauda equina. It is continuous above with the meningeal layer of dura covering the brain. At the foramen magnum, the meningeal layer of dura is firmly attached to the periosteal layer of the dura, thus closing off the spinal epidural space superiorly. Below, the spinal dura ends on the filum terminale at about the level of the lower border of the second sacral vertebra (*Fig. 5–14*). There is a slight variation in this level, from the middle of the first to the middle of the third sacral vertebra, the lower level being most common in infants. The dura mater extends along each nerve root and becomes continuous with connective tissue surrounding each spinal nerve **(epineurium)** at the intervertebral foramen. The inner surface of the dura mater is separated from the arachnoid mater by the potential **subdural space.** The dural sheath lies loosely in the vertebral canal, except in the lumbar region, where a few bands of fibrous tissue connect it to the posterior longitudinal ligament. The sheath is separated from the walls of the canal by the extradural space (epidural space). This contains loose fatty areolar tissue and the internal vertebral venous plexus.

Arachnoid Mater

The arachnoid mater is a delicate nonvascular impermeable membrane that lies within the dura and outside the pia (*Fig. 5–14*). It is separated from the dura by the subdural space that contains a thin film of tissue fluid. The arachnoid is separated from the pia mater by a wide space, the **subarachnoid space,** which is filled with **cerebrospinal fluid.** The arachnoid is continuous above through the foramen magnum with the arachnoid covering the brain. Inferiorly, it ends on the filum termi-

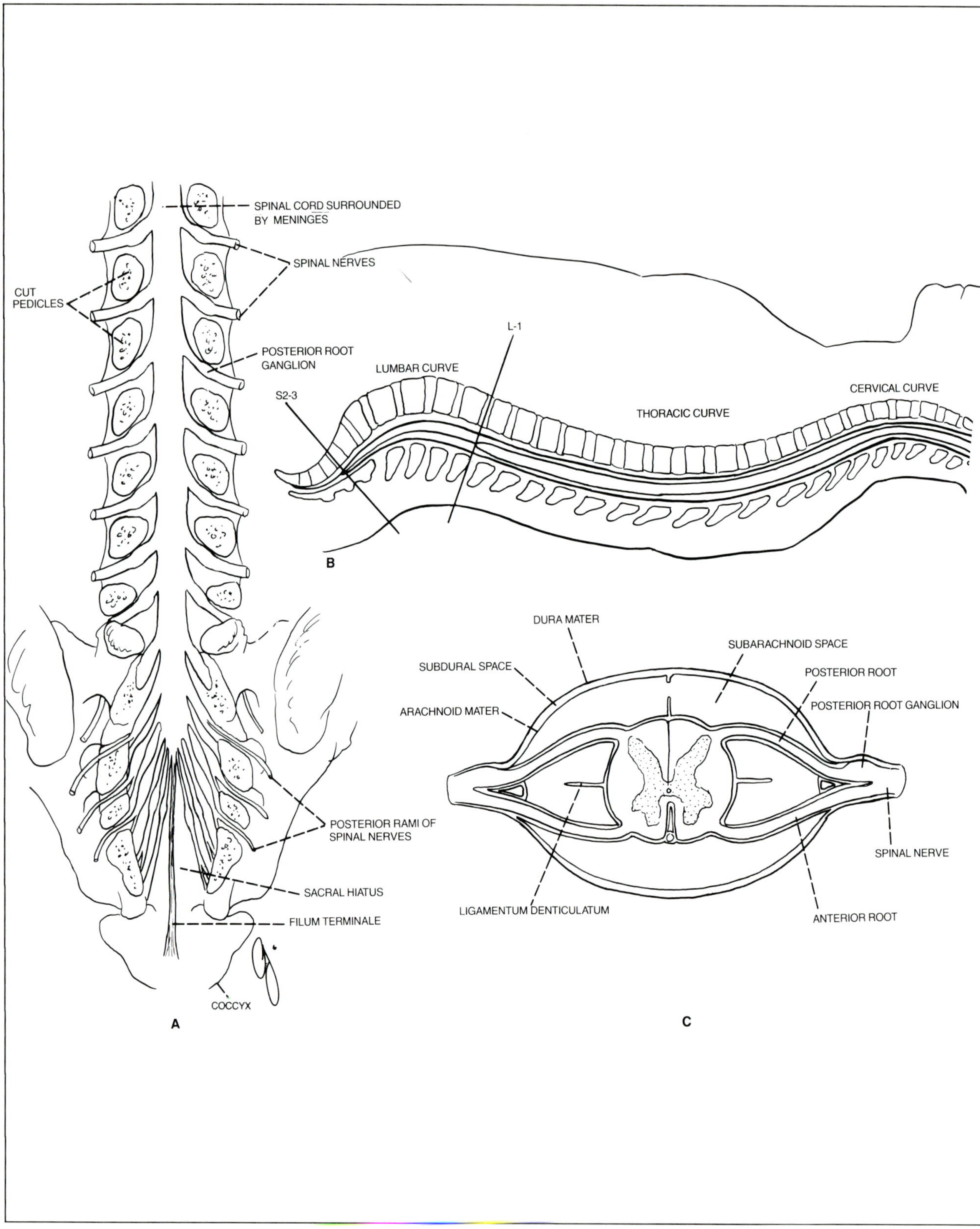

Figure 5–14. A. The dural sheath around the spinal cord and spinal nerves in the lumbosacral region; the laminae have been removed. **B.** Sagittal section of the spinal subarachnoid space showing its extent and curvatures in a patient in the reclining position. **C.** Cross section of the spinal cord showing the arrangement of the meninges. Note the lateral extensions of the subarachnoid space along the spinal nerve roots.

nale at about the level of the lower border of the second sacral vertebra. Between the levels of the conus medullaris and the lower end of the subarachnoid space lie the nerve roots of the cauda equina bathed in cerebrospinal fluid. The arachnoid continues along the spinal nerve roots, forming small lateral extensions of the subarachnoid space. In the cervical and thoracic regions in the adult, the distance between the arachnoid and the pia covering the spinal cord is only about 3 mm.

Pia Mater

The pia mater is a vascular membrane that closely covers the spinal cord (*Fig. 5–14*). It has lateral extensions on either side of the spinal cord that lie between the nerve roots and form the **ligamentum denticulatum.** This ligament has tooth-like processes, which pass laterally to adhere to the arachnoid mater and the dura mater, and suspends the spinal cord within the dural sheath. The pia mater extends along each nerve root as far as the spinal nerve. Inferiorly, it becomes the **filum terminale.**

Nerve Supply of the Meninges

The anterior surface is supplied by meningeal branches of the spinal nerves which reenter the vertebral canal through the intervertebral foramina.

Cerebrospinal Fluid

The cerebrospinal fluid is a clear, colorless fluid formed mainly by the **choroid plexus** of the lateral, third, and fourth ventricles. The fluid circulates through the ventricular system and enters the subarachnoid space through the three foramina in the roof of the fourth ventricle (*Fig. 5–15*). It ascends in the subarachnoid space over the outer surface of the cerebrum. Some of the fluid moves inferiorly in the subarachnoid space around the spinal cord and the cauda equina. The cerebrospinal fluid is mainly absorbed through the **arachnoid villi** that project into the dural venous sinuses, in particular the **superior sagittal venous sinus.** Groups of arachnoid villi are known as **arachnoid granulations.** There is no evidence that the slow circulation of the cerebrospinal fluid in the spinal subarachnoid space greatly influences the distribution of anesthetic solutions injected into it.

Composition of Cerebrospinal Fluid. The cerebrospinal fluid has inorganic salts in solution similar to those in the blood plasma. The glucose content is 40 to 80 mg per 100 ml, i.e., about half that of blood, and there is only a trace of protein (15 to 45 mg per 100 ml). Only a few cells are present and these are lymphocytes. The normal lymphocyte count is 0 to 3 cells/mm^3.

In the lateral recumbent position the cerebrospinal fluid pressure, as measured by lumbar puncture, is about 60 to 150 mm water. This pressure may be easily raised by straining, coughing, or compressing the internal jugular veins in the neck. The total volume of cerebrospinal fluid in the subarachnoid space and within the ventricles is about 130 ml.

Functions of the Cerebrospinal Fluid. The cerebrospinal fluid, which bathes the external and internal surfaces of the brain and the spinal cord, serves as a cushion between the central nervous system and the surrounding bones, thus protecting it against mechanical trauma. The close relationship of the fluid to the nervous tissue and the blood enables it to serve as a reservoir and assist in the regulation of the contents of the skull. For example, if the brain volume or the blood volume increases, the cerebrospinal fluid volume decreases. Because the cerebrospinal fluid is an ideal physiologic substrate, it probably plays an active part in the nourishment of the nervous tissue; it almost certainly assists in the removal of products of neuronal metabolism. It is possible that the secretions of the pineal gland influence the activities of the pituitary gland by circulating through the cerebrospinal fluid in the third ventricle.

Formation of the Cerebrospinal Fluid. The cerebrospinal fluid is formed in the choroid plexuses of the lateral, third, and fourth ventricles; some may originate as tissue fluid formed in the brain substance. As much as 500 ml of cerebrospinal fluid may be produced in 24 hours.

The choroid plexuses have a multifolded surface with each fold consisting of a core of vascular connective tissue covered with cuboidal epithelium from the ependyma. Electron microscope examination of the epithelial cells shows that their free surfaces are covered with microvilli. The blood of the capillaries is separated from the ventricular lumen by endothelium, a basement membrane, and the surface epithelium. The epithelial cells are fenestrated and permeable to large molecules.

There is considerable evidence that the choroid plexuses actively secrete cerebrospinal fluid. It has been shown experimentally in animals that the passage of sodium is stopped when a carbonic anhydrase inhibitor is applied to the choroid plexus. Furthermore, it has been shown that the concentrations of electrolytes, particularly potassium, calcium, and magnesium, differ in the cerebrospinal fluid as compared with those of the blood plasma.

Absorption of the Cerebrospinal Fluid. The main sites for the absorption of the cerebrospinal fluid are the arachnoid villi that project into the dural venous sinuses, especially the superior sagittal sinus. Structurally, each arachnoid villus is a diverticulum of the subarachnoid space that pierces the dura mater. The arachnoid diverticulum is capped by a thin cellular layer, which in turn is covered by the endothelium of the venous sinus. The arachnoid granulations increase in number and size with age and tend to become calcified with advanced age.

The absorption of cerebrospinal fluid into the venous sinuses occurs when the cerebrospinal fluid pressure ex-

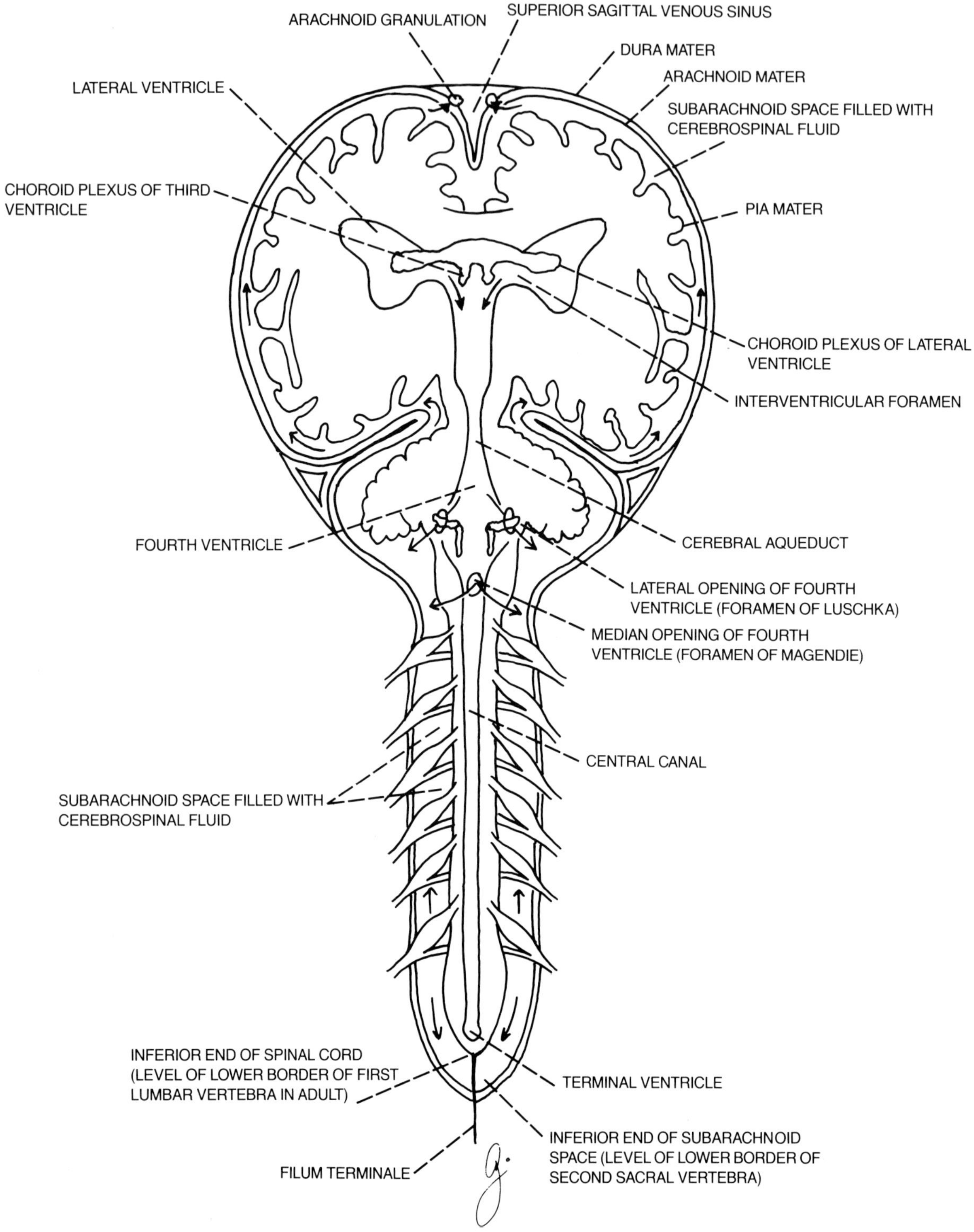

Figure 5–15. Diagram showing the origin and circulation of the cerebrospinal fluid.

ceeds that in the sinus. Electron microscope studies of the arachnoid villi seem to indicate that fine tubules lined with endothelium permit a direct flow of fluid from the subarachnoid space into the lumen of the venous sinuses. Should the venous pressure rise and exceed the cerebrospinal fluid pressure, compression of the tips of the villi closes the tubules and prevents the reflux of blood into the subarachnoid space.

Some of the cerebrospinal fluid probably is absorbed directly into the veins in the subarachnoid space and some possibly escapes through the perineural lymphatic vessels of the cranial and spinal nerves.

CLINICAL NOTES

CONGENITAL ANOMALIES OF THE SPINAL CORD

Spina Bifida

In spina bifida, the spines and the arches of one or more adjacent vertebrae fail to develop. The anomaly occurs most commonly in the lower thoracic, lumbar, and sacral regions. Beneath this defect the meninges and the spinal cord may or may not be involved to varying degrees. The spina bifida is a result of failure of the embryonic mesenchyme, which grows in between the neural tube and the surface ectoderm, to form the vertebral arches in the affected region. Some authorities believe that local overgrowth of the neural plate in the early stages of development might be a factor in the production of spina bifida. The types of spina bifida are as follows:

1. SPINA BIFIDA OCCULTA. The spines and the arches of one or more vertebrae, usually in the lumbar region, are absent, and the vertebral canal remains open posteriorly (*Fig. 5–16*). The spinal cord and the nerve roots usually are normal. The defect is covered by the postvertebral muscles and cannot be seen from the surface. A small tuft of hair or a fatty tumor may be present over the defect. Most cases are symptomless and are diagnosed by chance when the vertebral column is x-rayed.
2. MENINGOCELE. The meninges project through the defect in the vertebral arches, forming a cystic swelling beneath the skin and contains cerebrospinal fluid that communicates with the subarachnoid space (*Fig. 5–16*). The spinal cord and the nerves usually are normal.
3. MENINGOMYELOCELE. The normal spinal cord, or cauda equina, lies within the meningeal sac, which projects through the vertebral arch defect (*Fig. 5–16*). The spinal cord or the nerve roots are adherent to the inner wall of the sac.
4. MYELOCELE. The neural tube fails to close in the region of the defect (*Fig. 5–16*). An oval raw area is visible on the skin surface; this represents the neural groove that is not united. The central canal discharges clear cerebrospinal fluid onto the surface.
5. SYRINGOMYELOCELE. This condition is rare. A meningomyelocele is present, and in addition, the central canal of the spinal cord at the level of the bony defect is grossly dilated (*Fig. 5–16*).

Spina bifida occulta is the most common defect. The next most common defect is myelocele, and many afflicted infants are born dead. If the child is born alive, death from infection of the spinal cord occurs within a few days.

Most cases of spina bifida occulta require no treatment. A meningocele should be removed surgically within a few days of birth. Cases of meningomyelocele should also be treated surgically. The sac is opened and the spinal cord or nerves are freed and carefully replaced in the vertebral canal. The meninges are sutured over the cord and the postvertebral muscles are approximated.

HEMISECTION OF THE SPINAL CORD (BROWN-SÉQUARD SYNDROME)

Hemisection of the spinal cord may be caused by trauma or an expanding tumor. Incomplete hemisection is common; complete hemisection is rare. The following characteristic clinical features (*Fig. 5–17*) are seen in patients with a complete hemisection of the cord after the period of spinal shock has ended:

1. Ipsilateral lower motor neuron paralysis in the segment of the lesion and muscular atrophy. These conditions are caused by damage to the neurons in the anterior gray column and possibly by damage to the nerve roots of the same segment.
2. Ipsilateral spastic paralysis below the level of the lesion. There is an ipsilateral Babinski sign present, and, depending on the segment of the cord damaged, there are an ipsilateral loss of superficial abdominal reflexes and an ipsilateral loss of the cremasteric reflex. All these signs are caused by the loss of the corticospinal tracts on the side of the lesion. The spastic paralysis is produced by the interruption of the descending tracts other than the corticospinal tracts.
3. Ipsilateral band of cutaneous anesthesia in the segment of the lesion. This is caused by the destruction of the posterior root and its entrance into the spinal cord at the level of the lesion.
4. Ipsilateral loss of tactile discrimination and of vibratory and proprioceptive sense below the level of the lesion. These signs are caused by the destruction of the ascending tracts in the posterior white column on the same side as the lesion.
5. Contralateral loss of pain and temperature sense below the lesion. This is caused by the destruction of the crossed lateral spinothalamic tracts on the same side as the lesion. Because the tracts cross obliquely, the sensory loss occurs one or two segments below the lesion distally.
6. Contralateral but not complete loss of tactile sense below the lesion. This condition is brought about by the destruc-

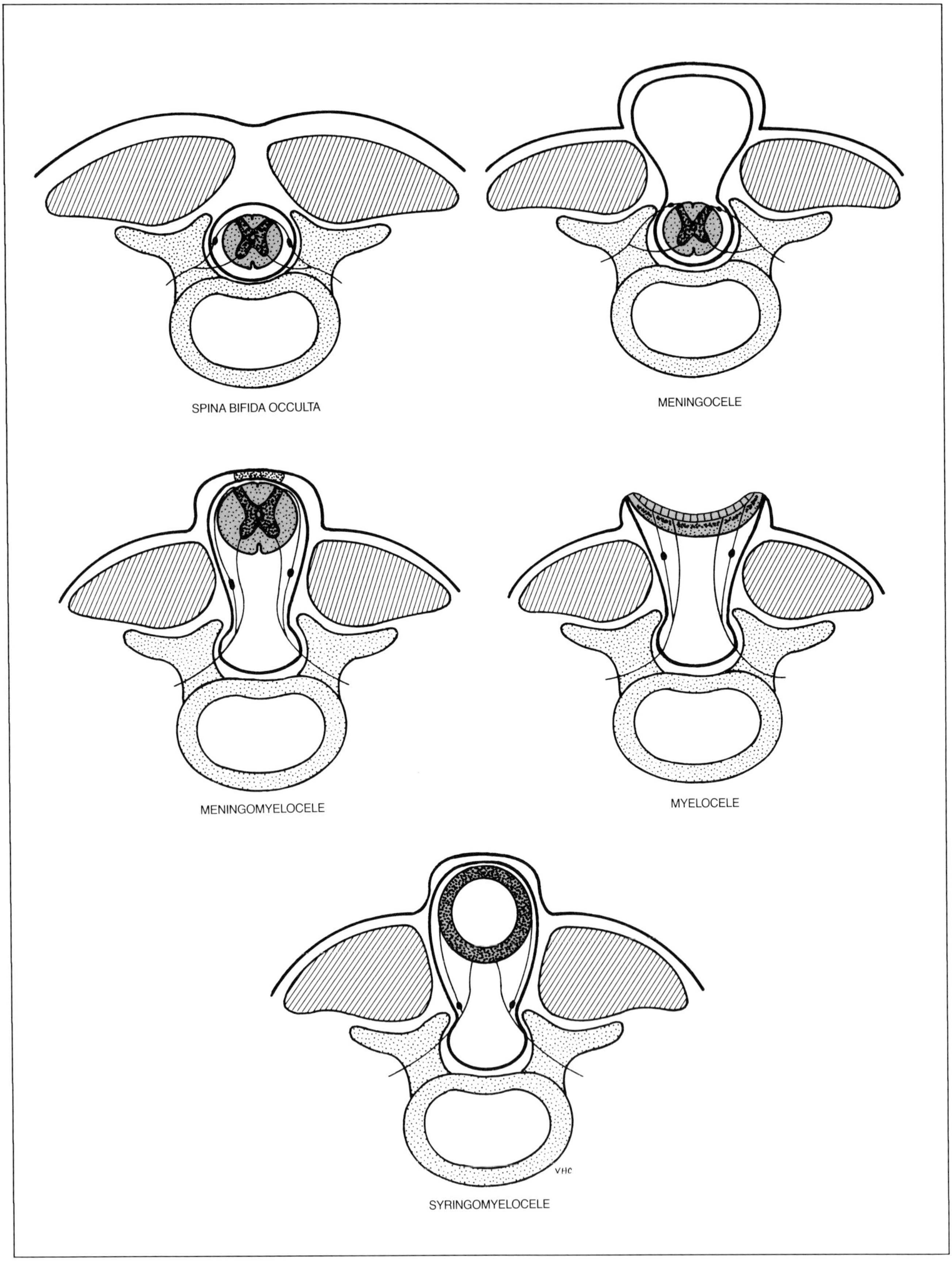

Figure 5–16. The different types of spina bifida.

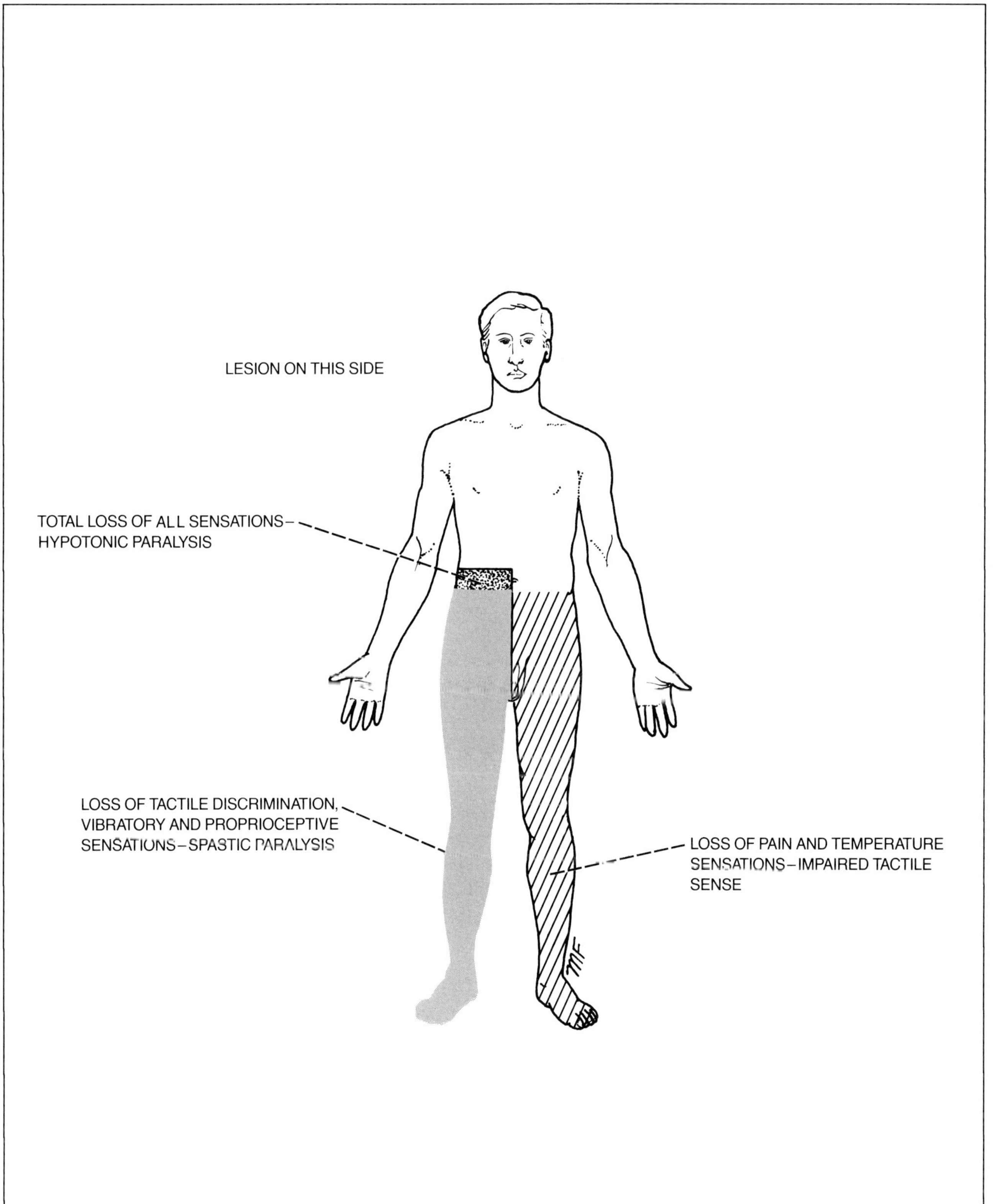

Figure 5–17. Brown-Séquard syndrome with spinal cord lesion at right tenth thoracic segmental level.

tion of the crossed anterior spinothalamic tracts on the side of the lesion. Here again, because the tracts cross obliquely, the sensory impairment occurs two or three segments below the lesion distally.

COMPLETE TRANSECTION OF THE SPINAL CORD

Complete transection of the spinal cord results in complete loss of all sensibility and voluntary movement below the level of the lesion. It may be caused by a fracture dislocation of the vertebral column, other trauma, or an expanding tumor. The following characteristic clinical features will be seen after the period of spinal shock has ended:

1. Bilateral lower motor neuron paralysis in the segment of the lesion and muscular atrophy. This is caused by damage to the neurons in the anterior gray columns and possibly damage to the nerve roots of the same segment.
2. Bilateral spastic paralysis below the level of the lesion. A bilateral Babinski sign is present and, depending on the level of the segment of the spinal cord damaged, there is bilateral loss of the superficial abdominal and cremasteric reflexes. All these signs are caused by an interruption of the corticospinal tracts on both sides of the cord. The bilateral spastic paralysis is produced by the cutting of the descending tracts other than the corticospinal tracts.
3. Bilateral loss of all sensation below the level of the lesion. The loss of tactile discrimination and vibratory and proprioceptive sense is caused by bilateral destruction of the ascending tracts in the posterior white columns. The loss of pain, temperature, and light touch senses is caused by the section of the lateral and anterior spinothalamic tracts on both sides. Because these tracts cross obliquely, the loss of thermal and light touch sensations occurs two or three segments below the lesion distally.
4. Bladder and bowel functions are not under voluntary control, which causes serious difficulties in the management of these patients.

ANATOMY OF SPINAL ANESTHESIA

Between the levels of the lower end of the spinal cord (vertebral levels L1–2 in the adult) and the lower end of the subarachnoid space (vertebral levels S2–3), lie the spinal nerve roots of the cauda equina bathed in cerebrospinal fluid. The nerve roots are destined for body levels below L-1. Because at this level of the vertebral column the spinal cord is absent and the spinal nerve roots are mobile and easily pushed aside, it is in this region that spinal needles are introduced into the subarachnoid space to induce spinal anesthesia. The induction of spinal anesthesia at higher levels is possible, but the presence of the spinal cord and the great obliquity of the thoracic spines requires considerable technical expertise. Moreover, the distance between the pia mater covering the spinal cord and the arachnoid mater in the cervical and thoracic regions is only 3 mm. Because of the fusion of the laminae of adjacent vertebrae in the sacrum, the only way to perform spinal anesthesia in the sacral region is through the sacral hiatus.

CLINICAL PROBLEMS

1. A 54-year-old man visited his physician because of increasing difficulty in walking. Three weeks ago he had been able to walk with the help of a cane. Since that time walking had become increasingly difficult. He had complete control of micturition and defecation. On examination, the hand grip was weak on both sides but power was normal in the proximal segments of the upper extremities. The tendon reflexes of the upper limbs and the sensory function were normal. Both lower limbs showed muscular weakness with increased muscle tone, especially on the right side. The knee jerks and ankle jerks in both lower limbs were exaggerated and there were bilateral extensor plantar responses. There was a loss of sensation of pain below the sixth thoracic dermatome on both sides of the body. Postural sense was impaired in both great toes and vibration sense was absent below the sixth thoracic segmental level. A myelogram revealed a complete block at the lower border of the fifth thoracic vertebra. Using your knowledge of anatomy of the spinal cord, make a diagnosis. Name the tracts in the spinal cord that are responsible for the conduction of the sensation of pain. What is the position of these tracts in the spinal cord? Which tracts are responsible for the conduction of postural sense and vibration sense within the cord? Why did the patient have increasing difficulty in walking? Why were the tendon reflexes in the lower limbs exaggerated, and why did the patient exhibit bilateral extensor plantar responses?

2. What is the normal pressure of the cerebrospinal fluid with the patient lying on the side? What is the normal concentration of protein and glucose in the cerebrospinal fluid?

3. An orthopedic surgeon was about to fuse the first and second lumbar vertebral bodies through an anterior approach. He was aware of the complicated anatomy in this region but he was particularly concerned about compromising the blood supply to the spinal cord. What is the arterial supply to the spinal cord? Is the blood supply to the spinal cord particularly vulnerable in the lumbar region?

4. When performing a spinal anesthetic why is the fourth lumbar spine commonly used as a bony landmark? What level does the spinal cord commonly end inferiorly in a 4-year-old child? What level does the cord end inferiorly in an adult?

5. When introducing the spinal needle into the subarachnoid space before performing a spinal anesthesia, occasionally a paresthesia occurs. Has the needle hit (a) a spinal nerve? (b) a spinal nerve root? (c) a ramus of a spinal nerve? or (d) the spinal cord?

6. The following anatomical facts are important to an anesthesiologist who is about to perform a spinal anesthetic or a spinal extradural block. Define the following (a) filum terminale, (b) ligamentum denticulatum, (c) internal vertebral venous plexus, (d) conus medullaris, and (e) epidural space.

7. A 40-year-old woman was hospitalized for investigation. She had symptoms of analgesia and thermoanesthesia on the medial side of the right hand of 8 months duration. Five weeks before her admittance, she had burnt her right little finger on a hot stove. Although she could smell the burning skin she was initially unaware that the burn had occurred. On examination she was found to have reduced pain and temperature sense of the eighth cervical and first thoracic dermatomes of the right hand. Her tactile sense was normal in these areas. Examination of the left upper limb showed similar dissociated sensory loss. The fingers on both hands showed several small burn scars. In the context of this chapter what is your diagnosis? Name the tracts involved in the disease process.

8. After drinking heavily at a party, an 18-year-old student crashed his car into a bridge abutment. On examination in the emergency room, he was found to have

a fracture dislocation of the tenth thoracic vertebra with signs and symptoms of severe injury to the spinal cord. On physical examination, he had an upper motor neuron paralysis of the right leg. He also had loss of muscle joint sense of the right leg. He had a band of cutaneous hyperesthesia extending around the abdominal wall on the right side just below the level of the umbilicus. A short distance below this, he had a narrow band of anesthesia and analgesia. On the left side, there was total analgesia, thermoanesthesia, and partial loss of tactile sense of the abdominal wall skin from just below the umbilicus to involve the entire left leg. At what level was the spinal cord damaged? Was the spinal cord completely sectioned? If not, on which side did the hemisection occur? Explain the sensory losses found on examination in this patient.

9. After a spinal anesthetic for a hysterectomy, a patient complained of a persistent headache, 24 hours after the operation, which was made worse by sitting up. The possible cause of the headache was discussed with the first year anesthetic resident. Could prolonged hypotension during the course of the operation be a possible cause? In anatomic terms explain why patients having a spinal anesthetic are prone to hypotension. Could the headache be caused by a persistent dural leak? When is the cerebrospinal fluid pressure highest in the lumbar region, lying down or sitting up? Is the epidural pressure higher or lower than the cerebrospinal fluid pressure?

ANSWERS TO CLINICAL PROBLEMS

1. A laminectomy of the fourth, fifth, and sixth thoracic vertebrae was performed on this patient. At the level of the fifth thoracic vertebra a small swelling was seen on the posterior surface of the spinal cord, which was attached to the dura mater. Histologic examination showed that it was a meningioma. The tumor was easily removed. The lateral spinothalamic tracts are responsible for the conduction of pain impulses up the spinal cord. These tracts are situated in the lateral white columns of the spinal cord. Postural sense and vibration sense are conducted up the spinal cord in the posterior white column through the fasciculus cuneatus from the upper limbs and the upper part of the thorax, and in the fasciculus gracilis from the lower part of the trunk and the lower limb. The increasing difficulty in walking was caused by pressure on the corticospinal tracts in the lateral white column. The exaggeration of the tendon reflexes of the lower limbs and the bilateral extensor plantar responses were caused by the pressure on the descending tracts at the level of the tumor.

2. The normal cerebrospinal fluid pressure in a patient in a lateral recumbent position is 60 to 150 mm water. The protein concentration is 15 to 45 mg per 100 ml, and the glucose concentration is 40 to 80 mg per 100 ml.

3. The arterial supply to the spinal cord is described on page 170. In the lower thoracic or upper lumbar vertebral levels a large and important radicular artery arises from an intersegmental branch of the descending aorta on the left side. This artery may be the main source of blood to the lower two-thirds of the spinal cord. For further details on the blood supply to the spinal cord, see page 170.

4. The spine of the fourth lumbar vertebra can easily be identified as it lies on a line joining the highest points of the iliac crests. In a 4-year-old child the spinal cord ends below at the level of the upper border of the third lumbar vertebra. In an adult, the spinal cord ends below at the level of the lower border of the first lumbar vertebra (for variations, see page 153).

5. The spinal needle on entering the subarachnoid space below the lower limit of the spinal cord commonly touches, deflects, or rarely impinges on an anterior or a posterior root of a spinal nerve belonging to the cauda equina. Paresthesias will also occur if the periosteum is touched.

6. (a) Filum terminale. This is a prolongation of the pia mater that extends from the apex of the conus medullaris as a fibrous cord and is attached inferiorly to the posterior surface of the coccyx. It contains no nervous tissue. (b) Ligamentum denticulatum. This ligament is present on either side of the spinal cord. It is a thickening of the pia mater that passes laterally to adhere to the arachnoid and dura. Its tooth-like processes are attached to the arachnoid and dura between adjacent spinal nerves (*see Fig.* 5–2). (c) Internal vertebral venous plexus. This plexus lies within the vertebral canal in the epidural space. It receives the venous drainage of the spinal cord, the meninges, and the vertebral bodies and is drained by intervertebral veins into the vertebral, posterior intercostal, lumbar, and lateral sacral veins. (d) Conus medullaris. This is the cone-shaped lower end of the spinal cord. (e) Epidural space (extradural space). This space lies between the dural sheath covering the spinal cord and the cauda equina and the bones and ligaments that form the vertebral canal. It contains loose fatty areolar tissue and the internal vertebral venous plexus. The spinal nerves cross the epidural space as they leave the dura mater to exit the vertebral canal through the intervertebral foramina.

7. This patient has the early signs and symptoms of syringomyelia. The cavitation had interrupted the lateral and anterior spinothalamic tracts as they decussated in the gray and white commissures of the spinal cord at the level of the eighth cervical and first thoracic segments of the spinal cord. The condition was worse on the right side. Tactile discrimination was normal in both upper limbs indicating that the fasciculus cuneatus was not affected.

8. Fracture dislocation of the tenth thoracic vertebra resulted in damage to the 11th thoracic segment of the spinal cord. The unequal sensory and motor losses on the two sides indicate a right hemisection of the cord. The band of hyperesthesia on the right side was caused by the irritation of the cord immediately above the level of the lesion. The band of anesthesia and analgesia was caused by the destruction of the cord on the right side at the level of the 11th thoracic segment, i.e., all afferent fibers entering the cord at that point were interrupted. The loss of pain and thermal sensibilities and the loss of light touch below the umbilicus on the left side were caused by the interruption of the lateral and anterior spinothalamic tracts on the right side of the cord.

9. Hypotension is one of the common accompaniments of spinal anesthesia, but is usually not responsible for postoperative headache. Hypotension associated with spinal anesthesia is caused by the blockade of the preganglionic sympathetic fibers present in the anterior nerve roots of segments T-1 to L-2. In very high thoracic spinal anesthesia (not in this patient), the cardiac accelerator fibers are also blocked leading to bradycardia.

   Puncture of the dura and arachnoid, in performing spinal anesthesia may result in a persistent cerebrospinal fluid leak. Such a leak may lead to a headache in the occipital region, which is accentuated by attempting to sit upright. In the upright position, the cerebrospinal fluid pressure is greatest in the lumbar region, therefore further leakage may occur. The epidural pressure is negative in the majority of patients and is probably caused by transmission of the negative pressure from the thorax through the intervertebral foramina. The negative pressure in the epidural space may facilitate the leakage of cerebrospinal fluid into the extradural space after lumbar puncture. The cerebrospinal fluid pressure in the lumbar region, in the lateral recumbent position, is between 60 and 150 mm of water and sitting up, it may reach 200 to 300 mm of water.

CHAPTER 6

Nerve Fibers and Spinal Nerves

NERVE FIBERS

A nerve fiber is the name given to the processes of a nerve cell. Nerve fibers may be divided into dendrites and axons.

The **dendrites** are the short processes of the cell body (*Fig. 6–1*). Their diameter narrows as they extend from the cell body and they often branch profusely. In many neurons the smaller branches bear large numbers of small projections called **dendritic spines.** The cytoplasm of the dendrites closely resembles that of the cell body and contains Nissl granules, mitochondria, microtubules, microfilaments, ribosomes, and agranular endoplasmic reticulum. Essentially dendrites are extensions of the cell body and they conduct the nerve impulse toward the cell body.

The **axon** is the longest process of the cell body. It arises from a conical elevation on the cell body, devoid of Nissl granules, called the **axon hillock** (*Fig. 6–1*). An axon tends to have a smooth contour and is uniform in diameter.

Axons give off collateral branches along their length but they tend not to branch close to the cell body. Just before their termination, axons commonly branch profusely. The diameter of axons varies considerably with different neurons. Those of larger diameter conduct impulses rapidly and those of smaller diameter conduct slowly.

The plasma membrane bounding the axon is called the **axolemma.** The cytoplasm of the axon is termed the **axoplasm.** Axoplasm differs from the cytoplasm of the cell body in possessing no Nissl granules or Golgi apparatus. The sites for the production of protein, namely RNA and ribosomes, are absent.

It is commonly stated that an axon always conducts impulses away from the cell body. The axons of the sensory posterior root ganglion cells are an exception. Here, the long neurite, which is identical to an axon, carries the impulse toward the cell body.

Bundles of nerve fibers found in the central nervous system are referred to as **nerve tracts;** those in the peripheral nervous system are called **peripheral nerves.** Two types of nerve fibers are present in the central and peripheral parts of the nervous system, namely myelinated and nonmyelinated fibers.

Myelinated Nerve Fibers

A myelinated nerve fiber is one that is surrounded by a myelin sheath. The myelin sheath is not part of the neuron but is formed by a supporting cell (*Fig. 6–2*). In the central nervous system the supporting cell is the oligodendrocyte; in the peripheral nervous system, it is the Schwann cell.

The myelin sheath is a segmented, discontinuous layer interrupted at regular intervals by the **nodes of Ranvier.** Each segment of the myelin sheath measures approximately 0.5 to 1 mm in length. In the central nervous system, each oligodendrocyte may form and maintain myelin sheaths for as many as 60 nerve fibers (axons). In the peripheral nervous system, there is usually only one Schwann cell for each segment of a nerve fiber.

Formation of Myelin. Myelin sheaths begin to form during the latter part of fetal development and during the first year postnatally.

In the **peripheral nervous system,** the nerve fiber or axon first indents the side of a Schwann cell (*Fig. 6–3*). Later, as the axon sinks farther into the Schwann cell, the external plasma membrane of the Schwann cell forms a **mesaxon,** which suspends the axon within the Schwann cell. Subsequently, it is thought, the Schwann cell rotates on the axon so that the plasma membrane becomes wrapped around and around the axon in a spiral. The direction of the spiral is clockwise in some segments and counterclockwise in others. At the beginning, the wrappings are loose, but gradually the cytoplasm between the layers of the cell membrane disappears leaving cytoplasm near the surface and in the region of the nucleus. The wrappings become tight with maturation of the nerve fiber. The thickness of the myelin depends on the number of spirals of Schwann cell membrane. Some nerve fibers

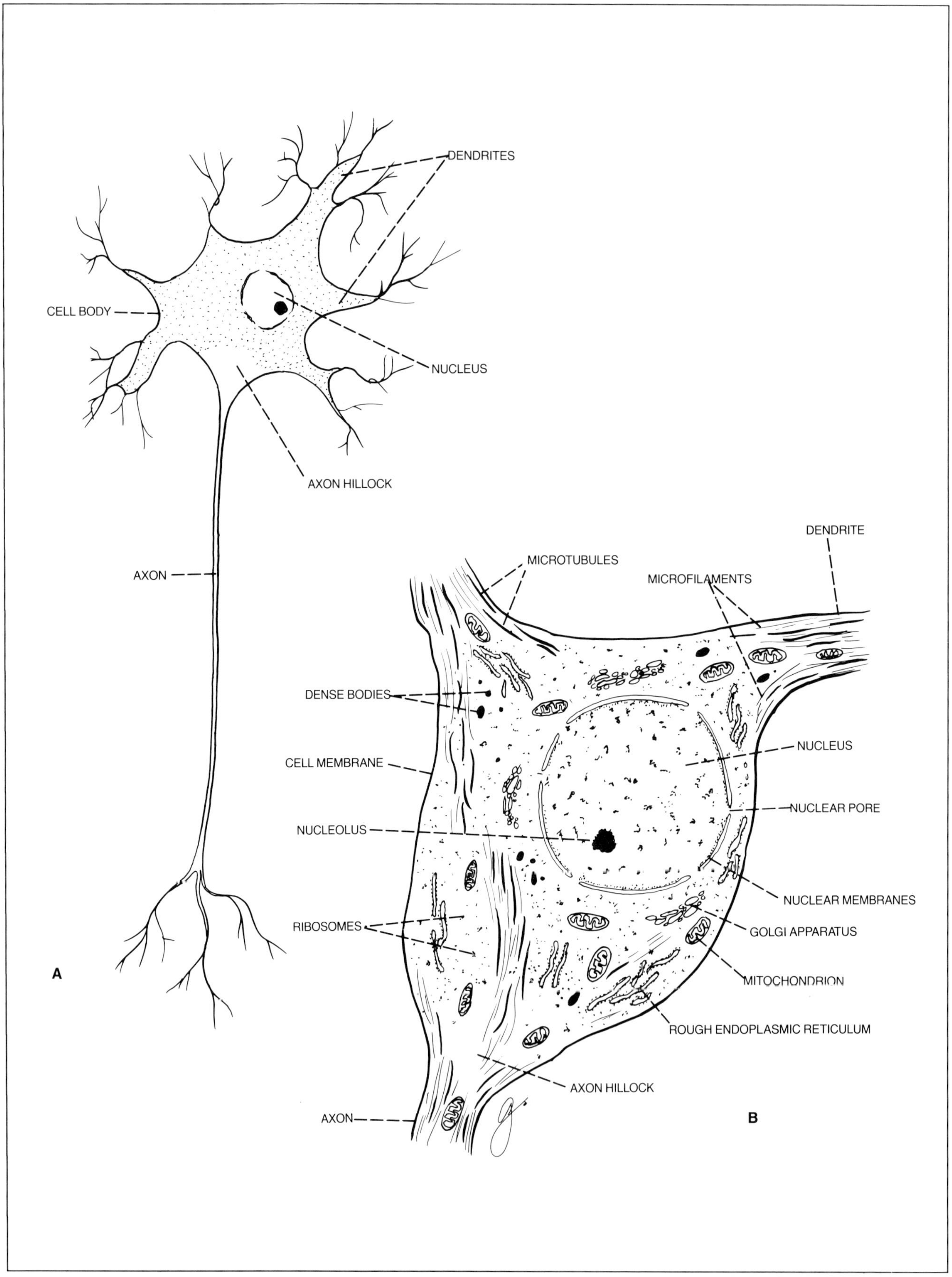

Figure 6–1. Diagrammatic representation of **(A)** a neuron; **(B)** the fine structure of a neuron.

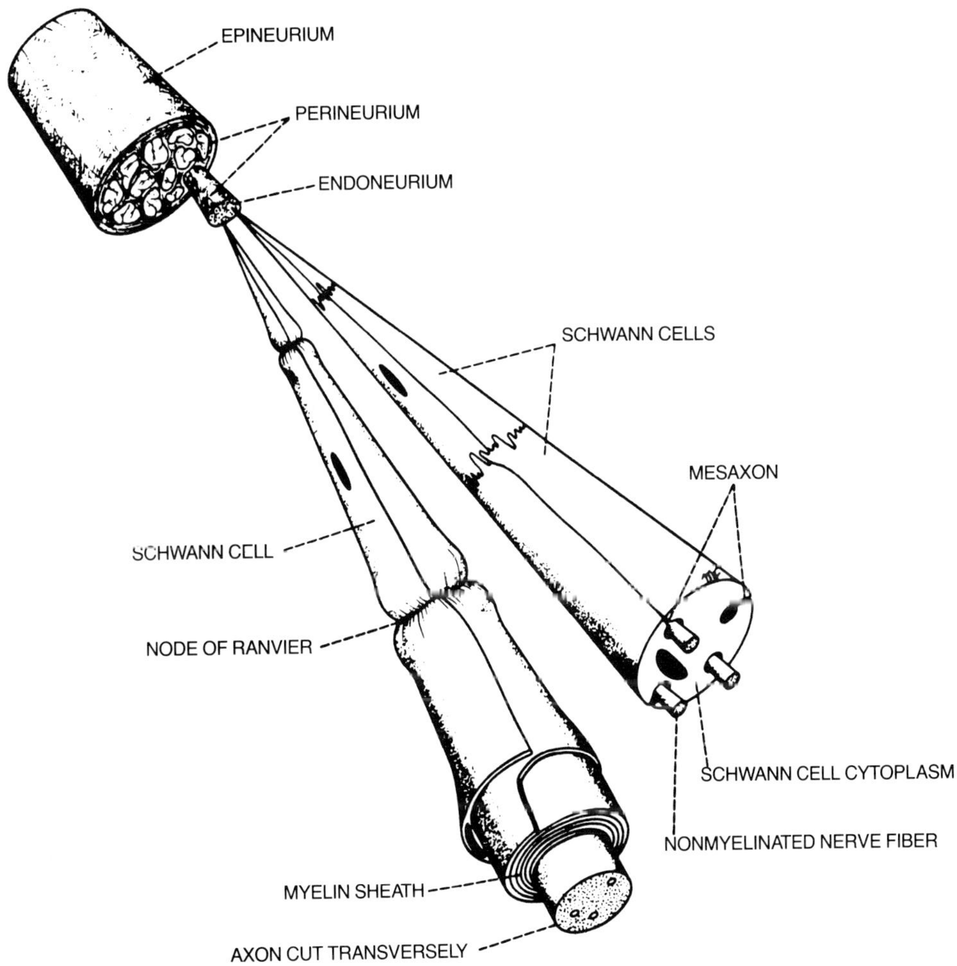

Figure 6–2. Exploded view of a peripheral nerve, showing the connective tissue sheaths and the structure of myelinated and nonmyelinated nerve fibers.

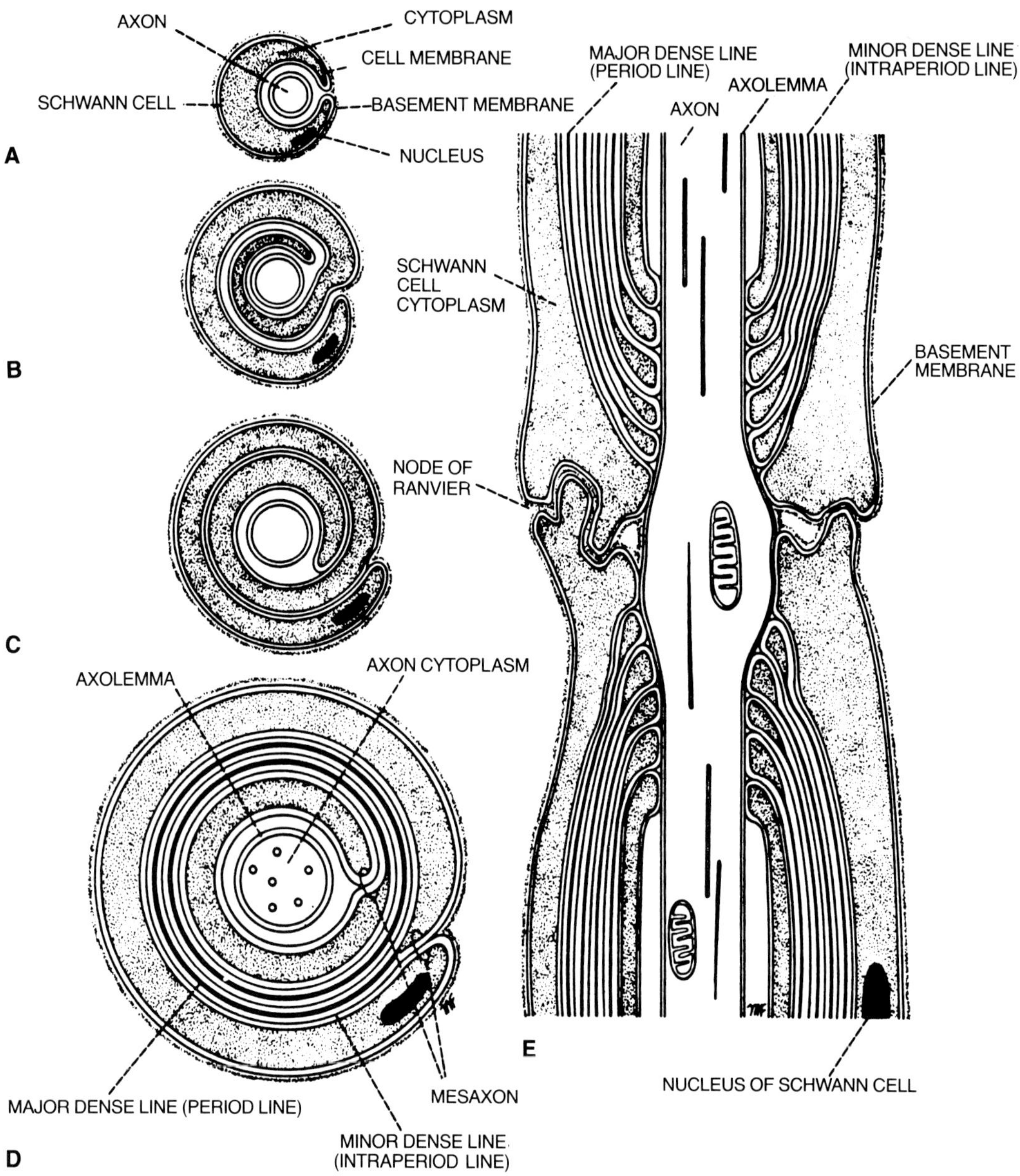

Figure 6–3. A myelinated nerve fiber in the peripheral nervous system. **A, B, C,** and **D** show cross sections of a nerve at different stages in the formation of the myelin sheath.

are surrounded by only a few turns of the membrane, whereas others have as many as 50 turns. In electron micrographs of cross sections of mature myelinated nerve fibers, the myelin is seen to be laminated (*Fig. 6–3*). Each lamella measures 13 to 18 nm thick. The dark major dense line, about 2.5 nm thick, consists of two inner protein layers of the plasma membrane that are fused together. The lighter minor dense line, about 10 nm thick, is formed by the approximation of the outer surfaces of adjacent plasma membranes and is made up of lipid. The fused outer protein layers of the plasma membranes are very thin and form a thin intraperiod line situated in the center of the lighter lipid layer (*Fig. 6–3*). At the nodes of Ranvier, the adjacent Schwann cells terminate and the plasma membrane of the axon is exposed.

The **incisures of Schmidt-Lanterman** are seen on longitudinal sections of myelinated nerve fibers. They represent areas where the dark major dense line is not formed as a result of the localized persistence of Schwann cell cytoplasm. This persistence of cytoplasm involves all the layers of the myelin, and thus, there is a continuous spiral of cytoplasm from the outermost region of the Schwann cell to the region of the axon. This spiral of cytoplasm may provide a pathway for the conduction of metabolites from the surface region of the Schwann cell to the axon.

In the **central nervous system,** oligodendrocytes are responsible for the formation of the myelin sheaths. The plasma membrane of the oligodendrocyte becomes wrapped around the axon and the number of layers will determine the thickness of the myelin sheath. The **nodes of Ranvier** are situated in the intervals between adjacent oligodendroctyes. A single oligodendrocyte may be connected to the myelin sheaths of as many as 60 nerve fibers. For these reasons, the process of myelination in the central nervous system cannot take place by rotation of the oligodendrocyte on the axon, as did the Schwann cell in the peripheral nervous system. It is possible that myelination in the central nervous system occurs by the growth in length of the mesaxon of the oligodendrocyte, the mesaxon wrapping itself around the axon. There are incisures of Schmidt-Lanterman in nerve fibers of the central nervous system.

Nonmyelinated Nerve Fibers

The smaller axons of the central nervous system, the postganglionic axons of the autonomic part of the nervous system, and some fine sensory axons associated with the reception of pain, are nonmyelinated.

In the **peripheral nervous system** each axon, which is usually less than 1 μ (1 μm) in diameter, indents the surface of the Schwann cell so that it lies within a trough. As many as 15 or more axons may share a single Schwann cell, each lying within its own trough or sometimes sharing a trough. In some situations the troughs are deep and the axons are embedded deep in the Schwann cells, forming a **mesaxon** from the Schwann cell plasma membrane (*Fig. 6–2*). There are series of Schwann cells along the lengths of these axons. There are no nodes of Ranvier.

In areas where there are synapses or where motor transmission occurs, the axon emerges from the trough of the Schwann cell for a short distance, thus exposing the active region of the axon.

In the **central nervous system,** the nonmyelinated nerve fibers are not particularly related to the oligodendrocytes.

Conduction Velocity in Nerve Fibers

The conduction velocity of a nerve fiber is proportional to the cross-sectional area of the axon, the thicker fibers conducting more rapidly than those of smaller diameter. In the large alpha fibers, the rate may be as high as 80 to 120 m/sec; the beta fibers 30 to 90 m/sec; the gamma fibers 6 to 60 m/sec, and the delta fibers 6 to 30 m/sec. The very small C fibers that are unmyelinated conduct at the rate of 0.5 to 2.0 m/sec. It is interesting to note that over 66 percent of all nerve fibers in peripheral nerves are C fibers.

In myelinated nerve fibers, the presence of a myelin sheath serves as an insulator, as a result of which, a myelinated nerve fiber can be stimulated only at the nodes of Ranvier. Thus, the action potential jumps from one node to the next (*Fig. 6–4*), a process known as **saltatory conduction.** In nonmyelinated nerve fibers, the action potential passes continuously along the axolemma, progressively exciting neighboring areas of membrane (*Fig. 6–4*). This is a much slower mechanism than is found in myelinated fibers.

PERIPHERAL NERVES

The peripheral nerves are the cranial and spinal nerves. Each peripheral nerve consists of parallel bundles of nerve fibers, which may be efferent or afferent axons, and may be myelinated or nonmyelinated. Each parallel bundle is surrounded by connective tissue sheaths (*Fig. 6–2*).

The nerve trunk is surrounded by a dense connective tissue sheath called the **epineurium** (*Fig. 6–2*). Within the sheath are bundles of nerve fibers, each of which is surrounded by a connective tissue sheath called the **perineurium.** Between the individual nerve fibers is a loose, delicate connective tissue referred to as the **endoneurium.** The connective tissue sheaths serve to support the nerve fibers and their associated blood vessels and lymphatic vessels.

Peripheral nerve fibers may be classified into three groups, depending on their size and speed of conduction:

- **Group A fibers** are 1 to 20 μ in diameter and they conduct at the rate of 5 to 120 m/sec. They are myelinated somatic efferent and afferent fibers.

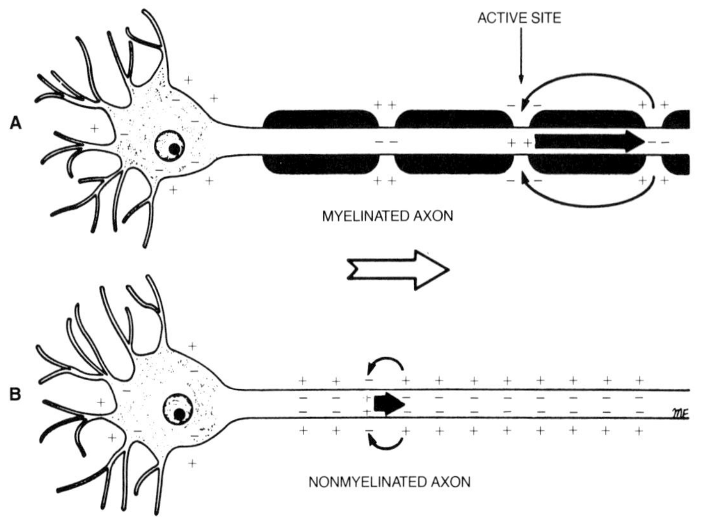

Figure 6–4. The electrical changes that occur in: **(A)** a stimulated myelinated axon and **(B)** a stimulated nonmyelinated axon.

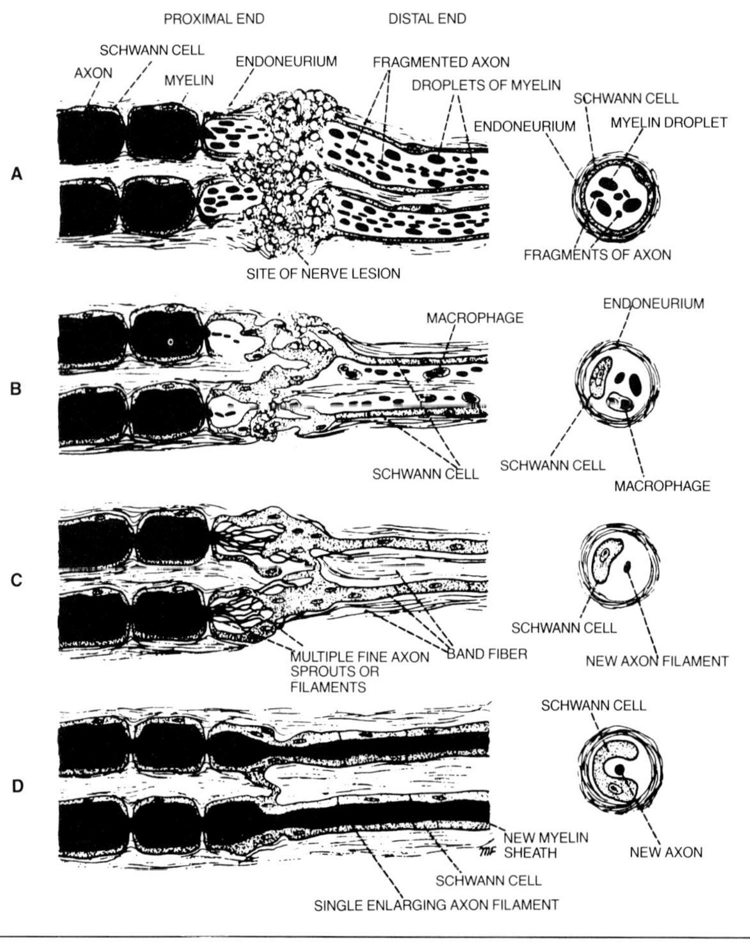

Figure 6–5. Degeneration and regeneration in a divided peripheral nerve.

- **Group B fibers** are 1 to 3 μ in diameter and they conduct at 3 to 15 m/sec. They are myelinated, efferent, preganglionic, autonomic fibers.
- **Group C fibers** are 0.5 to 2 μ in diameter and they conduct at 0.5 to 2 m/sec. They are unmyelinated afferent or efferent postganglionic autonomic fibers.

CLINICAL NOTES

Responses of Nerve Fibers to Injury

If the axon of the nerve cell is divided, degenerative changes will take place (1) in the distal segment that is separated from the cell body, (2) in a portion of the axon proximal to the injury, and (3) may occur in the cell body from which the axon arises.

Changes in the Distal Segment of the Axon. The changes spread distally from the site of the lesion (*Fig. 6–5*) and include its terminations; the process is referred to as **Wallerian degeneration.** The first day, the axon becomes swollen and irregular; by the third or fourth day, the axon is broken into fragments (*Fig. 6–5*) and the debris is digested by the surrounding Schwann cells and tissue macrophages. The entire axon is destroyed within a week.

Meanwhile, the myelin sheath slowly breaks down and lipid droplets appear within the Schwann cell cytoplasm (*Fig. 6–5*). Later, the droplets are extruded from the Schwann cell and subsequently, phagocytosed by tissue macrophages. The Schwann cells now begin to proliferate rapidly and become arranged in parallel cords within the persistent basement membrane. The endoneurial sheath and the contained cords of Schwann cells are sometimes referred to as a **band fiber.** If regeneration does not occur, the axon and the Schwann cells are replaced by fibrous tissue produced by local fibroblasts.

In the central nervous system, degeneration of the axons and myelin sheaths follows a similar course, and the debris is removed by the phagocytic activity of the microglial cells. Little is known about the role of oligodendrocytes in this process. The astrocytes now proliferate and replace the axons.

Changes in the Proximal Segment of the Axon. The changes in the proximal segment of the axon are similar to those that take place in the distal segment (*Fig. 6–5*), but only extend proximally above the lesion as far as the first node of Ranvier. The proliferating cords of Schwann cells in the peripheral nerves bulge from the cut surfaces of the endoneurial tubes.

Regeneration of the Axons

Regrowth of the axon is possible in peripheral nerves and appears to depend on the presence of endoneurial tubes and the special qualities possessed by Schwann cells. In the central nervous system, there is an attempt at regeneration of the axons, as evidenced by sprouting of the axons, but the process ceases after about 2 weeks. There is no evidence that restoration of function takes place. The regeneration process is aborted by the absence of endoneurial tubes (which are necessary to guide the regenerating axons), the failure of oligodendrocytes to serve in the same manner as Schwann cells, and the laying down of scar tissue by the active astrocytes. It has also been suggested that there is an absence of nerve growth factors in the central nervous system or that the neuroglial cells may produce nerve growth inhibiting factors.

Regeneration of Axons in Peripheral Nerves

If one assumes that the proximal and distal stumps of the severed nerve are in close apposition, the following regenerative processes take place (*Fig. 6–5*). The Schwann cells, having undergone mitotic division, now fill the space within the basal membrane of the endoneurial tubes of the proximal stump as far proximally as the next node of Ranvier, and in the distal stump as far distally as the end organs. Where a small gap exists between the proximal and distal stumps, the multiplying Schwann cells form a number of cords to bridge the gap.

Each proximal axon end now gives rise to multiple fine sprouts or filaments with bulbous tips. These filaments, as they grow, advance along the clefts between the Schwann cells and thus cross the interval between the proximal and distal nerve stumps. Many such filaments now enter the proximal end of each endoneurial tube and grow distally in contact with the Schwann cells (*Fig. 6–5*). It is clear that the filaments from many different axons may enter a single endoneurial tube. Only one filament persists, however, the remainder degenerating, and that one filament grows distally to reinnervate a motor or sensory end organ. While crossing the gap between the severed nerve ends, many filaments fail to enter an endoneurial tube and grow out into the surrounding connective tissue. It is interesting to note that the formation of multiple sprouts or filaments from a single proximal axon greatly increases the chances of a neuron's becoming connected to a sensory or motor ending. It is not known why one filament within a single endoneurial tube should be selected to persist whereas the remainder degenerate.

Once the axon has reached the end organ, the adjacent Schwann cells start to lay down a myelin sheath. This process begins at the site of the original lesion and extends in a distal direction. By this means, the nodes of Ranvier and the Schmidt-Lanterman incisures are formed.

Many months may elapse before the axon reaches its appropriate end organ, depending on the site of the nerve injury. The rate of growth has been estimated to be of the order of 2 to 4 mm daily. If, however, one takes into consideration the almost certain delay incurred by the axons as they cross the site of the injury, an overall regeneration rate of 1.5 mm daily is a useful figure to remember for clinical use. Even if all the difficulties outlined above are overcome and a given neuron reaches the original end organ, the enlarging axonal filament within the endoneurial tube reaches only about 80 percent of its original diameter. For this reason the conduction velocity will not be as great as the original axon. Moreover, a given motor axon tends to innervate more muscle fibers than formerly, therefore the control of the muscle is less precise.

SPINAL NERVES

There are 31 pairs of spinal nerves, which leave the spinal cord and pass through intervertebral foramina in the vertebral column. Each spinal nerve is connected to the spinal cord by two roots, the anterior and posterior roots (*Fig. 6–6*). The **anterior root** consists of bundles of nerve fibers carrying nerve impulses away from the central nervous system; these are called efferent or motor fibers. The **posterior root** consists of bundles of nerve fibers carrying nerve impulses to the central nervous system; these are the afferent sensory fibers. The cell bodies of these nerve fibers are situated in a swelling on the posterior root called the **posterior root ganglion.** Because a spinal nerve is formed by the union of an anterior root and a posterior root, it possesses both motor and sensory nerve fibers, and therefore, is a **mixed nerve.**

During development the spinal cord grows in length more slowly than the vertebral column. When growth ceases in the adult the lower end of the spinal cord only reaches inferiorly as far as the lower border of the first lumbar vertebra (see page 153). To accommodate for this disproportionate growth in length, the length of the roots increases progressively from above downward. In the upper cervical region, the spinal nerve roots are short and run almost horizontally, but the roots of the lumbar and sacral nerves below the level of the termination of the cord form a vertical bundle of nerves that resembles a horse's tail and is called the **cauda equina.**

Names of Spinal Nerves

The 31 pairs of spinal nerves are named according to the regions of the vertebral column with which they are associated: 8 cervical, 12 thoracic, 5 lumbar, 5 sacral, and 1 coccygeal nerve. Note that there are 8 cervical nerves and only 7 cervical vertebrae and that there is 1 coccygeal nerve and 4 coccygeal vertebrae.

Branches of Spinal Nerves

After emerging from the intervertebral foramen, each spinal nerve immediately divides into a large **anterior ramus** and a smaller **posterior ramus,** each containing both motor and sensory nerve fibers. The posterior ramus passes posteriorly around the vertebral column to supply the muscles and the skin of the back of the neck and the trunk. The anterior ramus supplies the muscles and the skin over the anterolateral body wall and the muscles and the skin of the limbs.

In addition to the anterior and posterior rami, the spinal nerves give off a small **meningeal branch,** which supplies the vertebrae, the vertebral joints, and the meninges. The spinal nerves also have branches called the **rami communicantes,** which are associated with the sympathetic part of the autonomic nervous system (see page 257).

Plexuses

At the root of the limbs the anterior rami join one another to form complicated nerve plexuses, the **cervical** and **branchial plexuses** of the upper limbs and the **lumbar** and **sacral plexuses** of the lower limbs.

Cervical Plexus. The cervical plexus is formed by the anterior rami of the first four cervical nerves (*Fig. 6–7*). The rami are joined by connecting branches, which form loops that lie in front of the origins of the levator scapulae and the scalenus medius muscles. The cervical plexus supplies the skin and the muscles of the head, the neck, and the shoulders.

Phrenic Nerve

The phrenic nerve (C_3, C_4, and C_5) is a large and very important branch of the cervical plexus; it is the only motor nerve supply to the diaphragm. The phrenic nerve runs vertically downward through the neck on the scalenus anterior muscle (*Fig. 6–8*). Because of the obliquity of the scalenus anterior muscle, the nerve crosses the muscle from its lateral to its medial border. The nerve enters the thorax by passing anterior to the subclavian artery and it crosses the internal thoracic artery from lateral to medial.

The **right phrenic nerve** descends along the right side of the superior vena cava and in front of the root of the right lung. It then passes along the right side of the pericardium to the diaphragm. The **left phrenic nerve** descends along the left side of the left subclavian artery and crosses the left side of the aortic arch and the left vagus nerve. It passes in front of the root of the left lung and descends on the pericardium to the diaphragm.

In addition to the motor fibers to the diaphragm, the phrenic nerve contains sensory fibers from the pericardium and the mediastinal parietal pleura and the pleura and the peritoneum covering the upper and lower surfaces of the central part of the diaphragm.

Table 6–1 summarizes the branches of the cervical plexus and their distribution.

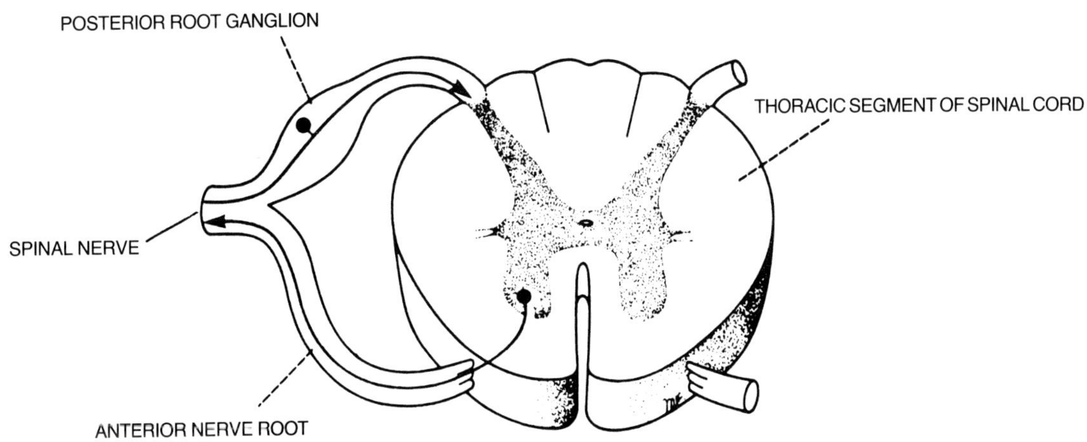

Figure 6–6. Transverse section of the thoracic region of the spinal cord, showing the formation of a spinal nerve from the union of an anterior and a posterior nerve root.

CLINICAL NOTES

CERVICAL PLEXUS BLOCK

Blockade of the main roots of the cervical plexus, formed by the anterior rami of the first four cervical nerves, is referred to as a **deep cervical plexus block.** If only cutaneous analgesia is required, the superficial cutaneous nerves that emerge from the posterior border of the sternocleidomastoid muscle may be blocked; this is referred to as a **superficial cervical plexus**

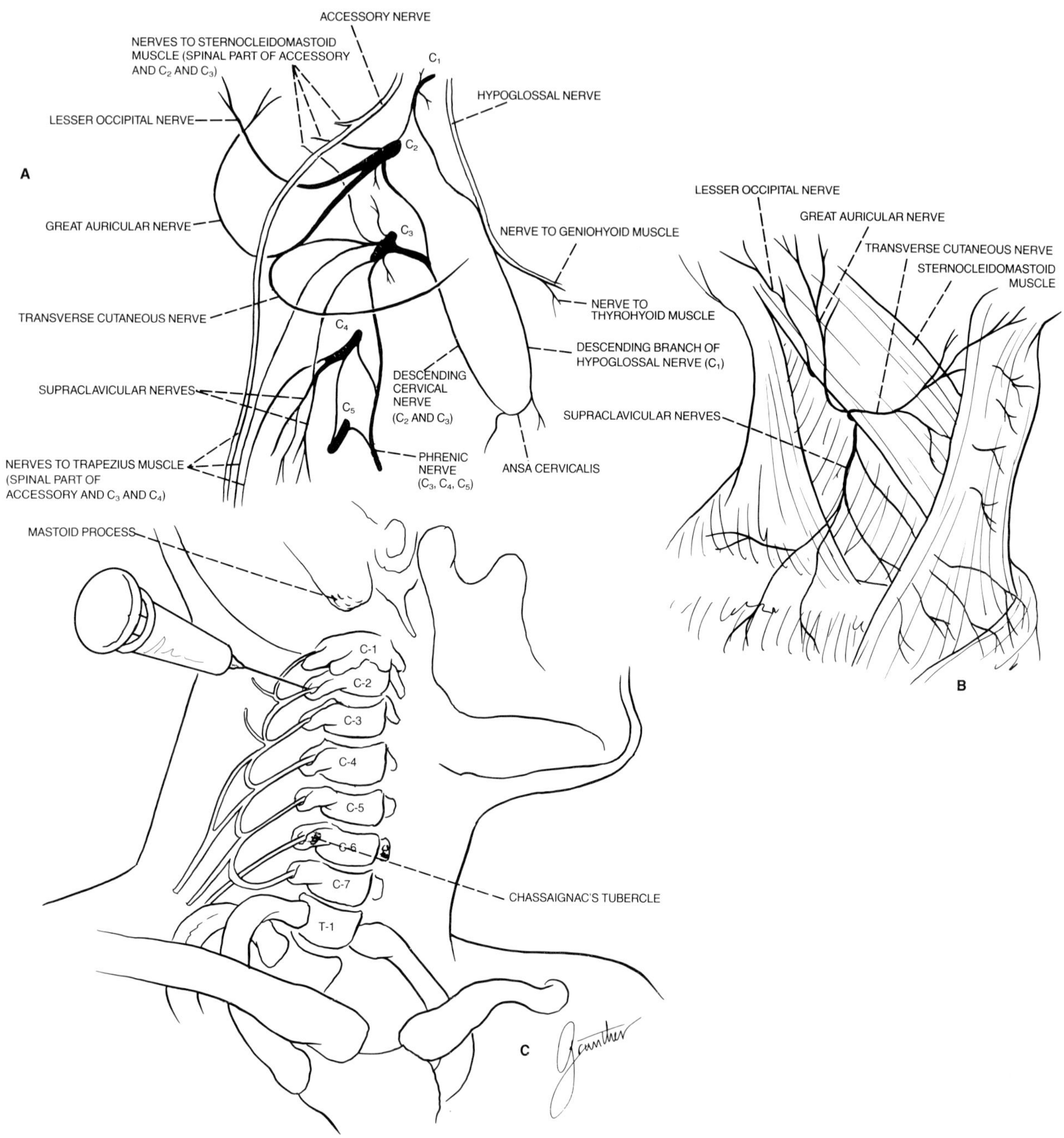

Figure 6–7. A. The cervical plexus and its main branches. **B.** The superficial branches of the cervical plexus in the neck. **C.** Important bony landmarks used in cervical plexus nerve blocks.

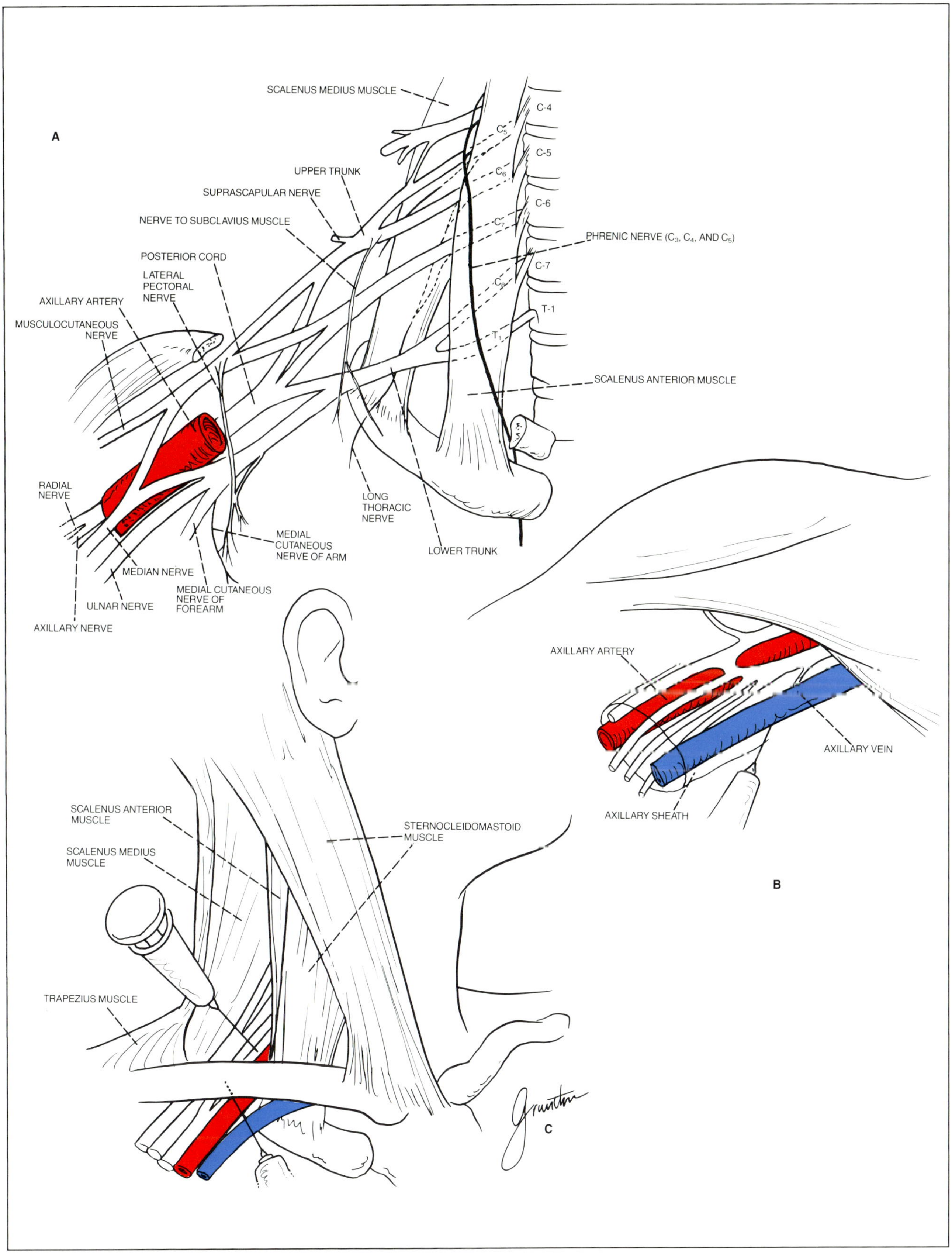

Figure 6–8. **A.** The brachial plexus and its main branches. **B.** The axillary sheath and contents. **C.** Important bony and muscular landmarks used in brachial plexus nerve blocks.

TABLE 6–1. BRANCHES OF THE CERVICAL PLEXUS AND THEIR DISTRIBUTION

Branches	Distribution
Cutaneous	
Lesser occipital	Skin of scalp behind ear
Greater auricular	Skin over parotid gland, auricle, and angle of jaw
Transverse cutaneous	Skin over side and front of neck
Supraclavicular	Skin over upper part of chest and shoulder
Muscular	
Segmental	Prevertebral muscles, levator scapulae
Ansa cervicalis (C_1, C_2, and C_3)	Omohyoid, sternohyoid, sternothyroid
C_1 fibers via hypoglossal nerve	Thyrohyoid, geniohyoid
Phrenic nerve (C_3, C_4, and C_5)	Diaphragm (most important muscle of respiration)
Sensory to Pericardium, Pleura, and Peritoneum	
Phrenic nerve (C_3, C_4, and C_5)	Pericardium, mediastinal parietal pleura, pleura and peritoneum covering central diaphragm

block. For details of techniques of nerve blocks of the cervical plexus and its branches, the reader is referred to the **Atlas of Regional Anesthesia** by Jordan Katz.

Anatomy of Deep Cervical Plexus Block

The tip of the mastoid process, the prominent anterior tubercle (Chassaignac's tubercle) of the sixth cervical vertebra, and the posterior margin of the clavicular origin of the sternocleidomastoid muscle are identified. Note that all of these structures are in line with one another. The anterior ramus of C_2 may be blocked by inserting a needle on the line about 1 inch below the tip of the mastoid process (*Fig. 6–7*). This should contact the tip of the transverse process of the second cervical vertebra and injection of the anesthetic agent can then be made. The insertion of similar needles 0.5 to 1 inch apart on the same line below the first needle, will enable one to block C_3 and C_4 anterior rami, respectively. Note that the vertebral vessels lie between the transverse processes, therefore, avoid passing the needle between the processes.

Anatomy of Superficial Cervical Plexus Block

The posterior margin of the sternocleidomastoid muscle is identified (*Fig. 6–7*). Halfway down the posterior margin of this muscle the great auricular nerve runs upward and backward, the transverse cutaneous nerve runs forward, and the supraclavicular nerves run downward and backward. The introduction of an anesthetic solution at this point will infiltrate the nerves.

Anatomy of Phrenic Nerve Block

The phrenic nerve (C_3, C_4, and C_5) in the lower part of the neck descends on the anterior surface of scalenus anterior muscle (*Fig. 6–8*). A block needle is inserted just posterior to the posterior border of the sternocleidomastoid muscle at about 1 inch above the clavicle. It is directed forward and medially, and the anesthetic solution is infiltrated into the tissues around the nerve.

Brachial Plexus. The brachial plexus is formed in the posterior triangle of the neck by the union of the anterior rami of the fifth, sixth, seventh, and eighth cervical and the first thoracic spinal nerves (*Fig. 6–8*). The plexus may be divided up into **roots, trunks, divisions, and cords.** The roots of C_5 and C_6 unite to form the upper trunk, the root of C_7 continues as the **middle trunk,** the roots of C_8 and T_1 unite to form the **lower trunk.** Each trunk now divides into **anterior and posterior divisions.** The anterior divisions of the upper and middle trunks unite to form the **lateral cord,** the anterior division of the lower trunk continues as the **medial cord,** and the posterior divisions of all three trunks join to form the **posterior cord.**

The roots of the brachial plexus enter the posterior triangle at the base of the neck between the scalenus anterior and scalenus medius muscles and behind the sternocleidomastoid muscle. Here, they are surrounded by a sheath of fascia, the **axillary sheath,** which is derived from the prevertebral layer of the deep cervical fascia. The trunks and divisions cross the posterior triangle of the neck and the cords become arranged around the axillary artery within the axillary sheath in the axilla (*Fig. 6–8*).

Branches of the brachial plexus arise from the roots, the trunks, and the cords. The courses of the more important nerves and their branches are as follows:

Musculocutaneous Nerve. The musculocutaneous nerve (*Fig. 6–8*) arises from the lateral cord of the brachial plexus (C_5, C_6, and C_7). It pierces the coracobrachialis muscle and descends between the biceps and the brachialis muscles. It pierces the deep fascia in the region of the elbow and is distributed along the lateral side of the forearm as the **lateral cutaneous nerve of the forearm.** The musculocutaneous nerve supplies the coracobrachialis, both heads of the biceps, and the greater part of the brachialis muscles.

Median Nerve. The median nerve (*Fig. 6–8*) arises from the medial and lateral cords of the brachial plexus (C_5, C_6, C_7, C_8, and T_1). The nerve descends on the lateral side of the axillary artery and at first on the lateral side of the brachial artery. At the middle of the arm, it crosses the brachial artery to reach its medial side. The nerve enters the forearm between the two heads of pronator teres muscle and descends on the posterior surface of the flexor digitorum superficialis muscle lying on the flexor digitorum profundus muscle. At the wrist, it lies posterior to the tendon of the palmaris longus and between the tendons of the flexor carpi radialis and flexor digitorum superficialis muscles. The median nerve enters the palm by passing **behind** the flexor retinaculum and through the carpal tunnel.

BRANCHES OF THE MEDIAN NERVE IN THE FOREARM

1. **Muscular branches.** Pronator teres, flexor carpi radialis, palmaris longus, flexor digitorum superficialis
2. **Articular branches.** Elbow joint
3. **Anterior interosseous nerve**
   a. **Muscular branches.** Flexor pollicis longus, pronator quadratus, lateral half of flexor digitorum profundus
   b. **Articular branches.** Wrist and carpal joints
4. **Palmar branch.** Distributed to the skin over the lateral part of the palm

BRANCHES OF THE MEDIAN NERVE IN THE PALM

1. **Muscular branches.** Abductor pollicis brevis, flexor pollicis brevis, opponens pollicis, and the first and second lumbrical muscles
2. **Cutaneous branches** to the palmar aspect of the lateral three and one-half fingers and the distal half of the dorsal aspect of each finger

Ulnar Nerve. The ulnar nerve (*Fig. 6–8*) arises from the medial cord of the brachial plexus (C_8, T_1). It descends along the medial side of the axillary and brachial arteries in the anterior compartment of the arm. At the middle of the arm it pierces the medial intermuscular septum and passes down **behind** the medial epicondyle of the humerus. It then enters the anterior compartment of the forearm by passing between the two heads of flexor carpi ulnaris muscle. It descends behind the flexor carpi ulnaris medial to the ulnar artery. At the wrist, it passes **anterior** to the flexor retinaculum, lateral to the pisiform bone. It then divides into superficial and deep terminal branches.

BRANCHES OF THE ULNAR NERVE IN THE FOREARM

1. **Muscular branches.** Flexor carpi ulnaris, medial half of flexor digitorum profundus
2. **Articular branches.** Elbow joint
3. **Palmar branch.** Supplies skin over hypothenar eminence
4. **Dorsal cutaneous branch.** Supplies skin over medial side of back of hand and back of medial one and one-half fingers over the proximal phalanges

BRANCHES OF THE ULNAR NERVE IN THE HAND

Superficial terminal branch descends into the palm and gives off the following branches:

1. **Muscular branches.** Palmaris brevis
2. **Cutaneous branches.** Supplies skin over palmar aspect of medial one and one-half fingers including their nail beds

Deep terminal branch runs backward between the abductor digiti minimi and the flexor digiti minimi muscles and pierces the opponeus digiti minimi. It gives off the following branches:

1. **Muscular branches.** Abductor digiti minimi, flexor digiti minimi, opponens digiti minimi, all palmar and all dorsal interossei, third and fourth lumbricals, adductor pollicis
2. **Articular branches.** Carpal joints

Radial Nerve. The radial nerve (*Fig. 6–8*) arises from the posterior cord of the brachial plexus (C_5, C_6, C_7, C_8, and T_1). It descends behind the axillary and brachial arteries and passes posteriorly between the long and medial heads of the triceps muscle to enter the posterior compartment of the arm. The radial nerve then winds round in the spiral goove on the back of the humerus with the profunda brachii vessels. Piercing the lateral intermuscular septum just above the elbow, it continues downward into the cubital fossa between the brachialis and the brachioradialis muscles. At the level of the lateral epicondyle it divides into superficial and deep branches.

BRANCHES OF THE RADIAL NERVE IN THE AXILLA

1. **Muscular branches.** Long and medial heads of the triceps
2. **Cutaneous branch.** Posterior cutaneous branch of the arm

BRANCHES OF THE RADIAL NERVE IN THE SPIRAL GROOVE

1. **Muscular branches.** Lateral and medial heads of the triceps, anconeus
2. **Cutaneous branches.** Lower lateral cutaneous nerve of the arm, posterior cutaneous nerve of the forearm

BRANCHES OF THE RADIAL NERVE IN THE ANTERIOR COMPARTMENT OF THE ARM

1. **Muscular branches.** Brachialis, brachioradialis, extensor carpi radialis longus
2. **Articular branches.** Elbow joint

Superficial Branch of the Radial Nerve. The superficial branch of the radial nerve is the direct continuation of the radial nerve. It descends under cover of the brachioradialis muscle on the lateral side of the radial artery. In the lower part of the forearm it passes backward under the tendon of the brachioradialis muscle and descends on to the back of the hand.

BRANCHES OF THE SUPERFICIAL BRANCH OF THE RADIAL NERVE

1. **Cutaneous branches.** Lateral two-thirds of the posterior surface of the hand (variable area) and posterior surface of the lateral two and one-half fingers over the proximal phalanxes

Deep Branch of the Radial Nerve. The deep branch of the radial nerve winds around the lateral side of the neck of the radius in the supinator muscle to enter the posterior compartment of the forearm. It descends between the superficial and deep layers of muscle and reaches the back of the interosseous membrane.

BRANCHES OF THE DEEP BRANCH OF THE RADIAL NERVE

1. **Muscular branches.** Extensor carpi radialis brevis, supinator, extensor digitorum, extensor digiti minimi, extensor carpi ulnaris, abductor pollicis longus, extensor pollicis brevis, extensor pollicis longus, extensor indicis
2. **Articular branches.** Wrist and carpal joints

Axillary Nerve. The axillary nerve arises from the posterior cord of the brachial plexus (C_5 and C_6). It passes backward through the quadrilateral space below the shoulder joint in company with the posterior circumflex humeral vessels. The main trunk supplies the shoulder joint and its **anterior terminal branch** winds around the surgical neck of the humerus beneath the deltoid muscle; the anterior branch supplies the deltoid muscle and the skin that covers its lower half. A **posterior terminal branch** supplies the teres minor muscle, the deltoid muscle, and then becomes the **upper lateral cutaneous nerve of the arm,** which supplies the skin over the lower part of the deltoid muscle.

A summary of the branches of the brachial plexus and their distribution is shown in Table 6–2.

CLINICAL NOTES

BRACHIAL PLEXUS BLOCK

The brachial plexus in the neck occupies the lower part of the posterior triangle (*Fig. 6–8*). It lies below and anterior to a line connecting the cricoid cartilage to the midpoint of the clavicle. In the axilla, the brachial plexus and its branches are arranged within the axillary sheath around the axillary artery, which can be palpated.

Anatomy of Interscalene Block

At the level of the cricoid cartilage (C-6), the posterior border of the sternocleidomastoid muscle can be palpated. With the head turned laterally and upward from the side of the block, the palpating finger can feel the groove between the scalenus anterior and the scalenus medius muscles, just lateral to the sternocleidomastoid muscle (*Fig. 6–8*). The blocking needle is inserted into the interval between the scalene muscles and the roots of the upper part of the brachial plexus can be blocked.

Anatomy of Supraclavicular Block

The trunks of the brachial plexus can be blocked as they cross the first rib and enter the axilla. The third part of the subclavian artery is first palpated in the lower anterior corner of the posterior triangle. The posterior border of the sternocleidomastoid muscle is then felt. A blocking needle is inserted at the level of the cricoid cartilage between the scalenus anterior and the scalenus medius muscles and directed caudally behind the subclavian artery toward the upper surface of the first rib (*Fig. 6–8*). It is here that the brachial plexus is very compact consisting of the upper, middle, and lower trunks. The obvious possible complication of this approach is a pneumothorax.

Anatomy of Infraclavicular Block

The middle of the clavicle is identified. A blocking needle is inserted 1 inch (2.5 cm) inferior to it and directed laterally in the direction of the head of the humerus and toward the subclavian artery. The anesthetic solution is infiltrated around the trunks of the brachial plexus (*Fig. 6–8*). The tip of the needle should not enter either of the subclavian vessels.

Anatomy of Axillary Approach

With the arm abducted to an angle greater than 90 degrees, the axillary artery within the axillary sheath may be palpated high up in the axilla (*Fig. 6–8*). The artery is compressed and a blocking needle is inserted just proximal to the point of compression into the axillary sheath. The cords and branches of the plexus lie within the sheath along with artery. The disadvantage of this approach is the difficulty in blocking the musculocutaneous nerve. The object of compressing the artery distal to the point of injection is to close off the sheath distally so that the anesthetic agent may rise in the sheath to the musculocutaneous nerve.

Anatomy of Musculocutaneous Nerve Block

The musculocutaneous nerve may be blocked with the rest of the brachial plexus as it arises from the lateral cord of the brachial plexus (see previous discussion). It may also be blocked as it emerges between the biceps and the brachialis muscles just above the lateral epicondyle of the humerus; it is here that it becomes the lateral cutaneous nerve of the forearm (*Fig. 6–9*).

TABLE 6–2. BRANCHES OF THE BRACHIAL PLEXUS AND THEIR DISTRIBUTION

Branches	Distribution
Roots	
Dorsal scapular nerve (C_5)	Rhomboid minor, rhomboid major, levator scapulae muscles
Long thoracic nerve (C_5, C_6, and C_7)	Serratus anterior muscle
Upper Trunk	
Suprascapular nerve	Supraspinatus and infraspinatus muscles
Nerve to subclavius	Subclavius muscle
Lateral Cord	
Lateral pectoral	Pectoralis major muscle
Musculocutaneous nerve (C_5, C_6, and C_7)	Coracobrachialis, biceps brachii, brachialis muscles and supplies skin along lateral border of forearm when it becomes the lateral cutaneous nerve of forearm
Lateral root of median nerve	See median nerve, medial root
Posterior Cord	
Upper subscapular nerve	Subscapularis muscle
Thoracodorsal nerve	Latissimus dorsi muscle
Lower subscapular nerve	Subscapularis and teres major muscles
Axillary nerve (C_5 and C_6)	Deltoid and teres minor muscles; upper lateral cutaneous nerve of arm supplies skin over lower half of deltoid muscle
Radial nerve (C_5, C_6, C_7, C_8, and T_1)	Triceps, anconeus, part of brachialis, extensor carpi radialis longus; via deep radial nerve branch supplies extensor muscles of forearm: supinator, extensor carpi radialis brevis, extensor carpi ulnaris, extensor digitorum, extensor digiti minimi, extensor indicis, abductor pollicis longus, extensor pollicis longus, extensor pollicis brevis; skin, lower lateral cutaneous nerve of arm, posterior cutaneous nerve of arm, and posterior cutaneous nerve of forearm; skin on lateral side of dorsum of hand and dorsal surface of lateral 3½ fingers. Articular branches to elbow, wrist, and carpal joints.
Medial Cord	
Medial pectoral	Pectoralis major and minor muscles
Medial cutaneous nerve of arm joined by intercostal brachial nerve from second intercostal nerve	Skin of medial side of arm
Medial cutaneous nerve of forearm	Skin of medial side of forearm
Ulnar nerve (C_8 and T_1)	Flexor carpi ulnaris and medial half of flexor digitorum profundus, flexor digiti minimi, abductor digiti minimi, opponens digiti minimi, adductor pollicis, third and fourth lumbricals, interossei, palmaris brevis, skin of medial half of dorsum of hand and palm, skin of palmar and dorsal surfaces of medial 1½ fingers
Medial root of median nerve (with lateral root) forms median nerve (C_5, C_6, C_7, C_8, and T_1)	Pronator teres, flexor carpi radialis, palmaris longus, flexor digitorum superficialis, abductor pollicis brevis, flexor pollicis brevis, opponens pollicis, first two lumbricals, by way of anterior interosseous branch flexor pollicis longus, flexor digitorum profundus (lateral half) pronator quadratus. Palmar cutaneous branch to lateral half of palm and digital branches to palmar surface lateral 3½ fingers. Articular branches to elbow, wrist, and carpal joints

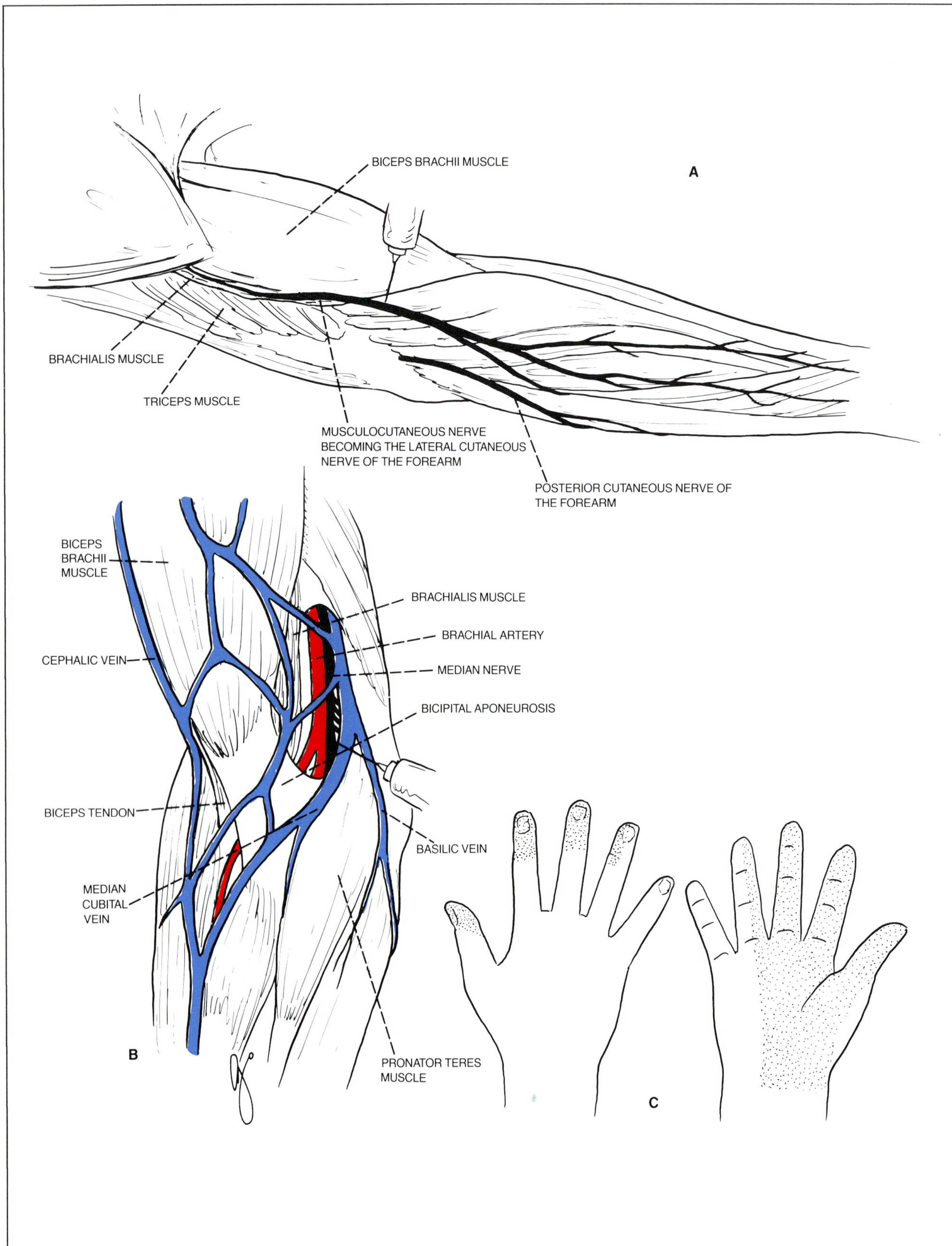

Figure 6–9. A. The musculocutaneous nerve in the region of the elbow joint. **B.** The median nerve entering the cubital fossa. **C.** The cutaneous distribution of the median nerve in the hand.

Anatomy of Median Nerve Block

The median nerve may be blocked at the elbow. The nerve enters the cubital fossa on the medial side of the brachial artery, which can easily be palpated (*Fig. 6–9*).

The median nerve can also be blocked in front of the wrist (*Fig. 6–10*). Here, the nerve lies between the tendons of the flexor carpi radialis (laterally) and the flexor digitorum superficialis (medially); it usually lies posterior to the tendon of the palmaris longus muscle (frequently absent). The nerve may be infiltrated here with local anesthetic just proximal to the flexor retinaculum.

Anatomy of Ulnar Nerve Block

At the elbow, the ulnar nerve enters the forearm between the olecranon process of the ulna and the medial epicondyle of the humerus (*Fig. 6–10*). Here, the nerve may be palpated and infiltrated with an anesthetic agent.

At the wrist, the ulnar nerve enters the hand anterior to the flexor retinaculum and lateral to the tendon of the flexor carpi ulnaris muscle and the pisiform bone (*Fig. 6–10*). The nerve may be infiltrated with an anesthetic agent at this site.

Anatomy of Radial Nerve Block

At the elbow, the radial nerve descends anterior to the lateral epicondyle of the humerus in the interval between the brachialis and the brachioradialis muscles (*Fig. 6–11*). The nerve may be infiltrated with an anesthetic agent at this point.

Just proximal to the wrist, the superficial branch of the radial nerve lies lateral to the radial artery. The nerve leaves the artery and passes laterally and backward under the tendon of the brachioradialis to reach the posterior surface of the wrist. At the level of the proximal flexor crease on the lateral side of the radial artery, the nerve may be infiltrated with an anesthetic agent.

Anatomy of Digital Nerve Block

The finger is supplied by four digital nerves at 2 o'clock, 5 o'clock, 7 o'clock, and 10 o'clock. The nerves are easily blocked with small volumes of local anesthetic.

Thoracic Spinal Nerves

There are 12 thoracic spinal nerves. The posterior rami supply the deep back muscles and the skin over the posterior surface of the back (*Fig. 6–12*). The anterior rami do not join together to form a plexus; instead, the anterior rami of the first 11 thoracic spinal nerves run forward between ribs in intercostal spaces and are known as **intercostal nerves.** The anterior ramus of the 12th thoracic nerve lies below the 12th rib in the abdomen and runs forward in the abdominal wall.

Intercostal Nerves. The intercostal nerves are the anterior rami of the first 11 thoracic spinal nerves. Each intercostal nerve enters an intercostal space (*Fig. 6–12*) and then runs forward inferiorly to the intercostal vessels in the subcostal groove of the corresponding rib, between the transverse thoracic and the internal intercostal muscles. The first six nerves are distributed within their intercostal spaces. The seventh to ninth intercostal nerves leave the anterior ends of their intercostal spaces by passing deep into the costal cartilages, to enter the abdominal wall. In the case of the 10th and 11th nerves, as the corresponding ribs are floating, these nerves pass directly into the abdominal wall.

Branches of the Intercostal Nerves

1. **Collateral branch.*** Runs forward below main nerve in lower part of intercostal space
2. **Lateral cutaneous branch.** Arises in midaxillary line and divides into anterior and posterior branches, which supply the skin
3. **Anterior cutaneous branch.** Terminal portion of main trunk. Reaches the skin near the midline and divides into a medial and a lateral branch
4. **Muscular branches.** Numerous muscular branches are given off by the main nerve and its collateral branch
5. **Pleural branches** to parietal pleura and **peritoneal branches** (7 through 11 intercostal nerves only) to parietal peritoneum. These are sensory nerves.

The first six intercostal nerves give off numerous branches that supply: (1) the skin and the parietal pleura covering the outer and inner surface of each intercostal space, respectively, and (2) the intercostal muscles of each intercostal space. In addition, the 7th to 11th intercostal nerves supply (1) the skin and the parietal peritoneum covering the outer and inner surfaces of the abdominal wall, respectively, and (2) the anterior abdominal muscles, which include the external oblique, internal oblique, transverse abdominal, and rectus abdominis muscles. It should be noted that the first and second intercostal nerves are an exception.

The **first intercostal nerve** is joined to the brachial plexus by a large branch that is equivalent to the lateral cutaneous branch of typical intercostal nerves. The remainder of the first intercostal nerve is small.

The **second intercostal nerve** is joined to the medial cutaneous nerve of the arm by the **intercostobrachial nerve** (*Fig. 6–12*). The second intercostal nerve, therefore, supplies the skin of the armpit and the upper medial side of the arm. In coronary artery heart disease, pain is referred along this nerve to the medial side of the arm.

---

* According to Siddiqi and Mullick (J Anat, 69: 350, 1935), this branch is present in only 10 percent of individuals.

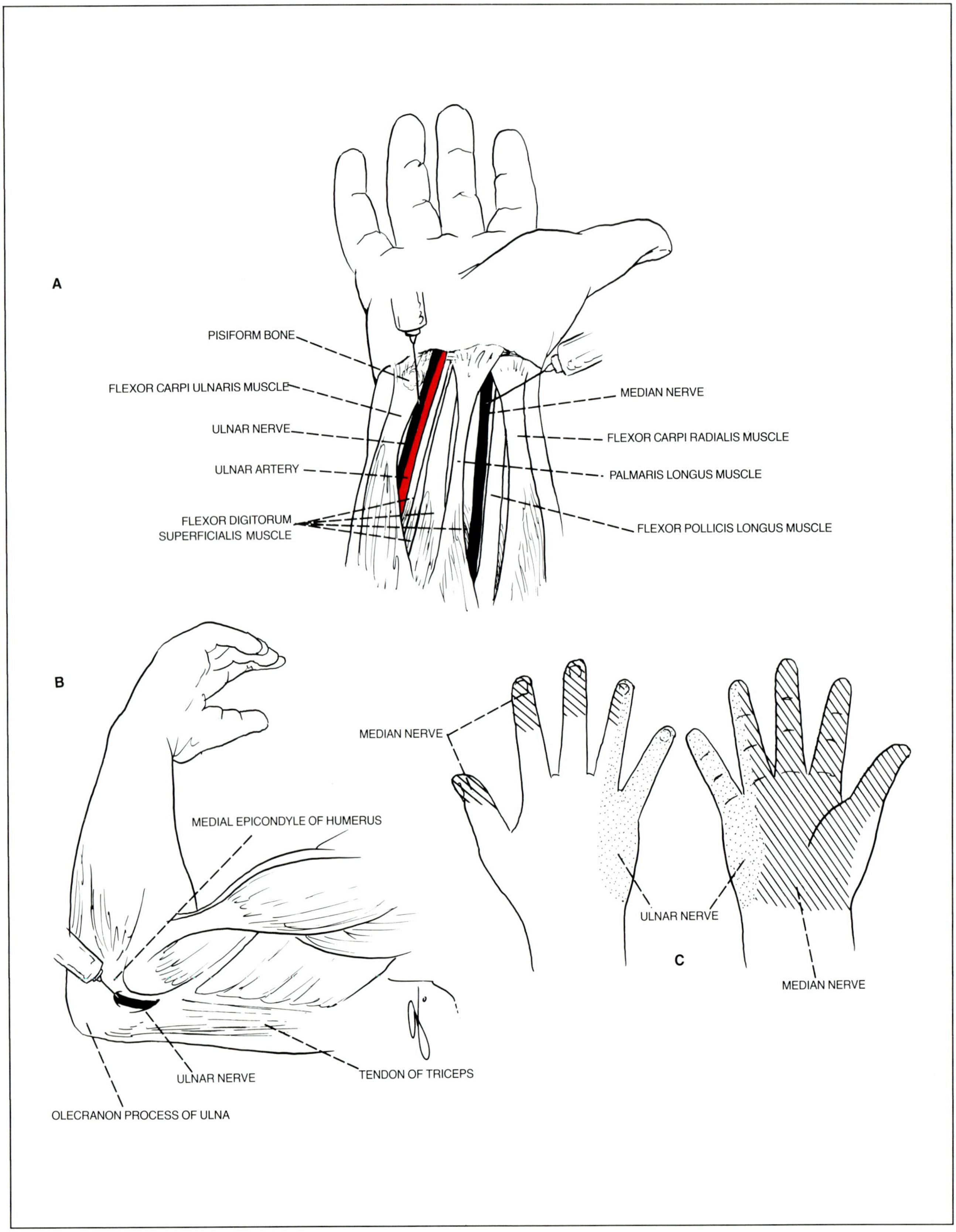

Figure 6–10. A. The median and ulnar nerves in the region of the wrist. **B.** The ulnar nerve as it passes posterior to the medial epicondyle of the humerus. **C.** The cutaneous distribution of the median and ulnar nerves in the hand.

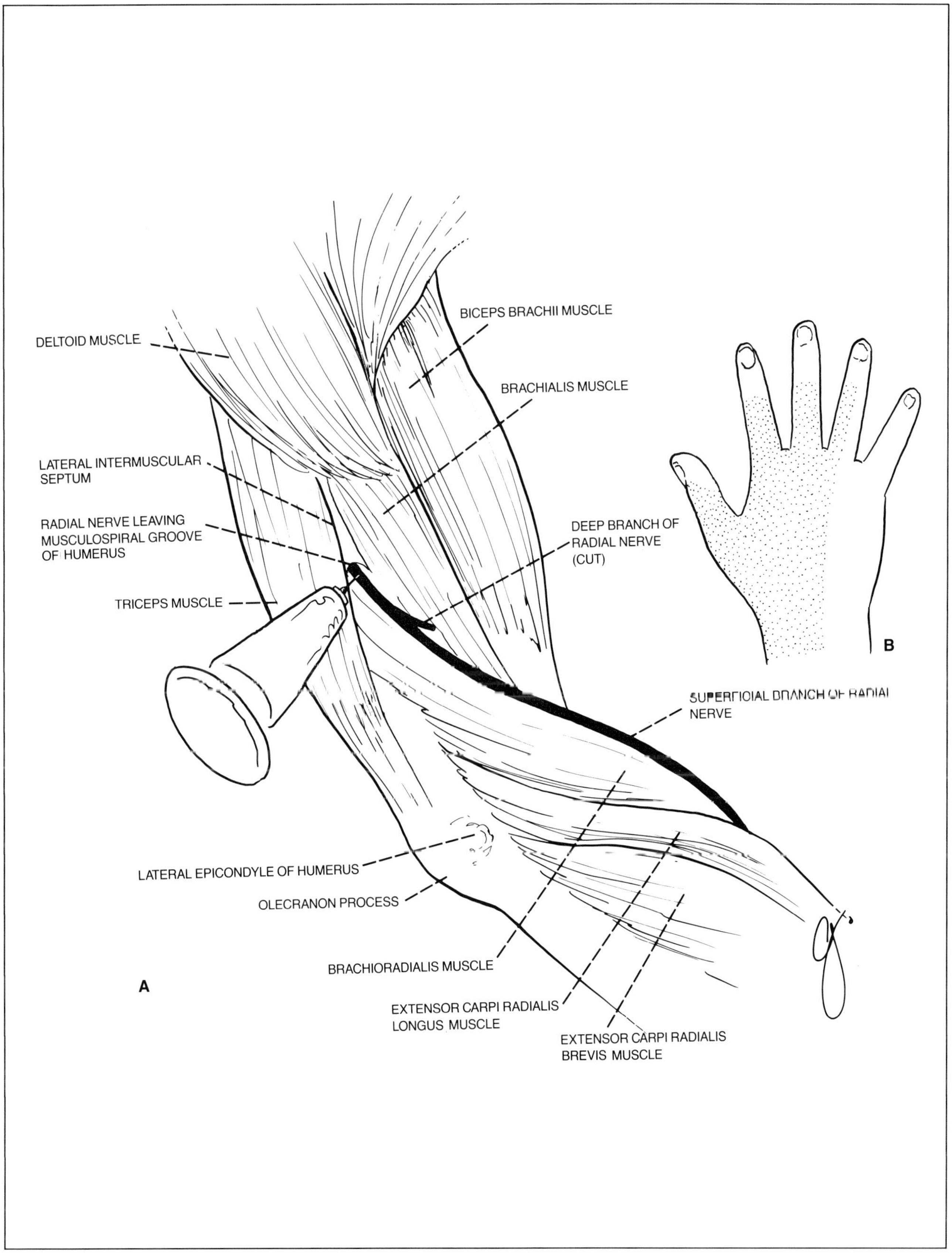

Figure 6–11. A. The radial nerve as it passes forward in the region of the elbow joint. **B.** The cutaneous distribution of the radial nerve in the hand.

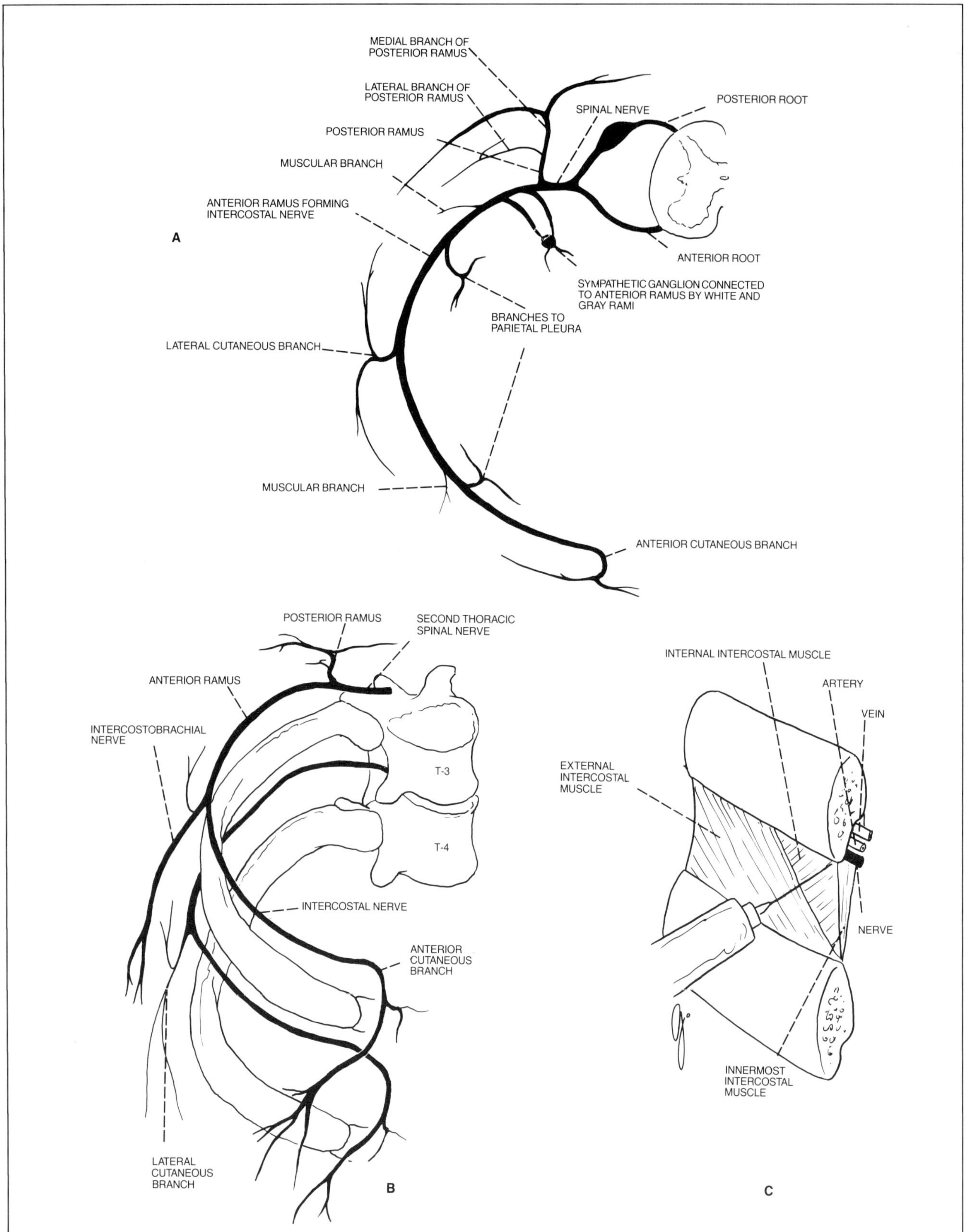

Figure 6–12. A. The origin and distribution of a thoracic spinal nerve. **B.** The distribution of two intercostal nerves relative to the rib cage. **C.** Section through an intercostal space showing the positions of the intercostal nerve, artery, and vein relative to the intercostal muscles.

CLINICAL NOTES

THORACIC SPINAL NERVE BLOCK

Anatomy of Thoracic Paravertebral Nerve Block

The site of emergence of the thoracic spinal nerve from the intervertebral foramen into the paravertebral space is the location of the nerve block. The thoracic spine is identified, which lies immediately superior to the nerve to be blocked. The anesthetic needle is inserted lateral to the spine, which is over the transverse process of the immediately inferior thoracic vertebra. The needle is now passed inferior to the transverse process and is located just lateral to the intervertebral foramen.

Anatomy of Intercostal Nerve Block

To produce analgesia of the anterior and lateral thoracic and abdominal walls, the intercostal nerve should be blocked before the lateral cutaneous branch arises at the midaxillary line. The ribs may be identified by counting up from the 12th or down from above. The needle is directed toward the rib near the lower border (*Fig. 6–12*) and the tip comes to rest near the subcostal groove. Here, the local anesthetic is infiltrated around the nerve. Remember that the order of structures lying in the neurovascular bundle from above downward is: intercostal vein, artery, and nerve, and that these structures are situated between the posterior intercostal membrane of the internal intercostal muscle and the parietal pleura. Further laterally the nerve lies between the internal intercostal muscle and the subcostals and the innermost intercostals (i.e., transverse thoracic muscle).

Anatomy of Anterior Abdominal Nerve Block

The nerves of the anterior and lateral abdominal walls are the anterior rami of the 7th through the 12th thoracic nerves and the first lumbar nerve. The anterior ends of the intercostal nerves T_7 through T_{11} enter the abdominal wall by leaving the intercostal spaces and passing posterior to the costal cartilages (*Fig. 6–13*). An abdominal field block is most easily carried out along the lower border of the costal margin and then infiltrating the nerves as they emerge between the xiphoid process and the 10th or 11th rib along the costal margin.

Lumbar Plexus. The lumbar plexus, which is one of the main nervous pathways supplying the lower limb, is formed inside the psoas muscle from the anterior rami of the upper four lumbar nerves (*Fig. 6–14*). The branches of the plexus emerge from the lateral and medial borders of the muscle and from its anterior surface. The courses of the more important nerves and their branches are as follows:

Femoral Nerve. This is the largest branch of the lumbar plexus and arises from L_2, L_3, and L_4 lumbar nerves. It emerges from the lateral border of the psoas muscle in the abdomen and descends between the psoas and the iliacus muscles. It enters the thigh behind the inguinal ligament and lies lateral to the femoral vessels and the femoral sheath. About 1.5 inches (4 cm) below the inguinal ligament, it terminates by dividing into anterior and posterior divisions.

BRANCHES OF THE FEMORAL NERVE IN THE ABDOMEN

1. Muscular branches to iliacus

BRANCHES OF THE FEMORAL NERVE IN THE THIGH

1. **Cutaneous branches.** The **medial cutaneous nerve of the thigh** supplies the skin on the medial side of the thigh. The **intermediate cutaneous nerve of the thigh** supplies the skin on the anterior surface of the thigh. The **saphenous nerve** descends into the adductor canal crossing the femoral artery on its anterior surface. The nerve emerges on the medial side of the knee joint between the tendons of the sartorius and gracilis muscles. It descends along the medial side of the leg in company with the great saphenous vein. It passes **anterior** to the medial malleous and along the medial border of the foot as far as the ball of the big toe.
2. **Muscular branches.** Sartorius, pectineus, quadriceps femoris. The muscular branch to the rectus femoris also supplies the hip joint; the branches to the three vasti muscles also supply the knee joint.

Obturator Nerve. The obturator nerve arises from the lumbar plexus (L_2, L_3, and L_4). It emerges on the medial border of the psoas muscle within the abdomen. The nerve descends and crosses the pelvic brim anterior to the sacroiliac joint and posterior to the common iliac vessels. It then runs downward and forward on the lateral wall of the pelvis in the angle between the internal and external iliac vessels. Here, it is accompanied by the obturator vessels. On reaching the obturator canal (upper part of obturator foramen), it divides into anterior and posterior divisions.

BRANCHES OF THE OBTURATOR NERVE

1. **Parietal peritoneum.** Sensory fibers to parietal peritoneum on lateral wall of pelvis
2. **Anterior division.** Descends into thigh anterior to obturator externus and adductor brevis
   a. **Muscular branches.** Gracilis, adductor brevis, adductor longus, and sometimes pectineus
   b. **Sensory branch.** Skin on medial side of thigh
   c. **Articular branch.** Hip joint
3. **Posterior division.** Descends through obturator ex-

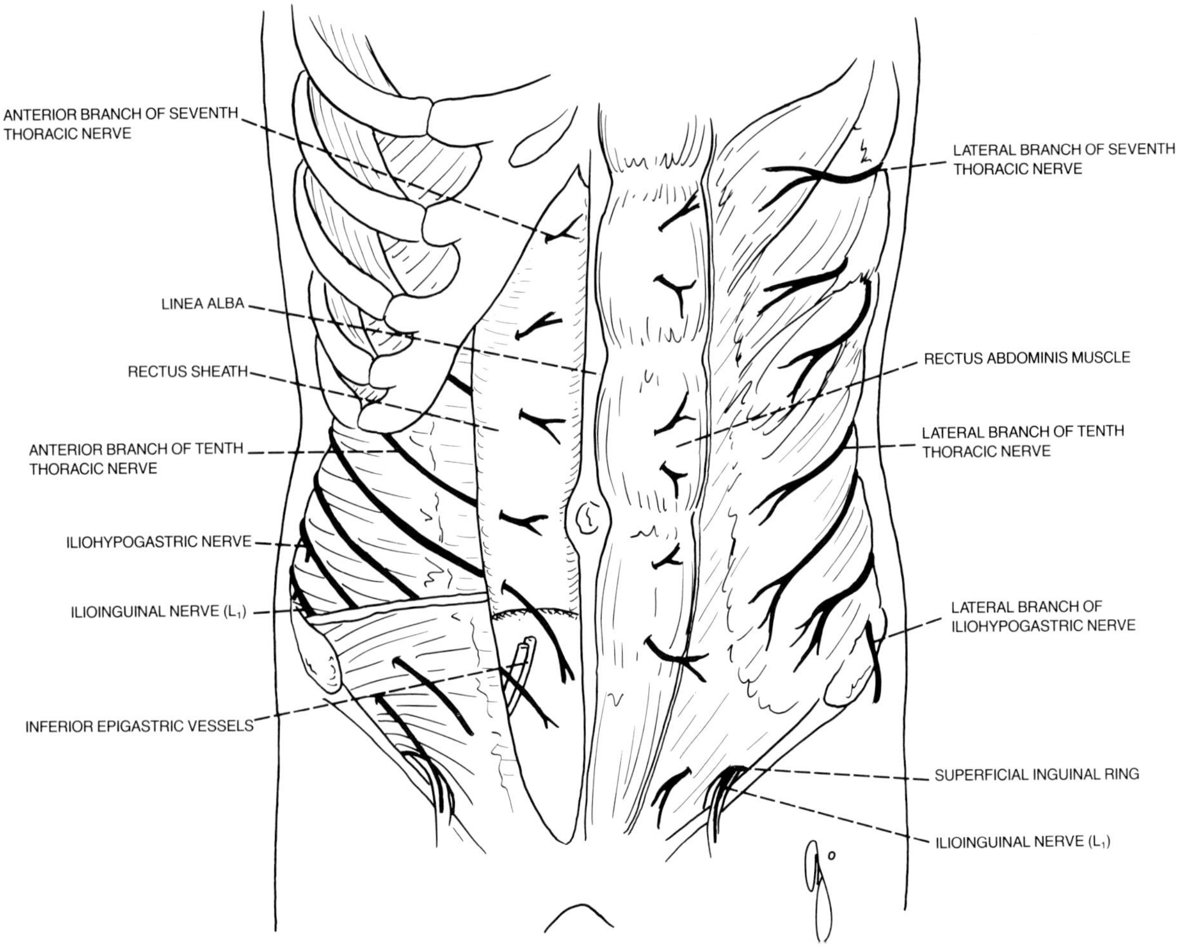

Figure 6–13. The distribution of the thoracic and lumbar nerves in the anterior abdominal wall.

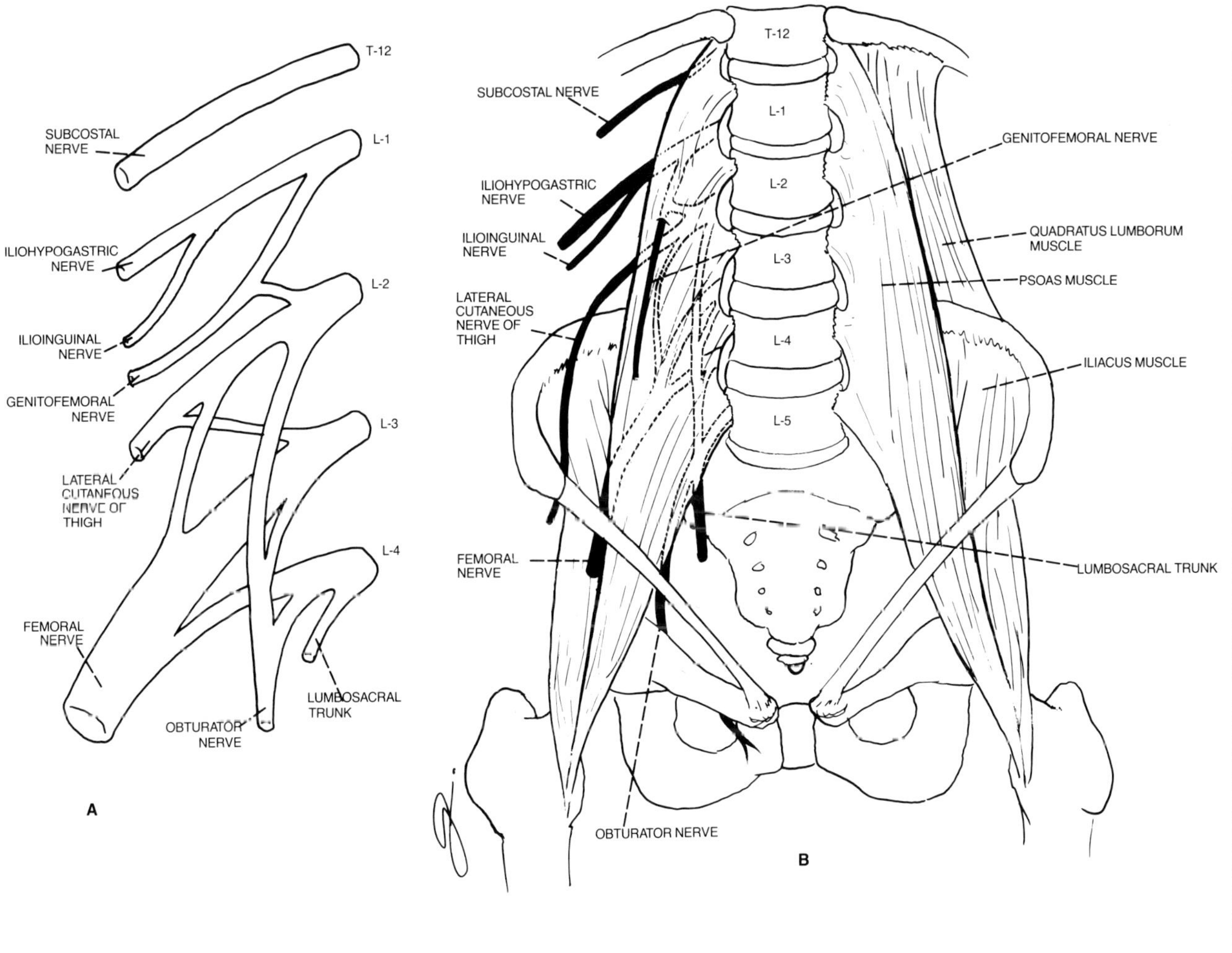

Figure 6–14. A. The lumbar plexus and its main branches. **B.** The lumbar plexus and its branches on the posterior abdominal wall.

TABLE 6–3. BRANCHES OF THE LUMBAR PLEXUS AND THEIR DISTRIBUTION

Branches	Distribution
Iliohypogastric nerve	External oblique, internal oblique, transversus abdominis muscles of anterolateral abdominal wall; skin over lower anterolateral abdominal wall and buttock
Ilioinguinal nerve	External oblique, internal oblique, transversus abdominis muscles of anterolateral abdominal wall; skin of upper medial aspect of thigh, root of penis and scrotum in male, mons pubis and labia majora in female
Lateral cutaneous nerve of the thigh	Skin of anterior and lateral surfaces of the thigh
Genitofemoral nerve (L_1 and L_2)	Cremaster muscle in scrotum of male; skin over anterior surface of thigh. Nervous pathway for cremasteric reflex
Femoral nerve (L_2, L_3, and L_4)	Iliacus muscle, pectineus, sartorius, quadriceps femoris, by medial and intermediate cutaneous branches to the skin of the anterior surface of the thigh and by saphenous branch to skin of medial side of the leg and medial side of the foot. Articular branches. Hip and knee joints
Obturator nerve (L_2, L_3, and L_4)	Gracilis, adductor brevis, adductor longus, obturator externus, ? pectineus, adductor magnus (adductor portion), and skin on medial surface of thigh; Articular branches: hip and knee joints
Segmental branches	Quadratus lumborum and psoas muscles

ternus and passes behind adductor brevis and in front of adductor magnus

a. **Muscular branches.** Obturator externus, adductor magnus (adductor part) and sometimes adductor brevis
b. **Articular branch.** Knee joint

A summary of the branches of the lumbar plexus and their distribution is shown in Table 6–3.

CLINICAL NOTES

LUMBAR SPINAL NERVE BLOCK

Anatomy of Lumbar Paravertebral Block

Each lumbar spinal nerve exits through the intervertebral foramen, which is located about 1 inch (2.5 cm) deep to the caudal edge of the transverse process. The correct vertebral spine is identified (L-4 spine lies on the line joining the highest points on the iliac crests). The caudal edge of the lumbar spine lies at the level of the cephalad edge of the transverse process of the next lower vertebra. The blocking needle is inserted lateral to the caudal edge of the lumbar spine. Once the transverse process is contacted, the needle is reinserted superior to the transverse process in a medial direction. The spinal nerve is then infiltrated with local anesthetic.

Anatomy of Ilioinguinal and Iliohypogastric Nerve Blocks

The ilioinguinal and iliohypogastric nerves (L_1) arise from the lumbar plexus within the psoas muscle. There may also be a small contribution from T_{12}. Having passed around the abdominal wall between the transversus abdominis and the internal oblique muscles, the iliohypogastric nerve finally pierces the external oblique aponeurosis above the superficial inguinal ring to supply the skin (*Fig. 6–13*). The ilioinguinal nerve passes forward through the inguinal canal and emerges through the superficial ring to supply the skin of the groin and part of the scrotum or labium majus.

The two nerves are easily blocked by inserting the anesthetic needle 1 inch (2.5 cm) above the anterior superior iliac spine on the spinoumbilical line.

Anatomy of Genitofemoral Nerve Block

The genitofemoral nerve (L_1 and L_2) arises from the lumbar plexus within the psoas muscle. Having emerged from the psoas muscle, the femoral branch runs under the inguinal ligament with the external iliac artery, at a point halfway between the anterior superior iliac spine and the symphysis pubis. It supplies a small area of skin below the inguinal ligament. The genital branch in the male supplies the cremaster muscle and part of the skin of the scrotum. In the female, the nerve supplies a small area of skin below the inguinal ligament and part of the skin of the labium majus. The terminal branches are most easily blocked by infiltrating the subcutaneous tissue through a needle inserted through the skin lateral to the pubic tubercle.

Anatomy of Femoral Nerve Block

The femoral nerve (L_2, L_3, and L_4) arises within the substance of the psoas muscle. It enters the thigh from beneath the inguinal ligament at a point midway between the anterior superior iliac spine and the pubic tubercle (*Fig. 6–15*). Here, it lies lateral to the femoral artery. It quickly divides into a leash of terminal branches that lie deep to the deep fascia of the thigh.

The nerve may be blocked by introducing the an-

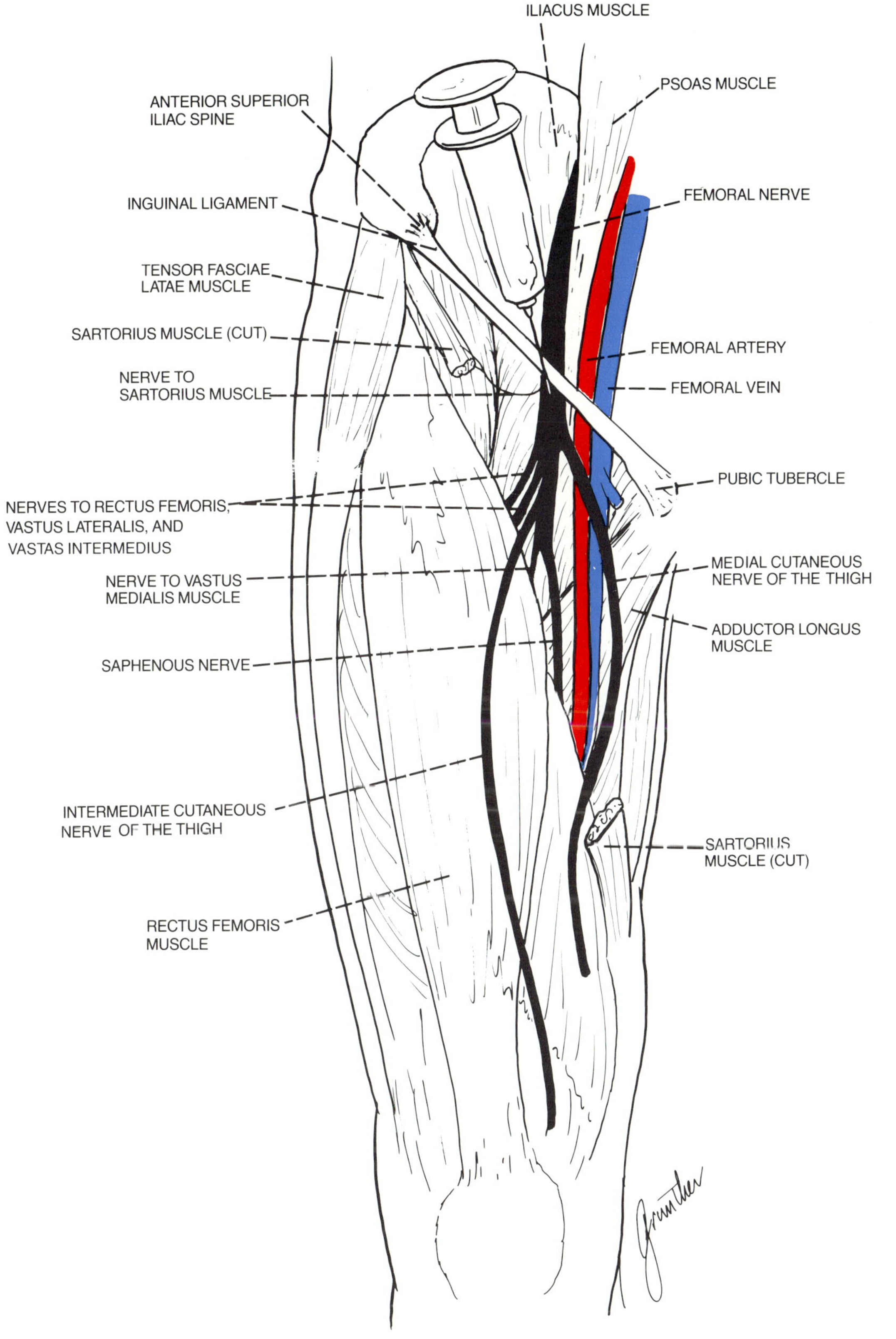

Figure 6–15. The femoral nerve and its branches in the front of the thigh.

esthetic needle just below the midpoint of the inguinal ligament and lateral to the femoral artery.

Anatomy of Lateral Cutaneous Nerve of the Thigh Block

The lateral cutaneous nerve of the thigh (L_2 and L_3) arises from the lumbar plexus within the substance of the psoas muscle. It enters the thigh behind (or through) the lateral end of the inguinal ligament just medial to the anterior superior iliac spine (*Fig. 6–14*). It then descends anterior or through the sartorius muscle and divides into its terminal anterior and posterior branches.

The nerve may be blocked by inserting the anesthetic needle just inferior to the inguinal ligament about 0.5 inches (1.3 cm) medial to the anterior superior iliac spine.

Anatomy of Obturator Nerve Block

The obturator nerve (L_2, L_3, and L_4) arises from the lumbar plexus within the substance of the psoas muscle. Having descended into the pelvis, it runs downward and forward along the lateral wall and enters the thigh through the obturator canal. It terminates by dividing into anterior and posterior divisions (*Fig. 6–16*).

The nerve may be blocked by inserting an anesthetic needle 1 inch (2.5 cm) below and lateral to the pubic tubercle so that the tip of the needle comes to lie just outside the obturator canal.

Anatomy of Saphenous Nerve Block

The saphenous nerve is the longest terminal branch of the femoral nerve and it is entirely sensory. It becomes superficial on the medial side of the knee after emerging between the tendons of sartorius and gracilis muscles. It supplies the skin down the medial side of the leg and the medial border of the foot, and terminates in the region of the ball of the great toe. It accompanies the great saphenous vein and passes anterior to the medial malleolus at the ankle.

The nerve may be blocked by inserting the anesthetic needle on the medial side of the knee joint either over the medial femoral condyle or lower down over the condyle of the tibia (*Fig. 6–19*). Be careful to avoid the great saphenous vein. The nerve may also be blocked at the ankle where it passes anterior to the medial malleolus.

Sacral Plexus. The sacral plexus is formed from the anterior rami of the fourth and fifth lumbar nerves and the anterior rami of the first, second, third, and fourth sacral nerves (*Fig. 6–17*). Note that the contribution from the fourth lumbar nerve joins the fifth lumbar nerve to form the **lumbosacral trunk.** The lumbosacral trunk passes down into the pelvis and joins the sacral nerves as they emerge from the anterior sacral foramina. Parasympathetic branches arise from the second, third, and fourth sacral nerves to form the **pelvic splanchnic nerves.**

The courses of the most important nerves and their branches are as follows:

Sciatic Nerve. The sciatic nerve (*Fig. 6–17*) is the largest nerve in the body and arises from L_4, L_5, S_1, S_2, and S_3. It passes out of the pelvis into the gluteal region through the greater sciatic foramen. The nerve appears below the piriformis muscle (*Fig. 6–18*) and curves downward and laterally lying on the superior gemellus, obturator internus, inferior gemellus, and quadratus femoris muscles to reach the back of the adductor magnus muscle. It is covered by the gluteus maximus and the long head of the biceps femoris muscles. In the lower third of the thigh (occasionally at a higher level), it ends by dividing into the tibial and common peroneal nerves.

BRANCHES OF THE SCIATIC NERVE

1. **Muscular branches.** Biceps femoris (long head), semitendinosus, semimembranosus, and the hamstring part of adductor magnus
2. **Articular branches.** Hip joint

Tibial Nerve. The tibial nerve (L_4, L_5, S_1, S_2, and S_3) runs downward through the popliteal fossa crossing the popliteal artery from the lateral to the medial side. It then passes deep to the gastrocnemius and soleus muscles and descends to the interval between the medial malleolus and the heel. It is covered here by the flexor retinaculum and divides into the medial and lateral plantar nerves.

BRANCHES OF THE TIBIAL NERVE

1. **Cutaneous branches.** The **sural nerve** is usually joined by the sural communicating branch of the common peroneal nerve. It supplies the skin of the calf and the back of the leg and accompanies the small saphenous vein behind the lateral malleolus to supply the skin of the lateral border of the foot and the lateral side of the little toe. The **medial calcaneal branch** supplies the skin over the medial surface of the heel.
2. **Muscular branches.** Gastrocnemius, plantaris, soleus, popliteus, flexor digitorum longus, flexor hallucis longus, tibialis posterior
3. **Articular branches.** Knee joint, ankle joint

Medial Plantar Nerve. The medial plantar nerve runs forward deep to the abductor hallucis muscle with the medial plantar artery.

BRANCHES OF THE MEDIAL PLANTAR NERVE

1. **Cutaneous branches.** Medial part of sole and plantar digital nerves supply medial three and one-half toes and the nail beds
2. **Muscular branches.** Abductor hallucis, flexor digitorum brevis, flexor hallucis brevis, first lumbrical

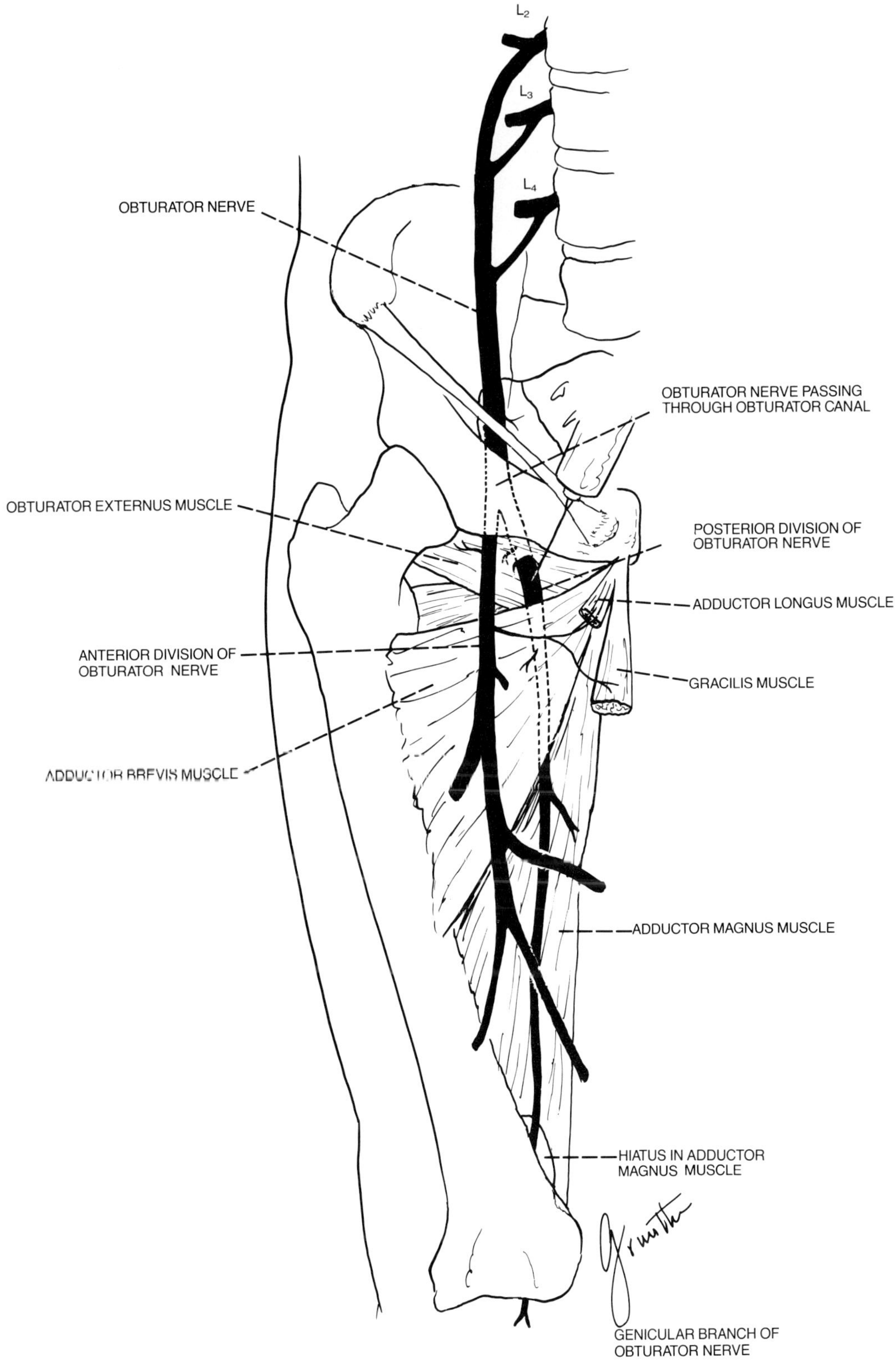

Figure 6–16. The origin, course, and distribution of the obturator nerve.

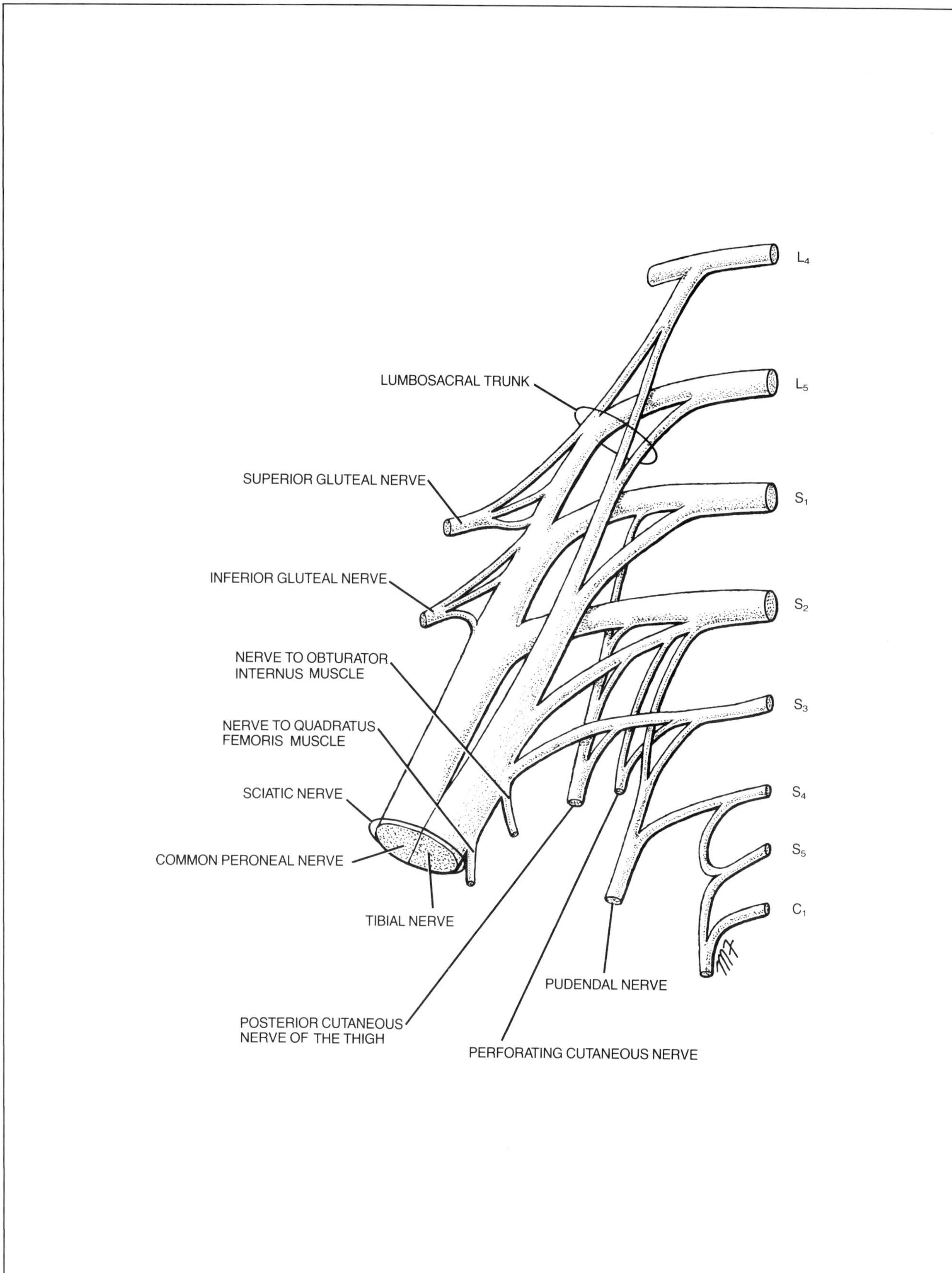

Figure 6–17. The sacral plexus and its main branches.

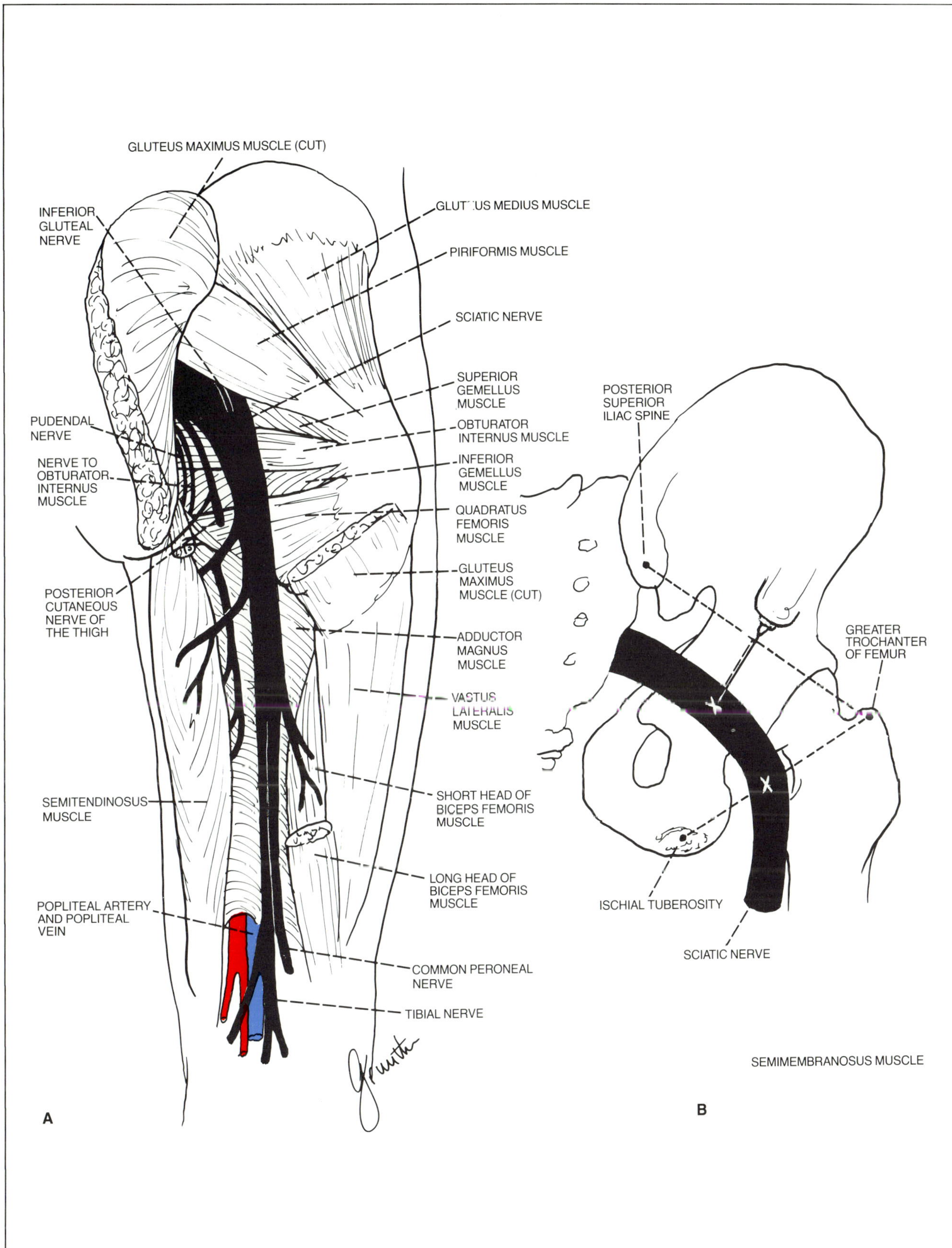

Figure 6–18. A. The sciatic nerve in the gluteal region and the back of the thigh. **B.** Important bony landmarks used when performing a sciatic nerve block.

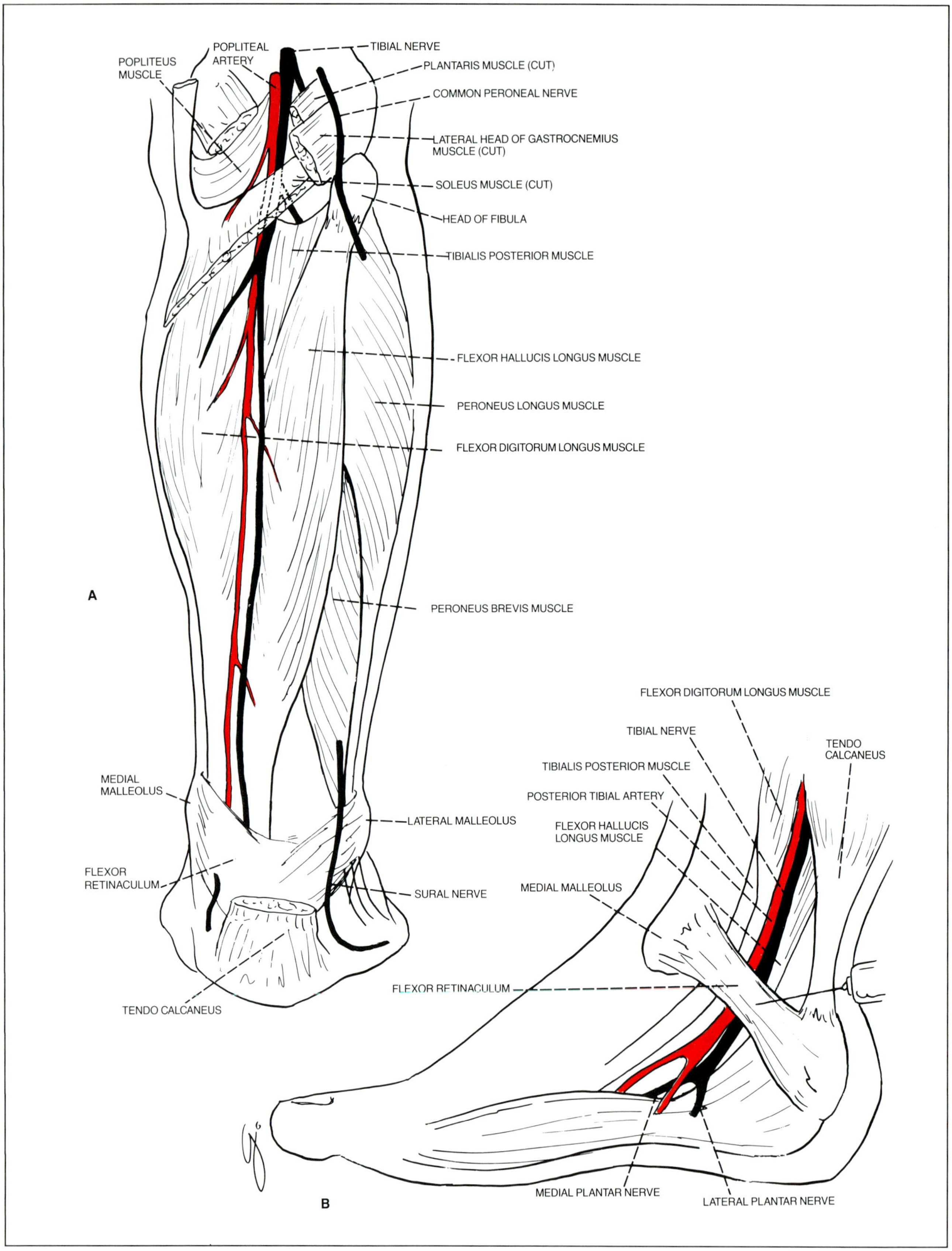

Figure 6–19. A. The tibial nerve in the posterior compartment of the leg. **B.** The tibial nerve as it descends posterior to the medial malleolus.

Lateral Plantar Nerve. The lateral plantar nerve runs forward deep to the abductor hallucis and flexor digitorum brevis in company with the lateral plantar artery.

BRANCHES OF THE LATERAL PLANTAR NERVE

1. **Cutaneous branches.** Skin of lateral part of sole and plantar digital branches to lateral one and one-half toes and the nail beds
2. **Muscular branches.** Flexor digitorum accessorius, abductor digiti minimi, flexor digiti minimi brevis, adductor hallucis, interosseous muscles, second, third, and fourth lumbricals

Common Peroneal Nerve. The common peroneal nerve (L_4, L_5, S_1, and S_2) runs downward through the popliteal fossa following the tendon of the biceps femoris muscle. It then passes laterally around the neck of the fibula, pierces the peroneus longus muscle, and divides into the superficial peroneal nerve and the deep peroneal nerve.

BRANCHES OF THE COMMON PERONEAL NERVE

1. **Cutaneous branches. Sural communicating branch** joins sural nerve. **Lateral cutaneous nerve of the calf** supplies the skin on the lateral side of the back of the leg
2. **Muscular branch.** Short head of the biceps femoris
3. **Articular branches.** Knee joint
4. **Terminal branches.**

Superficial Peroneal Nerve. The superficial peroneal nerve descends between the peroneus longus and brevis muscles and then between the peroneus brevis and the extensor digitorum longus muscles.

BRANCHES OF THE SUPERFICIAL PERONEAL NERVE

1. **Cutaneous branches.** Medial and lateral branches are distributed to skin on front of leg and dorsum of foot except cleft between big and second toes
2. **Muscular branches.** Peroneus longus and brevis

Deep Peroneal Nerve. The deep peroneal nerve (S_2, S_3, and S_4) leaves the peroneus longus muscle and enters the anterior compartment of the leg. It descends deep to the extensor digitorum longus muscle and is accompanied by the anterior tibial vessels. The nerve enters the dorsum of the foot on the lateral side of the dorsalis pedis artery and divides into medial and lateral terminal branches.

BRANCHES OF THE DEEP PERONEAL NERVE

1. **Cutaneous branch.** Adjacent side of big and second toes
2. **Muscular branches.** Tibialis anterior, extensor digitorum longus, peroneus tertius, extensor hallucis longus, extensor digitorum brevis
3. **Articular branches.** Ankle joint, tarsal joints

Pudendal Nerve. The pudendal nerve arises from S_2, S_3, and S_4 spinal nerves. It leaves the pelvis through the greater sciatic foramen and after a brief course in the gluteal region, enters the perineum through the lesser sciatic foramen. The nerve passes forward in the **pudendal canal** with the internal pudendal vessels on the lateral wall of the ischiorectal fossa. It ends by dividing into the perineal nerve and the dorsal nerve of the penis (clitoris).

BRANCHES OF THE PUDENDAL NERVE

1. **Inferior rectal nerve.** Crosses ischiorectal fossa to anal canal
   (a) **Sensory branch.** To mucous membrane of lower half of anal canal and perianal skin
   (b) **Muscular branch.** External anal sphincter
2. **Perineal nerve.**
   (a) **Cutaneous branches. Posterior scrotal (labial) nerves** to posterior surface of scrotum or labia majora
   (b) **Muscular branches.** Superficial and deep transverse perineal muscles, bulbospongiosus, ischiocavernosus, sphincter urethrae, levator ani
3. **Dorsal Nerve of the penis (Clitoris).** Supplies skin and deeper structures of penis (clitoris)

The branches of the sacral plexus and their distribution are shown in Table 6–4.

Coccygeal Nerve. The anterior ramus of the coccygeal nerve joins with branches of the anterior rami of the fourth and fifth sacral nerves to form the **coccygeal plexus.** Branches from this plexus supply the skin over the coccyx.

CLINICAL NOTES

SACRAL SPINAL NERVE BLOCK

Anatomy of Sciatic Nerve Block

The sciatic nerve (L_4, L_5, S_1, S_2, and S_3) arises from the sacral plexus. It emerges from the pelvis through the lower part of the greater sciatic foramen and curves downward and laterally through the gluteal region under cover of the gluteus maximus muscle (*Fig. 6–18*). It lies approximately midway between the ischial tuberosity and the tip of the greater trochanter of the femur.

The posterior superior iliac spine and the tip of the great trochanter are palpated. A line joining these two points is bisected at right angles by another line. At a location 2 inches (5 cm) caudad on this second

TABLE 6–4. BRANCHES OF THE SACRAL PLEXUS AND THEIR DISTRIBUTION

Branches	Distribution
Superior gluteal nerve	Gluteus medius, gluteus minimus, and tensor fasciae latae muscles
Inferior gluteal nerve	Gluteus maximum muscle
Nerve to piriformis	Piriformis muscle
Nerve to obturator internus	Obturator internus and superior gemellus muscles
Nerve to quadratus femoris	Quadratus femoris and inferior gemellus muscles
Perforating cutaneous nerve	Skin over medial aspect of buttock
Posterior cutaneous nerve of thigh	Skin over posterior surface of thigh and popliteal fossa, also skin over lower part of buttock, scrotum, or labium majus
Sciatic nerve (L_4, L_5, S_1, S_2, and S_3)	
Tibial portion	Hamstring muscles [semitendinosus, semimembranosus, biceps femoris (long head), adductor magnus (hamstring part)], gastrocnemius, soleus, plantaris, popliteus, tibialis posterior, flexor digitorum longus, flexor hallucis longus, and via medial and lateral plantar branches to muscles of sole of foot. Sural branch supplies skin on lateral side of leg and foot
Common peroneal portion	Biceps femoris muscle (short head) and via deep peroneal branch: tibialis anterior, extensor hallucis longus, extensor digitorum longus, peroneus tertius, and extensor digitorum brevis muscles. Skin over cleft between first and second toes. Superficial peroneal branch supplies the peroneus longus and brevis muscles and skin over lower third of anterior surface of leg and dorsum of foot
Pudendal nerve	Muscles of perineum including the external anal sphincter, mucous membrane of lower half of anal canal, perianal skin, skin of penis, scrotum, clitoris, labia majora and minora

line, an anesthetic needle may be inserted through the gluteus maximus muscle and the tissue around the sciatic nerve infiltrated (*Fig. 6–18*).

The sciatic nerve may also be blocked as it is leaving the gluteal region, by inserting the needle at a point halfway between the ischial tuberosity and the tip of the greater trochanter (*Fig. 6–18*).

Anatomy of Tibial Nerve Block

The tibial nerve (L_4, L_5, S_1, S_2, and S_3) is the largest terminal branch of the sciatic nerve. It is deeply placed and descends through the back of the thigh, the popliteal fossa, and the back of the leg. At the ankle, the nerve, accompanied by the posterior tibial artery, becomes superficial. It lies behind the medial malleolus, between the tendons of the flexor digitorum longus and the flexor hallucis longus muscles, and is covered by the flexor retinaculum (*Fig. 6–19*).

The tibial nerve is commonly blocked as it lies behind the medial malleolus. By careful palpation, the pulsations of the posterior tibial artery can be felt midway between the medial malleolus and the heel. The nerve lies immediately lateral to the artery and the blocking needle can be inserted at this location.

Anatomy of Common Peroneal Nerve Block

The common peroneal nerve (L_4, L_5, S_1, and S_2) is the smaller of the terminal branches of the sciatic nerve. It descends through the back of the thigh and the popliteal fossa (*Fig. 6–19*). It then winds laterally around the neck of the fibula to pierce the peroneus longus muscle (*Fig. 6–20*). It is here that it divides into the superficial and deep peroneal nerves. As the nerve lies on the lateral side of the neck of the fibula, it is subcutaneous and can easily be rolled against the bone.

The common peroneal nerve can be blocked by first palpating the nerve below the head of the fibula and infiltrating the tissue around the nerve with a local anesthetic agent.

Anatomy of Superficial Peroneal Nerve Block

The superficial peroneal nerve, a branch of the common peroneal nerve, descends through the lateral compartment of the leg (*Fig. 6–20*). It splits into terminal branches that supply the skin on the dorsum of the foot and the toes (except for the cleft between the first and second toes, which is innervated by the deep peroneal nerve and the lateral side of the little toe, which is supplied by the sural nerve).

The superficial peroneal nerve is easily blocked by infiltrating the anesthetic in the subcutaneous tissue along a transverse line connecting the medial and lateral malleoli (*Fig. 6–20*).

Anatomy of Deep Peroneal Nerve Block

The deep peroneal nerve descends in the anterior compartment of the leg lying deep to the extensor digitorum

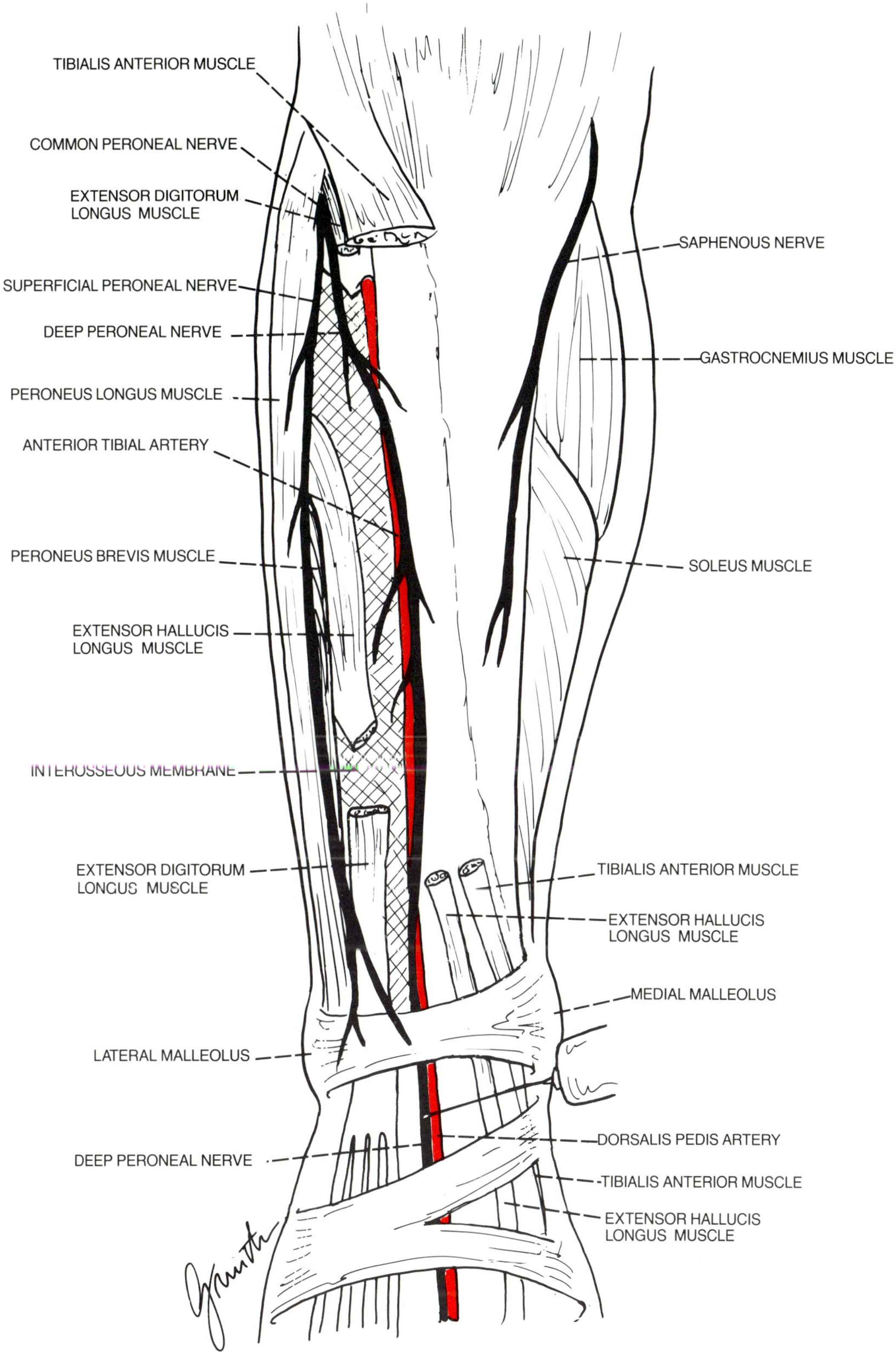

Figure 6–20. The common peroneal, the superficial peroneal, and the deep peroneal nerves in the leg. Note the relationships of the deep peroneal nerve as it lies on the dorsum of the foot.

longus muscle and accompanying the anterior tibial vessels (*Fig. 6–20*). At the ankle, it passes onto the dorsum of the foot on the lateral side of the dorsalis pedis artery and is superficially placed between the tendons of extensor digitorum longus and the extensor hallucis longus muscles (*Fig. 6–20*).

The dorsalis pedis artery can be palpated, midway between the medial and lateral malleoli. With the foot actively dorsiflexed, the tendons of extensor digitorum longus and extensor hallucis longus muscles can be seen. The nerve lies on the lateral side of the artery between these tendons (*Fig. 6–20*). The needle may then be inserted over the nerve and the surrounding tissues infiltrated with anesthetic.

Anatomy of Sural Nerve Block

The sural nerve is a branch of the tibial nerve in the popliteal fossa. It descends superficially in the calf accompanied by the small saphenous vein. It passes behind the lateral malleolus and is distributed to the skin along the lateral border of the foot and little toe (*Fig. 6–19*). The nerve may be blocked by inserting a needle midway between the lateral malleolus and the tendo calcaneus (Achillis) and infiltrating the subcutaneous tissue with local anesthetic.

Anatomy of Pudendal Nerve Block

The pudendal nerve (S_2, S_3, and S_4) leaves the pelvis through the greater sciatic foramen. It crosses the ischial spine with the pudendal vessels and enters the perineum through the lesser sciatic foramen. It then passes forward along the lateral wall of the ischiorectal fossa (*Fig. 6–21*) in a sheath of deep fascia called the **pudendal canal** (Alcock's canal).

The pudendal nerve may be blocked by inserting the needle into the pudendal canal along the medial side of the ischial tuberosity.

Dermatomes. The skin covering the whole body is supplied segmentally by spinal nerves. The area of skin supplied by a single spinal nerve, and therefore, a single segment of the spinal cord, is called a **dermatome** (*Figs. 6–22 and 6–23*). On the trunk, adjacent dermatomes overlap considerably, therefore, to produce a region of complete anesthesia at least three adjacent spinal nerves have to be blocked. In the upper and lower limbs, the arrangement of the dermatomes is more complicated, and this is caused by the embryological changes that take place as the limbs grow out from the body wall. The skin of the face is largely supplied by the three divisions of the trigeminal cranial nerve. Here, there is little or no overlap of the areas of skin supplied by each of the divisions.

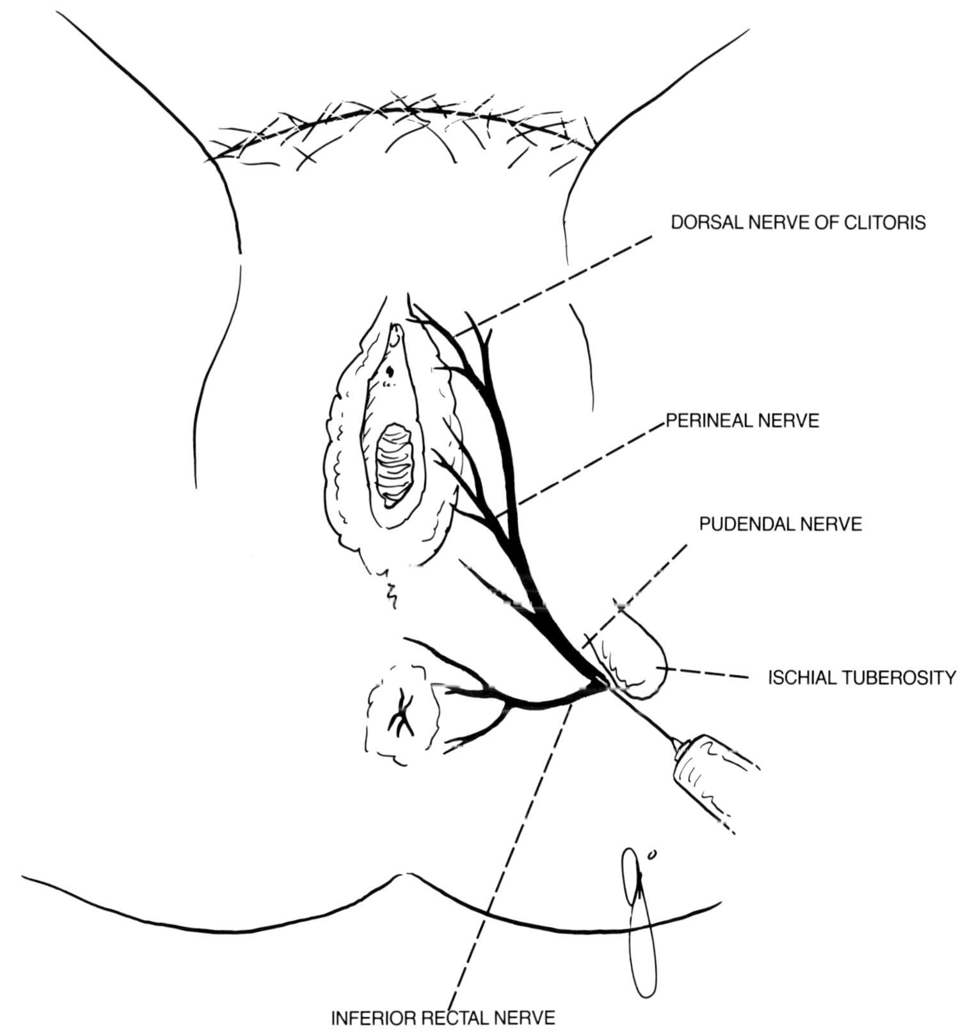

Figure 6–21. The pudendal nerve as it lies within the pudendal canal medial to the ischial tuberosity.

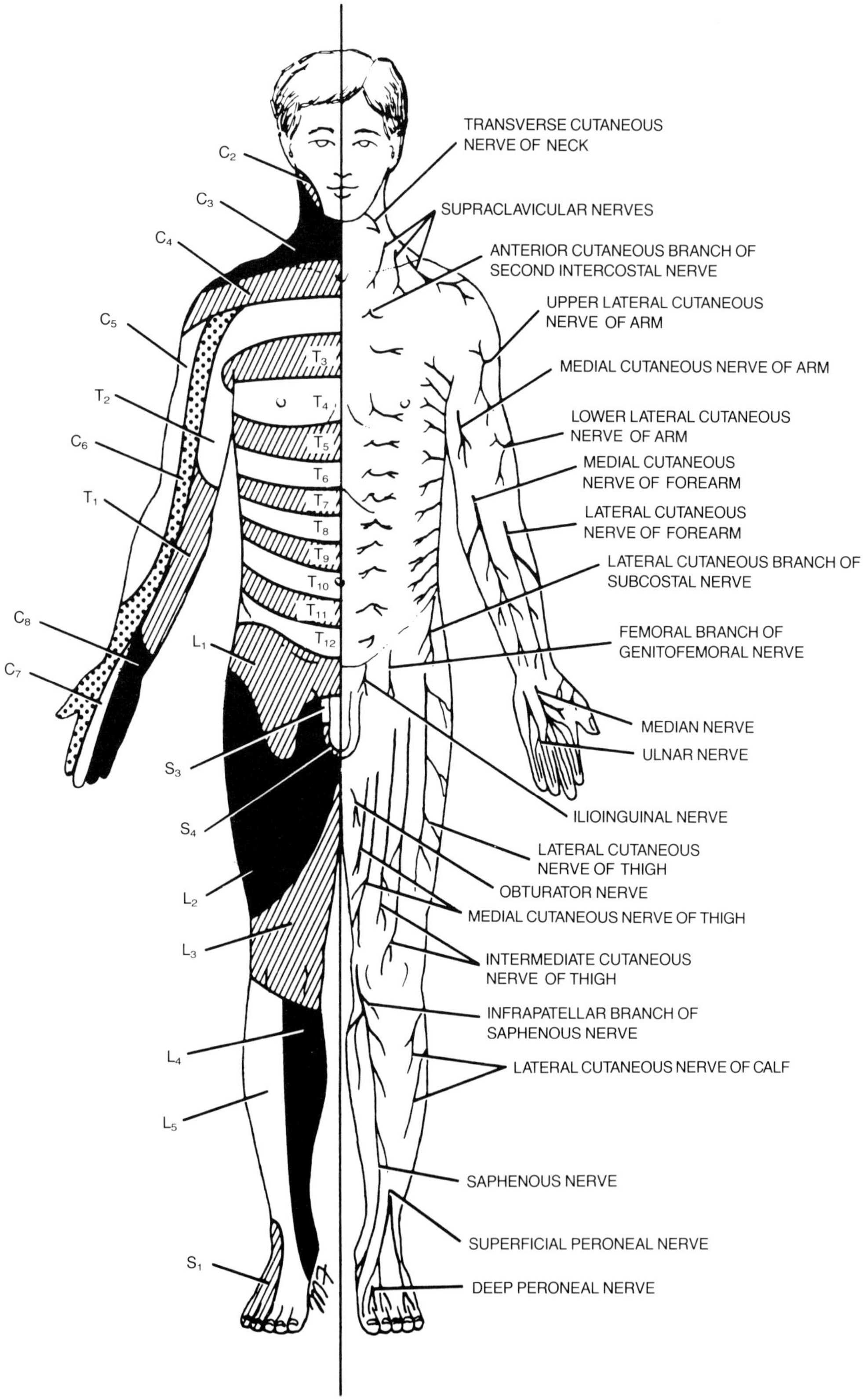

Figure 6–22. The dermatomes and distribution of cutaneous nerves on the anterior aspect of the body.

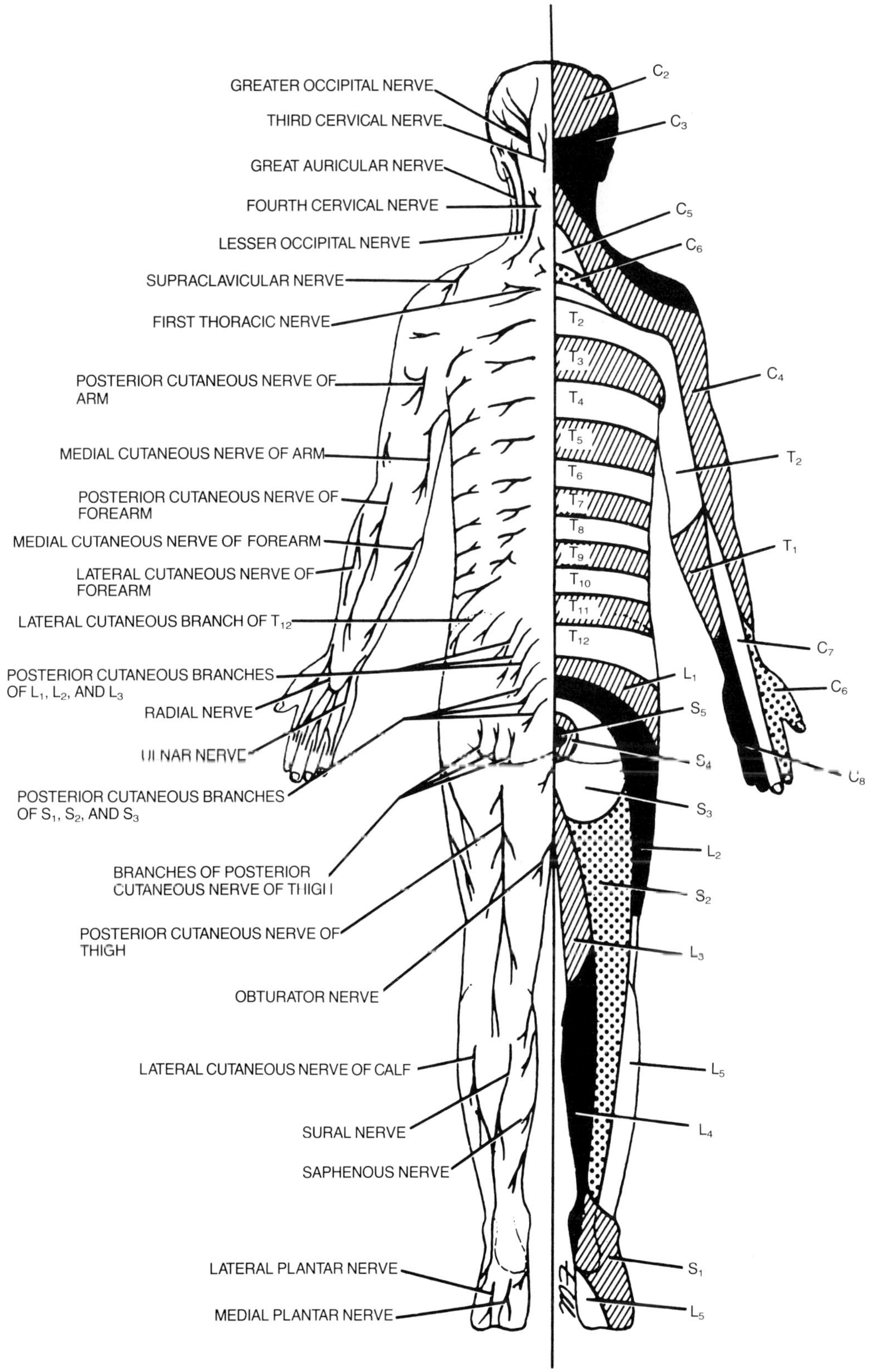

Figure 6–23. The dermatomes and distribution of cutaneous nerves on the posterior aspect of the body.

CLINICAL PROBLEMS

1. Demyelinating diseases of the nervous system, multiple sclerosis is a good example, have the common features of destruction of the myelin sheaths of nerve fibers. How does the normal process of myelination take place in a peripheral nerve and a nerve tract in the central nervous system? During development when does myelination normally occur?

2. The prognosis after peripheral nerve injury and repair is related to the following: (a) The more distal the injury to a nerve, the more successful the regeneration and the restoration of function will be. (b) The closer the lesion is to the cell body of the neuron, the more likely is the lesion to have a serious effect on the cell body. (c) Sensory nerve cells are more affected by retrograde effects than are motor nerve cells. (d) Young patients have a better return of function than elderly patients. Comment on these observations.

3. Define the following terms: (a) Schwann cells, (b) Schmidt-Lanterman incisure, (c) Wallerian degeneration, (d) node of Ranvier, and (e) band fiber.

4. A 35-year-old man was involved in a shoot-out in a liquor store. A bullet ricocheted off the wall and entered the gluteal region severing his right sciatic nerve. After operative reunion of the nerve the following observations were made. (a) The skin of the affected leg was redder and hotter than the normal limb. (b) Later, the skin of the affected leg and foot became blue and colder than normal, especially in cold weather. (c) The skin of the affected leg became dry and scaly and the nail growth was retarded. (d) Radiography of the leg on the affected side showed decalcification of the bones. Comment on these statements.

5. A 55-year-old woman with symptoms and signs of thyrotoxicosis was advised to have a partial thyroidectomy. A preoperative physical examination revealed that she suffered from chronic bronchitis and had atrial fibrillation. It was decided to perform the operation under a deep cervical plexus block. What is the difference between a superficial and a deep cervical plexus block? Name the nerves that form the cervical plexus. The anesthetic will produce analgesia over which dermatomes in the neck in a cervical plexus block?

6. A feeble 80-year-old man fell on his outstretched right hand and sustained a Colle's fracture at the distal end of his radius. Because of his poor physical condition, which included a severe bronchitis and emphysema, it was decided to reduce the fracture under a brachial plexus block. Which spinal nerves contribute to the brachial plexus? Are the roots of the plexus formed by anterior or posterior rami of the spinal nerves? What forms the axillary sheath? Is the axillary sheath a single compartment or do septa exist within the compartment? When using the axillary approach, which branch of the plexus is most likely to remain unblocked? Name the anatomical structures that are likely to be inadvertently invaded by the blocking needle and cause complications during an interscalene block when the needle is used in a horizontal direction? When the supraclavicular approach is made to block the brachial plexus, the entire upper limb is usually anesthetized except for a small area of skin on the medial side of the arm. What is the name of the nerve that supplies this area of skin? A supraclavicular brachial plexus block is sometimes complicated by a pneumothorax. Explain in anatomical terms how this can occur.

7. A patient is to undergo an extensive plastic reconstruction of the hand. Name the nerves that supply the skin and muscles of the hand and indicate where they may be blocked in the region of the wrist.

8. A 36-year-old man was involved in an automobile accident and sustained a fracture of his seventh right rib. It was decided to relieve his acute pain, caused by the broken rib, by performing an intercostal nerve block. Which intercostal nerve innervates the seventh rib? From which part of a spinal nerve is an intercostal nerve formed? What is the course of an intercostal nerve in an intercostal space? Name the order of the structures forming the neurovascular bundle in an intercostal space from above downward.

9. A 40-year-old woman with a history of gallstones was advised to have a cholecystectomy. Because of a preexisting heart disease secondary to rheumatic fever, it was decided to perform the operation under a local anesthetic. Describe the type of peripheral nerve block that could be used in this case.

10. Recently the use of nerve blocking as an anesthetic technique for inguinal hernia repair has been strongly advocated. The technique results in early hospital discharge and speeds up postoperative recovery. Name the nerves that innervate the inguinal region. From which segment of the spinal cord do they originate? Which important surface landmark would you use when introducing the blocking needle?

11. It was decided to use a skin graft from the anterior and lateral surfaces of the thigh to provide coverage for a burnt area on the chest wall. What is the sensory nerve supply to this region of the thigh. Which important surface landmarks would you use when blocking these nerves?

12. A sciatic nerve block and a saphenous nerve block were used in the reduction of a fracture dislocation of the ankle. Where may the sciatic nerve be conveniently blocked when using a posterior approach? Where may the saphenous nerve be blocked proximal to the ankle? Where does the saphenous nerve arise from?

ANSWERS TO CLINICAL PROBLEMS

1. The process of myelination in the peripheral nervous system takes place as the result of the axon indenting the Schwann cell. Later, the Schwann cell rotates on the axon so that the plasma membrane becomes wrapped around the axon in a spiral. In the central nervous system, the oligodendrocyte forms a myelin sheath in a similar manner; additional details are described on page 189. Myelination normally occurs during the latter part of fetal development and during the first year postnatally.

2. As the distal end of a nerve is approached, there are fewer structures to innervate; therefore, the chance of innervating the wrong structure during the process of regeneration is less. In addition, the greater the distance of the nerve injury from the nerve cell body, the less the metabolism of the nerve cell is affected by the injury. (b) A severe nerve injury close to the body of the nerve cell could result in the death of the entire neuron. (c) The metabolic activity of sensory neurons is more susceptible to change by retrograde phenomena than that of motor neurons. (d) All reparative processes within the body take place more readily in the young than in the elderly.

3. (a) Schwann cells are supporting cells for a peripheral nerve fiber. They are arranged in line along the course of the nerve fiber. In a myelinated nerve, it is the Schwann cell that is responsible for the formation of the myelin sheath. In a nonmyelinated nerve, as many as 15 or more axons indent the cell surface of a single Schwann cell. (b) A Schmidt-Lanterman incisure is seen on longitudinal sections of myelinated nerve fibers. It represents a localized persistence of Schwann cell cytoplasm. (c) Wallerian degeneration describes the changes that occur in the distal segment of a sectioned axon after injury (page 191). (d) A node of

Ranvier is a gap that exists between two adjacent Schwann cells (or oligodendrocytes, in the central nervous system) in a myelinated nerve; nodes of Ranvier are not present in nonmyelinated nerves. (e) A band fiber is the endoneurial sheath and the contained cords of Schwann cells that replace a degenerated peripheral nerve after nerve section. It will guide regenerating axons to their end organs.

4. Division of a peripheral nerve results in the section of postganglionic sympathetic fibers traveling in the nerve. The loss of vasomotor control causes the skin of the leg to become first red and hot. Later, because of the pooling of blood in the dermal capillaries, the blood becomes deoxygenated and blue in color and the skin becomes cold. The denervated sweat glands cease to function and the skin becomes dry and scaly. Because of poor peripheral circulation, the nail growth becomes retarded and the bones become decalcified; decalcification also follows disuse caused by paralysis.

5. A superficial cervical nerve block refers to blockade of the cutaneous nerves that emerge halfway down the posterior border of the sternocleidomastoid muscle, namely the lesser occipital, the great auricular, the transverse cutaneous, and the supraclavicular nerves. A deep cervical plexus block involves blocking all the nerves of the cervical plexus.

The cervical plexus is formed from the anterior rami of the first four cervical nerves. The dermatomes C_2, C_3, and C_4 involved in analgesia after a deep cervical plexus block, will extend from the chin down the front and sides of the neck and onto the chest as far as the sternal angle. Anesthesia will also extend to the skin covering the parotid region and over the angle of the jaw.

6. The brachial plexus is formed from the anterior rami of C_5, C_6, C_7, C_8, and T_1 spinal nerves. The axillary sheath is formed as a prolongation of the prevertebral fascia from the lateral margins of the scalenus anterior and scalenus medius muscles into the axilla; it encloses the axillary artery and the brachial plexus. Septa do exist within the sheath and may explain why axillary nerve blocks of the brachial plexus are sometimes incomplete. The musculocutaneous nerve is the most likely nerve to escape blockage during a brachial plexus block approached from the axilla.

If the needle is inserted horizontally when performing an interscalene block of the brachial plexus, it could pass between the bony transverse processes and pierce the vertebral vessels or enter the dural sheath around a cervical spinal nerve.

The intercostobrachial nerve, a branch of the second intercostal nerve, supplies with the medial cutaneous nerve of the arm, a small area of skin on the medial side of the arm. This nerve will not be blocked when the brachial plexus is blocked.

The brachial plexus in the root of the neck passes downward and laterally over the cervical dome of the pleura. If the vertical approach of introducing the needle is used it is possible for the needle to travel beyond the trunks of the plexus and pierce the parietal pleura.

7. The following nerves supply the skin of the hand. The lateral two-thirds of the palm and the anterior surface of the lateral three and one half fingers including the nail beds are innervated by the median nerve. The medial one-third of the palm and the anterior surface of the medial one and one half fingers including the nail beds are innervated by the ulnar nerve. On the posterior surface of the hand, the lateral two-thirds and the lateral three and one half fingers are innervated by the superficial branch of the radial nerve (the areas stated are variable). The medial one-third of the dorsal surface of the hand and the medial one and one half fingers are supplied by the ulnar nerve. All the small muscles of the hand are supplied by the deep branch of the ulnar nerve, except the three muscles of the thenar eminence and the first two lumbricals that are innervated by the median nerve.

The median nerve may be blocked just proximal to the flexor retinaculum, as it lies medial to the tendon of the flexor carpi radialis tendon and beneath the palmaris longus tendon (if present). The small palmar cutaneous branch of the median nerve that passes anterior to the flexor retinaculum may be blocked by subcutaneous infiltration in a plane anterior to the median nerve.

The ulnar nerve may be blocked as it lies on the lateral side of the tendon of the flexor carpi ulnaris at its point of insertion onto the pisiform bone. The dorsal cutaneous branch of the ulnar nerve that passes to the back of the hand and fingers is a large branch that arises from the ulnar nerve in the distal third of the forearm. It passes medially between the tendon of the flexor carpi ulnaris muscle and the ulna and may be blocked by subcutaneous infiltration about 2 inches (5 cm) above the wrist.

The superficial branch of the radial nerve may be blocked on the lateral side of the radial artery just proximal to the proximal wrist crease. To block all the small branches given off by the radial nerve proximal to this site, it is usual to inject a cuff of anesthetic on the lateral and posterior aspects of the wrist at this level.

8. The periosteum of the seventh rib is innervated by the seventh intercostal nerve.* Intercostal nerves are formed from the anterior rami of T_1 through T_{11} spinal nerves. An intercostal nerve passes forward around the thoracic wall first, between the posterior intercostal membrane and the parietal pleura, then between the internal intercostal muscle and the intercostalis intimus muscle (part of the transversus thoracis). The nerve lies in the subcostal groove of the rib of its own number. The intercostal vein, the intercostal artery, and the intercostal nerve lie within the subcostal groove in that order from above downward.

9. If the cholecystectomy is performed through the classical Kocher incision along the right costal margin or through a right supraumbilical paramedian incision, the right seventh, eighth, ninth, and tenth intercostal nerves may be blocked at the posterior axillary line or as a field block as they emerge into the anterior abdominal wall from beneath the costal cartilages.

10. The iliohypogastric and ilioinguinal nerves innervate the inguinal region. They originate from L-1 segment of the spinal cord. The anterior superior iliac spine and the umbilicus are the landmarks used for blocking these nerves. The needle is introduced about 1 inch (2.5 cm) medial along a line joining the spine and the umbilicus. Local spread of the anesthetic may also block the superficial terminal branches of T_{12} (T_{11}?) and the genital branch of the genitofemoral in nerve. In addition it is common to raise a skin wheal from the anterior superior iliac spine to the symphysis pubis. The neck of the hernia sac may also have to be infiltrated to block autonomic fibers.

11. The branches of the femoral nerve and the lateral cutaneous nerves of the thigh supply the skin of the anterior and lateral surfaces of the thigh. The surface landmarks used in blocking the femoral nerve are the anterior superior iliac spine and the pubic tubercle; the nerve emerges below the inguinal ligament halfway between these points and just lateral to the femoral artery. The lateral cutaneous nerve may be blocked 0.5 to 1 inch (1.3 to 2.5 cm) medial to the anterior superior iliac spine.

12. The sciatic nerve lies at a site 5 cm caudad on a line drawn at right angles to the midpoint of a line joining the posterior superior iliac spine to the tip of

---

* It is conceivable that in the individual who has a collateral branch of the sixth intercostal nerve, the periosteum of the seventh rib is also innervated by this nerve.

the greater trochanter. It may also be blocked lower down midway between the ischial tuberosity and the greater trochanter.

The saphenous nerve may be blocked on the medial side of the knee joint, either as it lies superficial over the medial femoral condyle or lower down over the medial tibial condyle.

The saphenous nerve arises from the femoral nerve.

CHAPTER 7

Cranial Nerves

There are 12 pairs of cranial nerves that leave the brain and pass through the foramina and the fissures in the skull. All the nerves are distributed in the head and the neck except the tenth, which also supplies structures in the thorax and the abdomen.

The cranial nerves are named and numbered as follows:

I. olfactory
II. optic
III. oculomotor
IV. trochlear
V. trigeminal
VI. abducent
VII. facial
VIII. vestibulocochlear
IX. glossopharyngeal
X. vagus
XI. accessory
XII. hypoglossal

The olfactory, the optic, and the vestibulocochlear nerves are entirely sensory. The oculomotor, trochlear, abducent, accessory, and hypoglossal nerves are entirely motor. The remaining cranial nerves are mixed nerves containing both sensory and motor fibers.

Anatomical Organization

The cranial nerves have central motor or sensory nuclei, or both, within the brain, and peripheral nerve fibers, that emerge from the brain and exit from the skull to reach their effector or sensory organs.

Motor Nuclei of the Cranial Nerves

Somatic Motor and Branchomotor Nuclei The somatic motor and branchomotor nerve fibers of a cranial nerve are the axons of nerve cells situated within the brain. These nerve cell groups form motor nuclei and they innervate striated muscle. Each nerve cell with its processes is referred to as a **lower motor neuron.** Such a nerve cell is, therefore, equivalent to the motor cells found in the anterior gray columns of the spinal cord.

The motor nuclei of the cranial nerves receive impulses from the cerebral cortex through the corticonuclear fibers. These fibers originate from the pyramidal cells in the inferior part of the precentral gyrus (area 4) and from the adjacent part of the postcentral gyrus. The corticonuclear fibers descend through the **corona radiata** and the **genu of the internal capsule.** They pass through the midbrain just medial to the corticospinal fibers in the **basis pedunculi** and end by synapsing either directly with the lower motor neurons within the cranial nerve nuclei or indirectly through the **internuncial neurons.**

The majority of the corticonuclear fibers to the motor cranial nerve nuclei cross the median plane before reaching the nuclei. Bilateral connections are present for all the cranial motor nuclei except for part of the facial nucleus that supplies the muscles of the lower part of the face and a part of the hypoglossal nucleus that supplies the genioglossus muscle.

The **general visceral motor nuclei** form the cranial outflow of the parasympathetic portion of the autonomic nervous system. They are the Edinger-Westphal nucleus of the oculomotor nerve, the superior salivatory nucleus of the facial nerve, the inferior salivatory nucleus of the glossopharyngeal nerve, and the dorsal motor nucleus of the vagus. These nuclei receive numerous afferent fibers including descending pathways from the hypothalamus.

Sensory Nuclei of the Cranial Nerves

The sensory nuclei of the cranial nerves include somatic and visceral afferent nuclei. The sensory or afferent parts of a cranial nerve are the axons of nerve cells outside the brain and are situated in ganglia on the nerve trunks (equivalent to posterior root ganglion of a spinal nerve) or may be situated in a sensory organ, such as the nose, eye, or ear. The central processes of these cells enter the brain and terminate by synapsing with cells forming the sensory nuclei. Axons from the nuclear cells now cross the midline and ascend to other sensory nuclei, e.g., the thalamus, where they synapse. The nerve cells of these nuclei send their axons to terminate in the cerebral cortex.

OLFACTORY NERVES

The olfactory nerves arise from olfactory receptor nerve cells in the olfactory mucous membrane located in the upper part of the nasal cavity above the level of the superior concha. Bundles of these nerve fibers pass through the openings of the cribriform plate of the ethmoid bone to enter the **olfactory bulb** inside the skull (*Fig. 7–1*), which, in turn, is connected to the **olfactory area of the cerebral cortex** by the olfactory tract.

CLINICAL NOTES

When testing for the sensation of smell, it must be remembered that the nose contains a right and a left nasal passage, therefore, corresponding olfactory nerves must be tested separately.

Fractures of the anterior cranial fossa or cerebral tumors of the frontal lobes may produce lesions of the olfactory nerves, with consequent anosmia. A lesion of the olfactory cortex on one side is unlikely to produce complete anosmia because fibers from each olfactory tract travel to both cerebral hemispheres.

OPTIC NERVE

The optic nerve is composed of axons from the ganglion cell layer of the retina. The nerve emerges from the back of the eyeball and leaves the orbital cavity of the skull through the optic canal. It unites with the optic nerve of the opposite side to form the **optic chiasma** (*Fig. 7–2*).

In the chiasma, the fibers from the medial (nasal) half of each retina cross the midline and enter the **optic tract** of the opposite side, whereas the fibers from the lateral (temporal) half of each retina pass posteriorly in the optic tract of the same side. Most of the fibers of the optic tract terminate by synapsing with nerve cells in the **lateral geniculate body,** which is a small projection from the posterior part of the thalamus. A few fibers pass to the pretectal nucleus and superior colliculus of the midbrain and are concerned with light reflexes (see page 279).

The axons from the lateral geniculate body pass posteriorly as the **optic radiation,** and terminate in the **visual cortex** of the cerebral hemisphere (*Fig. 7–2*).

CLINICAL NOTES

Visual Field Defects Associated with Lesions of the Optic Pathways

The various visual field defects are summarized in Figure 7–3. Note that circumferential blindness may be due to hysteria or optic neuritis (*Fig. 7–3*). Total blindness of one eye would follow complete section of one optic nerve (*Fig. 7–3*). Lesions of the optic tract and optic radiation (*Fig. 7–3*) produce the same hemianopia for both eyes, i.e., homonymous hemianopia. Bitemporal hemianopia is a loss of the lateral halves of the fields of vision of both eyes (*Fig. 7–3*). This condition is commonly caused by a tumor of the pituitary gland exerting pressure on the optic chiasma.

OCULOMOTOR NERVE

The oculomotor nerve emerges on the anterior surface of the midbrain. It passes forward in the middle cranial fossa in the lateral wall of the cavernous sinus. Here, it divides into a superior and an inferior ramus, which enter the orbital cavity through the superior orbital fissure (*Fig. 7–4*).

The oculomotor nerve supplies the following extrinsic muscles of the eye: levator palpebrae superioris, superior rectus, medial rectus, inferior rectus and inferior oblique. It also supplies through its branch to the ciliary ganglion and the short ciliary nerves, parasympathetic nerve fibers to the following intrinsic muscles: the constrictor pupillae of the iris and the ciliary muscles.

The oculomotor nerve is, therefore, entirely motor and is responsible for lifting the upper eyelid, turning the eye upward, downward, and medially, constricting the pupil, and accommodating the eye.

CLINICAL NOTES

In complete oculomotor paralysis the eye cannot be moved upward, downward, or inward. At rest the eye looks laterally (external strabismus) owing to the activity of the lateral rectus muscle and downward due to the activity of the superior oblique muscle. The patient has diplopia. There is ptosis of the upper eyelid because of paralysis of the levator palpebrae superioris muscle. The pupil is widely dilated because of the paralysis of the sphincter pupillae and the unopposed action of the dilator pupillae muscle (supplied by the sympathetic). Accommodation is impossible.

TROCHLEAR NERVE

The trochlear nerve, the most slender of the cranial nerves, leaves the posterior surface of the midbrain and immedi-

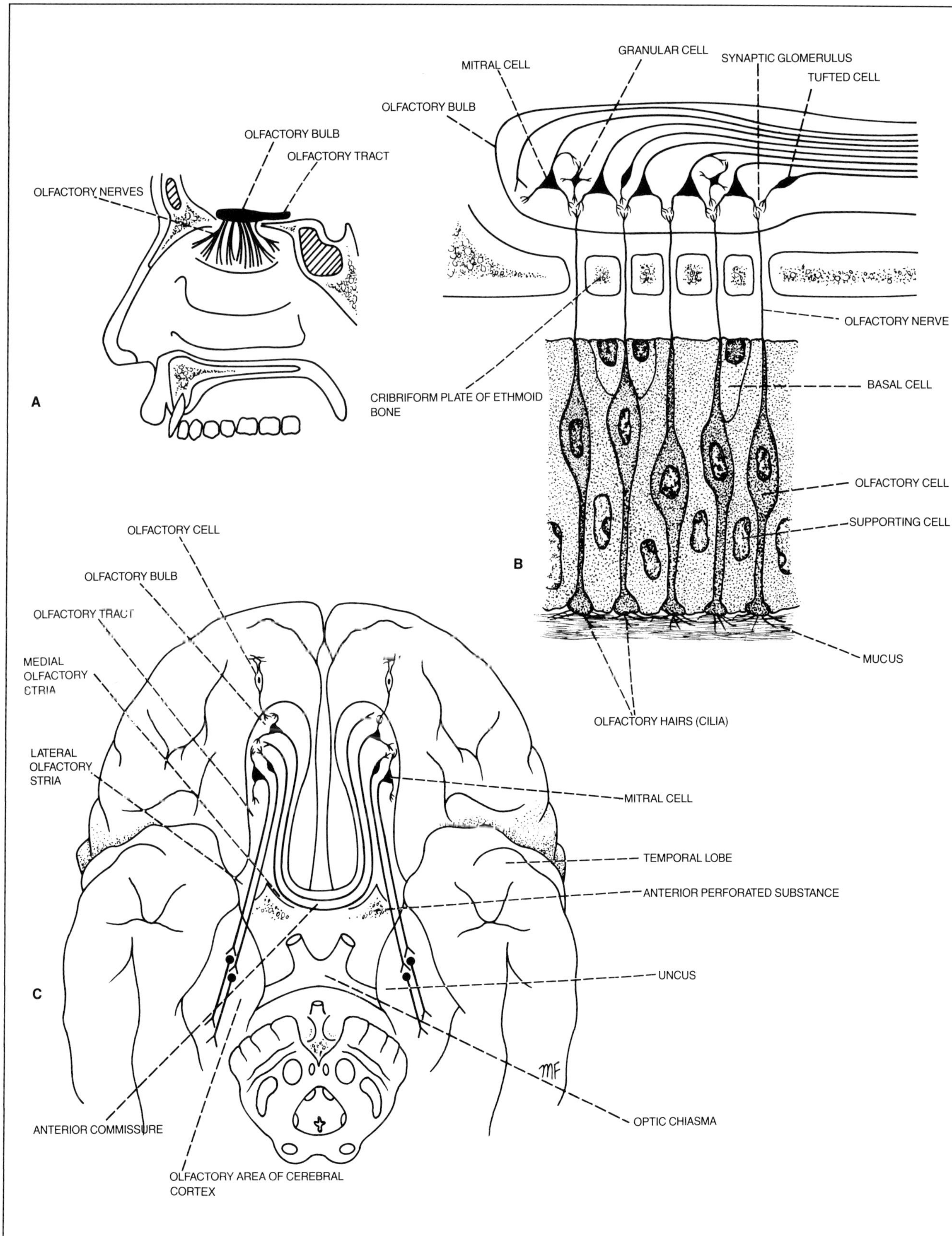

Figure 7–1. A. Distribution of the olfactory nerves on the lateral wall of the nose. **B.** Connections between the olfactory cells and neurons of the olfactory bulb. **C.** Connections between the olfactory cell and the rest of the olfactory system.

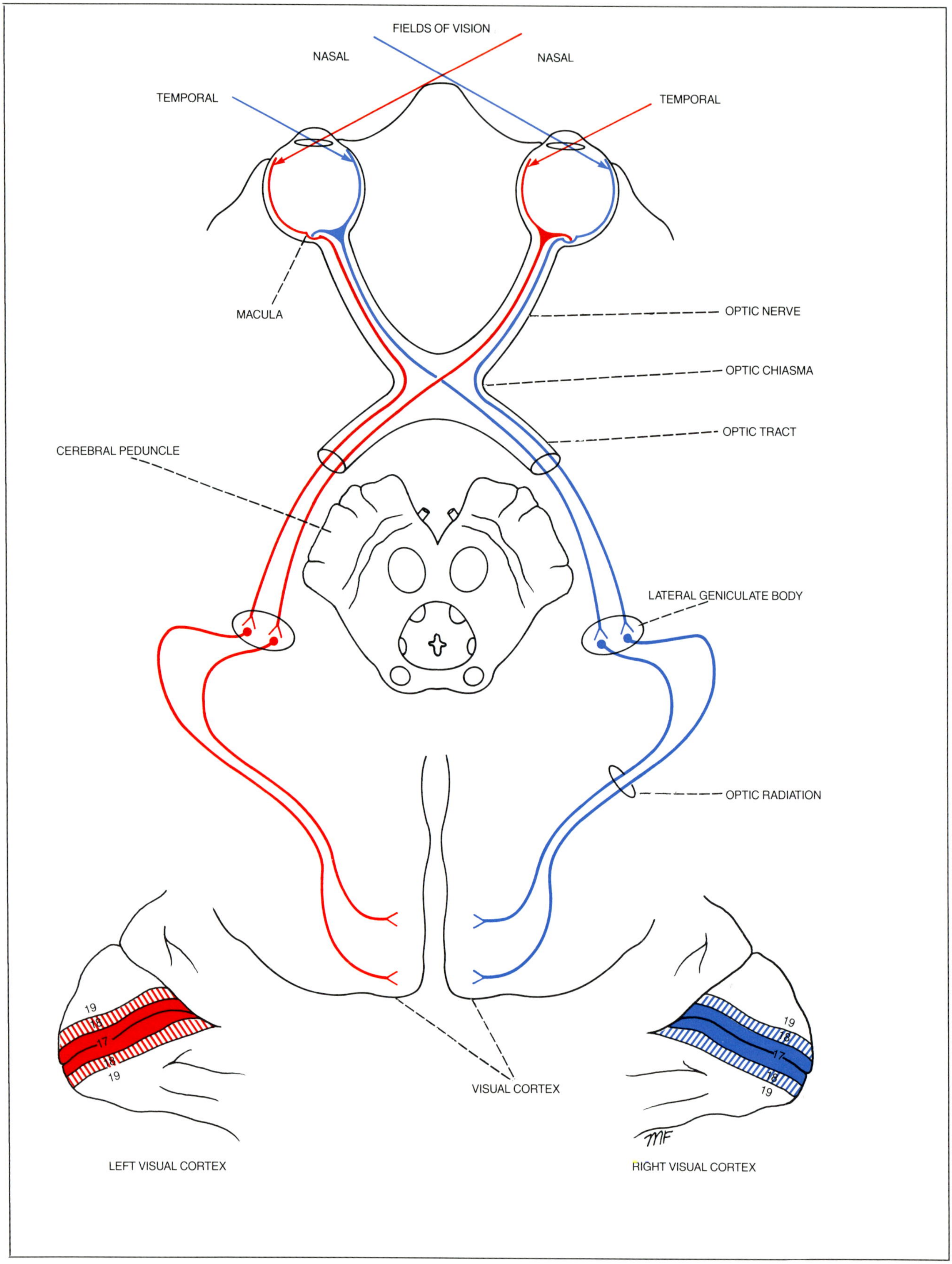

Figure 7–2. The optic pathway.

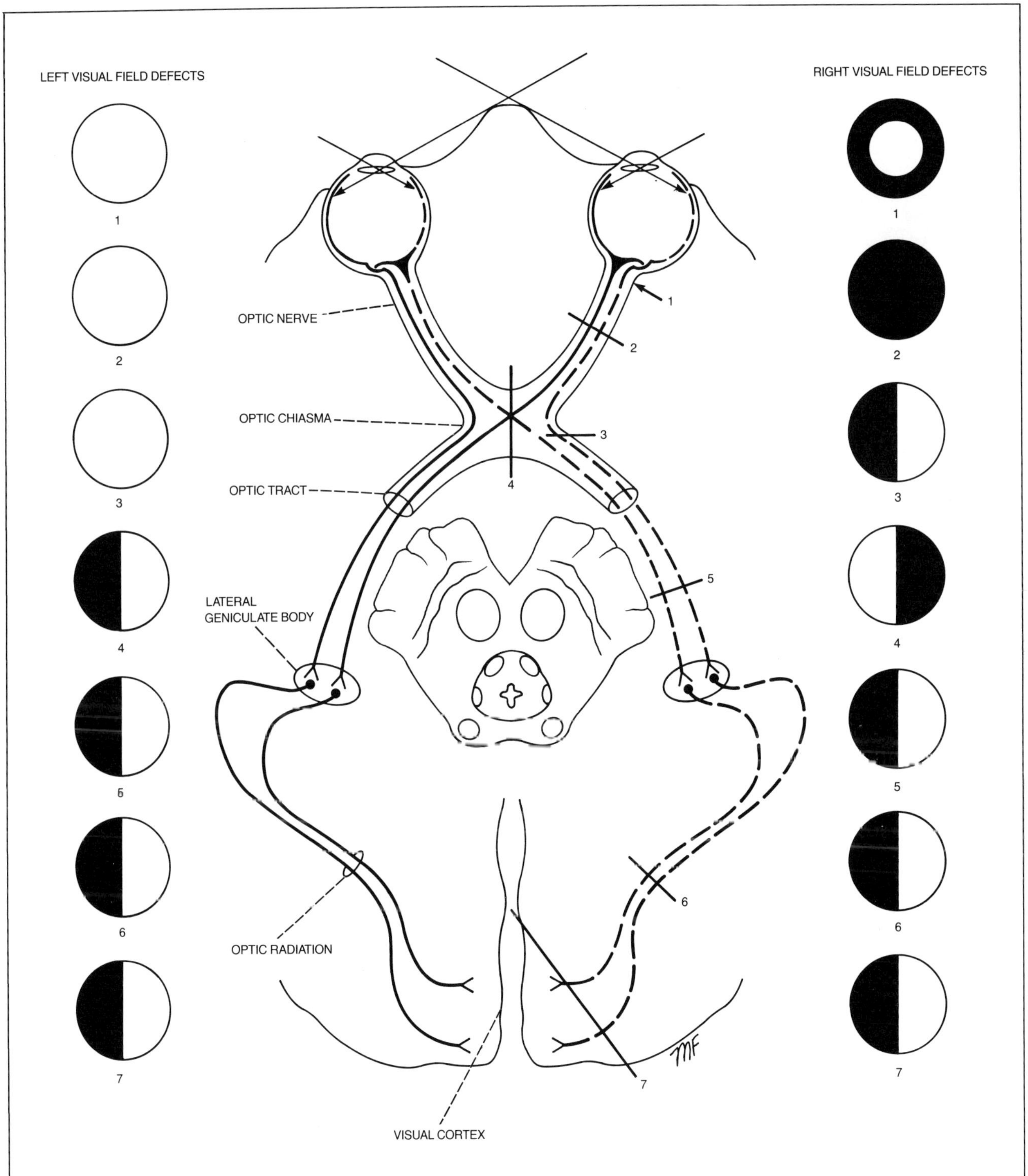

1. Right-sided circumferential blindness caused by retrobulbar neuritis.
2. Total blindness of right eye caused by division of right optic nerve.
3. Right nasal hemianopia caused by partial lesion of right side of optic chiasma.
4. Bitemporal hemianopia caused by complete lesion of optic chiasma.
5. Left temporal hemianopia and right nasal hemianopia caused by lesion of right optic tract.
6. Left temporal hemianopia and right nasal hemianopia caused by lesion of right optic radiation.
7. Left temporal hemianopia and right nasal hemianopia caused by lesion of right visual cortex.

Figure 7–3. Visual field defects associated with lesions of optic pathways.

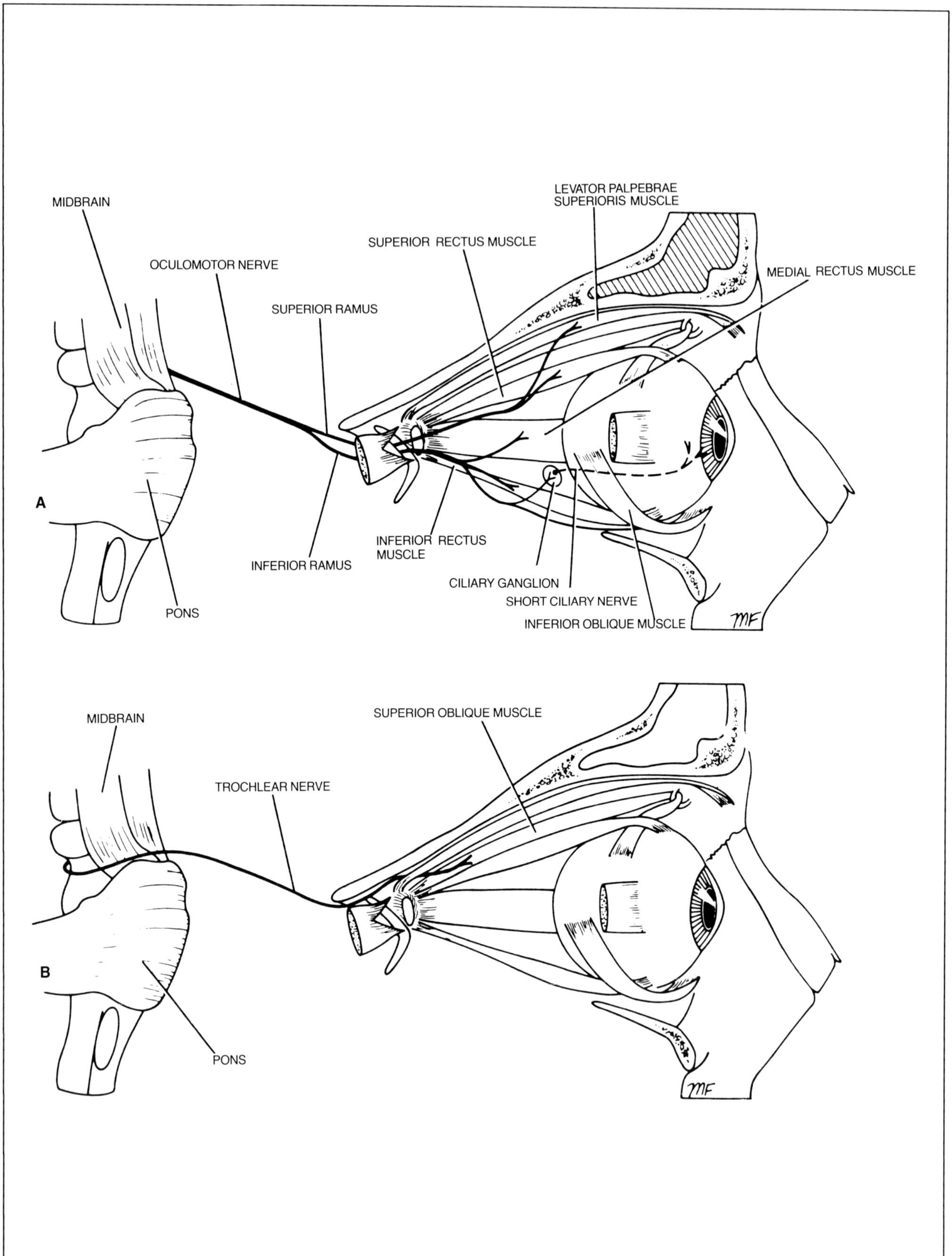

Figure 7–4. Origin and distribution of **(A)** the oculomotor nerve and **(B)** the trochlear nerve.

ately decussates with the nerve of the opposite side. The trochlear nerve passes forward through the middle cranial fossa in the lateral wall of the cavernous sinus and enters the orbit through the superior orbital fissure (*Fig. 7–4*). The nerve supplies the superior oblique muscle of the eyeball. The trochlear nerve is entirely motor and assists in turning the eye downward and laterally.

CLINICAL NOTES

In trochlear nerve paralysis, the patient complains of double vision when looking straight down. This is because the superior oblique muscle is paralyzed, and the eye turns medially as well as downward. In fact, the patient has great difficulty in turning the eye downward and laterally.

TRIGEMINAL NERVE

The trigeminal nerve is the largest cranial nerve. It is called trigeminal because it has three major branches; **ophthalmic (V_1), maxillary (V_2),** and **mandibular (V_3).** It is a mixed cranial nerve and contains both sensory and motor fibers; the sensory fibers are distributed mainly to the skin of the face (*Fig. 7–5*) and the motor fibers innervate the muscles of mastication.

The trigeminal nerve leaves the anterior aspect of the pons as a small motor root and a large sensory root. The nerve passes forward out of the posterior cranial fossa by crossing the superior border of the petrous part of the temporal bone. It then rests on the upper surface of the apex of the petrous part of the temporal bone in the middle cranial fossa (*Fig. 7–5*). The large sensory root now expands to form the crescent-shaped **trigeminal ganglion** that lies within a pouch of the dura mater, called the **trigeminal or Meckel's cave.** This covers the posterior two-thirds of the ganglion, which is bathed in cerebrospinal fluid. The ophthalmic, maxillary, and mandibular nerves arise from the convex anterior border of the ganglion.

The trigeminal ganglion has the following relations. Medially, it is related to the posterior part of the cavernous sinus and the internal carotid artery. Inferiorly, lies the motor root of the trigeminal nerve and the greater petrosal nerve, the apex of the petrous part of the temporal bone, and the cartilage filling in the foramen lacerum. Superiorly, lies the temporal lobe of the brain.

Ophthalmic Nerve (V_1)

The ophthalmic nerve is the smallest division of the trigeminal nerve and is entirely sensory. It runs forward in the lateral wall of the cavernous sinus in the middle cranial fossa and divides into three branches, the lacrimal, the frontal, and the nasociliary nerves, which enter the orbital cavity through the superior orbital fissure (*Fig. 7–5*).

The **lacrimal nerve** runs forward along the upper border of the lateral rectus muscle. It receives the zygomaticotemporal branch of the maxillary nerve, which contains the parasympathetic secretomotor branches to the lacrimal gland. The lacrimal nerve enters the lacrimal gland and is distributed to the conjunctiva and the skin of the upper eyelid.

The **frontal nerve** runs forward on the upper surface of the levator palpebrae superioris muscle and divides into the **supraorbital** and **supratrochlear nerves** (*Fig. 7–7*). The larger supraorbital nerve emerges from the orbital cavity by passing through the supraorbital notch, or foramen, and supplies the skin of the forehead and the scalp as far back as the vertex; it also supplies the frontal air sinus. The small supratrochlear nerve passes above the pulley for the superior oblique muscle and winds around the upper margin of the orbital cavity to supply the skin of the forehead medial to the area supplied by the supraorbital nerve.

The **nasociliary nerve** crosses the optic nerve and runs forward on the upper border of the medial rectus muscle (*Fig. 7–5*). It continues as the anterior **ethmoidal nerve** through the anterior ethmoidal foramen to enter the cranial cavity. It then descends through a slit at the side of the crista galli to enter the nasal cavity. It gives off two **internal nasal branches** and then continues as the **external nasal nerve.** The external nasal nerve appears on the face between the nasal bone and the upper lateral nasal cartilage (*Fig. 7–7*). It supplies the skin down as far as the tip of the nose.

Branches of the Nasociliary Nerve

1. **Sensory fibers** to the ciliary ganglion
2. **Long ciliary nerves** that contain sympathetic fibers to the dilator pupillae muscle and sensory fibers to the cornea
3. **Infratrochlear nerve** passes forward below the pulley of the superior oblique muscle and supplies the skin of the medial part of the upper eyelid and the adjacent part of the nose
4. **Posterior ethmoidal nerve** to the ethmoidal and sphenoid sinuses

Maxillary Nerve (V_2)

The maxillary nerve is purely sensory. It passes forward along the lower border of the lateral wall of the cavernous sinus and leaves the middle cranial fossa through the foramen rotundum. It then crosses the pterygopalatine fossa to enter the orbit through the inferior orbital fissure (*Fig. 7–5*). It continues as the **infraorbital nerve** in the infraorbital groove and emerges onto the face through the infraorbital foramen (*Fig. 7–9*). It gives sensory fibers to the skin of the face and side of the nose (*Fig. 7–5*).

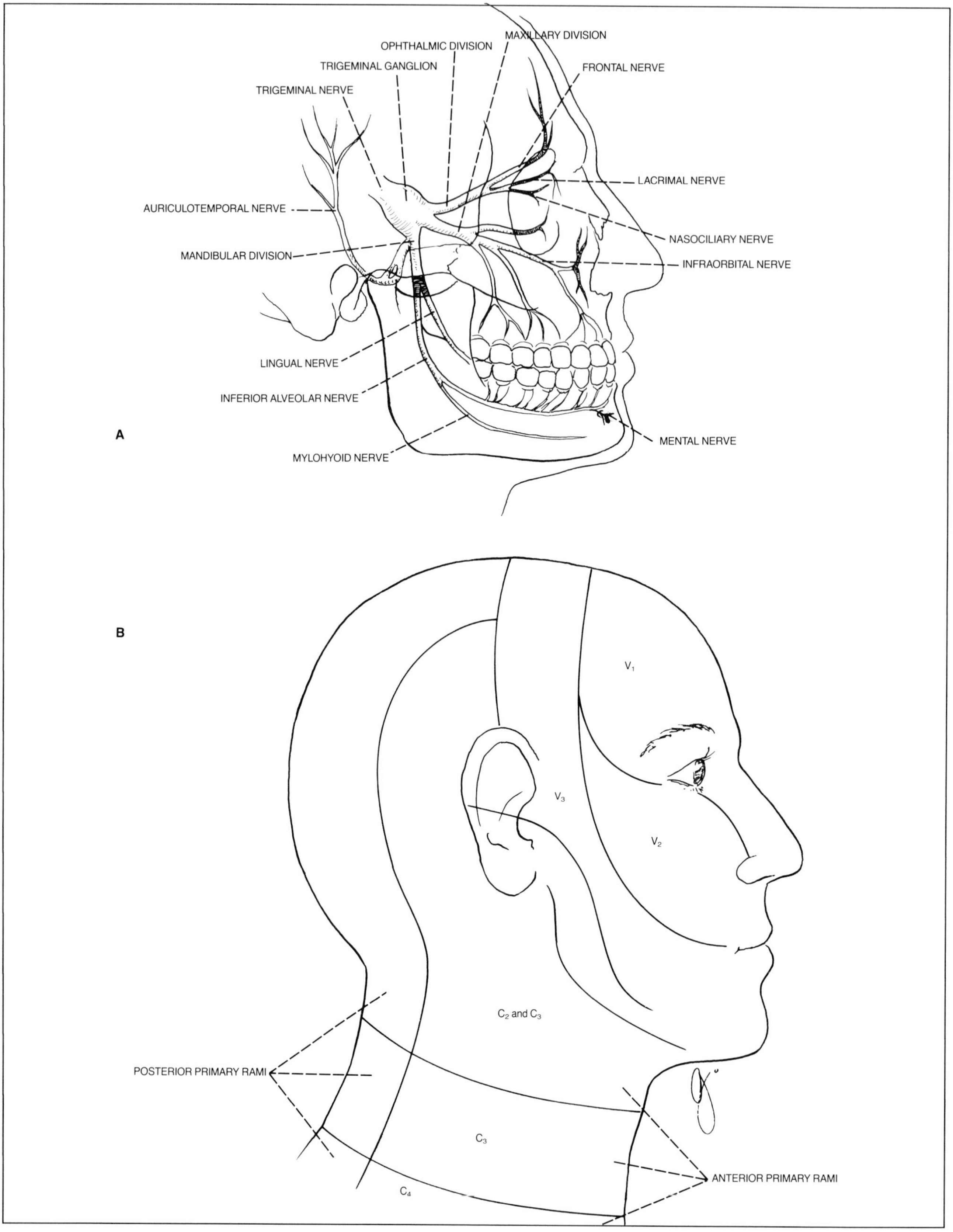

Figure 7–5. A. The distribution of the trigeminal nerve. **B.** The facial cutaneous distribution of the ophthalmic (V_1), the maxillary (V_2), and the mandibular (V_3) divisions of the trigeminal nerve. Note that the skin over the angle of the jaw is supplied by C_2 and C_3 segments of the spinal cord.

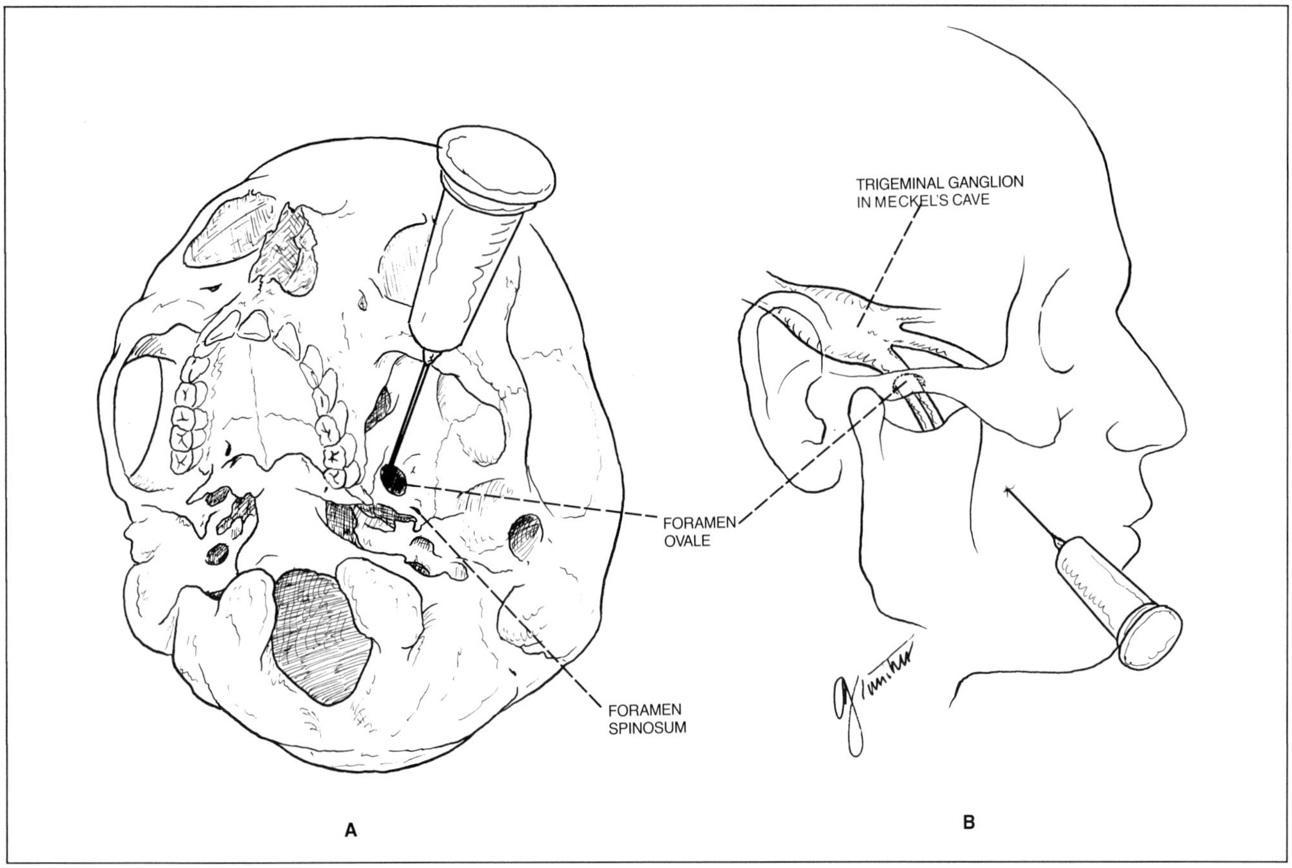

Figure 7–6. Anatomical landmarks for trigeminal ganglion block. **A.** Skull from inferior view. **B.** Lateral view.

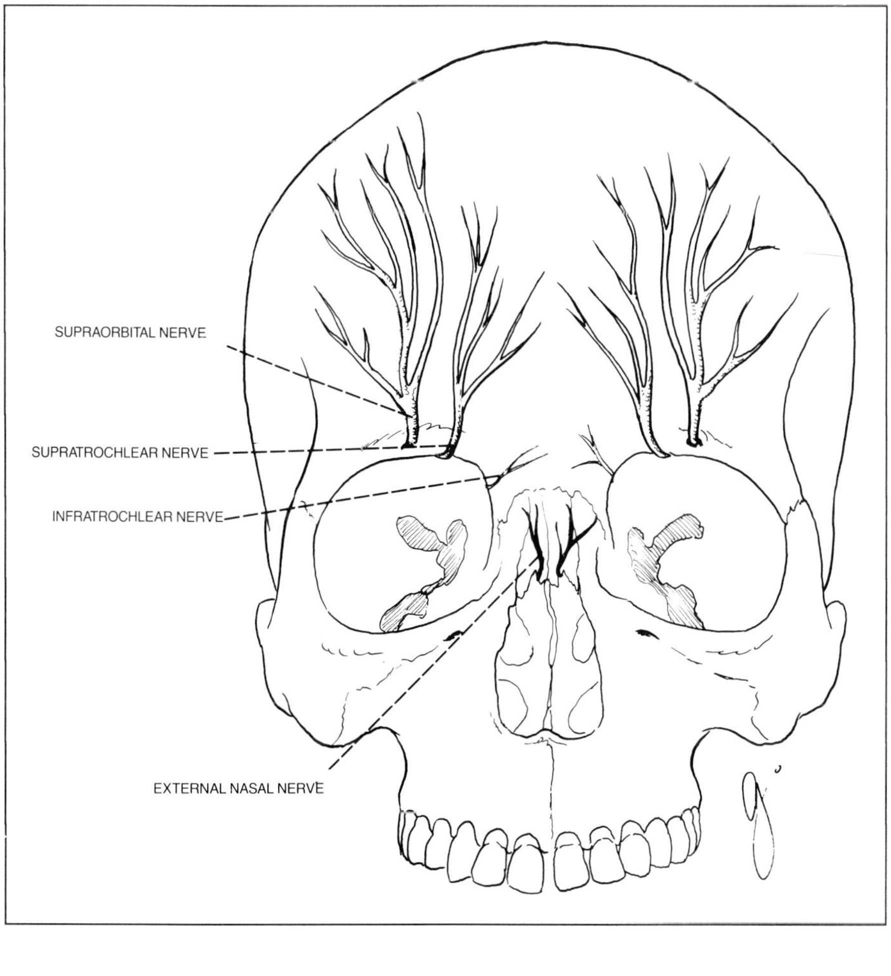

Figure 7–7. Anatomical landmarks for blocking branches of the ophthalmic nerve.

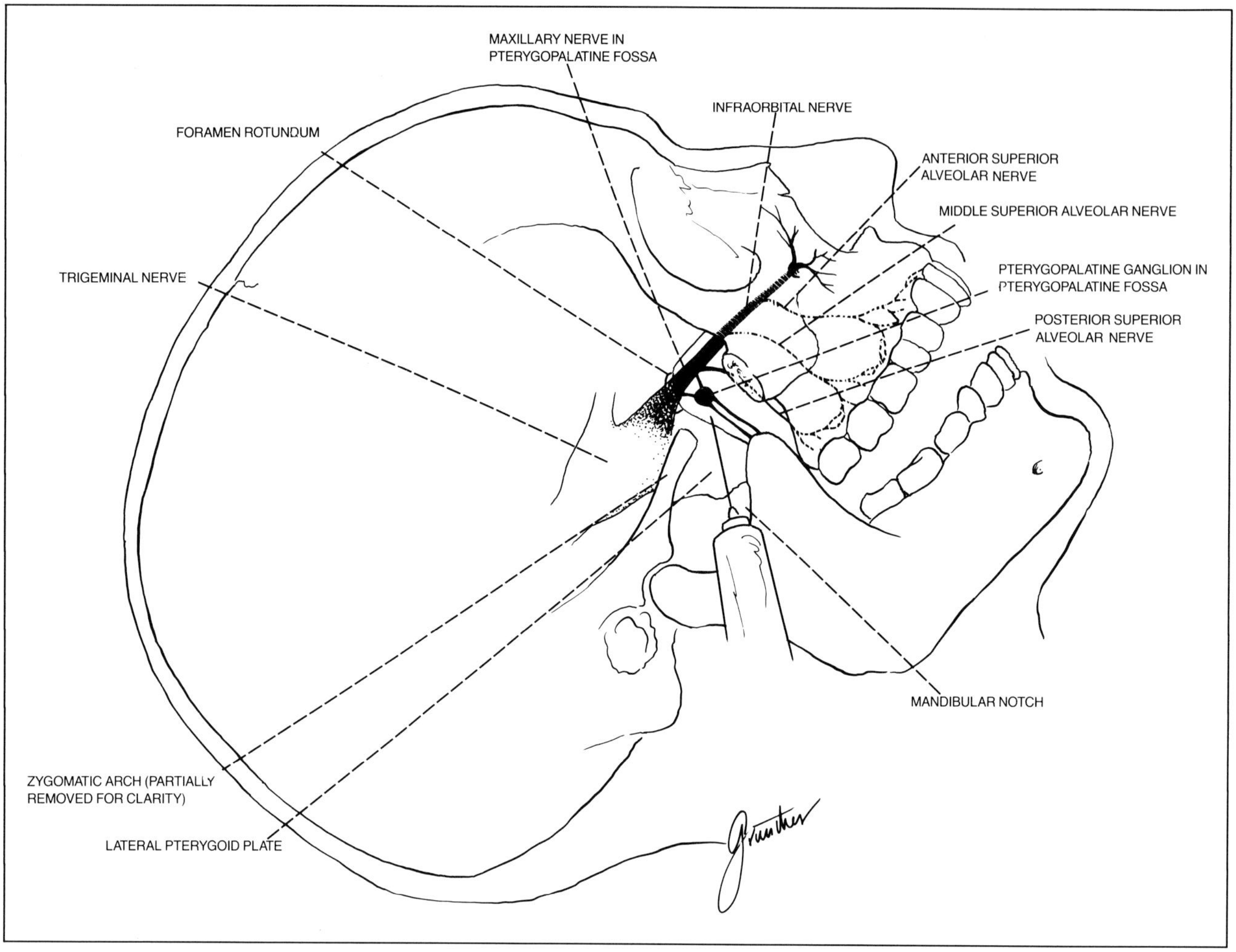

Figure 7–8. Anatomical landmarks for blocking the maxillary nerve.

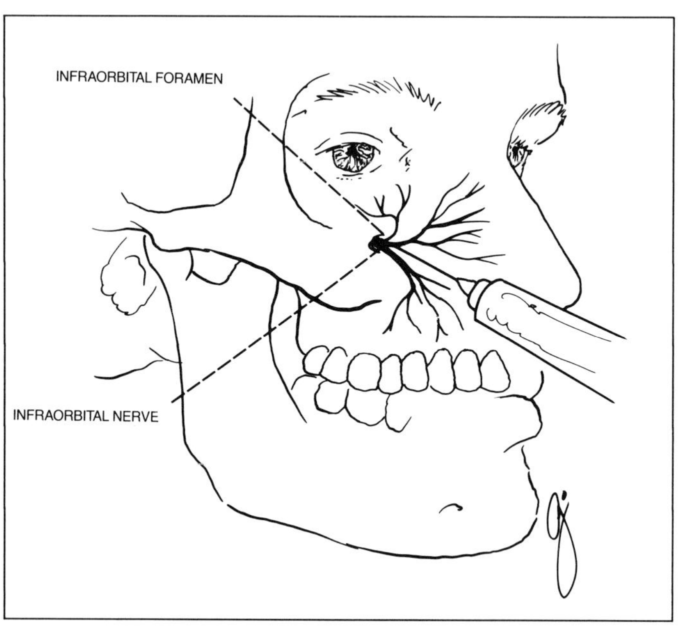

Figure 7–9. Anatomical landmarks for blocking the infraorbital nerve.

Branches of the Maxillary Nerve

1. **Meningeal branches.**
2. **Zygomatic branch.** This divides into the **zygomaticofacial and zygomaticotemporal branches.** The zygomaticofacial nerve passes onto the face through a small foramen on the lateral side of the zygomatic bone. It supplies the skin over the prominence of the cheek. The zygomaticotemporal nerve emerges in the temporal fossa through a small foramen on the posterior surface of the zygomatic bone and supplies the skin over the temple. The zygomaticotemporal branch gives parasympathetic secretomotor fibers to the lacrimal nerve for the lacrimal gland.
3. **Ganglionic branches.** Two short nerves suspend the pterygopalatine ganglion in the pterygopalatine fossa. They contain sensory fibers that have passed through the ganglion from the nose, the palate, and the pharynx. They also contain postganglionic parasympathetic fibers that are going to the lacrimal gland.
4. **Posterior superior alveolar nerve** innervates the maxillary sinus and the upper molar teeth and adjoining parts of the gum and the cheek.
5. **Middle superior alveolar nerve** goes to the maxillary sinus and the upper premolar teeth and the gums and the cheek.
6. **Anterior superior alveolar nerve** provides sensation to the maxillary sinus and upper canine and incisor teeth.

Pterygopalatine Ganglion

The pterygopalatine ganglion is a parasympathetic ganglion that is suspended from the maxillary nerve in the pterygopalatine fossa (*Fig. 7–10*). It is secretomotor to the lacrimal and nasal glands (see page 269).

Branches of the Pterygopalatine Ganglion

1. **Orbital branches** enter the orbit through the inferior orbital fissure and innervate the periosteum of the orbital cavity.
2. **Greater and lesser palatine nerves** go to the mucous membrane of the palate, the tonsil, and the nasal cavity.
3. **Nasal branches** enter the nose through the sphenopalatine foramen. They comprise **short medial** and **lateral posterior superior nasal nerves** and the **long nasopalatine nerve.** The latter nerve runs down on the nasal septum and supplies the hard palate by passing through the incisive foramen.
4. **Pharyngeal branch** innervates the roof of the nasopharynx; it passes backward through the palatovaginal canal.

Mandibular Nerve (V_3)

The mandibular nerve is the largest division of the trigeminal nerve and is motor and sensory. The sensory root leaves the trigeminal ganglion and passes out of the skull through the foramen ovale (*Fig. 7–5*). The motor root of the trigeminal nerve also leaves the skull through the foramen ovale and almost immediately joins the sensory root to form the trunk of the mandibular nerve. The nerve trunk then divides into a small anterior and a large posterior division.

Branches from the Main Trunk of the Mandibular Nerve

1. **Meningeal branch**
2. **Nerve to medial pterygoid** innervates the medial pterygoid, the tensor tympani, and the tensor veli palatini muscles

Branches from the Anterior Division of the Mandibular Nerve

1. **Masseteric nerve** to the masseter muscle
2. **Deep temporal nerves** to the temporalis muscle
3. **Nerve to the lateral pterygoid muscle**
4. **Buccal nerve** to the skin of the cheek and the mucous membrane lining the cheek

Branches from the Posterior Division of the Mandibular Nerve

1. **Auriculotemporal nerve** passes backward between the sphenomandibular ligament and the neck of the mandible (*Fig. 7–5*). It then ascends behind the temporomandibular joint on the medial surface of the parotid gland, and then crosses the root of the zygomatic arch posterior to the superficial temporal vessels. Its terminal branches supply the skin of the auricle, the external auditory meatus, the temporomandibular joint, and the scalp. The nerve conveys postganglionic parasympathetic secretomotor fibers from the otic ganglion to the parotid salivary gland.
2. **Lingual nerve** descends in front of the inferior alveolar nerve (*Fig. 7–5*) and enters the mouth by passing under the superior constrictor muscle. It runs forward on the side of the tongue to reach its tip. In its course it crosses the submandibular duct. At the lower border of the lateral pterygoid muscle it is joined by the chorda tympani nerve. It ends by giving branches to the mucous membrane of the anterior two-thirds of the tongue and the floor of the mouth. The fibers from the chorda tympani provide important preganglionic parasympathetic secretomotor fibers to the submandibular ganglion. Other fibers of the chorda tympani subserve taste from the anterior two-thirds of the tongue (except for the vallate papillae that are supplied by the glossopharyngeal nerve).
3. **Inferior alveolar nerve** descends posterior to the lingual nerve (*Fig. 7–5*) and enters the mandibular canal to supply the teeth of the lower jaw. It gives off numerous branches to the molar and premolar teeth, and the adjacent gingiva. It ends by dividing into an **incisive** and **mental branch.** The incisive branch inner-

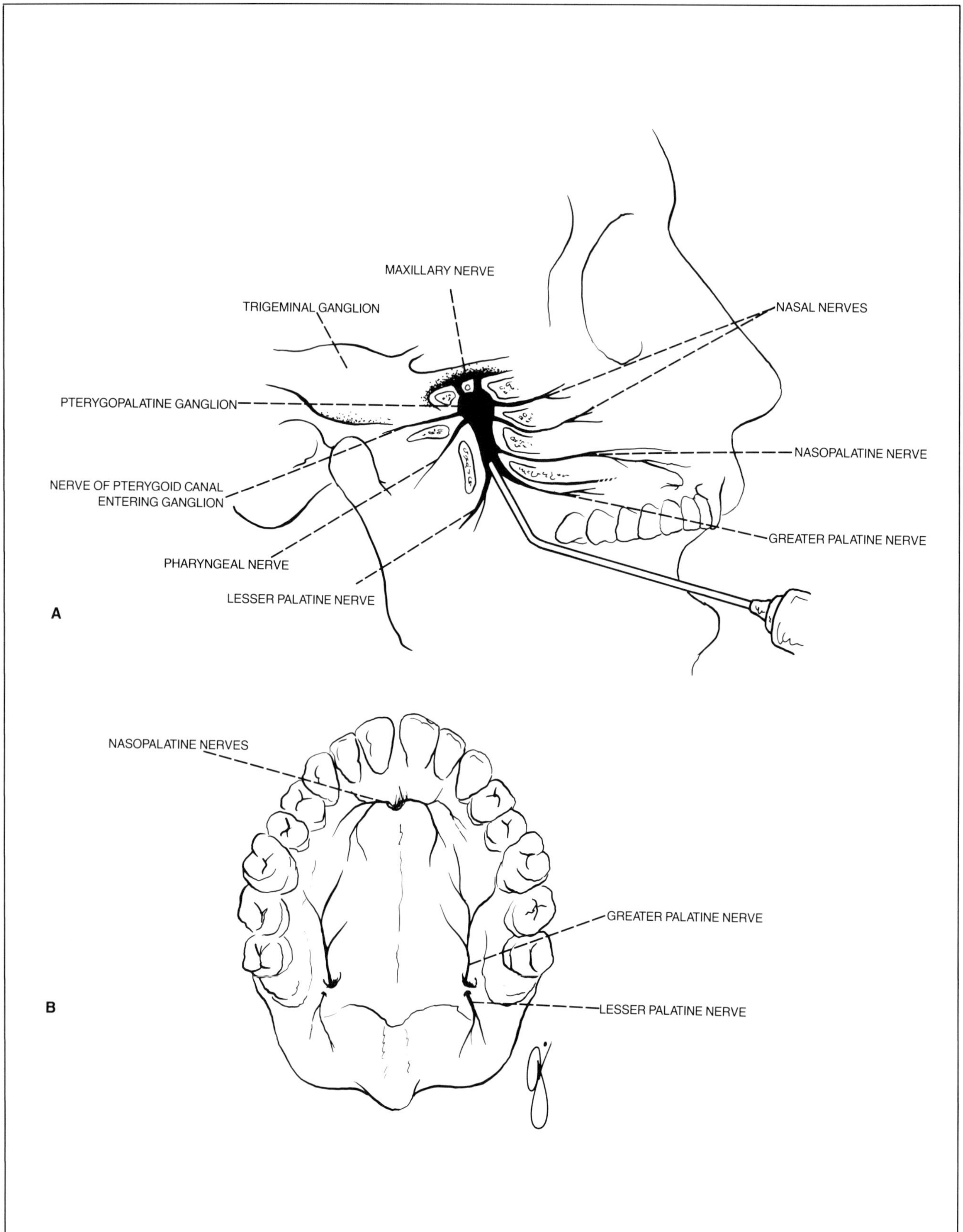

Figure 7–10. Anatomical landmarks for blocking the pterygopalatine ganglion. Note the distribution of the branches of the maxillary nerve through the ganglion to the walls of the nose and the palate.

vates the canine and incisive teeth and the mental branch emerges through the mental foramen as the **mental nerve** (*Fig. 7–5*) to supply the skin of the face over the chin. The inferior alveolar nerve gives off the **mylohyoid nerve** that innervates the mylohyoid muscle and the anterior belly of the digastric muscle (*Fig. 7–5*).

4. **Communicating branch** frequently runs from the inferior alveolar nerve to the lingual nerve.

Otic Ganglion

The otic ganglion is a small parasympathetic ganglion that is located medial to the mandibular nerve just below the foramen ovale. The ganglion is adherent to the nerve to the medial pterygoid muscle. The preganglionic fibers originate in the glossopharyngeal nerve and reach the ganglion via the lesser petrosal nerve (see page 270). The postganglionic secretomotor fibers reach the parotid salivary gland via the auriculotemporal nerve.

Submandibular Ganglion

The submandibular ganglion is a small parasympathetic ganglion that lies between the submandibular salivary gland and the hyoglossus muscle. It is connected to the lingual nerve and the salivary gland by several branches. Preganglionic parasympathetic fibers reach the ganglion via the chorda tympani and the lingual nerves. Postganglionic secretomotor fibers pass to the submandibular and sublingual salivary glands.

CLINICAL NOTES

ANATOMY OF TRIGEMINAL NERVE BLOCKS*

Trigeminal Ganglion Block

The blocking needle is introduced through the cheek lateral to the lateral margin of the mouth (*Fig. 7–6*). It is directed toward the midpoint of the zygomatic arch as seen from the lateral aspect. A guiding finger within the vestibule of the mouth cavity ensures that the needle does not enter the mouth. The long needle on reaching the greater wing of the sphenoid bone is passed through the foramen ovale for about 1 cm to the region of the trigeminal ganglion. It is important that the dural sheath of Meckel's cavity is not penetrated. The anesthetic may then be injected. The possible complications of injecting an anesthetic into the cavity will depend on the surrounding anatomic structures affected. The presence of the anesthetic in the cerebrospinal fluid may produce unconsciousness. A concentrated solution of anesthetic spilling over into the posterior cranial fossa may cause paralysis of the 6th to the 12th cranial nerves. The 3rd and 4th cranial nerves, as they pass forward in the lateral wall of the cavernous sinus, may also be affected.

Ophthalmic Nerve Block

The terminal branches of the ophthalmic division of the trigeminal nerve that emerge onto the face and the nose, include the supraorbital, supratrochlear, infratrochlear, external nasal, and lacrimal nerves (*Fig. 7–7*). The latter nerve gives off only a few branches to the skin and is seldom blocked.

Supraorbital Nerve Block

The supraorbital nerve emerges from the orbital cavity in the same vertical plane as the pupil when the patient is looking straight ahead. If the nerve passes through the supraorbital notch, this can easily be palpated on the supraorbital margin. If the nerve passes through a foramen, however, it is small and difficult to feel.

Supratrochlear Nerve Block

The supratrochlear nerve winds around the supraorbital margin about a finger's breadth medial to the supraorbital nerve.

Infratrochlear Nerve Block

The infratrochlear nerve emerges from the orbital cavity at the junction of the superior and medial walls.

Anterior Ethmoidal Nerve and External Nasal Nerve Blocks

The anterior ethmoidal nerve leaves the medial wall of the orbital cavity through the anterior ethmoidal foramen about 1 inch (2.5 cm) posterior to the orbital margin. Because the orbital veins are close and may be damaged by needle insertion, extreme care is required. The external nasal nerve is blocked at the point where it emerges between the nasal bone and the upper lateral nasal cartilage.

Maxillary Nerve Block

The maxillary nerve may be blocked as it crosses the pterygopalatine fossa (*Fig. 7–8*). The needle is inserted through the skin just below the midpoint of the zygomatic arch. With the mouth open slightly, the needle is directed medially and superiorly through the masseter muscle and the mandibular notch into the infratemporal fossa. The needle is advanced until it encounters the lateral pterygoid plate. The direction of the needle is now changed so that the tip passes anteriorly to the lateral pterygoid plate and enters the pterygopalatine fossa. The local anesthetic is then injected into the fossa around the nerve. Complications of this approach include damage to the maxillary artery and vein and their branches and spread of the anesthetic into the orbital cavity to the optic nerve through the infraorbital fissure causing temporary blindness.

Infraorbital Nerve Block

The infraorbital nerve emerges from the infraorbital foramen as a direct continuation of the maxillary nerve (*Fig. 7–9*). The opening of the foramen faces downward

* For details concerning the techniques of nerve blocks, the reader is referred to the *Atlas of Regional Anesthesia* by Jordan Katz.

and medially. The opening of the foramen can usually be palpated with gentle pressure just below the infraorbital margin, in line with the pupil.

Pterygopalatine (Sphenopalatine) Ganglion Block

The pterygopalatine ganglion is a parasympathetic ganglion that is small in size and suspended from the lower border of the maxillary nerve in the pterygopalatine fossa (*Fig. 7–10*). Passing through the ganglion without interruption are the sensory fibers from the orbit, the nose, the hard and soft palate, the gums, and the tonsillar region of the pharynx. The ganglion and, therefore, the sensory fibers may be blocked by inserting a long-angled needle into the greater palatine foramen with the mouth wide open. The greater palatine foramen leads superiorly into the pterygopalatine fossa. Injection of the anesthetic blocks the greater and lesser palatine nerves, the orbital nerves, the nasal nerves, and the pharyngeal nerve.

Mandibular Nerve Block

The mandibular nerve (V_3) is blocked at the point where it emerges through the foramen ovale (*Fig. 7–11*). The needle is inserted through the skin just below the midpoint of the zygomatic arch. With the mouth open slightly, the needle is directed medially and superiorly through the masseter muscle and the mandibular notch into the infratemporal fossa. On encountering the lateral pterygoid plate, the needle direction is changed so that the tip is directed posteriorly toward the nerve. The complications of mandibular nerve block include piercing the lateral wall of the oral pharynx (superior constrictor muscle) posterior to the lateral pterygoid plate.

Auriculotemporal Nerve Block

The auriculotemporal nerve is easily blocked as it ascends in front of the auricle over the posterior root of the zygoma, behind the superficial temporal artery (*Fig. 7–11*).

Lingual Nerve and Inferior Alveolar Nerve Blocks

Both the lingual and the interior alveolar nerves may be blocked as they pass downward and forward in the infratemporal fossa on the lateral surface of the medial pterygoid muscle, and on the medial surface of the ramus of the mandible (*Fig. 7–12*). With the patient's mouth wide open, the anterior border of the ramus of the mandible is palpated just above the third molar tooth. The blocking needle is advanced until the tip lies in close proximity to the mandibular foramen. The anesthetic solution will then infiltrate around both nerves.

Mental Nerve Block

The mental nerve may be blocked as it emerges from the mental foramen on the body of the mandible (*Fig. 7–12*). The foramen lies on the same vertical line that passes through the supraorbital notch, the infraorbital foramen, and the first premolar tooth.

ABDUCENT NERVE

The abducent nerve is a small nerve that emerges from the anterior surface of the brain between the pons and the medulla oblongata. It passes forward through the cavernous sinus in the middle cranial fossa lying below and lateral to the internal carotid artery. The nerve then enters the orbit through the superior orbital fissure (*Fig. 7–13*). The abducent nerve is entirely a motor nerve and supplies the lateral rectus muscle and is, therefore, responsible for turning the eye laterally.

CLINICAL NOTES

ANATOMY FOR ANESTHESIA FOR INTRAOCULAR SURGERY

The eyeball receives its sensory innervation from the nasociliary nerve, a branch of the ophthalmic division of the trigeminal nerve. The nasociliary nerve gives off two or three long ciliary nerves that pass forward to the back of the eyeball close to the attachment of the optic nerve.

The ciliary muscle and the constrictor pupillae muscle receive their parasympathetic supply by a branch from the nerve to the inferior oblique muscle from the lower division of the oculomotor nerve. The preganglionic fibers synapse in the ciliary ganglion and the postganglionic fibers pass forward to the eyeball in the short ciliary nerves. The sympathetic innervation of the dilator pupillae muscle and the smooth muscle portion of the levator palpebrae superioris muscle arises from the superior cervical sympathetic ganglion. The postganglionic fibers reach the orbit via the carotid plexus of nerves. Those fibers passing to the dilator pupillae muscle travel in the long ciliary nerves. Fibers passing to the levator palpebrae superioris muscle travel in the oculomotor nerve.

The extraocular muscles are innervated by the oculomotor nerve (superior rectus, medial rectus, inferior rectus and inferior oblique muscles), the trochlear nerve (superior oblique muscle), and the abducent nerve (lateral rectus). Vision is transmitted by the optic nerve.

Retrobulbar Injection

The lower lateral part of the orbital margin is palpated and the needle is inserted through the lower lid at this site and directed backward below the eyeball toward the apex of the orbital cavity. Before injecting the anesthetic, aspiration should be attempted to be certain that the needle is not in a blood vessel. Recently, a superior lateral approach has been advocated.

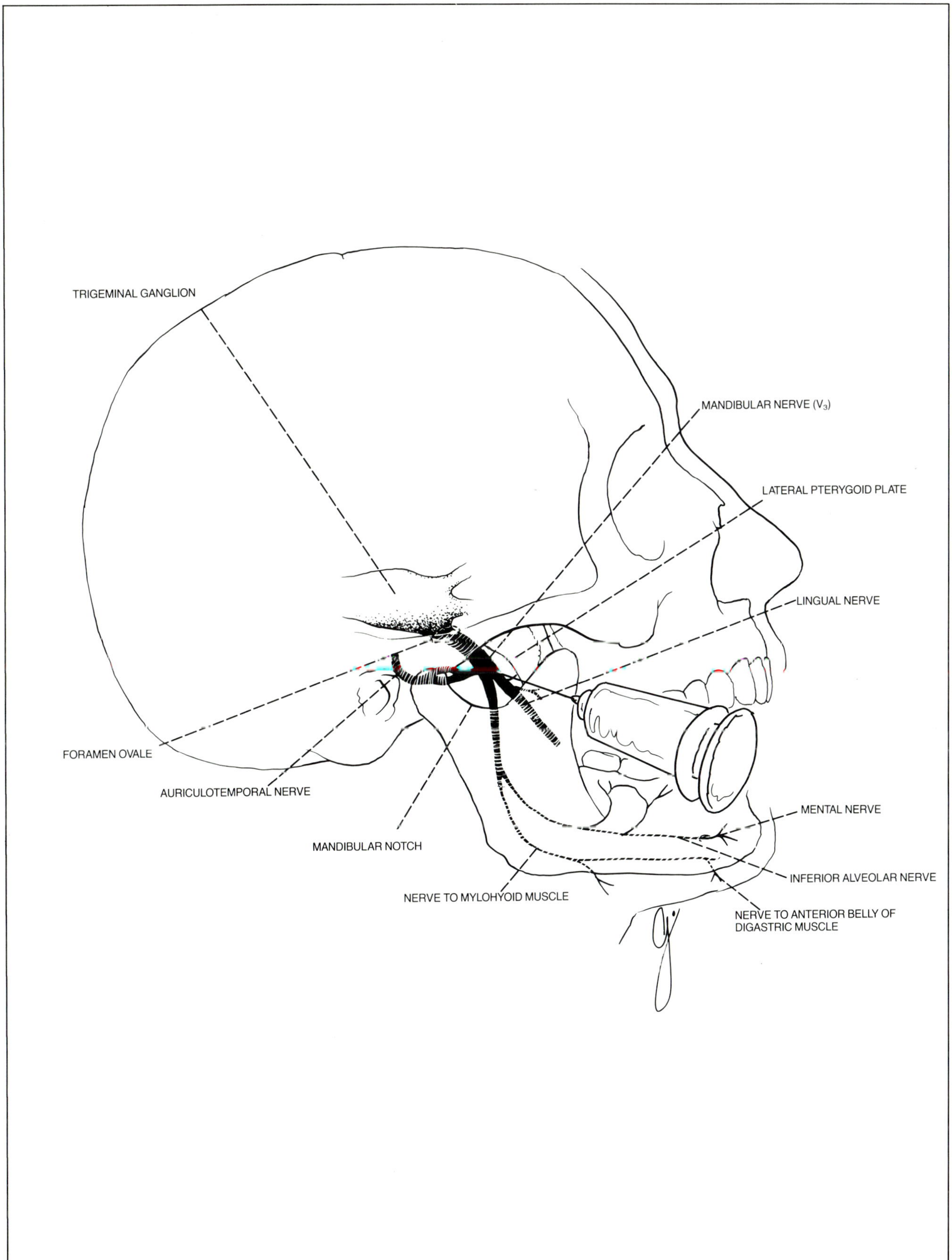

Figure 7–11. Anatomical landmarks for blocking the mandibular nerve.

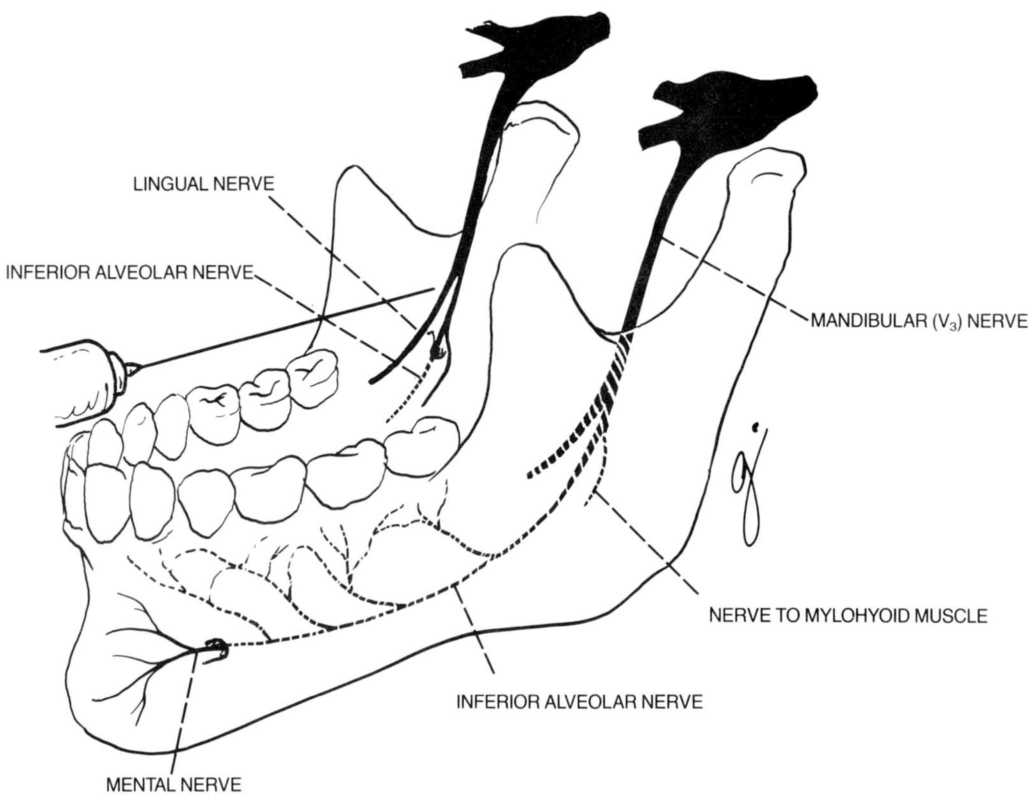

Figure 7–12. Anatomical landmarks for blocking the lingual and the inferior alveolar nerves.

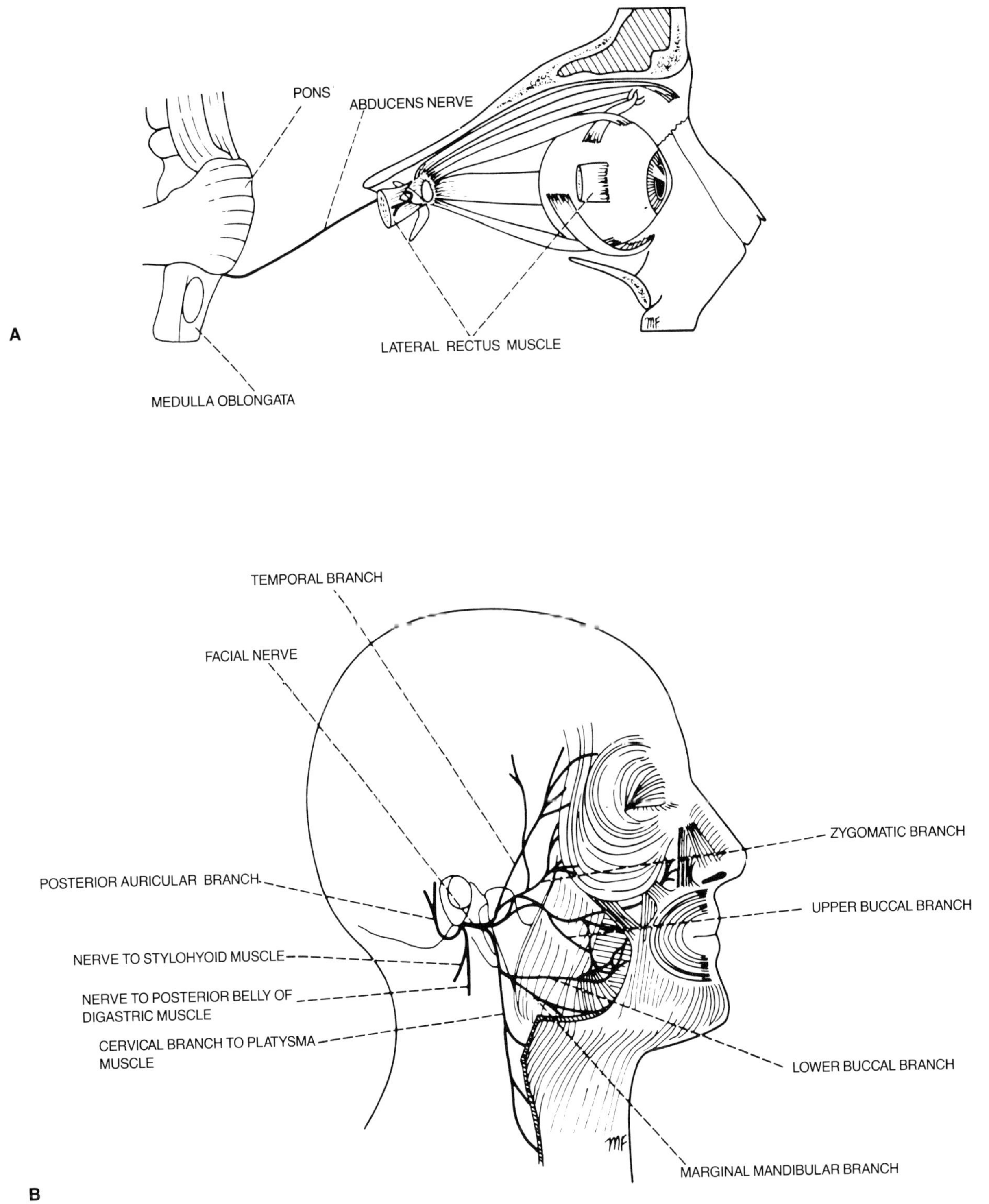

Figure 7–13. The origin and distribution of **(A)** the abducent nerve and **(B)** the facial nerve.

Facial Nerve Block

For all intraocular surgery, the orbicularis oculi muscle must be paralyzed to relax the eyelids. (See later discussion for anatomy.)

FACIAL NERVE

The facial nerve has a medial motor root and a lateral sensory root, the **nervus intermedius.** The two roots emerge from the anterior surface of the brain between the pons and the medulla oblongata. They pass laterally in the posterior cranial fossa with the vestibulocochlear nerve and enter the internal acoustic meatus in the petrous part of the temporal bone. At the bottom of the meatus, the nerve enters the facial canal and runs laterally through the inner ear. On reaching the medial wall of the tympanic cavity, the nerve expands to form the sensory **geniculate ganglion** and turns sharply backward above the promontory. At the posterior wall of the tympanic cavity, the facial nerve turns downward on the medial side of the aditus of the mastoid antrum, descends behind the pyramid, and emerges from the stylomastoid foramen.

Important Branches of the Facial Nerve

1. The **greater petrosal nerve** arises from the facial nerve at the geniculate ganglion. It contains preganglionic parasympathetic fibers that synapse in the pterygopalatine ganglion. The postganglionic fibers are secretomotor to the lacrimal gland and glands of the nose and the palate. The greater petrosal nerve also contains taste fibers from the mucous membrane of the palate (see page 237).
2. **Nerve to the stapedius** muscle arises from the facial nerve in the facial canal. It innervates the stapedius muscle within the pyramid.
3. **Chorda tympani** arises from the facial nerve in the facial canal. It runs forward over the medial surface of the upper part of the tympanic membrane and crosses the root of the handle of the malleus. The nerve leaves the tympanic cavity through the **petrotympanic fissure** and enters the infratemporal fossa to join the lingual nerve. The nerve contains preganglionic parasympathetic secretomotor fibers to the submandibular and sublingual salivary glands as well as taste fibers from the anterior two-thirds of the tongue and the floor of the mouth.
4. **Nerves to the posterior auricular muscle, posterior belly of the diagastric muscle, and nerve to the stylohoid muscle.** These muscular branches are given off from the main trunk as it emerges from the stylomastoid foramen (*Fig. 7–13*).
5. **Five terminal branches to the muscles of facial expression.** As the facial nerve leaves the stylomastoid foramen it enters the substance of the parotid salivary gland lying between the superficial and deep parts. Here, it gives off the **temporal, zygomatic, buccal, mandibular,** and **cervical** branches that emerge from the parotid gland and are distributed to the scalp and the facial muscles. (*Fig. 7–13*). Note that the buccal branch supplies the buccinator muscle. The cervical branch supplies the platysma and the depressor anguli oris muscles.

CLINICAL NOTES

ANATOMY OF FACIAL NERVE BLOCK

The facial nerve emerges from the stylomastoid foramen deep to the mastoid process and passes forward in the substance of the parotid gland. The anesthetic needle is inserted behind the midpoint of the posterior border of the ramus of the mandible below the external auditory meatus (*Fig. 7–14*). The needle is directed medially around the anterior edge of the mastoid process to reach a region just below the stylomastoid foramen.

VESTIBULOCOCHLEAR NERVE

The vestibulocochlear nerve consists of two sets of sensory fibers, vestibular and cochlear. Both parts of the nerve leave the anterior surface of the brain between the lower border of the pons and the medulla oblongata (*Fig. 7–15*), and run laterally in the posterior cranial fossa entering the internal acoustic meatus with the facial nerve.

The **vestibular fibers** represent the central processes of nerve cells of the vestibular ganglion situated in the internal acoustic meatus. These fibers originate from the vestibule and the semicircular canals and are, therefore, concerned with the sense of position and movement of the head.

The **cochlear fibers** represent the central processes of nerve cells of the **spiral ganglion of the cochlea.** These fibers originate in the **spiral organ of Corti** and are, therefore, concerned with the sense of hearing.

GLOSSOPHARYNGEAL NERVE

The glossopharyngeal nerve is a motor and a sensory nerve. It emerges from the anterior surface of the medulla oblongata between the olive and the inferior cerebellar peduncle. It passes laterally in the posterior cranial fossa and leaves the skull by passing through the jugular foramen. Here, the superior and inferior glossopharyngeal ganglia are situated on the nerve. The nerve then descends through the upper part of the neck (*Fig. 7–15*) in company with the internal jugular vein and the internal carotid artery to reach the posterior border of the stylopharyngeus

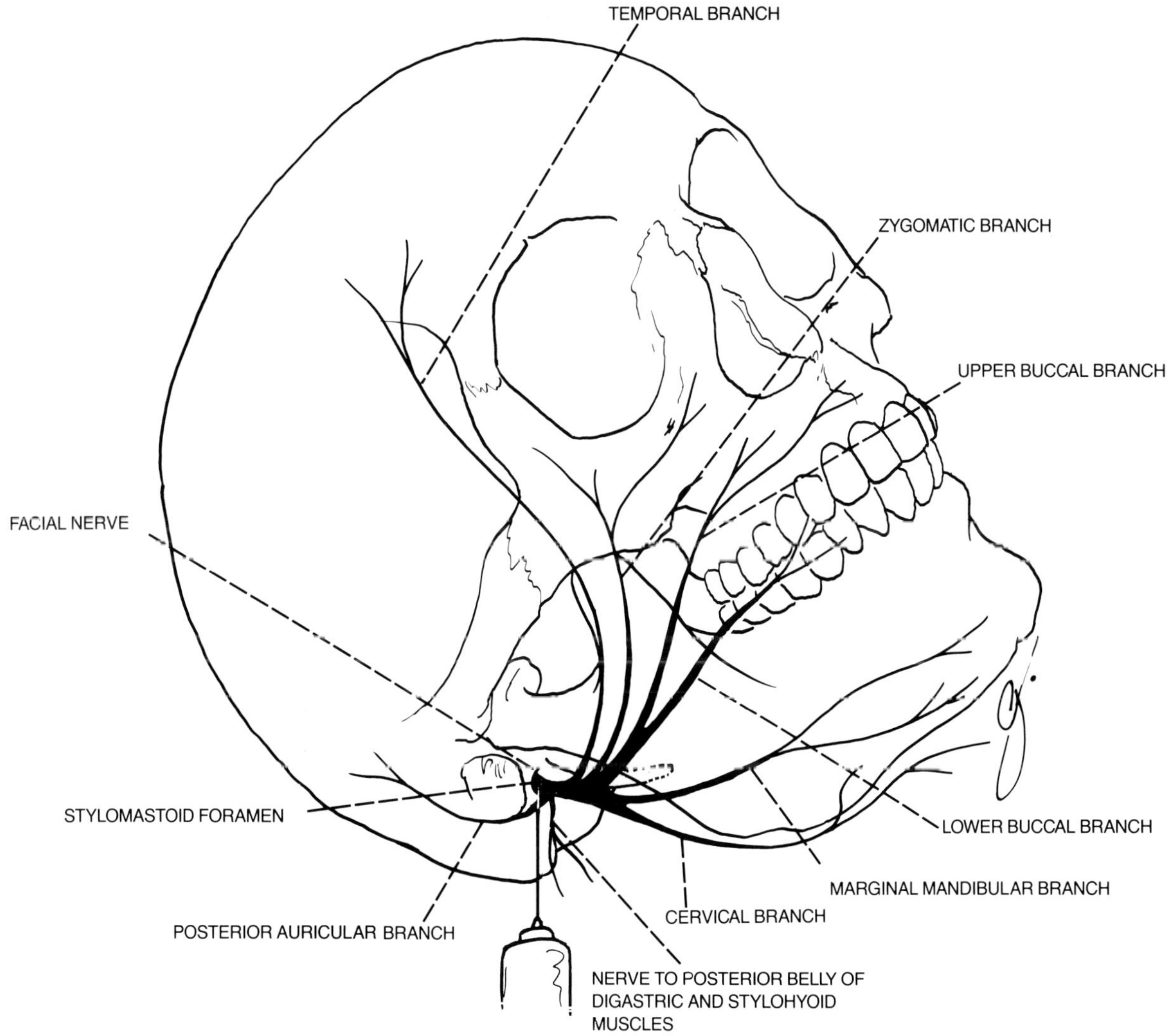

Figure 7–14. Anatomical landmarks for blocking the facial nerve.

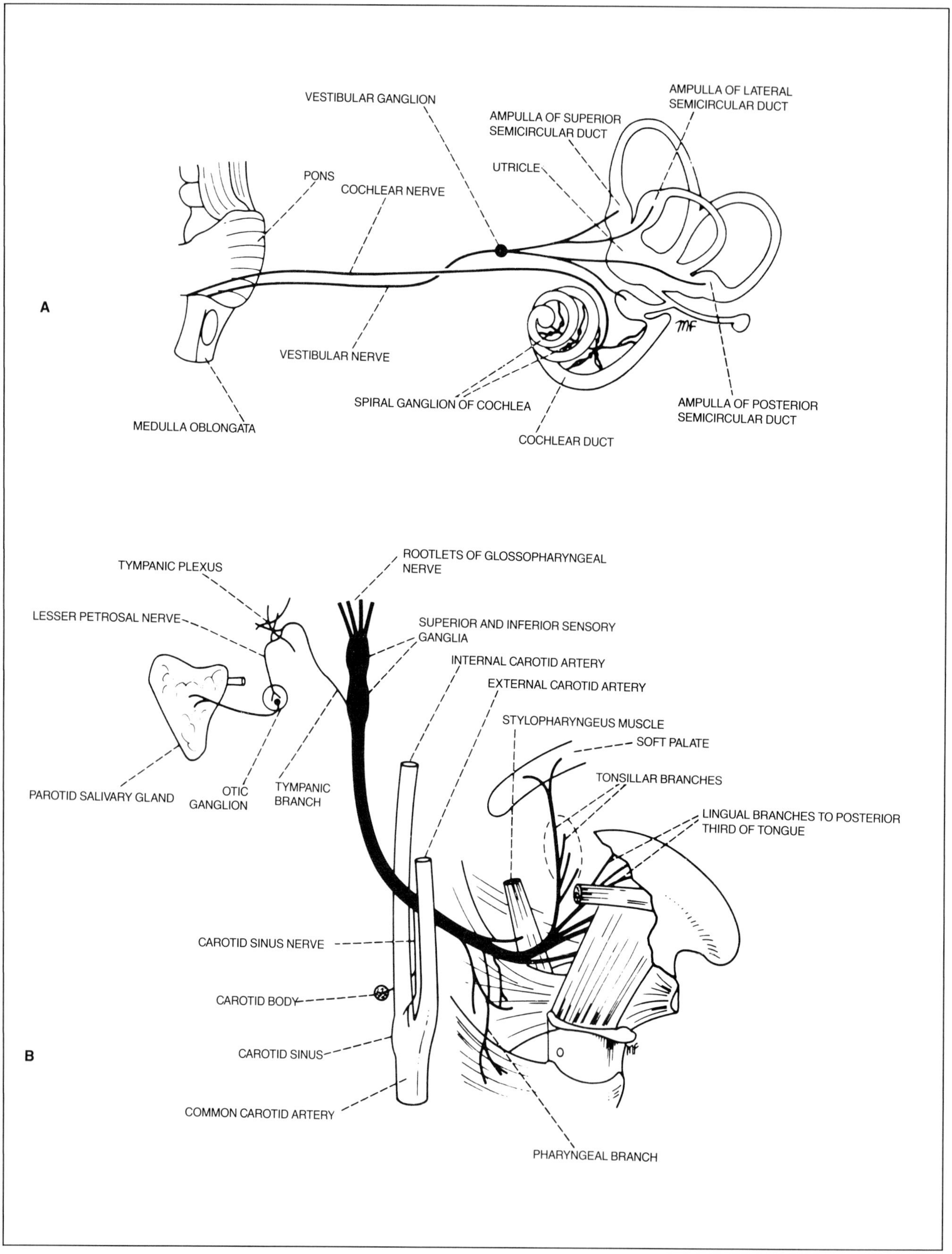

Figure 7–15. The origin and distribution of **(A)** the vestibulocochlear nerve and **(B)** the glossopharyngeal nerve.

muscle. It then passes forward between the superior and middle constrictor muscles to be distributed to the pharynx and back of the tongue.

Important Branches of the Glossopharyngeal Nerve

1. **Tympanic branch** passes through the floor of the tympanic cavity and contributes to the **tympanic plexus.** The secretomotor preganglionic parasympathetic fibers for the **parotid salivary gland** that originated in the glossopharyngeal nerve now leave the plexus as the **lesser petrosal nerve** and synapse in the **otic ganglion.**
2. **Carotid branch** contains sensory fibers from the carotid sinus (pressoreceptor mechanism for the regulation of blood pressure) and the carotid body (chemoreceptor mechanism sensitive to oxygen lack and carbon dioxide excess to regulate heart rate and respiration).
3. **Nerve to stylopharyngeus muscle.**
4. **Pharyngeal branches** that join the pharyngeal branch of the vagus and the pharyngeal branch of the sympathetic trunk to form **pharyngeal plexus.** The plexus gives sensory fibers to the mucous membrane of the pharynx.
5. **Lingual branch** supplies general sensation and special taste to the mucous membrane of the posterior third of the tongue and the vallate papillae.

CLINICAL NOTES

ANATOMY OF GLOSSOPHARYNGEAL NERVE BLOCK

The glossopharyngeal nerve leaves the skull through the middle of the jugular foramen along with the vagus and accessory nerves. It then comes to lie between the internal jugular vein and the internal carotid artery. Because of these important vascular relationships it is mandatory to perform aspiration tests before injecting the anesthetic.

The needle is inserted at a point midway between the angle of the jaw and the tip of the mastoid process and is directed medially toward the styloid process. The vagus and accessory nerves are blocked along with the glossopharyngeal nerve when the injection is made just deep to the styloid process.

The glossopharyngeal nerve can also be blocked by inserting the needle into the midpoint of the palatopharyngeal arch (posterior pillar of the fauces). This blocks the distribution of the nerve to the pharynx.

VAGUS NERVE

The vagus nerve is composed of motor and sensory fibers. It emerges from the anterior surface of the medulla oblongata between the olive and the inferior cerebellar peduncle. The nerve passes laterally through the posterior cranial fossa and leaves the skull through the jugular foramen. The vagus nerve possesses two sensory ganglia, a rounded **superior ganglion,** situated on the nerve within the jugular foramen, and cylindrical **inferior ganglion,** which lies on the nerve just below the foramen. Below the inferior ganglion the cranial root of the accessory nerve joins the vagus nerve and is distributed mainly in its pharyngeal and recurrent laryngeal branches.

The vagus nerve descends vertically in the neck within the carotid sheath with the internal jugular vein and the internal and common carotid arteries (*Figs. 3–16 and 7–17*).

The **right vagus nerve** (*Fig. 7–16*) crosses the anterior surface of the subclavian artery and enters the thorax posterolateral to the brachiocephalic artery, lateral to the trachea, and medial to the azygos vein. It passes posterior to the root of the right lung contributing to the pulmonary plexus. It then passes onto the posterior surface of the esophagus and contributes to the esophageal plexus. It enters the abdomen behind the esophagus through the esophageal opening of the diaphragm. The posterior vagal trunk (which is the name now given to the right vagus) is distributed to the posterior surface of the stomach and by a large celiac branch to the duodenum, liver, kidneys, and small and large intestine as far as the distal third of the transverse colon. This wide distribution is accomplished through the celiac, superior mesenteric, and renal plexuses.

The **left vagus nerve** (*Fig. 7–16*) enters the thorax between the left common carotid and the left subclavian arteries. It crosses the left side of the aortic arch and descends behind the root of the left lung contributing to the pulmonary plexus. The left vagus then descends on the anterior surface of the esophagus contributing to the esophageal plexus. It enters the abdomen through the esophageal opening of the diaphragm lying in front of the esophagus. The anterior vagal trunk (which is the name now given to the left vagus) divides into several branches that are distributed to the stomach, the liver, and the upper part of the duodenum and head of the pancreas.

Important Branches of the Vagus Nerve

1. **Meningeal and auricular branches.**
2. **Pharyngeal branch** contains nerve fibers from the cranial root of the accessory nerve. It joins the pharyngeal branches of the glossopharyngeal nerves and the sympathetic trunk to form the **pharyngeal plexus.** The pharyngeal nerve supplies all the muscles of the pharynx and soft palate except the stylopharyngeus and tensor veli palatini.
3. **Superior laryngeal nerve** divides into internal and external laryngeal nerves. The **internal laryngeal nerve** is sensory to the mucous membrane of the piri-

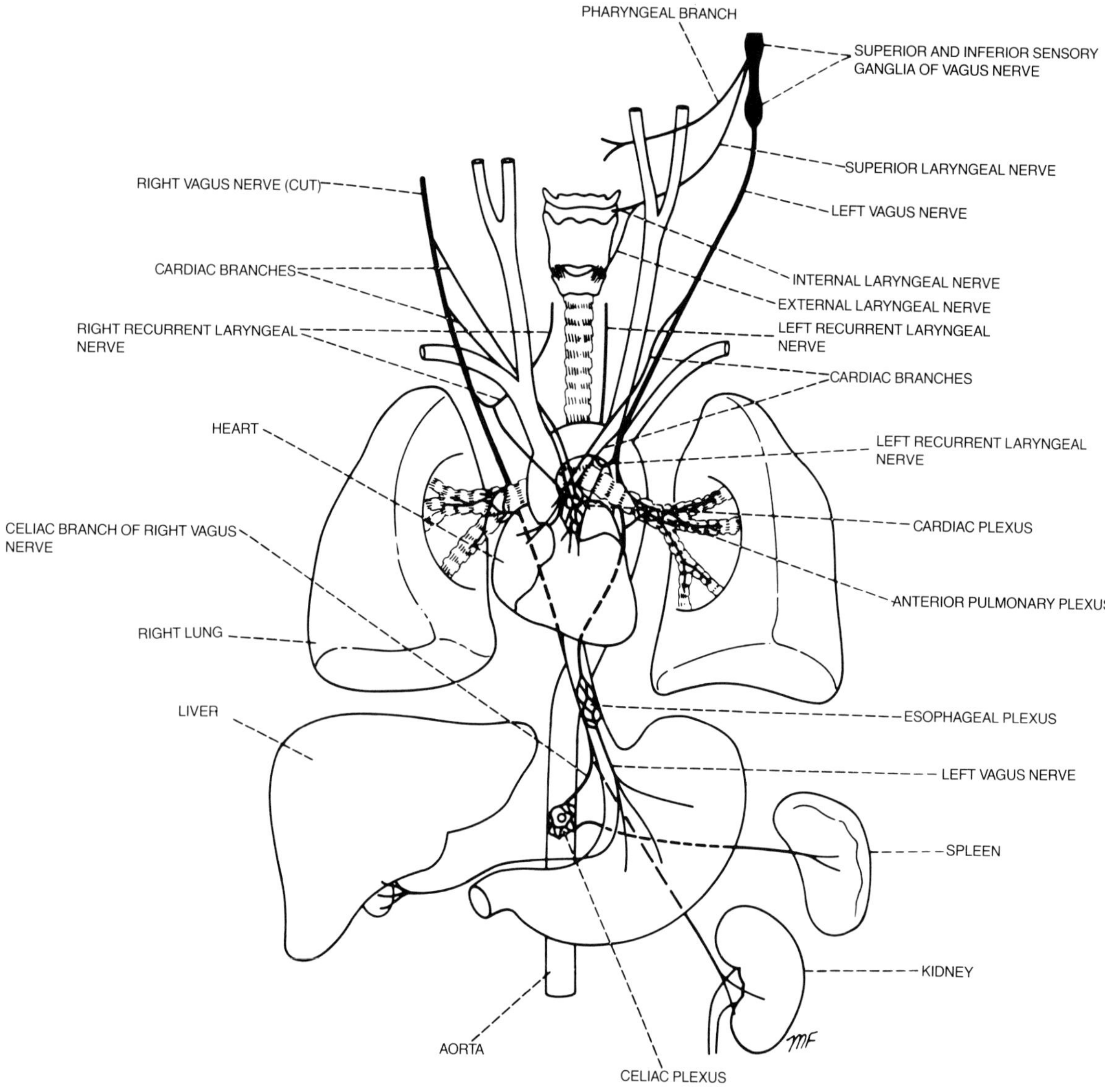

Figure 7–16. Distribution of the vagus nerve.

form fossa and the larynx down as far as the vocal folds. The **external laryngeal nerve** is motor and located close to the superior thyroid artery; it supplies the cricothyroid muscle.

4. **Recurrent laryngeal nerve.** On the right side, the nerve arises from the vagus as it crosses the first part of the subclavian artery (*Fig. 7–16*). The recurrent laryngeal nerve then hooks backward and upward behind the artery and ascends in the groove between the trachea and the esophagus. The nerve crosses either in front of or behind the inferior thyroid artery or may pass between its branches. It supplies all the muscles of the larynx except the cricothyroid and the mucous membrane of the larynx below the vocal folds and the mucous membrane of the upper part of the trachea. On the left side, the recurrent laryngeal nerve arises from the vagus as the latter crosses the arch of the aorta (*Fig. 7–16*). It hooks beneath the arch behind the ligamentum arteriosum and ascends into the neck in the groove between the trachea and the esophagus.
5. **Cardiac branches.** Two or three branches arise from the vagus in the neck. They descend into the thorax and end in the cardiac plexus.

The vagus nerve, thus, innervates the heart and the great vessels within the thorax, the larynx, the trachea, the bronchi and the lungs, and much of the alimentary tract from the pharynx to the distal part of the transverse colon. It also supplies glands associated with the alimentary tract such as the liver and the pancreas.

The vagus has the most extensive distribution of all the cranial nerves and supplies the structures named above with afferent and efferent fibers.

CLINICAL NOTES

ANATOMY OF VAGAL NERVE BLOCKS

The sensory and motor supply of the larynx and the trachea are the superior and the recurrent laryngeal nerves, branches of the vagus.

The superior laryngeal nerve is easily blocked as it passes forward below the palpable greater cornu of the hyoid bone. The block produces anesthesia from the laryngeal inlet down as far as the vocal folds (internal laryngeal branch of superior laryngeal nerve); it also paralyses the cricothyroid muscle that normally tenses the vocal folds (external laryngeal branch of superior laryngeal nerve).

The most effective method of blocking the terminal branches of the recurrent laryngeal nerve to the mucous membrane of the lower half of the larynx and the trachea is to insert the blocking needle through the cricothyroid membrane. The cricoid cartilage is easily felt and the needle is inserted through the membrane in the midline.

The permanent block of the recurrent laryngeal nerve is difficult because of the important neighboring anatomic structures. If such a block is to be carried out, however, the needle is inserted over the nerve as it ascends in the groove between the trachea and the esophagus.

ACCESSORY NERVE

The accessory nerve is a motor nerve that consists of cranial and spinal roots (*Fig. 7–17*). The **cranial root** emerges from the anterior surface of the medulla oblongata between the olive and the inferior cerebellar peduncle. The nerve runs laterally in the posterior cranial fossa and joins the spinal root, which arises from nerve cells in the anterior gray column of the upper five segments of the cervical spinal cord. The nerve ascends alongside the spinal cord and enters the skull through the foramen magnum. It then turns laterally to join the cranial root.

The two roots unite and leave the skull through the jugular foramen. The roots then separate (*Fig. 7–17*) and the cranial root joins the vagus nerve and is distributed in its pharyngeal and recurrent laryngeal branches to the muscles of the soft palate, the pharynx, and the larynx. The spinal root runs downward and laterally and enters the deep surface of the sternocleidomastoid muscle, which it supplies. The nerve then crosses the posterior triangle of the neck lying on the levator scapulae muscle. It then passes beneath the trapezius muscle, which it supplies.

The spinal accessory nerve thus brings about movements of the soft palate, the pharynx, and the larynx, and controls the movements of two large muscles in the neck.

CLINICAL NOTES

Anatomy of Accessory Nerve Block

The spinal part of the accessory nerve is occasionally blocked to paralyze the trapezius muscle.

The spinal part of the accessory nerve emerges from the posterior border of the sternocleidomastoid in the posterior triangle of the neck and runs downward and backward to pass beneath the anterior border of the trapezius. The course may be indicated by drawing a line from the angle of the mandible to the tip of the mastoid process and bisecting this line at right angles with a second line that runs downward across the posterior triangle; the second line indicates the course of the nerve.

HYPOGLOSSAL NERVE

The hypoglossal nerve is a motor nerve that emerges on the anterior surface of the medulla oblongata between

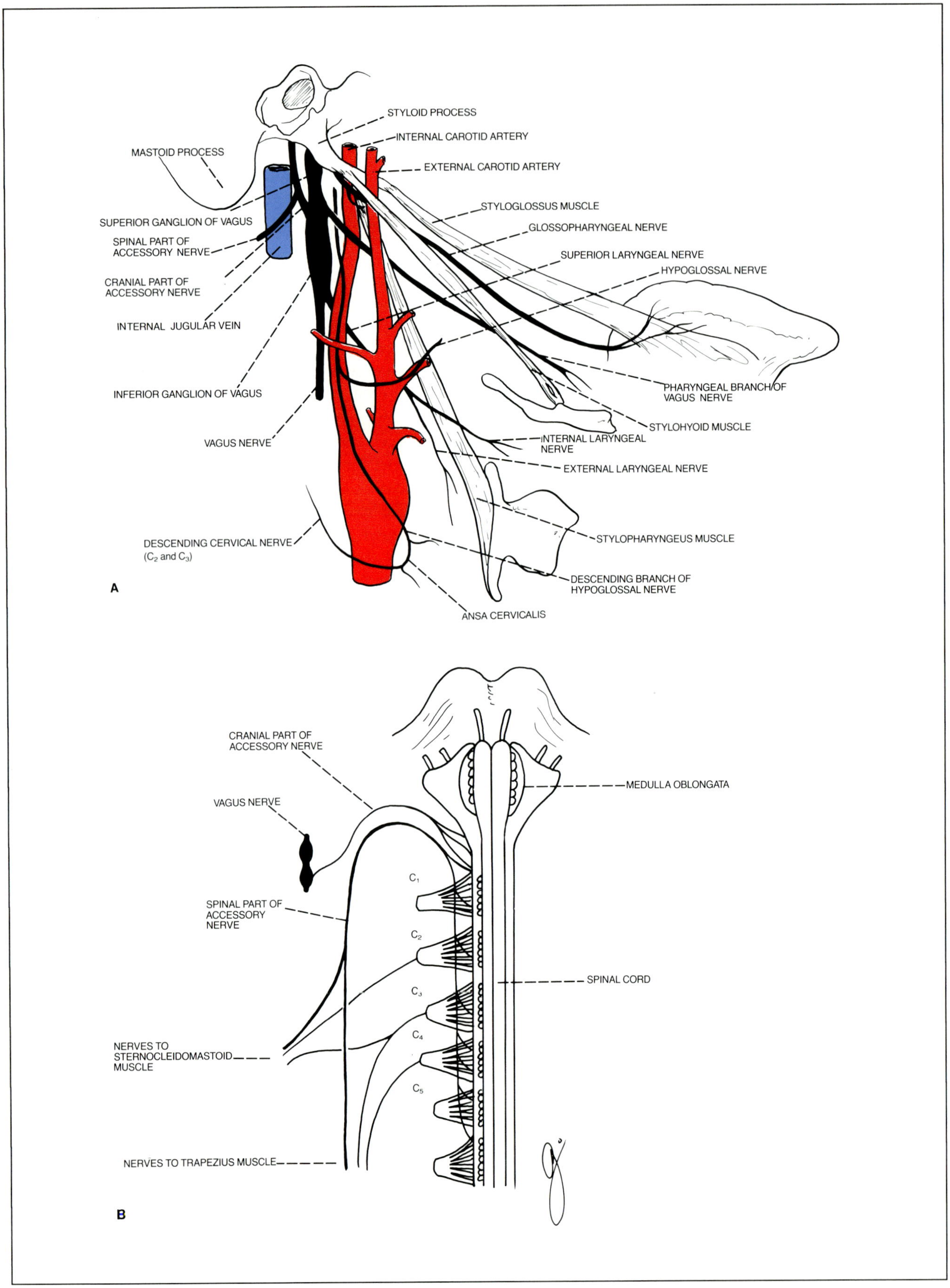

Figure 7–17. A. The main vessels and cranial nerves in the neck. **B.** The origin and distribution of the accessory nerve.

the pyramid and the olive. It crosses the posterior cranial fossa and leaves the skull through the hypoglossal canal. The nerve then passes downward and forward in the neck between the internal carotid artery and the internal jugular vein until it reaches the lower border of the posterior belly of the digastric muscle. Here, it turns forward and crosses the internal and external carotid arteries and the loop of the lingual artery. It passes deep to the posterior margin of the mylohyoid muscle lying on the lateral surface of the hyoglossus muscle (*Fig. 7–18*) and lies inferior to the submandibular duct and the lingual nerve. In the upper part of its course, the hypoglossal nerve is joined by C_1 fibers from the cervical plexus.

Important Branches of the Hypoglossal Nerve

1. **Meningeal branch.**
2. **Descending branch** (*Fig. 7–18*), consisting of C_1 fibers, descends in front of the internal and common carotid arteries. It is joined by the descending cervical nerve (C_2 and C_3) from the cervical plexus to form a loop called the **ansa cervicalis.** Branches from the loop supply the omohyoid, the sternohyoid, and the sternothyroid muscles.
3. **Nerve to the thyrohyoid muscle** (*Fig. 7–18*), consisting of C_1 fibers, innervates the thyrohyoid muscle.
4. **Muscular branches** (*Fig. 7–18*) to all the muscles of the tongue except the palatoglossus. It thus innervates the styloglossus, hyoglossus, genioglossus, and intrinsic muscles.
5. **Nerve to the geniohyoid muscle** (*Fig. 7–18*) consists of C_1 fibers.

The hypoglossal nerve proper thus innervates the majority of the muscles of the tongue and controls the shape and movements of the tongue.

CLINICAL NOTES

To test the integrity of the hypoglossal nerve, the patient is asked to extend his or her tongue. If there is a lower motor neuron lesion, it will be noted that the tongue deviates toward the paralyzed side. The tongue will be smaller on the side of the lesion owing to muscle atrophy, and fibrillation may accompany or precede the atrophy. If a patient has a lesion of the corticonuclear fibers, there will be no atrophy or fibrillation of the tongue, and on protrusion, the tongue will deviate to the side opposite the lesion. The explanation for these phenomena is as follows. The right and left genioglossus muscles pull the tongue forward in the midline. Should one be paralyzed (lower motor neuron), the tongue will be deviated to the side of the lesion. The hypoglossal nerve nucleus receives corticonuclear fibers from both cerebral hemispheres, but the portion of the nucleus that supplies the genioglossus muscle receives corticonuclear fibers only from the opposite cerebral cortex. It, therefore, follows that if the right corticonuclear fibers are sectioned, the left genioglossus muscle will be paralyzed (upper motor neuron) and the tongue will point to the left side.

The different components of the cranial nerves, their functions, and the openings in the skull through which the nerves leave the cranial cavity are summarized in Table 7–1.

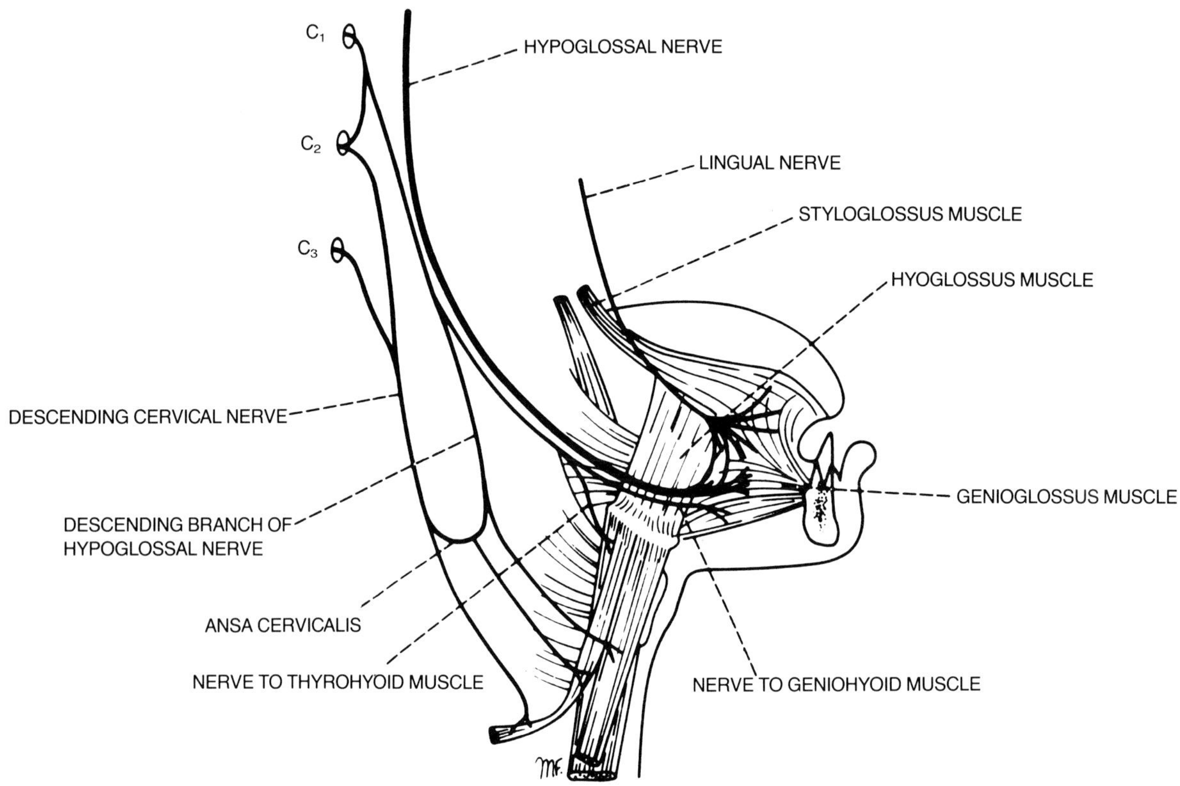

Figure 7–18. The distribution of the hypoglossal nerve.

TABLE 7–1. CRANIAL NERVES

Cranial Nerve	Components	Function	Opening in Skull
I Olfactory	Sensory	Smell	Openings in cribriform plate of ethmoid
II Optic	Sensory	Vision	Optic canal
III Oculomotor	Motor	Lifts upper eyelid; turns eyeball upward, downward, and medially; constricts pupil; accommodates eye	Superior orbital fissure
IV Trochlear	Motor	Assists in turning eyeball downward and laterally	Superior orbital fissure
V Trigeminal			
Ophthalmic division	Sensory	Cornea, skin of forehead, scalp, eyelids; also mucous membrane of paranasal sinuses and nasal cavity	Superior orbital fissure
Maxillary division	Sensory	Skin of face over maxilla; teeth of upper jaw; mucous membrane of nose, maxillary sinus, and palate	Foramen rotundum
Mandibular division	Motor	Muscles of mastication: mylohyoid, anterior belly of digastric, tensor veli palatini, and tensor tympani	Foramen ovale
	Sensory	Skin of cheek, skin over mandible and side of head, teeth of lower jaw and temporomandibular joint; mucous membrane of mouth and anterior part of tongue	
VI Abducent	Motor	Lateral rectus muscle turns eyeball laterally	Superior orbital fissure
VII Facial	Motor	Muscles of face and scalp: stapedius muscle, posterior belly of digastric, and stylohyoid muscles	Internal acoustic meatus, facial canal, stylomastoid foramen
	Sensory	Taste from anterior part of tongue, floor of mouth and palate	
	Secretomotor parasympathetic	Submandibular and sublingual salivary glands, lacrimal gland and glands of nose and palate	
VIII Vestibulocochlear			
Vestibular division	Sensory	Position and movement of head	Internal acoustic meatus
Cochlear division	Sensory	Hearing	
IX Glossopharyngeal	Motor	Stylopharyngeus muscle, assists in swallowing	
	Secretomotor parasympathetic	Parotid salivary	Jugular foramen
	Sensory	Posterior part of tongue and pharynx, carotid sinus, and carotid body	
X Vagus	Motor and sensory	Heart and great thoracic vessels; larynx, trachea, bronchi and lungs; alimentary tract from pharynx to splenic flexure of colon; liver, kidneys, and pancreas	Jugular foramen
XI Accessory			
Cranial root	Motor	Muscles of soft palate, pharynx, and larynx	Jugular foramen
Spinal root	Motor	Sternocleidomastoid and trapezius muscles	
XII Hypoglossal	Motor	Muscles of tongue (except palatoglossus) controlling its shape and movement	Hypoglossal canal

CLINICAL PROBLEMS

1. A 48-year-old woman woke up one morning to find the left side of her face paralyzed. Her physician found on examination that the left side of her face was completely paralyzed. She was also noted to have severe hypertension. The patient spoke with a slightly slurred speech. The physician told the patient that she had had a mild stroke and that she should be treated in bed. Do you think that the physician had made the correct diagnosis?

2. A feeble 65-year-old man suffering from bronchitis and emphysema was about to have surgery involving the lower jaw. It was decided to perform the surgery with a mandibular nerve block. Where does the mandibular nerve exit from the skull? In which bone is this foramen situated? What is the sensory distribution of the mandibular division of the trigeminal nerve? If the block is successful, which muscles would be paralyzed? Would such a nerve block interfere with salivary secretion? As the anesthetic needle is advanced medially across the infratemporal fossa, it encounters a bone on the medial wall. What is the name of that bone? What is the possible complication if the needle is advanced too deeply?

3. The visual pathway extends from the retina to the cortex of the occipital lobe of the brain. What are the effects of a lesion at the following sites along the visual pathway: (a) Midline section of the optic chiasma, (b) section of the left optic tract; (c) destruction of the cortex of the right occipital cortex; (d) section of the left optic radiation; and (e) section of the right optic nerve.

4. A 60-year-old man was diagnosed as having an advanced inoperable carcinoma of the oral pharynx with neoplastic infiltration extending laterally. How is it possible to test for the integrity of the 9th, 10th, and 11th cranial nerves? Because of severe chronic pain in the pharynx, it was decided to block the 9th and 10th cranial nerves with a neurolytic agent. Where may these nerves be blocked in this patient?

5. A 55-year-old man with a history of loss of weight and expectorating blood-stained sputum was found on radiographic examination to have a density in the middle lobe of the right lung. It was decided to perform a bronchoscopy and if possible take a biopsy for pathologic examination. What is the sensory innervation of the larynx and the trachea? Where may these nerves be blocked to perform a bronchoscopic examination?

6. A 35-year-old woman was found to have a small lipoma developing in the skin of her right forehead. For cosmetic reasons it was decided to remove it surgically under local anesthesia. What is the sensory innervation of the skin of the forehead? Where may these nerves be blocked?

7. An 83-year-old man with bilateral cataract was advised by his ophthalmologist to have surgical removal of his right lens. What is the innervation of the eyeball? How would you anesthetize the eyeball? It was considered important to paralyze the orbicularis oculi muscle. What is the motor innervation of the orbicularis oculi muscle? Where would you block this nerve?

8. A 33-year-old line backer from the Dulles Cowboys had repeatedly dislocated his left shoulder. His orthopedic surgeon advised surgery to tighten the joint capsule. It was decided to perform the surgery under an interscalenae brachial plexus block. In addition, it was proposed to paralyze the trapezius muscle to prevent the shoulder from being raised during surgery. What is the nerve supply to the trapezius muscle? Where may this nerve be conveniently blocked?

ANSWERS TO CLINICAL PROBLEMS

1. The physician made the diagnosis based on the facial paralysis, the slurred speech, and the hypertension. Unfortunately, he had forgotten the central connections of the motor nucleus of the facial nerve. A lesion of the corticonuclear fibers on one side of the brain will cause paralysis only of the muscles of the lower part of the opposite side of the face. The part of the facial nerve nucleus that innervates the upper part of the face receives corticonuclear fibers from both cerebral cortices. This patient had a complete paralysis of the entire left side of the face, which could only be due to a lesion of the lower motor neuron. The correct diagnosis was Bell's palsy, an inflammation of the connective tissue sheath of the facial nerve. The majority of the patients recover within 2 weeks.

2. The mandibular division of the trigeminal nerve exits from the middle cranial fossa through the foramen ovale in the greater wing of the sphenoid bone. The sensory distribution of this nerve is as follows: A small meningeal branch returns to the middle cranial fossa. From the anterior division, the buccal nerve supplies the skin and mucous membrane of the cheek. From the posterior division, the lingual nerve innervates the mucous membrane of the floor of the mouth and the anterior two-thirds of the tongue with general sensation. The inferior alveolar nerve supplies the teeth and gums of the lower jaw and the skin over the chin. The auriculotemporal nerve innervates the skin of the auricle, the external auditory meatus, the temporomandibular joint, and the scalp.

   A successful mandibular nerve block would result in the paralysis of the muscles of mastication (masseter, temporalis, medial and lateral pterygoids), the tensor veli palatini, and the tensor tympani.

   The otic ganglion is situated close to the medial side of the main trunk of the mandibular nerve just inferior to the foramen ovale and is very likely to be blocked along with the mandibular nerve. Paralysis of this ganglion would result in the blocking of the parasympathetic secretomotor fibers to the parotid salivary gland.

   The name of the bone on the medial wall of the infratemporal fossa is the lateral pterygoid plate.

   If the anesthetic needle is advanced too far behind the posterior border of the lateral pterygoid plate, it will pierce the superior constrictor and enter the pharynx.

3. (a) Bilateral hemianopia: (b) Right homonymous hemianopia: (c) Left homonymous hemianopia: (d) Right homonymous hemianopia: (e) Complete blindness of the right eye.

4. The glossopharyngeal nerve supplies the posterior one-third of the tongue with fibers that subserve common sensations and taste. These sensations may easily be tested. The vagus nerve, by means of its pharyngeal branch, supplies many muscles of the soft palate, including the levator veli palatini. These muscles may be tested by asking the patient to say "ah" and observing the uvula, which is normally elevated in the midline. Additional tests may be carried out by observing the movements of the vocal folds with a laryngoscope (see page 23).

   The spinal part of the accessory nerve may be tested by asking the patient to shrug his shoulders by using the trapezius muscles or to rotate his head so that he looks upward to the opposite side by contracting the sternocleidomastoid muscles. Both these muscles are innervated by the spinal part of the accessory nerve.

   The site for blocking the ninth and tenth cranial nerves is described on pages 247 and 249.

5. The upper half of the mucous membrane of the larynx, from the inlet down as far as the vocal folds, is innervated by the internal laryngeal branch of the superior

laryngeal branch of the vagus. The mucous membrane of the lower half of the larynx and the upper half of the trachea is innervated by the recurrent laryngeal branch of the vagus.

Local anesthesia for bronchoscopy must be combined with sedation. To permit easy unimpeded passage of the bronchoscope, the cough and gag reflexes must be blocked and the possibility of laryngeal spasm eliminated. Topical application of local anesthetic solution to the mucous membrane of the pharynx will block the gag reflex (glossopharyngeal nerve), the upper half of the larynx may be anesthetized with a bilateral superior laryngeal nerve block through the skin below the greater cornu of the hyoid bone (see page 249), or more commonly, by spraying the anesthetic onto the mucous membrane of the larynx. Tracheal anesthesia (recurrent laryngeal nerve) may be performed by transtracheal injection (see page 249) or spraying the mucous membrane under direct vision through a laryngoscope.

6. The skin of the forehead is innervated by the supraorbital and supratrochlear branches of the frontal nerve, a branch of the ophthalmic division of the trigeminal nerve. The nerves may be blocked as they emerge onto the forehead around the superior margin of the orbital cavity (see page 239).

7. The eyeball is innervated by sensory fibers from the long ciliary nerves, branches of the nasociliary nerve, and a branch of the ophthalmic division of the trigeminal nerve. Block of the nasociliary nerve and its branches (retrobulbar block) is described on page 240. The orbicularis oculi muscle is supplied by the 7th cranial nerve. This nerve may be blocked as it emerges from the stylomastoid foramen as described on page 244.

8. The trapezius muscle is innervated by the spinal part of the accessory nerve (it also receives proprioceptive fibers from C_3 and C_4). The course of the spinal part of the accessory nerve (and C_3 and C_4) in the posterior triangle of the neck may be marked out on the skin by drawing a line from the tip of the mastoid process to the angle of the mandible and bisecting it at right angles. The second line when extended downward across the posterior triangle indicates the course of the nerves. The spinal part of the accessory nerve may be blocked along its course.

CHAPTER 8

Autonomic Nervous System

The autonomic nervous system is concerned with the innervation of involuntary structures such as the heart, the smooth muscles, and the glands. It is distributed throughout the central and peripheral nervous systems. The autonomic system is divided into two parts, the **sympathetic** and the **parasympathetic,** and in both there are afferent and efferent nerve fibers.

SYMPATHETIC PART OF THE AUTONOMIC SYSTEM

The larger sympathetic part of the autonomic system is widely distributed throughout the body innervating the heart and the lungs, the muscle in the walls of many blood vessels, the hair follicles and the sweat glands, and many abdominopelvic viscera.

The function of the sympathetic system is to prepare the body for an emergency. The heart rate is increased, arterioles of the skin and the intestine are constricted, those of the skeletal muscles are dilated, and the blood pressure is raised. There is a redistribution of blood so that it leaves the skin and the gastrointestinal tract and passes to the brain, the heart, and the skeletal muscle. In addition, the sympathetic nerves dilate the pupils, inhibit the smooth muscle of the bronchi, the intestine, and the bladder wall, and close the sphincters. The hair is made to stand on end and sweating occurs.

The sympathetic system includes the efferent outflow from the spinal cord, two ganglionated sympathetic trunks, important branches, plexuses, and regional ganglia.

Efferent Nerve Fibers (Sympathetic Outflow)

The lateral gray columns (horns) of the spinal cord from the first thoracic segment to the second lumbar segment (sometimes third lumbar segment) possess the cell bodies of the sympathetic connector neurons (*Fig. 8–1*). The myelinated axons of these cells leave the cord with the anterior nerve roots and pass into the spinal nerve trunks. They then leave the nerve via the **white rami communicantes** to the **paravertebral ganglia** of the **sympathetic trunk.** Once these fibers (preganglionic) reach the ganglia in the sympathetic trunk, distribution is as follows:

1. The fibers synapse with an excitor neuron in the ganglion (*Figs. 8–1 and 8–2*). The gap between the two neurons is bridged by the neurotransmitter **acetylcholine.** The postganglionic nonmyelinated axons leave the ganglion and pass to the thoracic spinal nerves as **gray rami communicantes.** They are distributed in branches of the spinal nerves to smooth muscle in the blood vessel walls, sweat glands, and arrector pili muscles of the skin.
2. Axons may travel cephalad in the sympathetic trunk to synapse in ganglia in the cervical region (*Fig. 8–2*). The postganglionic nerve fibers pass via gray rami communicantes to join cervical spinal nerves. Many of the preganglionic fibers entering the lower part of the sympathetic trunk from the lower thoracic and upper two lumbar segments of the spinal cord travel caudal to synapse in ganglia in the lower lumbar and sacral regions. Here again, the postganglionic nerve fibers pass via gray rami communicantes to join the lumbar, sacral, and coccygeal spinal nerves (*Fig. 8–2*).
3. The fibers may pass through the ganglia of the sympathetic trunk without synapsing. These myelinated fibers leave the sympathetic trunk as the **greater splanchnic, lesser splanchnic,** and **lowest or least splanchnic nerves.** The greater splanchnic nerve is formed from branches of the fifth to the ninth thoracic ganglia. It descends obliquely on the side of the bodies of the thoracic vertebrae and pierces the crus of the diaphragm to synapse with excitor cells in the ganglia of the **celiac plexus,** the **renal plexus,** and the suprarenal medulla. The lesser splanchnic nerve is formed from branches of the 10th and 11th thoracic ganglia. It descends with the greater splanchnic nerve and pierces the diaphragm to join excitor cells in ganglia

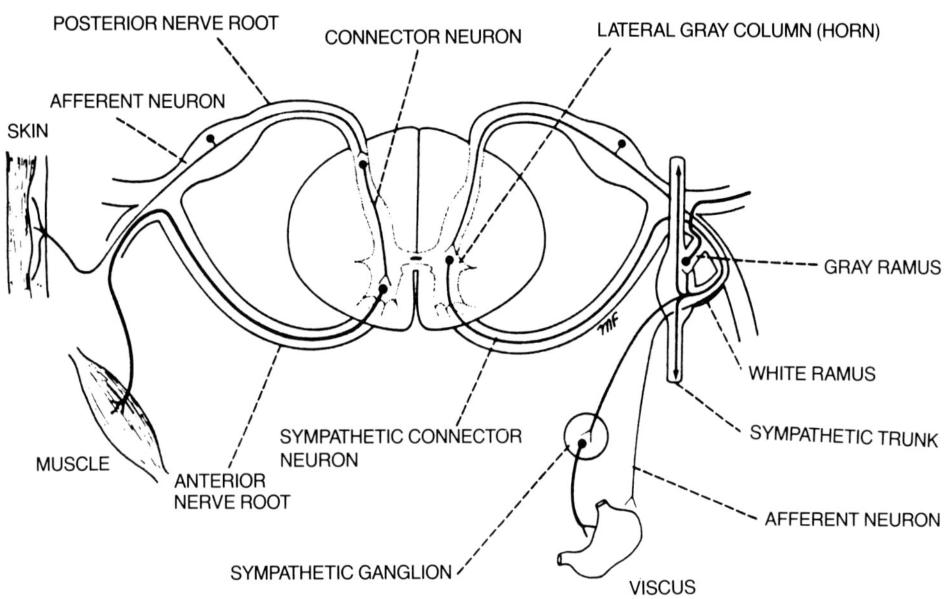

Figure 8–1. General arrangement of somatic part of nervous system (*left*) compared with autonomic part of nervous system (*right*).

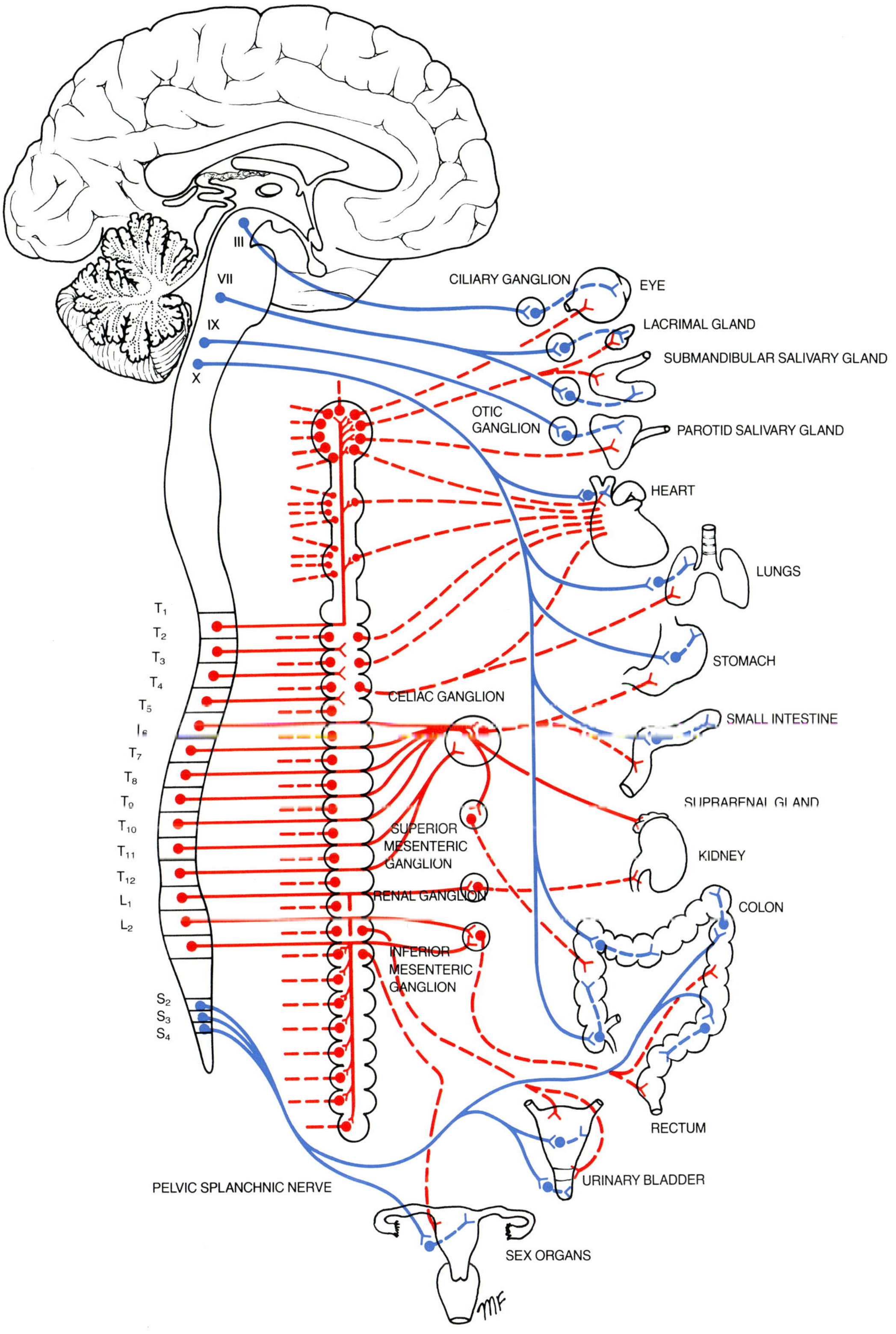

Figure 8–2. Efferent part of autonomic nervous system. Preganglionic parasympathetic fibers are shown in *solid blue* and postganglionic parasympathetic fibers in *interrupted blue.* Preganglionic sympathetic fibers are shown in *solid red* and postganglionic sympathetic fibers in *interrupted red.*

in the lower part of the **celiac plexus.** The lowest splanchnic nerve (when present) arises from the 12th thoracic ganglion, pierces the diaphragm, and synapses with excitor neurons in the ganglia of the **renal plexus.** The splanchnic nerves, therefore, are formed of preganglionic fibers. The postganglionic fibers arise from the excitor cells in the peripheral plexuses and are distributed to the smooth muscle and glands of the viscera. As previously noted, a few preganglionic fibers, traveling in the greater splanchnic nerve, end directly on the cells of the **suprarenal medulla.** These medullary cells may be regarded as modified sympathetic excitor neurons and are responsible for epinephrine and norepinephrine secretion.

Afferent Nerve Fibers

The afferent myelinated nerve fibers travel from the viscera through the sympathetic ganglia without synapsing. They pass to the spinal nerve via white rami communicantes and reach their cell bodies in the posterior root ganglion of the corresponding spinal nerve (*Fig. 8–1*). The central axons then enter the spinal cord and may form the afferent component of a local reflex arc or ascend to higher centers, such as the hypothalamus.

Sympathetic Trunks

The sympathetic trunks are two ganglionated nerve trunks that extend the whole length of the vertebral column (*Fig. 8–2*). In the neck, each trunk has three ganglia; in the thorax, there are 11 or 12; in the lumbar region, four or five, and in the pelvis, four or five. In the neck, the trunks lie anterior to the transverse processes of the cervical vertebrae; in the upper part of the thorax, they are anterior to the heads of the ribs; in the lower part of the thorax, they are on the sides of the vertebral bodies; in the abdomen, they are anterolateral to the bodies of the lumbar vertebrae; and in the pelvis, they are anterior to the sacrum and medial to the anterior sacral foramina. Below, the two trunks end by joining together to form a single ganglion, the **ganglion impar.**

Cervical Part of the Sympathetic Trunk. The cervical part of the sympathetic trunk extends upward to the base of the skull and below to the first rib, where it becomes continuous with the thoracic part of the sympathetic trunk. The trunk lies behind the carotid sheath. The cervical part of the trunk possesses three ganglia, the superior, middle, and inferior cervical ganglia (*Figs. 8–2 and 8–3*). This part of the trunk sends gray rami communicantes to the cervical spinal nerves, but there are no white rami communicantes.

Superior Cervical Ganglion. The superior cervical ganglion lies immediately below the skull and is joined to the middle ganglion by the connecting trunk.

BRANCHES

1. **Internal carotid nerve,** consisting of postganglionic nerve fibers, accompanies the internal carotid artery into the carotid canal. It divides into branches that form the **internal carotid plexus** around the artery.
2. **Gray rami communicantes,** consisting of postganglionic nerve fibers, pass laterally to the upper four cervical spinal nerves.
3. **Arterial branches** to the common and external carotid arteries.
4. **Cranial nerve branches** to the 9th, 10th, and 12th cranial nerves.
5. **Pharyngeal branches** that join the pharyngeal branches of the ninth and tenth cranial nerves to form the **pharyngeal plexus.**
6. **Superior cardiac branch** descends in the neck to end in the cardiac plexus in the thorax.

Middle Cervical Ganglion. The middle cervical ganglion lies at the level of the cricoid cartilage. It is the smallest of the three cervical ganglia and is connected to the inferior ganglion by two or more nerve bundles. The anterior bundle crosses anterior to the first part of the subclavian artery and is known as the **ansa subclavia.**

BRANCHES

1. **Gray rami communicantes** to the fifth and sixth cervical spinal nerves.
2. **Thyroid branches** that reach the thyroid gland along the inferior thyroid artery.
3. **Middle cardiac branch** descends in the neck to join the cardiac plexus in the thorax.
4. **Tracheal and esophageal branches.**

Inferior Cervical Ganglion. In the majority of individuals (80 percent) the inferior cervical ganglion is fused with the first thoracic ganglion to form the large **stellate ganglion.** It is located between the transverse process of the seventh cervical vertebra and the neck of the first rib. The vertebral artery and the vertebral veins lie anterior to it. Inferior to the ganglion lies the cervical dome of the pleura (*Fig. 8–4*).

BRANCHES

1. **Gray rami communicantes** to the seventh and eighth cervical nerves.
2. **Arterial branches** to the subclavian and vertebral arteries.
3. **Inferior cardiac branch** that descends to join the cardiac plexus.

The portion of the sympathetic trunk that connects the middle cervical ganglion to the inferior or stellate ganglion is in the form of two or more nerve bundles. The preganglionic fibers that supply the head and the

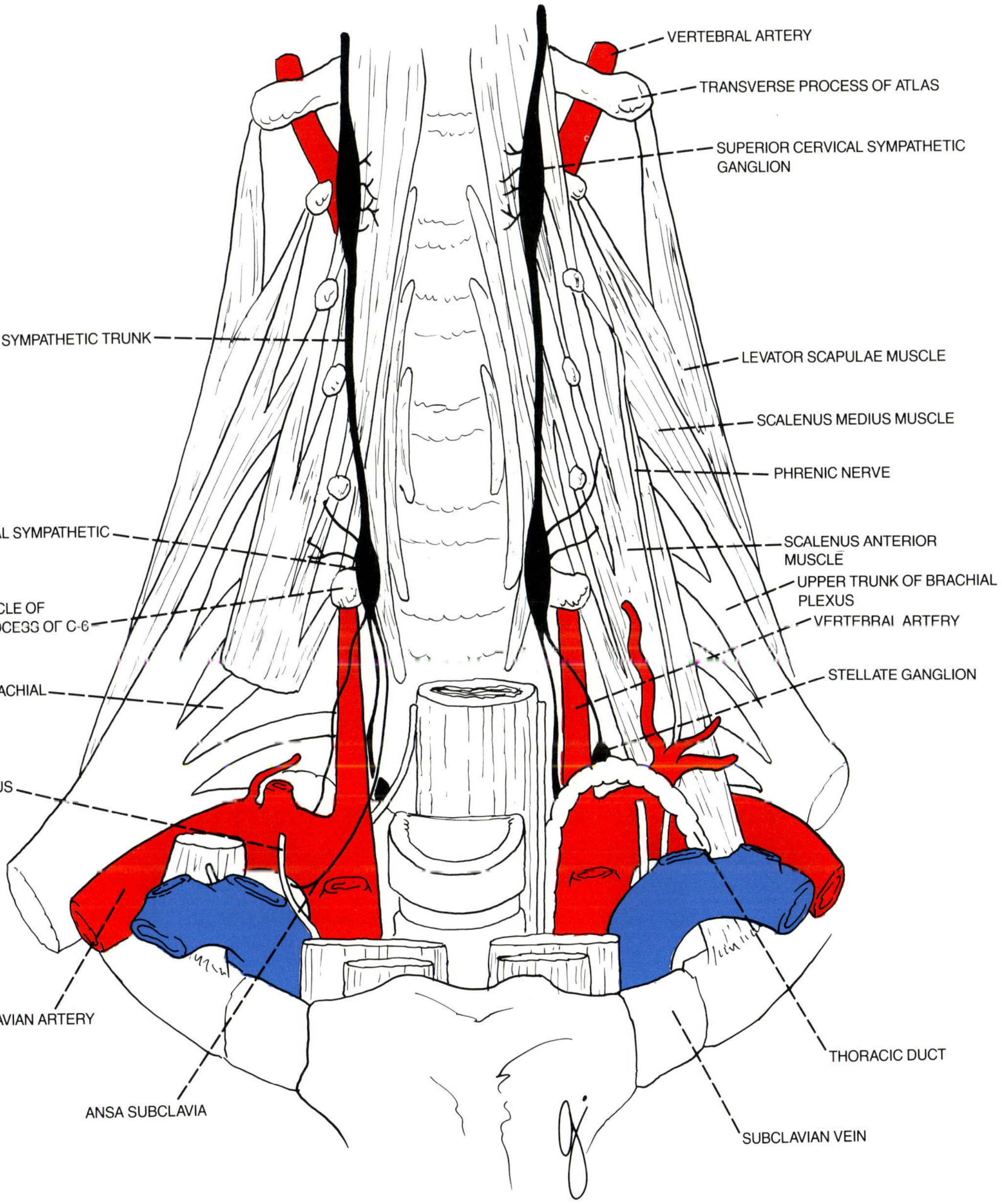

Figure 8–3. The root of the neck showing the cervical parts of the sympathetic trunks.

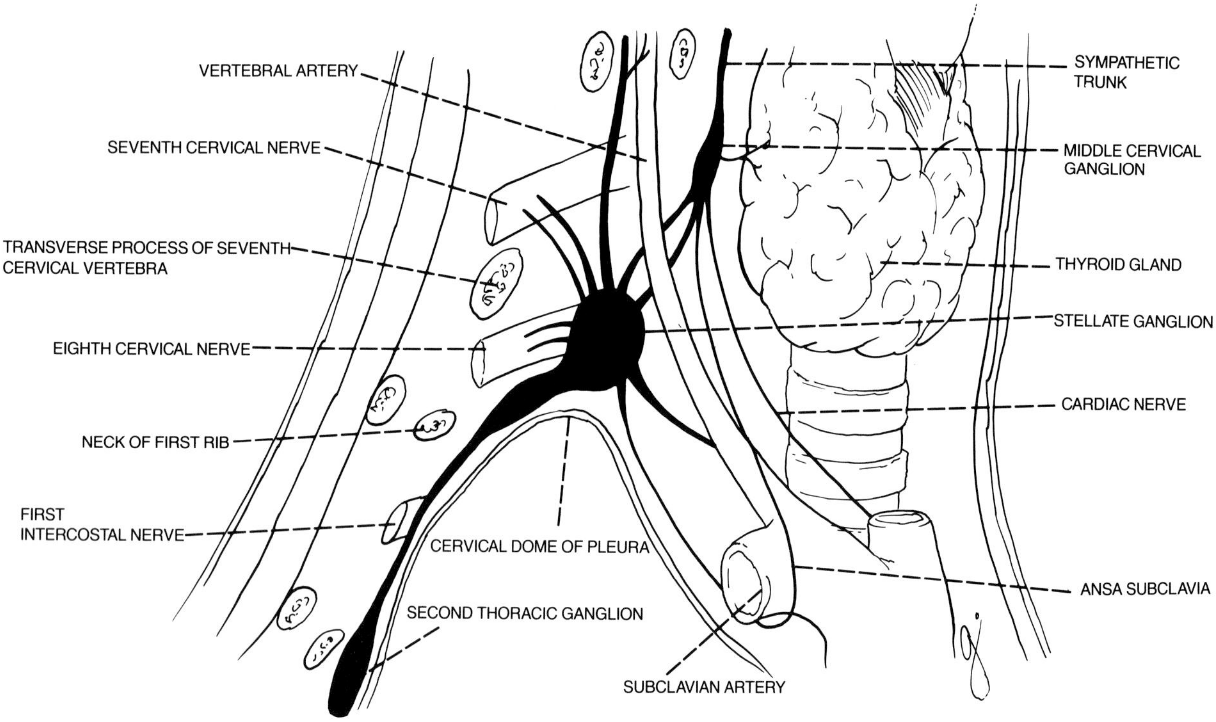

Figure 8–4. Parasagittal section of the neck as seen from the right side. Note the proximity of the stellate ganglion to the cervical dome of the pleura.

neck leave the spinal cord through the upper five thoracic nerves (mainly the upper two or three). They ascend the sympathetic trunk to synapse with cells in the cervical ganglia. The postganglionic fibers are distributed as branches of the ganglia as described previously.

Thoracic Part of the Sympathetic Trunk. The thoracic part of the sympathetic trunk first runs downward in front of the heads of the ribs and then passes medially to lie on the side of the vertebral bodies. It lies behind the costal pleura. It leaves the thorax by passing behind the medial arcuate ligament to become continuous with the lumbar part of the sympathetic trunk. The trunk has 11 or 12 segmentally arranged ganglia (*Figs. 8–2, 8–5, and 8–6*).

Branches of the Thoracic Part of the Sympathetic Trunk

1. **White rami communicantes** join each ganglion to a corresponding thoracic spinal nerve. A white ramus contains preganglionic nerve fibers and afferent sensory nerve fibers.
2. **Gray rami communicantes** join each ganglion to a corresponding thoracic spinal nerve. A gray ramus contains postganglionic nerve fibers.
3. The first five ganglia give postganglionic fibers to the heart, the aorta, the lungs, and the esophagus.
4. The lower seven ganglia give preganglionic fibers and receive afferent sensory fibers that are grouped together to form splanchnic nerves and supply the abdominal viscera. They enter the abdomen by piercing the crura of the diaphragm. The **greater splanchnic nerve** arises from ganglia 5 through 9, the **lesser splanchnic nerve** arises from ganglia 10 and 11, and the **lowest splanchnic nerve** arises from the last thoracic ganglion (*Figs. 8–5 and 8–6*).

Lumbar Part of the Sympathetic Trunk. The lumbar part of the sympathetic trunk runs downward on the bodies of the lumbar vertebrae in the retroperitoneal connective tissue along the medial border of the psoas muscle (*Fig. 8–7*). It is continuous above with the thoracic part of the trunk behind the medial arcuate ligament and below with the pelvic part of the trunk behind the common iliac vessels. The right trunk lies posterior to the right margin of the inferior vena cava and the left trunk lies close to the left margin of the aorta and the lateral aortic lymph nodes, anterior to the lumbar vessels. Each trunk has four or five segmentally arranged ganglia (*Fig. 8–2*).

Branches of the Lumbar Part of the Sympathetic Trunk

1. **White rami communicantes** join the first two ganglia to the first two lumbar spinal nerves. A white ramus contains preganglionic nerve fibers and afferent sensory nerve fibers.
2. **Gray rami communicantes** join each ganglion to a corresponding lumbar spinal nerve. A gray ramus contains postganglionic nerve fibers.
3. Fibers pass medially to the sympathetic plexuses on the abdominal aorta and its branches. (These plexuses also receive fibers from splanchnic nerves and the vagus nerve.)
4. Fibers pass downward and medially anterior to the common iliac vessels to enter the pelvis, where, together with branches from sympathetic nerves in front of the aorta, they form a large bundle of nerve fibers called the **superior hypogastric plexus.**

Pelvic Part of the Sympathetic Trunk. The pelvic part of the sympathetic trunk is continuous above, behind the common iliac vessels, with the lumbar part of the trunk (*Fig. 8–8*). Below, the two trunks come together in front of the coccyx to form the **ganglion impar.** Each trunk descends in the retroperitoneal connective tissue behind the rectum and in front of the sacrum and medial to the anterior sacral foramina. The sympathetic trunk has four or five segmentally arranged ganglia (*Fig. 8–8*).

Branches of the Pelvic Part of the Sympathetic Trunk

1. Gray rami communicantes to the sacral and coccygeal spinal nerves.
2. Fibers that join the hypogastric plexuses.

Medulla of Suprarenal Gland. Preganglionic sympathetic fibers descend to the gland in the **greater splanchnic nerve,** a branch of the thoracic part of the sympathetic trunk (*Fig. 8–2*). The nerve fibers terminate on the secretory cells of the medulla, which are comparable to postganglionic neurons. Acetylcholine is the transmitter substance between the nerve endings and the secretory cells, as at any other preganglionic endings. The sympathetic nerves stimulate the secretory cells of the medulla to increase the output of epinephrine and norepinephrine. There is no parasympathetic innervation.

A summary diagram showing the spinal segmental levels of the sympathetic connector neurons associated with different organs in the body is shown in Fig. 8–9.

PARASYMPATHETIC PART OF THE AUTONOMIC NERVOUS SYSTEM

The activities of the parasympathetic part of the autonomic system are directed toward conserving and restor-

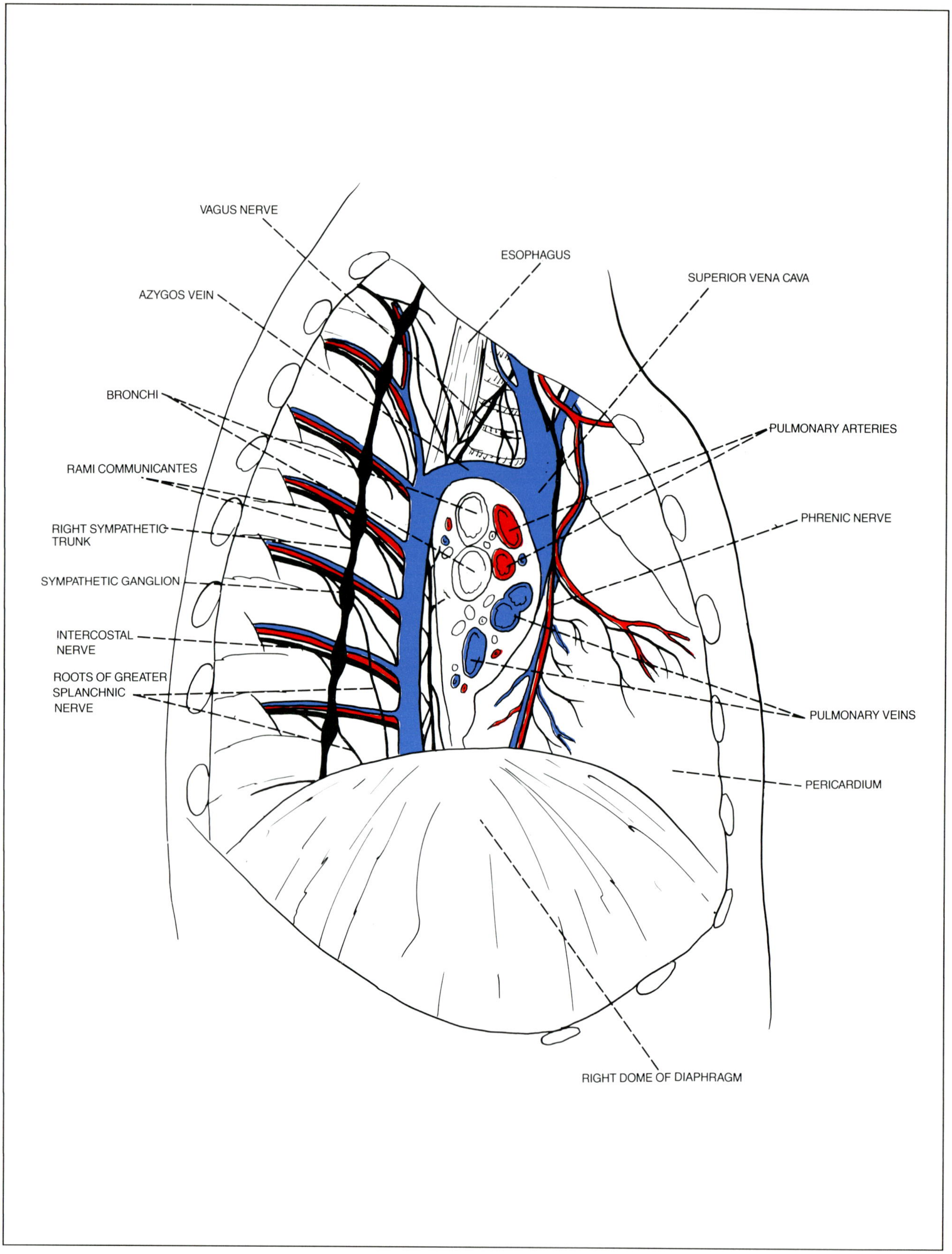

Figure 8–5. Right side of mediastinum showing the thoracic part of the sympathetic trunk and its main branches.

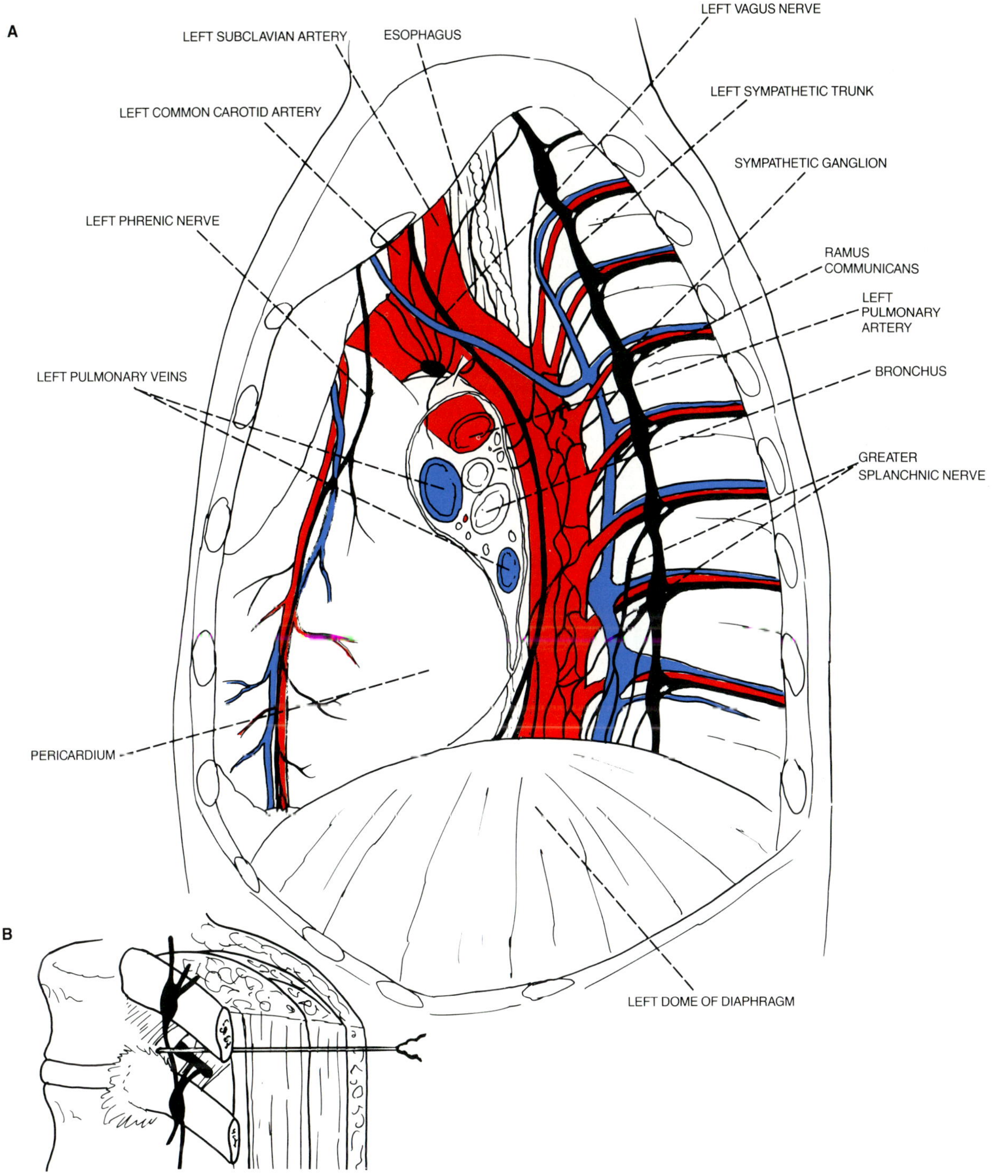

Figure 8–6. A. Left side of mediastinum showing the thoracic part of the sympathetic trunk and its main branches. **B.** Relationship of the sympathetic trunk to the vertebral bodies, the necks of the ribs, and the intercostal spaces.

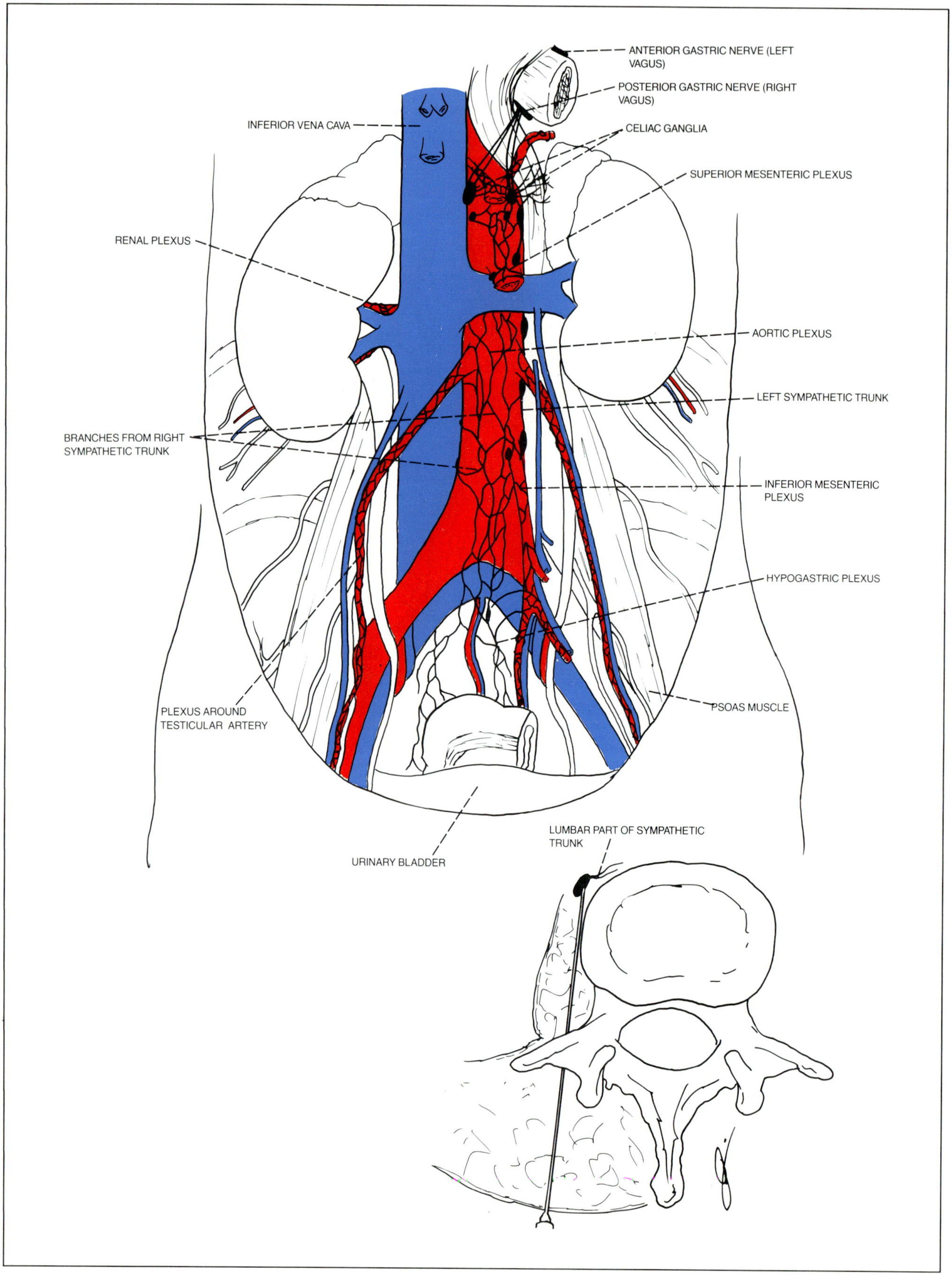

Figure 8–7. Posterior abdominal wall showing the aorta and related sympathetic plexuses.

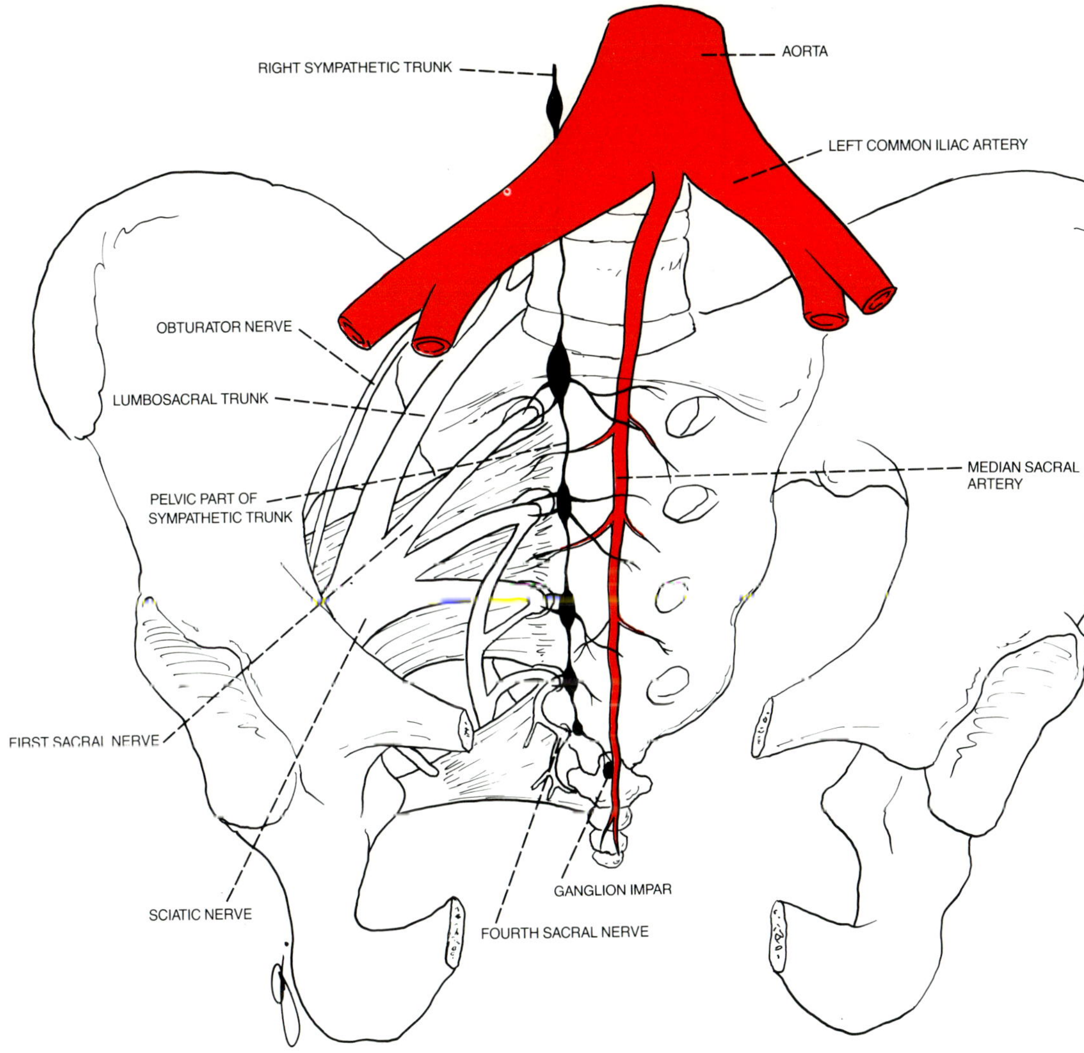

Figure 8–8. Posterior pelvic wall showing the right sacral plexus and the pelvic part of the right sympathetic trunk.

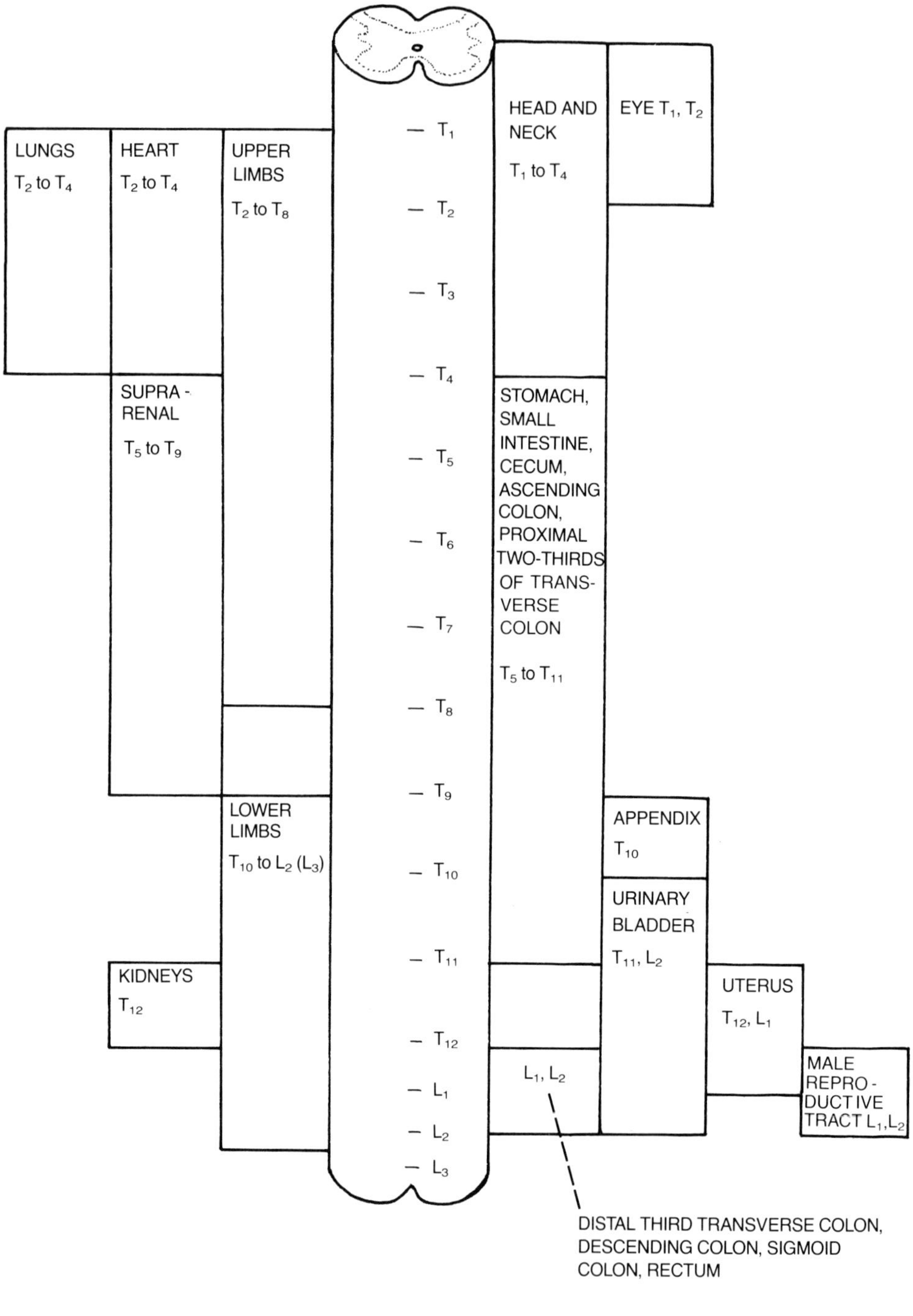

Figure 8–9. Diagram showing spinal segmental levels of sympathetic connector neurons associated with different organs in the body.

ing energy. The heart rate is slowed, pupils are constricted, peristalsis is increased, glandular activity is augmented, sphincters are opened, and the bladder wall is contracted.

Efferent Nerve Fibers (Craniosacral Outflow)

The parasympathetic part of the autonomic nervous system is located in the nuclei of the oculomotor (Edinger-Westphal nucleus), the facial (superior salivary and lacrimatory nuclei), the glossopharyngeal (inferior salivatory nucleus), and the vagus (dorsal nucleus) cranial nerves, and the second, third, and fourth sacral segments of the spinal cord (*Fig. 8–2*). The axons of the central connector nerve cells are myelinated and emerge from the brain within the cranial nerves. The sacral connector nerve cells give rise to myelinated axons that leave the spinal cord in the anterior nerve roots of the corresponding spinal nerves. They then leave the sacral nerves and form the **pelvic splanchnic nerves** (*Fig. 8–2*).

The efferent fibers of the craniosacral outflow are preganglionic and synapse in peripheral ganglia located close to the viscera they innervate. Here again, acetylcholine is the neurotransmitter. The cranial parasympathetic ganglia are the **ciliary,** the **pterygopalatine,** the **submandibular,** and the **otic.** In certain locations, the ganglion cells are placed in nerve plexuses, such as the **cardiac plexus, pulmonary plexus, myenteric plexus (Auerbach's plexus),** and **mucosal plexus (Meissner's plexus);** the last two plexuses are associated with the gastrointestinal tract. The pelvic splanchnic nerves synapse in ganglia in the hypogastric plexuses. The postganglionic parasympathetic fibers are nonmyelinated and short in length.

Parasympathetic Ganglia

Ciliary Ganglion. The ciliary ganglion is about the size of a pin's head and situated at the back of the orbital cavity between the optic nerve and the lateral rectus muscle.

BRANCHES

1. **Parasympathetic preganglionic fibers** that arise from the Edinger-Westphal nucleus of the oculomotor nerve. The fibers reach the ganglion via the oculomotor nerve and its branch to the inferior oblique muscle.
2. **Parasympathetic postganglionic fibers** that arise from the ganglion nerve cells and pass to the eyeball via the short ciliary nerves. The fibers innervate the ciliary muscle and the sphincter pupillae muscle.
3. **Sensory fibers from the eyeball** that reach the ganglion via the short ciliary nerves and pass through it without interruption to join the nasociliary branch of the ophthalmic division of the trigeminal nerve.
4. **Sympathetic postganglionic fibers** that arise from the superior cervical sympathetic ganglion and reach the ciliary ganglion via the internal carotid plexus. The fibers pass through the ganglion without interruption to the eyeball via the short ciliary nerves.

Pterygopalatine (Sphenopalatine) Ganglion. The pterygopalatine ganglion is located in the pterygopalatine fossa and is suspended from the lower border of the maxillary nerve.

BRANCHES

1. **Parasympathetic preganglionic fibers** that arise from the lacrimatory nucleus of the facial nerve. The fibers run in the nervous intermedius of the facial nerve and then travel in the **greater petrosal branch** of the nerve. The greater petrosal nerve joins with the **deep petrosal nerve** (sympathetic nerve from internal carotid plexus) to form the **nerve of the pterygoid canal.** The latter nerve joins the ganglion.
2. **Parasympathetic postganglionic fibers,** which arise from the ganglion nerve cells, pass to the maxillary nerve and its zygomatic branch. The zygomaticotemporal branch then gives off a communicating branch to the lacrimal nerve. The lacrimal nerve supplies the parasympathetic fibers to the lacrimal gland. Other postganglionic fibers travel with the **palatine** and **nasal branches** of the ganglion to the glands of the pharynx, the palate, and the nose.
3. **Sympathetic postganglionic orbital branches** reach the ganglion via the nerve of the pterygoid canal. They arise from the superior cervical sympathetic ganglion and pass without interruption through the pterygopalatine ganglion entering the orbit through the inferior orbital fissure innervating the **orbitalis muscle** (smooth muscle in back of orbital cavity).
4. **Palatine, nasal, and pharyngeal branches** are in reality sensory nerves of the maxillary division of the trigeminal nerve. They leave the maxillary nerve as two bundles that pass uninterrupted through the ganglion and exit as the **greater** and **lesser palatine nerves,** the **lateral posterior superior nasal nerves,** the **medial posterior superior nasal nerves,** the **nasopalatine nerves,** and the **pharyngeal nerve.**

Submandibular Ganglion. The submandibular ganglion lies on the side of the tongue suspended from the lingual nerve. Additional ganglion cells lie in the hilum of the submandibular salivary gland.

BRANCHES

1. **Parasympathetic preganglionic fibers,** which arise from the superior salivatory nucleus of the facial nerve, travel with the nerve, then its chorda tympani branch, which joins the lingual nerve. The fibers leave the lingual nerve to enter the ganglion.
2. **Parasympathetic postganglionic fibers,** from the ganglion nerve cells, pass to the submandibular and sublingual salivary glands.
3. **Postganglionic sympathetic fibers** pass without interruption through the ganglion to supply the blood vessels of the submandibular and sublingual salivary glands. These fibers originated in the superior cervical sympathetic ganglion.

Otic Ganglion. The otic ganglion lies just below the foramen ovale on the medial side of the mandibular division of the trigeminal nerve. It is attached to the nerve to the medial pterygoid muscle, a branch of the mandibular nerve.

BRANCHES

1. **Parasympathetic preganglionic nerve fibers,** which arise from the inferior salivatory nucleus of the glossopharyngeal nerve, pass via the tympanic branch of the glossopharyngeal nerve to the tympanic plexus in the middle ear. They leave the plexus via the **lesser petrosal nerve** to reach the ganglion.
2. **Parasympathetic postganglionic nerve fibers,** which arise from the ganglion nerve cells, travel via the auriculotemporal nerve to the parotid salivary gland.
3. **Postganglionic sympathetic fibers** pass without interruption through the ganglion to supply the blood vessels of the parotid salivary gland. These fibers originated in the superior cervical sympathetic ganglion.

Afferent Nerve Fibers

The afferent myelinated fibers leave the viscera and reach their cell bodies in the sensory ganglia of the cranial nerves or in posterior root ganglia of the sacral spinal nerves. The central axons then enter the central nervous system and form regional reflex arcs or ascend to higher centers such as the hypothalamus.

LARGE AUTONOMIC PLEXUSES

Large collections of sympathetic and parasympathetic efferent nerve fibers and their associated ganglia, together with visceral afferent fibers, form autonomic nerve plexuses in the thorax, the abdomen, and the pelvis. Branches from these plexuses innervate the viscera. In the abdomen, the plexuses are associated with the aorta and its branches. The subdivisions of these autonomic plexuses are named according to the branch of the aorta along which they are lying. The following plexuses are summarized: cardiac, pulmonary, celiac, superior mesenteric, inferior mesenteric, aortic, superior hypogastric, and inferior hypogastric plexus.

Cardiac Plexuses. Superficial and deep cardiac plexuses are found at the base of the heart. The superficial plexus lies below the arch of the aorta and the deep plexus lies in front of the bifurcation of the trachea. Sympathetic fibers cause cardiac acceleration, increased force of contraction of the cardiac muscle, and dilatation of the coronary arteries; parasympathetic fibers from the vagus cause slowing of the heart and constriction of the coronary arteries.

Pulmonary Plexuses. Anterior and posterior pulmonary plexuses lie in front and behind the hilus of each lung. Sympathetic fibers cause bronchodilatation and slight vasoconstriction; parasympathetic (vagal) fibers cause bronchoconstriction, slight vasodilatation, and bronchial secretion.

Celiac Plexus. A network of fibers and ganglion cells form two celiac ganglia around the celiac artery (*Fig. 8–7*). The network receives sympathetic fibers from the greater and lesser splanchnic nerves and parasympathetic fibers from the vagus. Branches are distributed along the branches of the celiac artery.

Superior Mesenteric Plexus. The superior mesenteric plexus is continuous above with the celiac plexus and consists of a network of fibers and ganglion cells (*Fig. 8–7*). It also receives a distribution from the right vagus. Branches are distributed along the branches of superior mesenteric artery.

Inferior Mesenteric Plexus. The inferior mesenteric plexus is connected with the aortic plexus and receives branches from the lumbar part of the sympathetic trunk (*Fig. 8–7*). It also receives parasympathetic branches from the pelvic splanchnic nerve.

Aortic Plexus. The aortic plexus is a diffuse plexus around the abdominal part of the aorta (*Fig. 8–7*). Neural concentrations around the origins of the celiac, renal, superior mesenteric, and inferior mesenteric arteries form the celiac plexus, renal plexus, superior mesenteric plexus, and inferior mesenteric plexus, respectively.

Superior Hypogastric Plexus. The superior hypogastric plexus is located at the bifurcation of the abdominal part of the aorta and extends down in front of the promontory of the sacrum (*Fig. 8–7*). It is formed as a continuation of the aortic plexus and from branches of the third and fourth lumbar ganglia. It divides inferiorly into right and left hypogastric nerves and contains sympathetic and sacral parasympathetic nerve fibers.

Inferior Hypogastric Plexus. The inferior hypogastric plexus lies in the extraperitoneal connective tissue on each side of the rectum. Each plexus has contributions from a hypogastric and a pelvic splanchnic nerve. Numerous branches pass to the pelvic viscera. Parasympathetic fibers from the pelvic splanchnic nerve ascend through the hypogastric nerve to the superior hypogastric plexus and the inferior mesenteric plexus to be distributed to the splenic flexure, and descending and sigmoid parts of the colon.

Structure of Autonomic Ganglia

Autonomic ganglia are situated along the course of efferent nerve fibers of the autonomic nervous system. Sympathetic ganglia form part of the sympathetic trunk or are prevertebral in position. Parasympathetic ganglia, on the other hand, are situated close to or within the walls of the viscera as parasympathetic plexuses.

An autonomic ganglion consists of a collection of multipolar neurons together with capsular or satellite cells and a connective tissue capsule. Nerve bundles are attached to each ganglion and consist of preganglionic nerve fibers that enter the ganglion, postganglionic nerve fibers that have arisen from neurons within the ganglion, and afferent and efferent nerve fibers that pass through the ganglion without synapsing. Preganglionic fibers are small and myelinated; postganglionic fibers are smaller and nonmyelinated.

An autonomic ganglion is the site where preganglionic fibers synapse on postganglionic neurons. However, the presence of small interneurons and collateral branches suggests that a ganglion may play a greater role than just relaying information and that it probably serves some integrative function.

The synaptic transmitter that excites the postganglionic neurons in both sympathetic and parasympathetic ganglia is acetylcholine. The action of acetylcholine in autonomic ganglia is terminated by hydrolysis by acetylcholinesterase. The small interneurons of ganglia contain dopamine, which is thought to be used as a transmitter. The electron microscopic structure of synapses in autonomic ganglia shows the characteristic membrane thickening and small clear vesicles. In addition, there are some larger granular vesicles. The smaller vesicles contain acetylcholine; the content of the granular vesicles is not known.

Structure of Postganglionic Nerve Endings

Postganglionic fibers terminate on the effector cells without special discrete endings. The axons run between the gland cells and the smooth and cardiac muscle cells and lose their covering of Schwann cells. At sites where transmission occurs, clusters of vesicles are present within the axoplasm. If the site on the axon is at some distance from the effector cell, transmission time may be slow. The diffusion of transmitter through large extracellular distances causes a given nerve to have an action on a large number of effector cells.

Higher Control of the Autonomic Nervous System

The sympathetic outflow in the spinal cord (T_1 through L_2 or L_3) and the parasympathetic craniosacral outflow (cranial nuclei III, VII, IX, and X, and spinal cord levels S_2, S_3, and S_4) are controlled by the hypothalamus. The hypothalamus appears to integrate the autonomic and neuroendocrine systems, thus preserving body homeostasis. It receives signals from all parts of the nervous system, afferent input from the viscera, and information concerning the hormone levels in the blood. This input is integrated within the hypothalamus and transmitted to the lower autonomic centers in the brainstem and spinal cord by descending tracts of the reticular formation. In a similar manner, **releasing factors** or **release-inhibiting factors** are liberated into the circulation effecting hormone levels and endocrine secretions, thus influencing organ activity.

Functions of the Autonomic Nervous System

The autonomic nervous system, along with the endocrine system, maintains body homeostasis. The endocrine control is slower and exerts its influence by means of bloodborne hormones.

The system should not be regarded as an isolated portion of the nervous system, for we know that it can play a role with somatic activity in expressing emotion, and that certain autonomic activities, such as micturition, can be brought under voluntary control. The various activities of the autonomic and endocrine systems are integrated within the hypothalamus.

For important anatomical, physiological, and pharmacologic differences between the sympathetic and the parasympathetic parts of the autonomic system see Table 8–1.

TABLE 8–1. ANATOMIC PHYSIOLOGIC AND PHARMACOLOGIC DIFFERENCES AND SIMILARITIES BETWEEN SYMPATHETIC AND PARASYMPATHETIC PARTS OF THE AUTONOMIC NERVOUS SYSTEM

	Sympathetic	Parasympathetic
Outflow	T_1 to $L_2(L_3)$	Cranial nerves III, VII, IX, and X; S_2, S_3, and S_4
Preganglionic fibers	Myelinated	Myelinated
Ganglia	Paravertebral (sympathetic trunks); prevertebral (e.g., celiac, superior mesenteric)	Small ganglia close to viscera (e.g., otic, ciliary) or ganglion cell in plexuses (e.g., cardiac, pulmonary)
Neurotransmitter within ganglia	Acetylcholine	Acetylcholine
Postganglionic fibers	Long nonmyelinated	Short nonmyelinated
Characteristic activity	Widespread due to many postganglionic fibers and liberation of epinephrine and norepinephrine from suprarenal medulla	Discrete action with few postganglionic fibers
Neurotransmitter at postganglionic endings	Norepinephrine at most endings and acetylcholine at few endings (sweat glands)	Acetylcholine at all endings
Higher control	Hypothalamus	Hypothalamus

ENTERIC NERVOUS SYSTEM

The so called enteric nervous system, consisting of the myenteric and mucosal plexuses of the gastrointestinal tract, has recently received considerable attention by researchers. No longer can these plexuses be regarded as simply parasympathetic plexuses containing preganglionic and postganglionic nerve fibers and nerve cells. It has been shown that contractions of the smooth muscle in the gut wall can occur in the absence of the myenteric plexus, but the coordinated purposeful contractions, as seen with peristalsis and segmental movements, require the presence of a nerve plexus, even though this may be isolated from the central nervous system.

Different types of neurons have been found in the plexuses. Some neurons are bipolar or unipolar and are thought to be sensory and involved in local reflex activity; other neurons send axons to the celiac and mesenteric plexuses. Preganglionic parasympathetic fibers synapse on nerve cells that give rise to postganglionic fibers that innervate the smooth muscle and the glands. Postganglionic sympathetic fibers have been seen to terminate on parasympathetic nerve cells and probably play an inhibitory role on parasympathetic activity. Internuncial neurons are also present. It has been suggested, that although the enteric plexuses can coordinate the activities of the gut wall, the parasympathetic and the sympathetic inputs modulate these activities.

SOME IMPORTANT AUTONOMIC INNERVATIONS

Eye

Upper Lid. The upper lid is raised by the levator palpebrae superioris muscle. The major part of this muscle is formed by skeletal fibers innervated by the oculomotor nerve. A small part is composed of smooth muscle fibers innervated by sympathetic postganglionic fibers from the superior cervical sympathetic ganglion (*Fig. 8–10*).

Iris. The smooth muscle fibers of the iris consist of circular fibers that form the sphincter pupillae muscle and radial fibers that form the dilator pupillae muscle. The sphincter pupillae muscle is supplied by parasympathetic fibers from the parasympathetic nucleus (Edinger-Westphal nucleus) of the oculomotor nerve (*Fig. 8–10*). After synapsing in the **ciliary ganglion,** the postganglionic fibers pass forward to the eyeball in the **short ciliary nerves.** (The ciliary muscle of the eye is also supplied by the short ciliary nerves; see p. 269).

The dilator pupillae muscle is supplied by postganglionic fibers from the superior cervical sympathetic ganglion (*Fig. 8–10*), which reach the orbit along the internal carotid and ophthalmic arteries. They pass uninterrupted through the ciliary ganglion and reach the eyeball in the **short and long ciliary nerves.**

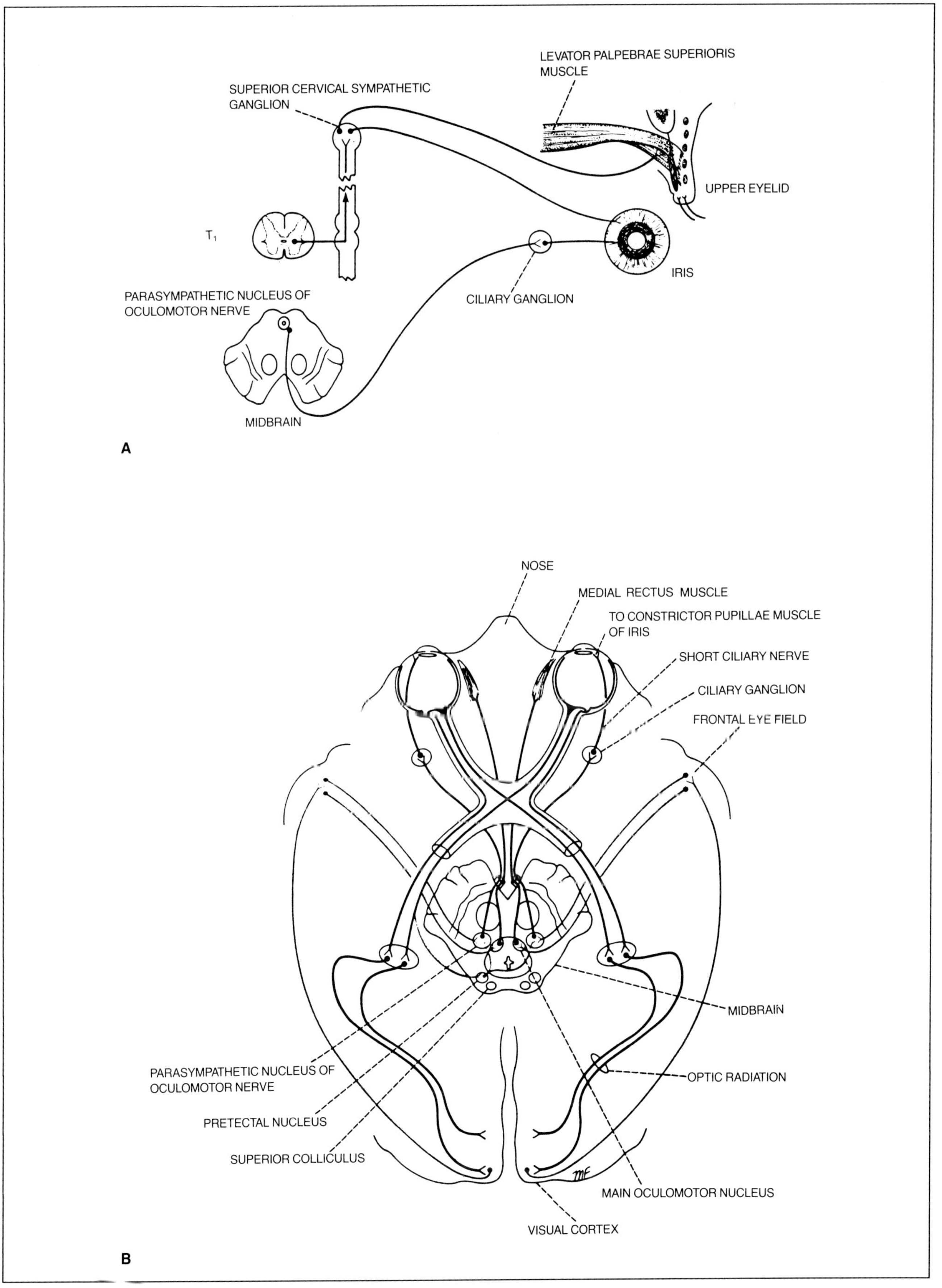

Figure 8–10. A. The autonomic innervation of the upper eyelid and the iris. **B.** The optic pathway and the visual reflexes.

Lacrimal Gland. The parasympathetic secretomotor nerve supply to the lacrimal gland originates in the lacrimatory nucleus of the facial nerve (*Fig. 8–2*). Preganglionic fibers reach the **pterygopalatine ganglion** through the **nervus intermedius** and its **great petrosal branch** and through the **nerve of the pterygoid canal.** The postganglionic fibers join the maxillary nerve and pass into its **zygomatic branch** and the **zygomaticotemporal nerve.** They reach the lacrimal gland via the **lacrimal nerve.**

The sympathetic postganglionic fibers arise from the superior cervical sympathetic ganglion and travel in the plexus of nerves around the internal carotid artery. They join the **deep petrosal nerve,** the **nerve of the pterygoid canal,** the **maxillary nerve,** the **zygomatic nerve,** the **zygomaticotemporal nerve,** and, finally, the lacrimal nerve.

Heart

Sympathetic postganglionic fibers arise from cervical and upper thoracic portions of the sympathetic trunks (*Fig. 8–11*). Postganglionic fibers reach the heart by way of the **superior, middle, and inferior cardiac branches** of the cervical portion of the sympathetic trunk and a number of **cardiac branches** from the thoracic portion of the sympathetic trunk. Fibers pass via the **cardiac plexuses** to terminate on the **sinoatrial** and **atrioventricular nodes,** the cardiac muscle fibers, and on the coronary arteries. Activation of these nerves results in cardiac acceleration, increased force of contraction of the cardiac muscle, and dilatation of the coronary arteries. Coronary dilatation, however, is mainly produced in response to local metabolic needs rather than by direct nerve stimulation.

The parasympathetic preganglionic fibers originate in the **dorsal nucleus of the vagus nerve** and descend into the thorax in the vagus nerves. The fibers terminate by synapsing with postganglionic neurons in the **cardiac plexuses.** Postganglionic fibers terminate on the **sinoatrial** and **atrioventricular nodes** and on the coronary arteries. Activation of these nerves results in a reduction in the rate and force of contraction of the myocardium and constriction of the coronary arteries. Here again, the coronary constriction is mainly produced by the reduction in local metabolic needs rather than by neural effects.

Lungs

Sympathetic postganglionic fibers arise from the second to fifth thoracic ganglia of the sympathetic trunk (*Fig. 8–11*). They pass through the pulmonary plexuses and enter the lung to form networks around the bronchi and the blood vessels. The sympathetic fibers produce bronchodilatation and are slightly vasoconstrictor.

Parasympathetic preganglionic fibers arise from the dorsal nucleus of the vagus and descend to the thorax within the vagus nerves. The fibers terminate by synapsing with postganglionic neurons in the pulmonary plexuses that enter the lung, where they form networks around the bronchi and the blood vessels. The parasympathetic fibers produce bronchoconstriction, slight vasodilatation, and increase glandular secretion.

Gastrointestinal Tract

Stomach and Intestine as Far as the Splenic Flexure. Preganglionic parasympathetic fibers enter the abdomen in the **anterior (left)** and **posterior (right)** vagal trunks (*Fig. 8–12*). The fibers are distributed to many abdominal viscera and the gastrointestinal tract from the stomach to the splenic flexure. Fibers that pass to the gastrointestinal tract terminate on postganglionic neurons in the **myenteric (Auerbach's)** and **submucous (Meissner's) plexuses.** The postganglionic fibers supply the smooth muscle and the glands. The parasympathetic nerves promote peristalsis and relax the sphincters; they also stimulate secretion.

Sympathetic preganglionic nerve fibers pass through the thoracic part of the sympathetic trunk and enter the **greater** and **lesser splanchnic nerves.** These descend into the abdomen and synapse with postganglionic neurons in the **celiac** and **superior mesenteric ganglia.** Postganglionic nerve fibers are distributed to the stomach and the intestine as nerve plexuses around the branches of the celiac and superior mesenteric arteries. The sympathetic nerves inhibit peristalsis and cause contraction of the sphincters; they also diminish secretion (see the enteric nervous system, page 272).

Descending Colon, Pelvic Colon, and Rectum. Preganglionic parasympathetic fibers originate in the gray matter of the spinal cord from the second to the fourth sacral segments (*Fig. 8–12*). Fibers pass through the **pelvic splanchnic nerves** and nerve plexuses around branches of the inferior mesenteric artery and terminate on postganglionic neurons in the myenteric (Auerbach's) and submucous (Meissner's) plexuses. Postganglionic fibers supply smooth muscle and glands. Parasympathetic nerves stimulate peristalsis and secretion.

Sympathetic preganglionic nerve fibers pass through the lumbar part of the sympathetic trunk and synapse with postganglionic neurons in the **inferior mesenteric plexus.** Postganglionic fibers are distributed to the bowel as nerve plexuses around the branches of the inferior mesenteric arteries. The sympathetic nerves inhibit peristalsis and secretion.

Involuntary Internal Sphincter of the Anal Canal. The circular smooth muscle coat is thickened at the upper end of the anal canal to form the involuntary internal sphincter. The sphincter is innervated by postganglionic sympathetic fibers from the **hypogastric plexuses**

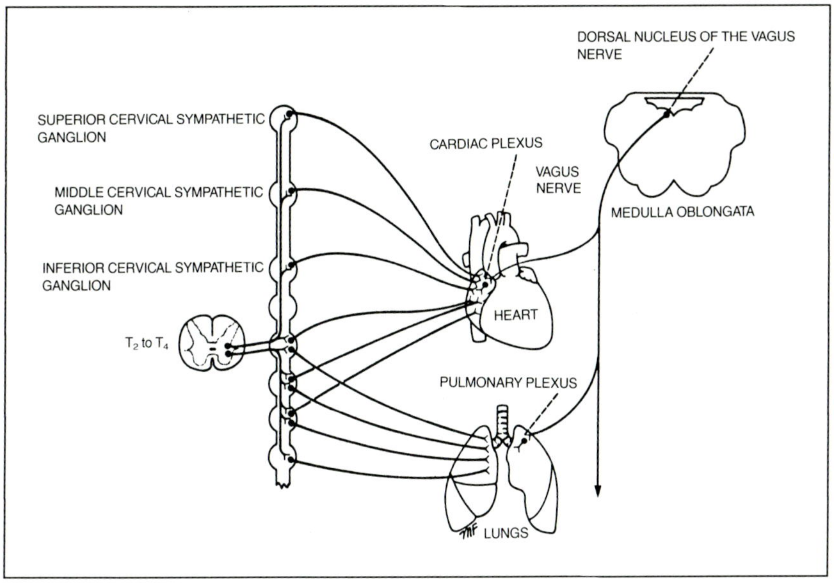

Figure 8–11. The autonomic innervation of the heart and the lungs.

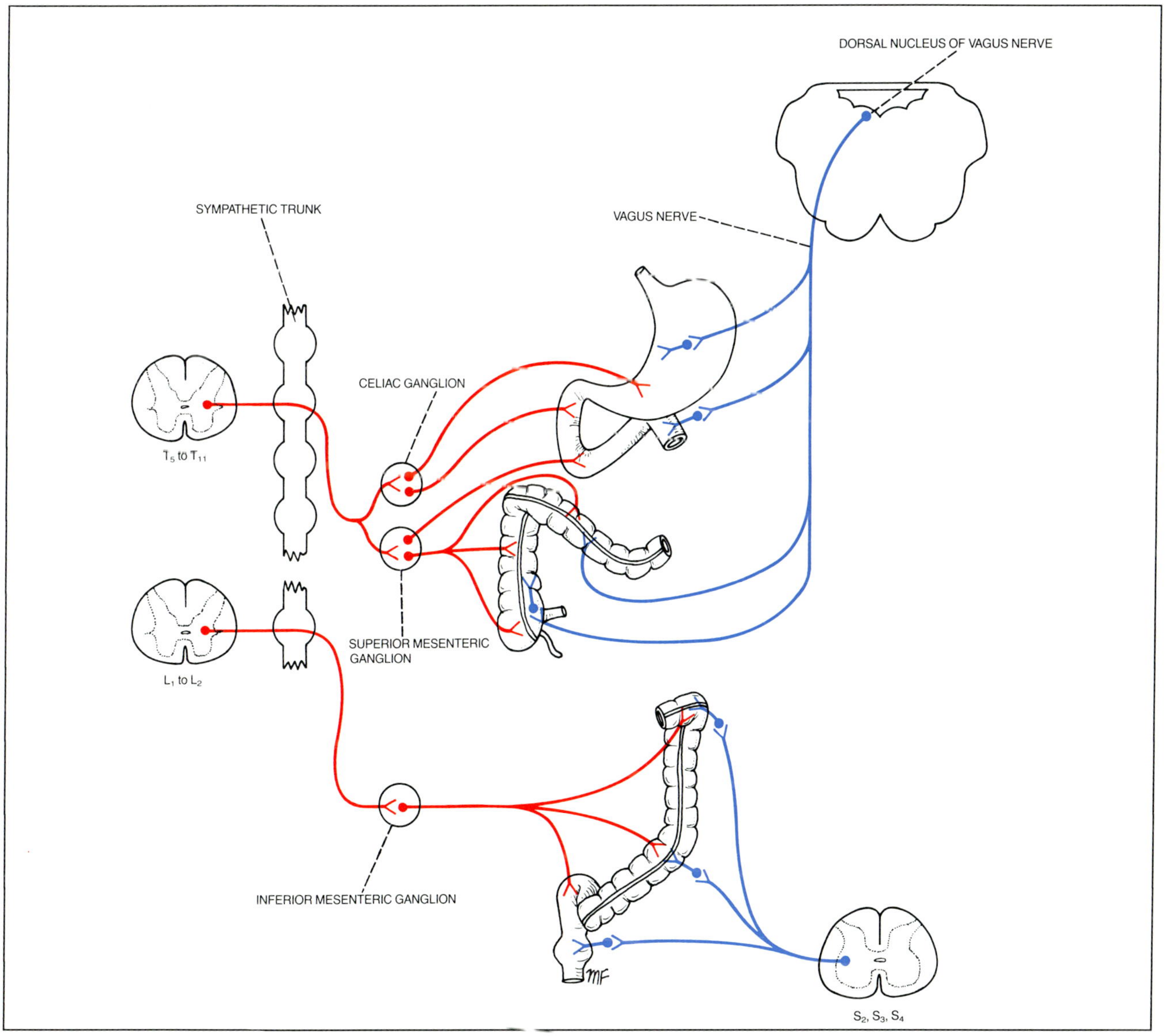

Figure 8–12. The autonomic innervation of the gastrointestinal tract.

(*Fig. 8–14*). Each hypogastric plexus receives sympathetic fibers from the **aortic plexus** and from the lumbar and pelvic parts of the sympathetic trunks. Sympathetic nerves cause the internal anal sphincter to contract.

Liver
The liver receives postganglionic parasympathetic fibers and sympathetic fibers derived from the celiac plexus. They accompany the hepatic artery and the portal vein as a network of nerves called the **hepatic plexus.** Within the liver, branches from the plexus are distributed with the blood vessels. Autonomic afferent fibers accompany the nerves.

Gallbladder and Biliary Ducts
The gallbladder and the biliary ducts receive postganglionic parasympathetic and sympathetic fibers from the hepatic plexus. Parasympathetic fibers derived from the vagus are thought to be motor to the smooth muscle of the gallbladder and the bile ducts and inhibitory to the sphincter of Oddi. Autonomic afferent fibers are also present. Some of these fibers are believed to leave the hepatic plexus and join the right phrenic nerve, thus explaining referred shoulder pain in gallbladder diseases.

Pancreas
The pancreas receives preganglionic parasympathetic fibers from the right vagus nerve and postganglionic sympathetic fibers from the ganglion cells of the celiac plexus. These efferent fibers, together with some afferent fibers, reach the organ along with its blood supply. The parasympathetic nerves synapse on ganglion cells distributed in the connective tissue within the gland. The blood vessels of the pancreas receive a sympathetic innervation, whereas the glands, including the islets of Langerhans, receive both parasympathetic and sympathetic fibers.

Kidney
Preganglionic sympathetic fibers pass through the lower thoracic part of the sympathetic trunk and the lowest thoracic splanchnic nerve to join the **renal plexus** around the renal artery (*Fig. 8–13*). Preganglionic fibers synapse with postganglionic neurons in the renal plexus and are distributed to the branches of the renal artery. Sympathetic nerves produce vasoconstriction within the kidney.

Preganglionic parasympathetic fibers from the vagus enter the renal plexus where they synapse with postganglionic neurons whose fibers are distributed to the kidney along the branches of the renal artery. The parasympathetic nerves are thought to cause vasodilation.

Ureter
The ureter receives parasympathetic and sympathetic fibers, which form a **ureteric plexus.** The upper part of the plexus receives nerves from the renal plexus, the middle part from the superior hypogastric plexus, and the terminal part from the inferior hypogastic plexus. Both efferent and afferent nerve fibers are present. T_{10} through L_1 and S_2, S_3, and S_4 segments of the spinal cord are involved in the innervation.

Urinary Bladder
The smooth muscle coat of the bladder wall is called the **detrusor muscle.** At the bladder neck the muscle is thickened to form the **sphincter vesicae.** The nerve supply of the smooth muscle is from the hypogastric plexuses (*Fig. 8–14*). Sympathetic postganglionic fibers originate in the first and second lumbar ganglia of the sympathetic trunk and travel to the hypogastric plexuses. The parasympathetic preganglionic fibers arise as the pelvic splanchnic nerves from the second, third, and fourth sacral nerves; they pass through the hypogastric plexuses to reach the bladder wall, where they synapse with postganglionic neurons.

The sympathetic nerves inhibit contraction of the detrusor muscle of the bladder wall and stimulate closure of the sphincter vesicae. The parasympathetic nerves stimulate the contraction of the detrusor muscle of the bladder wall and inhibit the action of the sphincter vesicae. Afferent nerve fibers from the bladder run with both the parasympathetic and sympathetic nerves.

Urethra
The mucous membrane of the urethra receives its sensory innervation from branches of the pudendal nerve (S_2, S_3, and S_4). The sphincter urethrae (striated muscle) is also supplied by the perineal branch of the pudendal nerve.

Testis
The testis is innervated by the **testicular plexus** that accompanies the testicular artery. The plexus is formed above by branches of the aortic and the renal plexuses, and receives additional nerves in the pelvis from the superior and inferior hypogastric plexuses. Nerve fibers derived from the aortic and renal plexuses are sympathetic postganglionic and sensory fibers. Nerve fibers from the hypogastric plexuses probably innervate the epididymis and vas deferens.

Prostate
The prostate receives its innervation from the lower part of the inferior hypogastric plexus. The nerves form a plexus, the **prostatic plexus,** consisting of postganglionic sympathetic fibers, preganglionic and postganglionic parasympathetic fibers, and nerve cells.

Uterus
The uterus is innervated by branches of the inferior hypogastric plexus situated in the base of the broad ligament (*Fig. 8–15*). This part of the plexus is often referred to as the **uterovaginal plexus.** The nerves pass medially

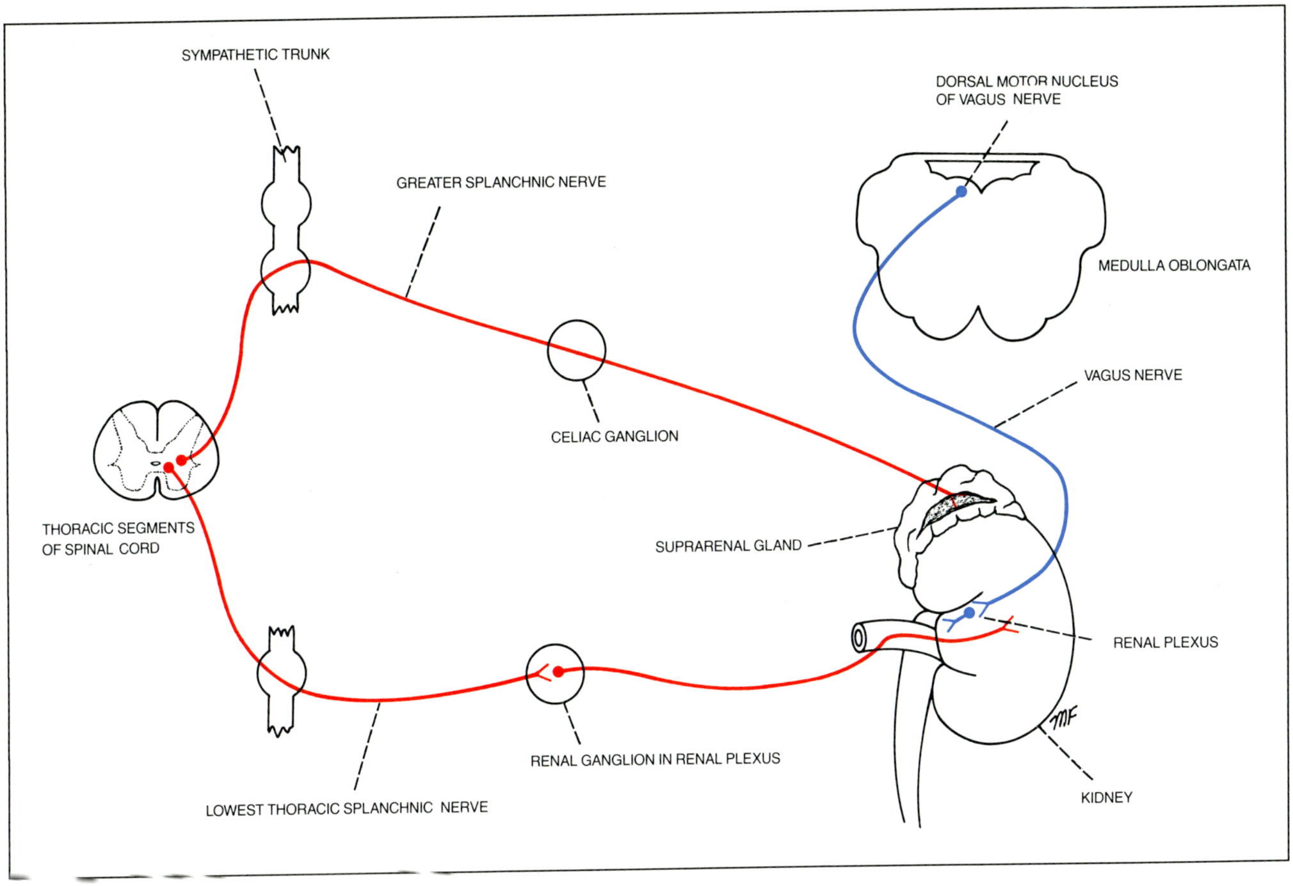

Figure 8–13. The autonomic innervation of the kidney and the suprarenal gland.

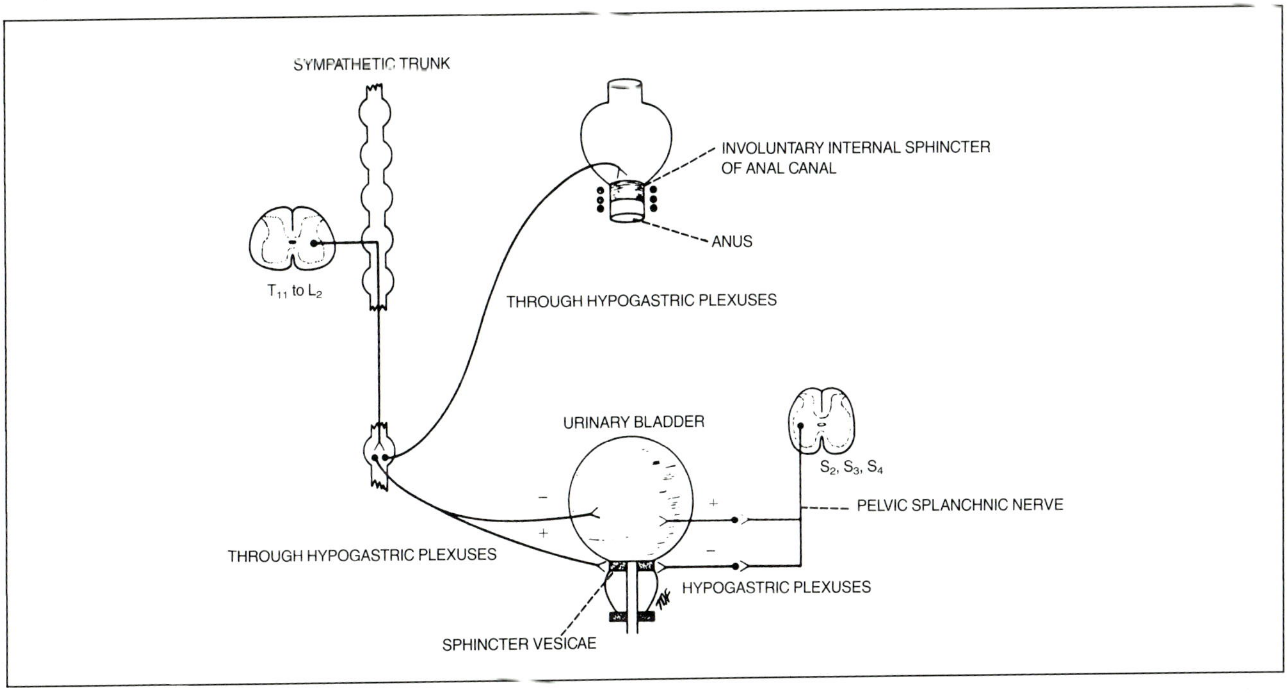

Figure 8–14. The autonomic innervation of the sphincters of the anal canal and the urinary bladder.

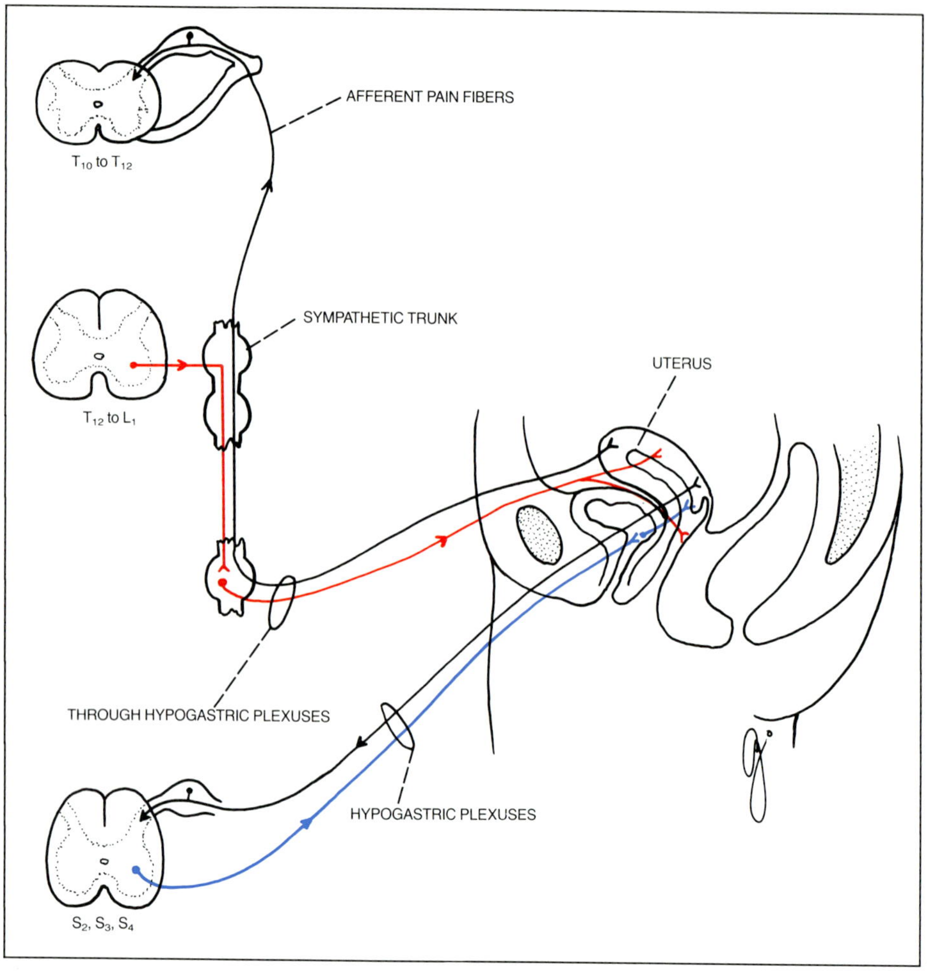

Figure 8–15. The autonomic innervation of the uterus; the pathway taken by the afferent sensory fibers is also shown.

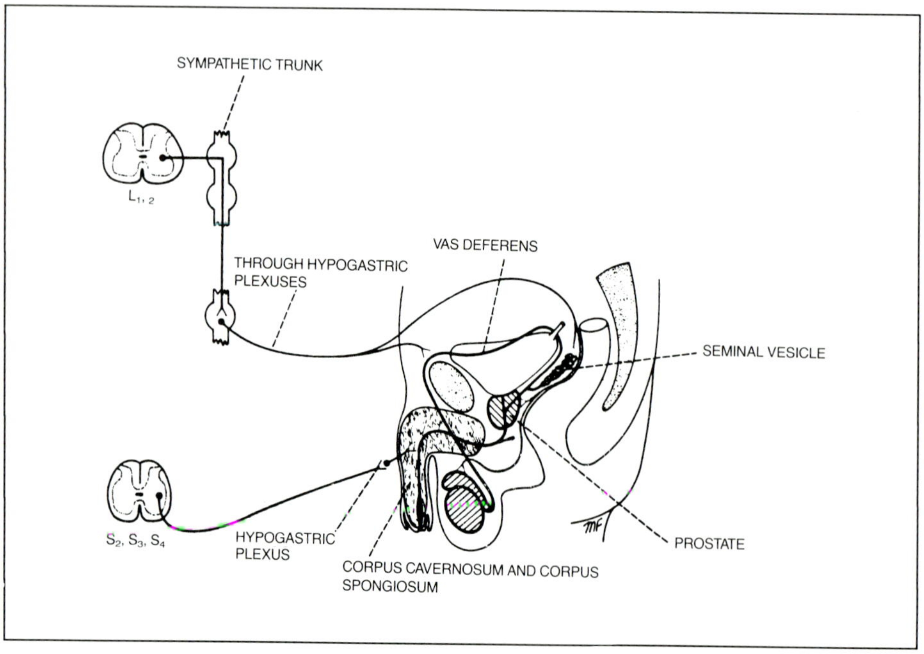

Figure 8–16. The autonomic innervation of the male reproductive tract.

accompanying the uterine artery to the cervix, body, and fundus, and inferiorly accompanying the vaginal artery to the vagina. Nerve fibers are sympathetic, parasympathetic, and afferent sensory. Sympathetic preganglionic fibers leave the spinal cord at segmental levels T_{12} and L_1 and are believed to synapse with ganglion cells in the uterovaginal plexus. Parasympathetic preganglionic fibers leave the spinal cord at levels S_2, S_3, and S_4 and relay in the uterovaginal plexus. Although it is recognized that the uterus is largely under hormonal control, sympathetic innervation may cause uterine contraction and vasoconstriction, whereas parasympathetic fibers have the opposite effect.

Afferent pain fibers from the fundus and the body of the uterus ascend to the spinal cord through the hypogastric plexuses, entering it through the posterior roots of the 10th, 11th, and 12th thoracic spinal nerves (*Fig. 8–15*). Fibers from the cervix run in the pelvic splanchnic nerves, and enter the spinal cord through the posterior roots of the second, third, and fourth sacral nerves.

Erection of the Penis and the Clitoris. The initial vascular engorgement is controlled by the parasympathetic part of the autonomic nervous system that originates in the gray matter of the second, third, and fourth sacral segments of the spinal cord (*Fig. 8–16*). The fibers enter the pelvic plexuses and synapse on the postganglionic neurons. Postganglionic fibers join the internal pudendal arteries and are distributed along their branches, which enter the erectile tissue. Parasympathetic nerves cause vasodilatation of the arteries and greatly increase the blood flow to the erectile tissue.

Ejaculation. The nervous impulses that pass to the genital organs are thought to leave the spinal cord at the first and second lumbar segments in the preganglionic sympathetic fibers (*Fig. 8–16*). Many of these fibers synapse with postganglionic neurons in the first and second lumbar ganglia. Other fibers may synapse in ganglia in the lower lumbar or pelvic parts of the sympathetic trunks. Postganglionic fibers are then distributed to the **vas deferens,** the **seminal vesicles,** and the **prostate** through the **hypogastric plexuses.** The sympathetic nerves stimulate the contractions of the smooth muscle in the walls of these structures and cause the spermatozoa, together with the secretions of the seminal vesicles and prostate, to be discharged into the urethra.

Arteries of the Upper Limb

The arteries of the upper limb are innervated by sympathetic nerves. Preganglionic fibers originate from cell bodies in the second to the eighth thoracic segments of the spinal cord (*Fig. 8–17*). They pass to the sympathetic trunk through white rami and ascend in the trunk to synapse in the middle cervical, inferior cervical, first thoracic, or stellate ganglia. The postganglionic fibers join the nerves that form the brachial plexus and are distributed to the arteries within the branches of the plexus. The sympathetic nerves cause vasoconstriction of cutaneous arteries and vasodilatation of arteries that supply skeletal muscle.

Arteries of the Lower Limb

The arteries of the lower limb are also innervated by sympathetic nerves (*Fig. 8–17*). Preganglionic fibers originate from cell bodies in the lower three thoracic and upper two or three lumbar segments of the spinal cord and pass to the lower thoracic and upper lumbar ganglia of the sympathetic trunk through white rami. Fibers synapse in the lumbar and sacral ganglia, and postganglionic fibers reach the arteries through branches of the lumbar and the sacral plexuses.

IMPORTANT AUTONOMIC REFLEXES

Visual Reflexes

Direct and Consensual Light Reflexes. Afferent nervous impulses travel from the retina through the optic nerve, the optic chiasma, and the optic tract (*Fig. 8–10*). A small number of fibers leave the optic tract and synapse on nerve cells in the **pretectal nucleus,** which lies close to the superior colliculus. The impulses are passed by axons of the pretectal nerve cells to the parasympathetic nuclei (Edinger-Westphal nuclei) of the oculomotor nerve on both sides. Here, the fibers synapse and travel through the oculomotor nerve to the **ciliary ganglion** in the orbit. Finally, postganglionic parasympathetic fibers pass through the **short ciliary nerves** to the eyeball and the constrictor pupillae muscle of the iris. Both pupils constrict in the consensual light reflex, as the pretectal nucleus sends fibers to the parasympathetic nuclei on both sides of the midbrain.

Accommodation Reflex. When the eyes are directed from a distant to a near object, contraction of the medial recti muscles brings about convergence of the ocular axes, the lens thickens to increase its refractive power by contraction of the ciliary muscle, and the pupils constrict to restrict the light waves to the thickest central part of the lens. The afferent impulses travel through the optic nerve, the optic chiasma, the optic tract, the lateral geniculate body, and the optic radiation to the visual cortex (*Fig. 8–10*). The latter is connected to the eye field of the frontal cortex. From here, cortical fibers descend via the internal capsule to the oculomotor nuclei in the midbrain. The oculomotor nerve travels to the medial recti muscles. Some of the descending cortical fibers synapse with the parasympathetic nuclei (Edinger-Westphal nuclei) of the oculomotor nerve on both sides. Here, the

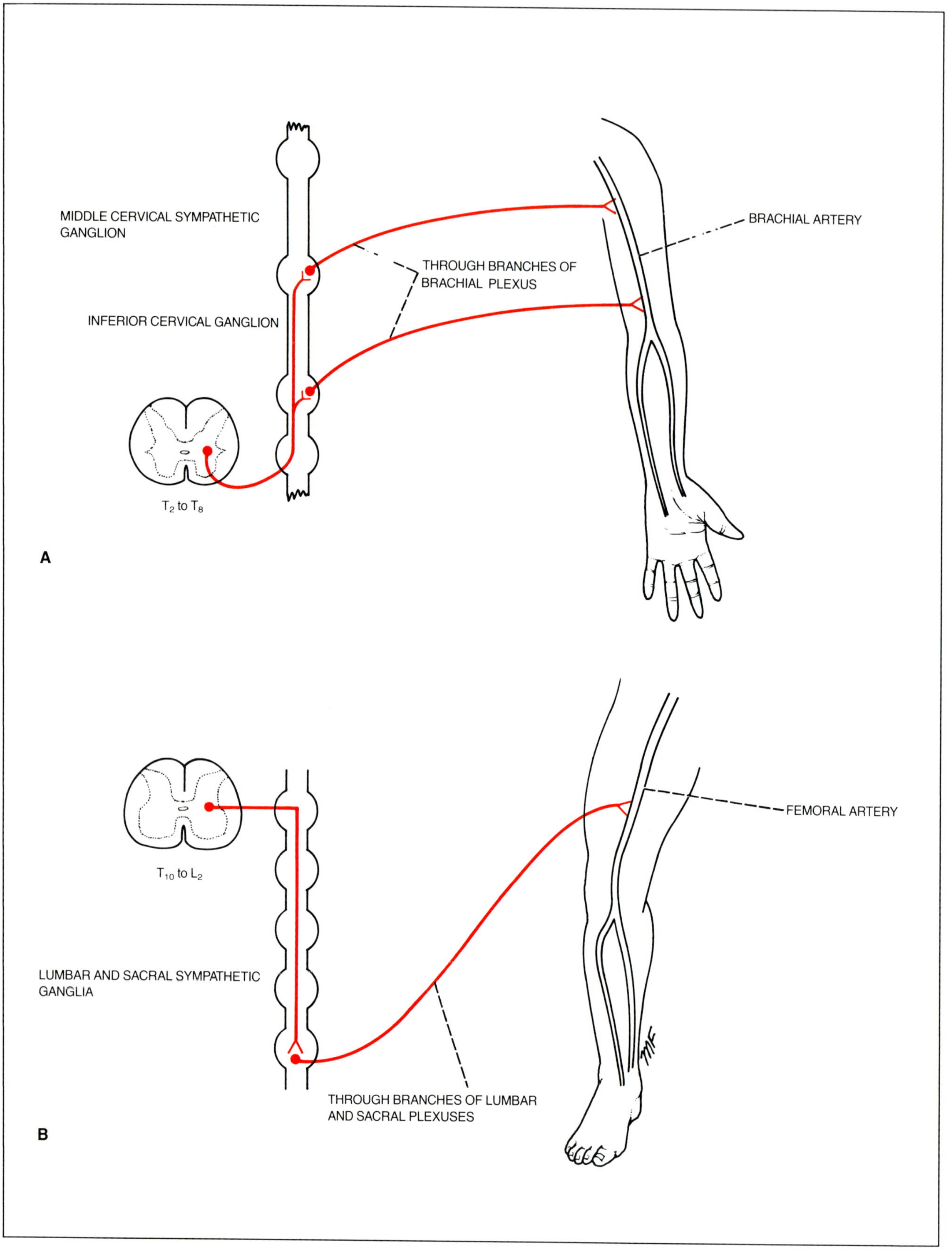

Figure 8–17. The sympathetic innervation of the arteries of **(A)** the upper limb and **(B)** the lower limb.

fibers synapse and the parasympathetic preganglionic fibers travel through the oculomotor nerve to the **ciliary ganglion** in the orbit. Finally, postganglionic parasympathetic fibers pass through the **short ciliary nerves** to the ciliary muscle and the constrictor pupillae muscle of the iris.

Cardiovascular Reflexes

The **carotid sinus** and the **aortic arch** serve as baroreceptors. As the blood pressure rises, nerve endings situated in the walls of these vessels are stimulated. Afferent fibers from the carotid sinus ascend in the glossopharyngeal nerve and terminate in the **nucleus solitarius.** Afferent fibers from the aortic arch ascend in the vagus nerve. Connector neurons in the medulla oblongata activate the parasympathetic nucleus (dorsal nucleus) of the vagus, which slows the heart rate. At the same time, reticulospinal fibers descend to the spinal cord and inhibit the preganglionic sympathetic outflow to the heart and cutaneous arterioles. The combined effect of stimulation of the parasympathetic action on the heart and inhibition of the sympathetic action on the heart and the peripheral blood vessels reduces the rate and force of contraction of the heart and the peripheral resistance. As a consequence, the blood pressure falls. Blood pressure of an individual is thus modified by afferent information received from baroreceptors and the modulator of the autonomic nervous system, namely, the hypothalamus; the latter, in turn, can be influenced by other, higher centers in the central nervous system.

The **Bainbridge right atrial reflex** is another example of a cardiovascular reflex. The nerve endings are situated in the wall of the right atrium and venae cavae. These endings are stimulated by a rise of venous pressure. Afferent fibers ascend in the vagus to the medulla oblongata and terminate on the **nucleus of the tractus solitarius.** Connector neurons inhibit the parasympathetic nucleus (dorsal) of the vagus, and reticulospinal fibers stimulate the thoracic sympathetic outflow to the heart, resulting in cardiac acceleration.

CLINICAL NOTES

Horner's Syndrome

Horner's syndrome includes: (1) constriction of the pupil, (2) ptosis, and (3) enophthalmos, and is caused by an interruption of the sympathetic nerve supply to the orbit. Pathologic causes include lesions in the brainstem or cervical part of the spinal cord that interrupt the reticulospinal tracts descending from the hypothalamus to the sympathetic outflow in the lateral gray column of the first thoracic segment of the spinal cord. Such lesions include **multiple sclerosis** and **syringomyelia.** Traction on the stellate ganglion caused by a **cervical rib,** or involvement of the ganglion in a metastatic lesion, may interrupt the peripheral part of the sympathetic pathway.

Argyll-Robertson Pupil

The Argyll Robertson pupil is a condition characterized by a small pupil that is of fixed size and does not react to light, but contracts with accommodation. It is usually caused by a lesion interrupting fibers that run from the pretectal nucleus to the parasympathetic nuclei (Edinger-Westphal nuclei) of the oculomotor nerve on both sides. The fact that the pupil constricts with accommodation implies that the connections between the parasympathetic nuclei and the constrictor pupillae muscle of the iris are intact.

Frey's Syndrome

Frey's syndrome is a complication that sometimes follows penetrating wounds of the parotid gland. During the process of healing, the postganglionic parasympathetic secretomotor fibers traveling in the auriculotemporal nerve grow out and join the distal end of the great auricular nerve, which supplies the sweat glands of the overlying facial skin. By this means, a stimulus intended to stimulate saliva flow instead produces sweat.

Hirschsprung's Disease (Megacolon)

Hirschsprung's disease is a congenital condition in which there is a failure of development of the myenteric plexus (Auerbach's plexus) in the distal part of the colon. The involved part of the colon possesses no parasympathetic ganglion cells and peristalsis is absent. This effectively blocks the passage of feces and the proximal part of the colon becomes enormously distended

Sympathetic Neural Blockade

Theoretically sympathetic blockade can be achieved at the following anatomic sites: (1) where the sympathetic preganglionic fibers leave the spinal cord (T_1 to $L_2(L_3)$) in the anterior roots of the spinal nerves (subarachnoid or epidural block); (2) at the site of synapses between the preganglionic fibers and the ganglionic neurons (ganglion block); (3) in a peripheral nerve where the postganglionic fibers travel to their destination (peripheral nerve block); (4) in a perivascular plexus (perivascular block); and (5) pharmacologic block at the ganglion or the terminal receptor sites.

Selective sympathetic blockade can best be accomplished by the blockade of sympathetic ganglia. Unfortunately, a sympathetic ganglion blockade is often accompanied by motor and sensory blockade. In the thorax, the ganglia of the sympathetic trunk are closely related to the intercostal nerves, therefore, differential blockade is difficult, but the cervical ganglia, the celiac ganglia, and the lumbar ganglia lie at some distance from the somatic nerves and are more easily dealt with.

The relay levels between the preganglionic fibers and the ganglionic neurons described in this text are those most often quoted in the literature. Variations do occur, however, and this may explain the poor results that sometimes occur from a technically successful block. The important anatomical landmarks used in sympathetic blockades are described later. Readers interested in the techniques of sympathetic blockades are referred to the **Atlas of Regional Anesthesia** by Jordan Katz.

Anatomy for Stellate Ganglion Blockade (Cervicothoracic Block)

As previously noted, in about 80 percent of individuals the inferior cervical ganglion is fused with the first thoracic ganglion of the sympathetic trunk to form the stellate ganglion. Because of the closeness of the ganglion to the dome of the pleura and the risk of pneumothorax, the anesthetic needle is inserted some distance above the ganglion. The anterior border of the sternocleidomastoid is palpated and the trachea is felt in the midline. Between these two landmarks can be felt the pulse of the common carotid artery. The needle is inserted and directed posteriorly at the level of the cricoid cartilage, between the cricoid medially, and the laterally displaced carotid artery until the tip contacts the transverse process of C-6. It is, then, withdrawn several millimeters until it lies anterior to the prevertebral layer of deep cervical fascia and the longus cervicis muscle. The local anesthetic is then injected after careful aspiration. Should the needle pass too far backward between the transverse processes it may impale the vertebral vessels.

Because in the remaining 20 percent of individuals, the inferior cervical ganglion and the first thoracic ganglion are separated by a short distance, it is common practice to sit the patient upright after the injection so that the anesthetic solution will diffuse inferiorly to the upper thoracic ganglia behind the cervical dome of the pleura and the costal pleura.

The anatomic signs of successful sympathetic block are ptosis, myosis, enophthalmus, flushing of facial skin and conjunctiva, and anhidrosis (Horner's syndrome), as well as vasodilation and increased warmth in the upper extremity.

Anatomy for Thoracic Sympathetic Block

The sympathetic trunk in the upper part of the thorax lies anterior to the necks of the ribs, close to the intercostal nerves, and behind the costal pleura. In the lower part of the thorax, the sympathetic trunk moves away from the ribs onto the sides of the bodies of the thoracic vertebrae.

In the upper part of the thorax, the needle is inserted a short distance lateral to the spinous process through the skin, the fascia, and the trapezius, rhomboid, and postvertebral muscles. The needle is then advanced below the transverse process and the local anesthetic is injected around the sympathetic trunk and the intercostal nerve.

In the middle and lower part of the thorax, the needle has to be advanced more anteriorly past the intercostal nerve to reach the sympathetic trunk on the side of the vertebral body.

The complications of thoracic sympathetic block are penetration of the costal pleura with resulting pneumothorax and passage of the needle through an intervertebral foramen into the subarachnoid space.

Anatomy for Lumbar Sympathetic Block

In the lumbar region, the sympathetic trunk lies on the anterolateral side of the lumbar vertebral bodies and is separated from the nerves of the lumbar plexus by the psoas fascia and the psoas muscle (the nerves lie within the muscle).

An appropriate lumbar spinous process is identified. The needle is advanced anteriorly and medially through the skin, the fascia, the postvertebral muscles, and between the transverse processes of adjacent vertebrae. When the needle contacts the side of the lumbar vertebral body it is withdrawn and reinserted so as to pass on the anterolateral surface of the bone. The anesthetic solution is injected around the sympathetic trunk.

Rare complications include the passage of the needle into an intervertebral foramen, the subarachnoid space, and major vessels.

Anatomy for Celiac Plexus Nerve Block

The celiac plexus surrounds the origin of the celiac artery in front of the abdominal part of the aorta, anterior to the body of the first lumbar vertebrae. The right celiac ganglion lies posteromedial to the inferior vena cava, and the left ganglion lies medial to the left kidney and the left suprarenal gland, and posterior to the pancreas.

The needle is inserted lateral to the spine of the first lumbar vertebra beneath the lower border of the 12th rib. The needle is advanced cephalomedially toward the body of the first lumbar vertebra piercing the skin, the fascia, and the postvertebral muscles. The needle passes inferiorly to the transverse process of the first lumbar vertebra and should come to lie lateral to the body of the first lumbar vertebra. The left side is preferred for the block because the aorta can be avoided by detecting the aortic pulsations along the needle. On the right side, the thin-walled nonpulsating inferior vena cava lies just lateral to the point of the needle and must be avoided. The anesthetic solution is then injected around the celiac ganglia.

Alternate techniques include direct transabdominal approaches using computed tomography scan or ultrasound to guide the needle tip to the origin of the celiac artery on the aorta.

CLINICAL PROBLEMS

1. A 38-year-old man was employed to remove some trees in the preparation of a piece of ground for a building site. Thirty feet from the ground he was sawing off the limb of a tree when he lost his footing and slipped. As he began to fall, he grabbed at a branch with his right hand and held on until he was rescued some minutes later. Careful examination of his right upper limb showed paralysis of the flexor carpi ulnaris, the flexor digitorum profundus, and weakness of the palmar and dorsal intercossei, and the thenar and hypothenar muscles. There was also some loss of sensation on the medial side of the forearm and the hand. It was noted that the pupil of the right eye was constricted and there was ptosis of the right upper lid. A slight degree of right-sided enophthalmus was also present. The skin of the right cheek felt warmer and drier than the left cheek. How can you account for these widespread physical signs?

2. A 23-year-old woman visited her physician with a history of experiencing severe attacks of painful discoloration of the fourth and fifth fingers of both hands especially during cold weather. A diagnosis of Raynaud's disease was made. In view of the severity of the symptoms and the possibility of gangrene of the fingertips, it was decided to recommend a cervicodorsal preganglionic sympathectomy. Before the operation the surgeon recommended that the anesthetist perform a cervicodorsal ganglion blockade to determine whether the blood flow to the affected fingers was increased. What important surface landmarks would you use at the site of injection? Why is the injection given at the level of the sixth cervical vertebra when the ganglia involved are located in the root of the neck? What are the important anatomic relations of the stellate ganglion? What is the innervation of the arteries of the upper limb?

3. A 16-year-old boy was admitted to the emergency room after an explosion in the basement of his home. The boy and his friends had found some sticks of dynamite while exploring a quarry, and were attempting to make a pipe bomb. The boy had a deep wound in his lower back and radiographic examination revealed the presence of a metal fragment in the vertebral canal at the level of the second lumbar vertebra. An extensive neurologic examination revealed a complete lesion of the cauda equina. Is this patient going to have a normal bladder function? What is the autonomic innervation to the bladder?

4. Explain the following anatomical facts. (a) The presence of sympathetic nerve fibers in a parasympathetic ganglion. (b) The large intestine is innervated by the vagus nerves down as far as the splenic flexure and yet the vagal nerve trunks apparently come to an end very soon after piercing the diaphragm with the esophagus. (c) How does the sacral parasympathetic outflow reach the splenic flexure of the colon? (d) The enteric system appears to be a special or separate part of the autonomic nervous system.

5. Two of the main complications of thoracic sympathetic block are pneumothorax and injection of the anesthetic solution into the subarachnoid space. In anatomical terms explain how these complications can take place.

6. After a neurolytic lumbar sympathetic blockade, the patient may experience discomfort for several weeks in the first lumbar dermatome. How can you explain this in terms of the anatomy of the area of injection.

7. Celiac plexus blockade combined with intercostal nerve block is occasionally used when surgery is contemplated on the upper abdominal viscera. Explain this statement in anatomical terms.

8. Neurolytic celiac plexus block is one of the best methods of relieving pain to carcinoma of the pancreas. Which nerve pathways are blocked?

9. When using a celiac plexus nerve block, a surgeon often comments on the smallness of the small intestine and the ease with which he is able to contain it within the abdomen when closing the peritoneal cavity. Explain this observation in anatomical terms.

ANSWERS FOR CLINICAL PROBLEMS

1. This man had sustained a severe traction injury of the eighth cervical and first thoracic roots of the brachial plexus. The various paralyzed muscles and the sensory loss are characteristic of Klumpke's paralysis. In addition the white ramus communicans passing from the first thoracic nerve to the stellate ganglion was torn, cutting off the preganglionic sympathetic fibers to the head and the neck, producing a Horner's syndrome.

2. The cricoid cartilage, the trachea, the anterior border of the sternocleidomastoid muscle, and the pulsating common carotid artery are important surface landmarks. The needle is directed posteriorly at the level of the cricoid cartilage (sixth cervical vertebral level) between the cricoid medially and the laterally displaced common carotid artery.

   The inferior cervical and the first thoracic ganglion or the stellate ganglion are located behind the cervical dome of the pleura. To avoid the possibility of producing a pneumothorax by inserting the needle low in the neck, the anesthetic solution is injected at a higher level and allowed to diffuse downward with the patient in the sitting position.

   The stellate ganglion lies in the interval between the transverse process of the seventh cervical vertebra and the neck of the first rib. It lies medial to the scalenus anterior muscle and is covered anteriorly by the vertebral vessels and the cervical dome of the pleura.

   The arteries of the upper limbs are innervated by sympathetic nerves. The preganglionic fibers originate from the nerve cells in the lateral gray columns of the second to the eighth thoracic segments of the spinal cord. They reach the sympathetic trunk via the white rami communicantes and ascend in the trunk to synapse in the middle cervical, inferior cervical, and first thoracic or stellate ganglia. The postganglionic fibers join the spinal nerves that form the brachial plexus traveling in the gray rami communicantes. The sympathetic fibers are distributed to the digital arteries within the branches of the brachial plexus.

3. The urinary bladder receives its sympathetic innervation from the first and second lumbar segments of the spinal cord and its parasympathetic fibers from the second, third, and fourth sacral segments of the cord. In this patient, with a complete section of the cauda equina at the level of the second lumbar vertebra, the preganglionic sympathetic fibers that descend in the anterior root of the first lumbar nerve were left intact, as they leave the vertebral canal to form the first lumbar nerve above the level of the metal fragment. The preganglionic parasympathetic fibers were, however, sectioned as they descended in the vertebral canal within the anterior roots of the second, third, and fourth sacral nerves. The patient would, therefore, have an autonomous bladder and would be without any external reflex control. The bladder would fill to capacity and then overflow. The patient could activate micturition by powerful contraction of the abdominal muscles, assisted by manual pressure on the lower anterior abdominal wall.

4. (a) It is not uncommon to find sympathetic preganglionic and postganglionic fibers passing through a parasympathetic ganglion without interruption. The nerve fibers are merely using the ganglion as a conduit en route to their destination. Visceral sensory nerve fibers travel in a similar manner. (b) The vagal nerve trunks on reaching the abdominal cavity split up after a short course with the esophagus

into their terminal branches. The posterior vagal trunk (right vagus) gives off an important branch that passes to the celiac and superior mesenteric plexuses. The terminal fibers are distributed with the branches of the celiac and superior mesenteric arteries to the small and large intestine as far as the splenic flexure. (c) The sacral outflow of the parasympathetic (S_2, S_3, and S_4) leave the anterior rami of the sacral nerves as the pelvic splanchnic nerve. These preganglionic fibers pass through the hypogastric and aortic plexuses to reach the inferior mesenteric plexus. The fibers are then distributed to the splenic flexure and descending colon along with the branches of the inferior mesenteric artery. (d) The enteric system consists of a nerve plexus within the wall of the intestine that coordinates the muscle movements of the bowel wall and the parasympathetic and sympathetic inputs modify these activities.

5. In the thorax the sympathetic trunk is covered anteriorly by the costal parietal pleura. If the anesthetic needle is advanced too far from the midline, the membrane will be perforated. In the lower part of the thorax, the sympathetic trunk is medially placed on the lateral surfaces of the thoracic vertebral bodies. If the needle is directed too much medially too early as it is advanced forward, it will enter an intervertebral foramen and possibly pierce the meninges to enter the subarachnoid space.

6. The use of excessively large volumes of solution when performing a neurolytic lumbar sympathetic blockade may damage the first lumbar nerve, and give rise to neuralgia in the distribution of the iliohypogastric (L_1), ilioinguinal (L_1), and genitofemoral (L_1, L_2) nerves. Also sensory nerve fibers that run in the rami communicantes that connect the upper lumbar sympathetic ganglia with the somatic nerves may be irritated or damaged.

7. Afferent visceral nerve fibers from the stomach and small intestine travel through the superior mesenteric and then the celiac plexuses on their way to the spinal cord via the greater, lesser, and least splanchnic nerves. A celiac plexus block will produce a visceral anesthesia. An intercostal nerve block produces anesthesia of the skin and parietal peritoneum of the anterior abdominal wall and relaxation of the abdominal wall musculature.

8. Afferent nerve fibers from the pancreas travel through the superior mesenteric and celiac plexuses and the greater, lesser, and least splanchnic nerves to reach the spinal cord. A neurolytic celiac block can be very affective in relieving chronic neoplastic pain from the pancreas.

9. A celiac plexus block paralyzes the sympathetic inhibition of the enteric nerve plexus in the wall of the intestines. The result is that the tone of the smooth muscle is increased with a reduction in the size of the intestine. Although the vagal parasympathetic fibers traveling through the celiac plexus are also presumably blocked, this does not seem to have much influence on the enteric plexus.

CHAPTER 9

Reception of Pain and Control of Muscular Activity

Pain is the sensation of discomfort, distress, or agony resulting from damage to tissue. Pain is essentially a protective mechanism that alerts the individual so that he or she may react and remove the pain stimulus.

PAIN RECEPTORS

Free nerve endings are the pain receptors (*Fig. 9–1*). They are found between the epithelial cells of the skin, the cornea, and the alimentary tract, and in connective tissues, including dermis, fascia, ligaments, joint capsules, tendons, periosteum, perichondrium, haversian systems of bone, tympanic membrane, and dental pulp; they are also present in muscle. The terminal endings are devoid of a myelin sheath and there are no Schwann cells covering their tips.

The pain receptors are stimulated by mechanical damage, by excessive heat or cold, and by chemicals. They are unique among sensory receptors in having very little ability to adapt and in fact they become increasingly sensitive if the stimulus is continued. The process of depolarization of the terminal nerve fibers by the injury is probably brought about by the release of substances, such as bradykinin, arachidonic acid, or prostaglandins.

CONDUCTION OF PAIN IMPULSES IN PERIPHERAL NERVES

Nerve impulses concerned with pain are conducted by small delta type A fibers (6 to 30 m/sec) and by type C fibers (0.5 to 2 m/sec). It has been shown that the delta type A fibers, which are thinly myelinated, are responsible for the acute ***sharp*** pain sensation, whereas the nonmyelinated type C fibers conduct the sensation of an aching, less intense pain. Because of the difference in conduction rates the sharp pain is felt first to be followed a short time later by the ***slow*** pain. The fast conducting fibers alert the individual to the site of tissue damage, whereas the slow conducting fibers are associated with the dull uncomfortable sensation of after pain and may be associated with the emotional aspects of pain.

Posterior Roots and Posterior Root Ganglia of Spinal Nerves

The cell bodies of pain afferent fibers are located in the posterior root ganglia. Some of these cells have been shown to contain the peptide, substance P. The axons from a single posterior root ganglion enter the spinal cord through a posterior root and from there may extend up and down the cord for several segments. A given area of skin is innervated by three adjoining segments of the spinal cord.*

Posterior Gray Column

The axons entering the spinal cord from the posterior root ganglion proceed to the tip of the posterior gray column and divide into ascending and descending branches (*Fig. 9–2*). These branches travel for a distance of one to two segments of the spinal cord and form the **posterolateral tract of Lissauer,** which is situated just posterior to the posterior gray column (*Fig. 9–2*).

The acute, fast pain fibers synapse on neurons in lamina I† and lamina V in the posterior gray column. The chronic, slow pain fibers synapse on neurons in laminae II and III, the area known as the substantia gelatinosa. Most of the neurons in laminae II and III have short axons and synapse with neurons in lamina V.

The axons of the neurons in laminae I and V and a few from laminae II and III now cross obliquely to

* This statement must not be confused with the definition of a dermatome, which is the area of skin supplied by a single segment of the spinal cord (or a single spinal nerve). Thus, a given area of skin is innervated by three overlapping dermatomes.

† In thick sections of the spinal cord the gray matter appears to have a laminar appearance. Rexed (1952) described ten layers of neurons in the cat spinal cord based on the arrangement of nerve cells in the gray matter.

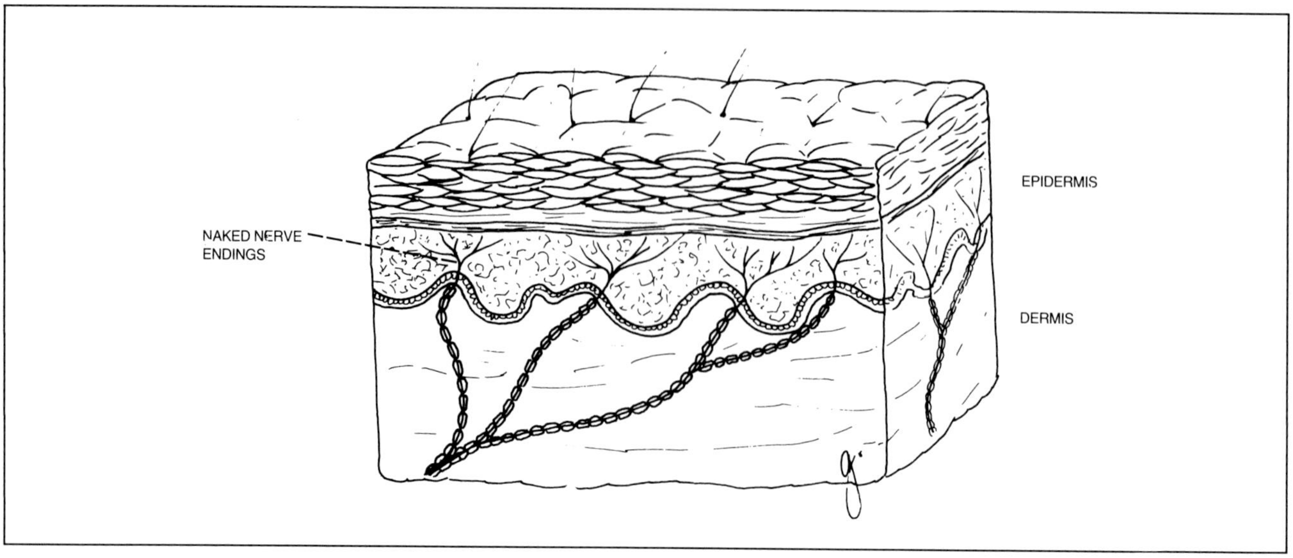

Figure 9–1. Free nerve endings in the skin. Note that the nerve fibers in the epidermis are naked.

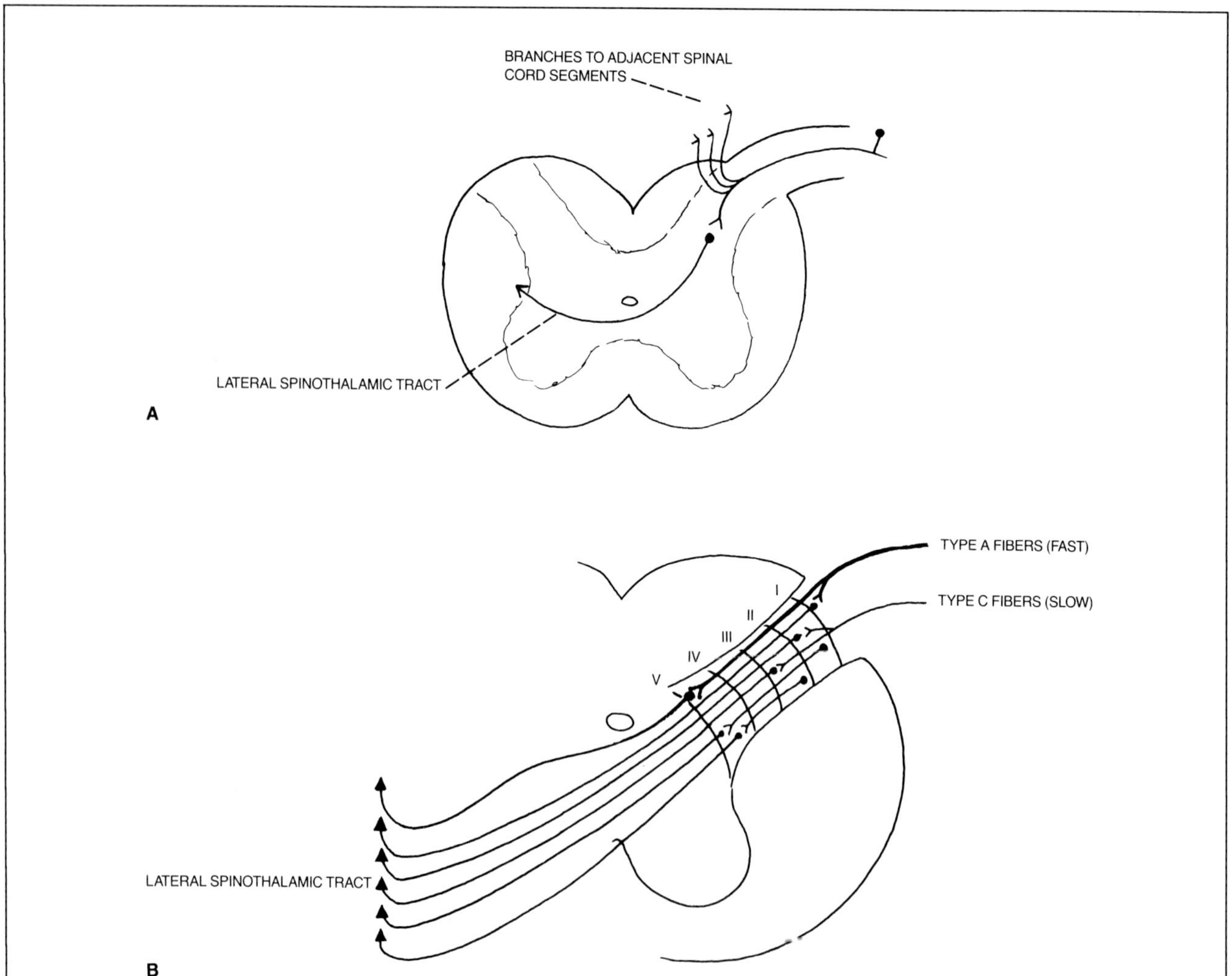

Figure 9–2. A. The pain fibers entering the spinal cord and dividing into ascending and descending branches forming the Lissauer tract. **B.** The termination of the acute, fast pain and the chronic, slow pain fibers in the various laminae of the posterior gray column.

the opposite side in the anterior gray and white commissures within one spinal segment of the cord, and ascend in the contralateral white column as the lateral spinothalamic tract (*Fig. 9–2*). As this tract ascends through the spinal cord, new axons are added to the anteromedial aspect, so that in the upper cervical segments of the cord the sacral fibers are posterolateral and the cervical segments are anteromedial.

The lateral spinothalamic tract continues to ascend through the medulla oblongata and the pons (*Fig. 9–3*). In the midbrain, it lies in the tegmentum lateral to the medial lemniscus. The great majority of the ascending fibers concerned with conducting acute sharp pain terminate by synapsing with cells in the reticular formation of the medulla, the pons, and the midbrain (*Fig. 9–3*). From there the neurons send fibers to the ventral posterolateral nucleus of the thalamus and the hypothalamus. The remaining acute pain fibers pass directly to the thalamus.

The axons of the thalamic neurons now pass via the posterior limb of the internal capsule and the corona radiata, to reach the somesthetic area in the postcentral gyrus of the cerebral cortex. The contralateral half of the body is represented as inverted, with the hand and the mouth situated inferiorly and the leg situated superiorly, and with the foot and the anogenital region on the medial surface of the hemisphere.

The axons ascending in the lateral spinothalamic tract concerned with conducting the sensation of slow chronic pain terminate in the reticular formation and the intralaminar nuclei of the thalamus (*Fig. 9–3*). It is these impulses, which are very poorly localized, that arouse the nervous system and create a sense of urgency.

Thalamus and Cerebral Cortex in the Reception of Pain

The thalamus and the other lower nerve centers enable the individual to be conscious of pain. The cerebral cortex is necessary, however, in understanding the quality of pain.

Localization of Pain

Suprisingly, fast, acute pain is poorly localized and an individual may only be able identify the stimulated area to within about 15 cm. Slow, chronic pain is even more poorly localized and the site may only be localized to a large region, such as a limb. The ability to precisely localize the source of pain is probably made possible because of the concurrent stimulation of tactile receptors.

PAIN CONTROL IN THE CENTRAL NERVOUS SYSTEM

The Analgesia System

It has been shown experimentally that stimulation of certain areas of the brainstem can reduce or block the sensations of pain. These areas include the periventricular area of the diencephalon, the periaqueductal gray matter of the midbrain, and the midline nuclei of the brainstem. It is believed that fibers of the reticulospinal tract pass down to the spinal cord and synapse on cells concerned with pain sensation in the posterior gray column. The analgesic system can suppress both acute, fast pain and chronic, slow, burning pain sensations.

The Gating Theory

Melzack and Wall* suggested that at the site where afferent pain fibers entered the central nervous system, inhibition could occur, and that this modulation of pain could be brought about by internuncial neurons (interneurons) located in the posterior gray column. The so called gate to the pain impulses could be closed by the internuncial neurons exerting presynaptic inhibition on primary afferent synapses and postsynaptic inhibition on the spinothalamic relay cells. It is now clear that gating is influenced not only by local control but also by large afferent fibers and by supraspinal controls.

It has been shown experimentally that stimulation of large myelinated afferent fibers carrying nonpainful information of touch and pressure from an area from which pain is being initiated, reduces the pain. Collateral branches from these large afferent fibers enter the substantia gelatinosa and inhibit the afferent fibers directly or through internuncial neurons, thus closing the gate. Once the nonpainful tactile stimulation ceases, however, the gate is opened and the painful stimuli ascend the spinothalamic tracts.

Supraspinal gate control has already been alluded to in the analgesic system and inhibitory impulses descend to the spinal cord in the reticulospinal tracts. Gate control of pain is summarized in Figure 9–4.

Afferent Pain Fibers in the Trigeminal Nerve

The sensation of pain travels along the axons whose cell bodies are situated in the trigeminal sensory (gasserian) ganglion (*Fig. 9–5*). The central processes of these cells form the large sensory root of the trigeminal nerve. About half the fibers divide into ascending and descending branches when they enter the pons; the remainder ascend or descend without division (*Fig. 9–5*). The ascending branches terminate in the main sensory nucleus and the descending branches terminate in the spinal nucleus. The sensations of touch and pressure are conveyed by nerve fibers that terminate in the main sensory nucleus. The sensations of pain and temperature pass to the spinal nucleus (*Fig. 9–5*). It should be noted that pain and temperature fibers from the ophthalmic division of the trigeminal nerve terminate in the inferior part of the spinal nucleus; fibers from the maxillary division termi-

* Melzack R, Wall PD: Pain mechanisms: A new theory. Science 150:971, 1965.

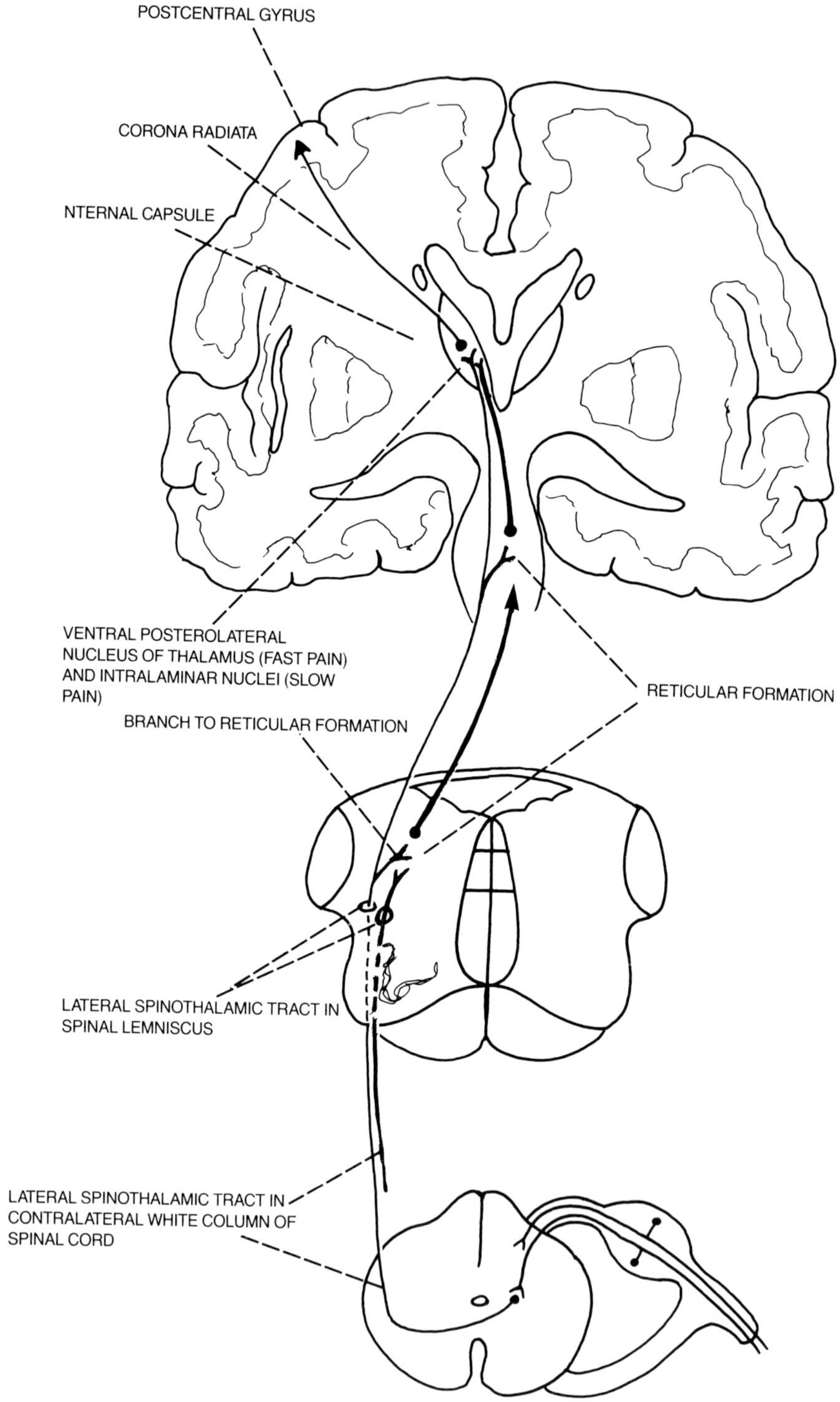

Figure 9–3. The final destinations of the fast, acute pain and the chronic, slow pain fibers in the lateral spinothalamic tract.

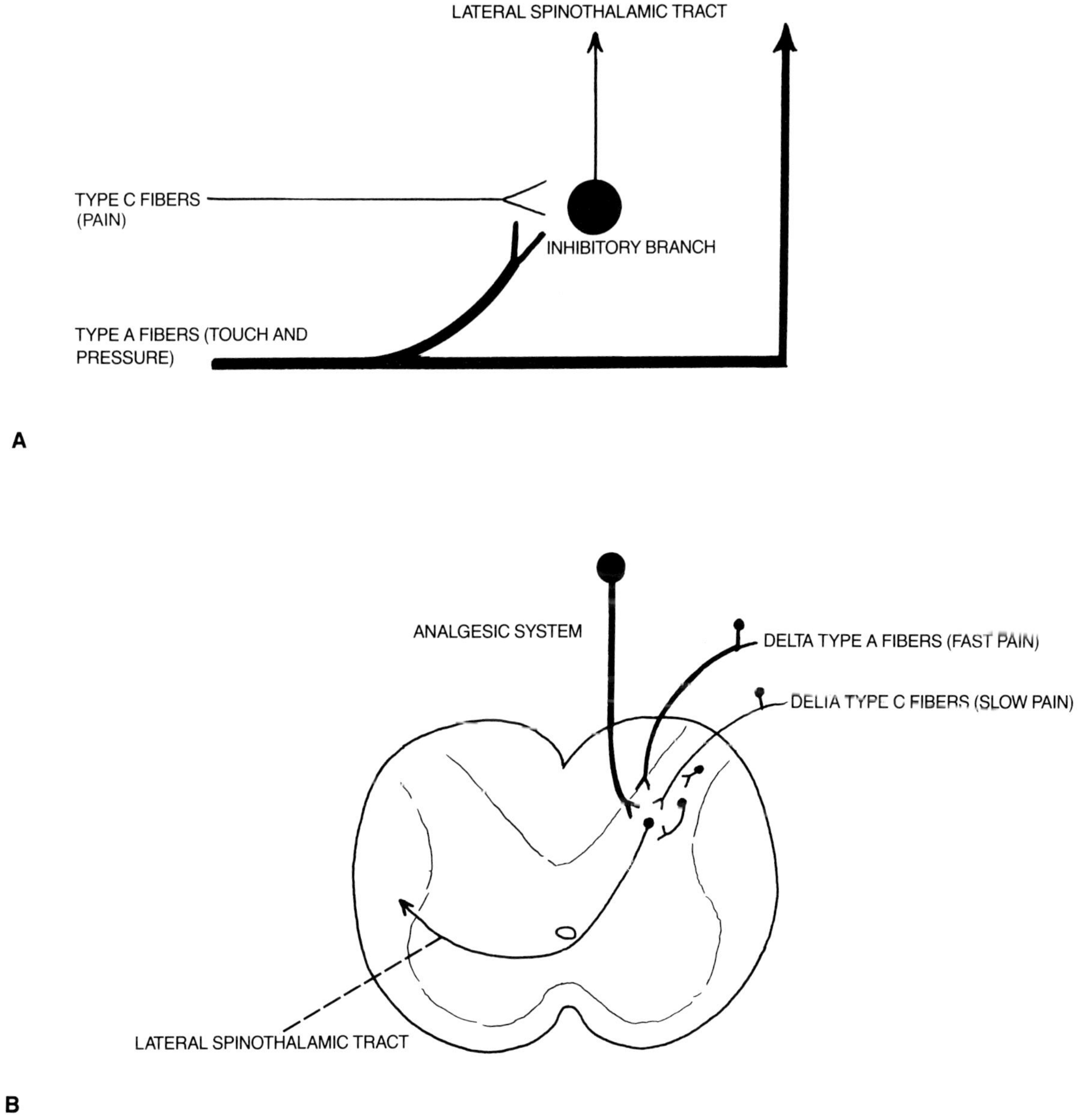

Figure 9–4. Diagrams showing the influence of large touch and pressure afferent fibers and supraspinal neurons of the analgesic system on the gate control.

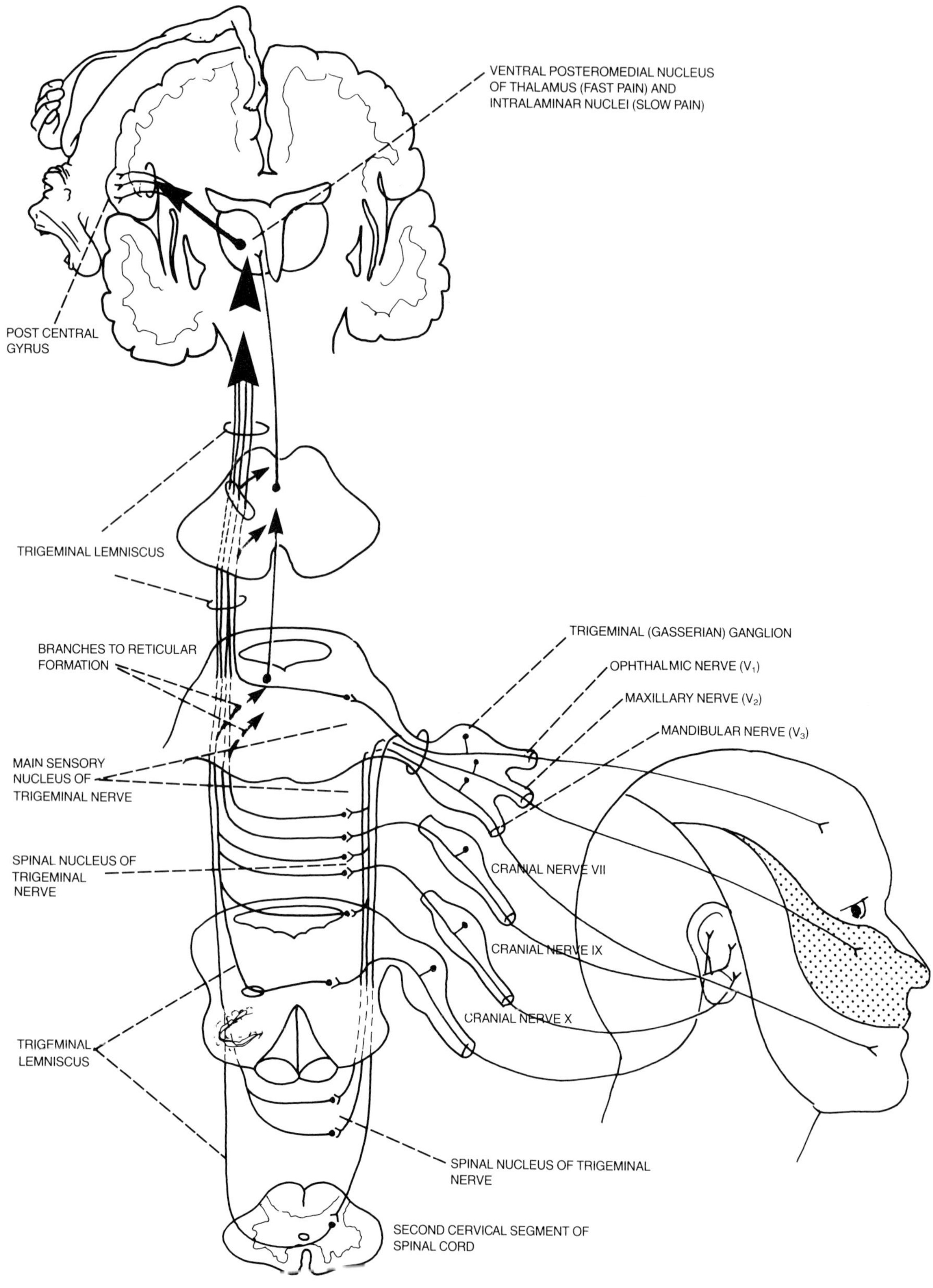

Figure 9–5. Diagram showing the pain pathways from the face. Also shows afferent pain pathways from the facial, glossopharyngeal, and vagus nerves.

nate in the middle of the spinal nucleus; and fibers from the mandibular division end in the superior part of the spinal nucleus.

The axons of the second-order neuron in the spinal nucleus now cross the median plane and ascend as the **trigeminal lemniscus.** It is believed that the acute, fast pain pathway and the chronic, slow, burning pain pathway exist, as seen with afferent fibers entering the central nervous system through the spinal nerves. The first pathway would thus end in the reticular formation and the ventral posteromedial nucleus of the thalamus and be relayed to the postcentral gyrus. The second pathway would end in the reticular formation and the intralaminar nuclei of the thalamus.

Afferent Pain Fibers in the Glossopharyngeal and Vagus Nerves

The pain fibers entering the medulla oblongata originate in the superior sensory ganglia of the glossopharyngeal and vagus nerves and end in the spinal nucleus of the trigeminal nerve. The fast and slow pain pathways of the second-order neurons in the spinal nucleus ascend to the reticular formation and the thalamus in a manner similar to those of the trigeminal nerve.

Visceral Pain Pathways

The causes of visceral pain include ischemia, chemical damage, spasm of smooth muscle, and distention. The sensations that arise in viscera located in the thorax and the abdomen accompany the sympathetic nerves and enter the spinal cord through the posterior roots. The cell bodies of the first-order neuron are situated in the posterior root ganglia. The peripheral processes of these cells are small type C fibers and receive nerve impulses from pain and stretch receptor endings in the viscera. They conduct only the burning and aching types of pain. The central processes, having entered the spinal cord, synapse with second-order neurons in the gray matter, probably in the posterior or lateral gray columns.

The axons of the second-order neurons are believed to join the spinothalamic tracts and ascend to terminate on the third-order neurons in the reticular formation and the ventral posterolateral nucleus of the thalamus. The final destination of the axons of the third-order neurons has not been established, but some may terminate in the postcentral gyrus of the cerebral cortex.

It has been reported that sensations from a full rectum and a full urinary bladder experienced before defecation and micturition are carried by ascending tracts located in the posterior white columns of the spinal cord.

Many of the visceral afferent fibers that enter the spinal cord branch and participate in reflex activity.

Referred Pain

Most viscera are innervated only by autonomic nerves. It follows, therefore, that visceral pain is conducted along small type C afferent autonomic nerves. Visceral pain, which is a burning aching type of pain, is diffuse and poorly localized, whereas somatic pain is sharp, intense, and discretely localized. Visceral pain frequently is referred to skin areas that are innervated by the same segments of the spinal cord as the painful viscus (*Fig. 9–6*). The explanation for referred pain is not known. One theory is that the nerve fibers from the viscus and the dermatome ascend in the central nervous system along a common pathway and the cerebral cortex is incapable of distinguishing between the sites of origin. Another theory is that under normal conditions the viscus does not give rise to painful stimuli, whereas the skin area repeatedly receives noxious stimuli. Because both afferent fibers enter the spinal cord at the same segment, the brain interprets the information as coming from the skin rather than the viscus. Pain arising from the gastrointestinal tract is referred to the midline. This can probably be explained on the basis that the tract arises embryologically as a midline structure and receives a bilateral nerve supply.

CLINICAL NOTES

REFERRED PAIN

Cardiac Pain

Inadequate blood supply to the myocardium (myocardial ischemia) results in the patient's experiencing severe pain over the middle of the sternum, often spreading to one or both arms, the root of the neck, and even the jaw. The pain is assumed to be caused by the accumulation of metabolites and oxygen deficiency, which stimulate the sensory nerve endings in the myocardium. The afferent nerve fibers ascend to the central nervous system through the cardiac branches of the sympathetic trunk and enter the spinal cord through the posterior roots of the upper four thoracic nerves. Note that the cardiac pain is not felt in the heart, but is referred to the skin areas supplied by the corresponding spinal nerves. The skin areas supplied by the upper four intercostal nerves and by the intercostobrachial nerve (T_2) are, therefore, affected. A certain amount of spread of nervous information must occur within the central nervous system, for the pain is sometimes felt in the neck and the jaw.

Appendicitis

Another common example of referred pain is found in the early stages of an acute appendicitis. Initially, visceral pain in the appendix is produced by distention of its lumen or spasm of its muscle. The afferent pain fibers enter the spinal cord at the level of the tenth thoracic segment, having ascended through the superior mesenteric plexus and the lesser splanchnic nerve. A vague referred pain is felt in the region of the umbili-

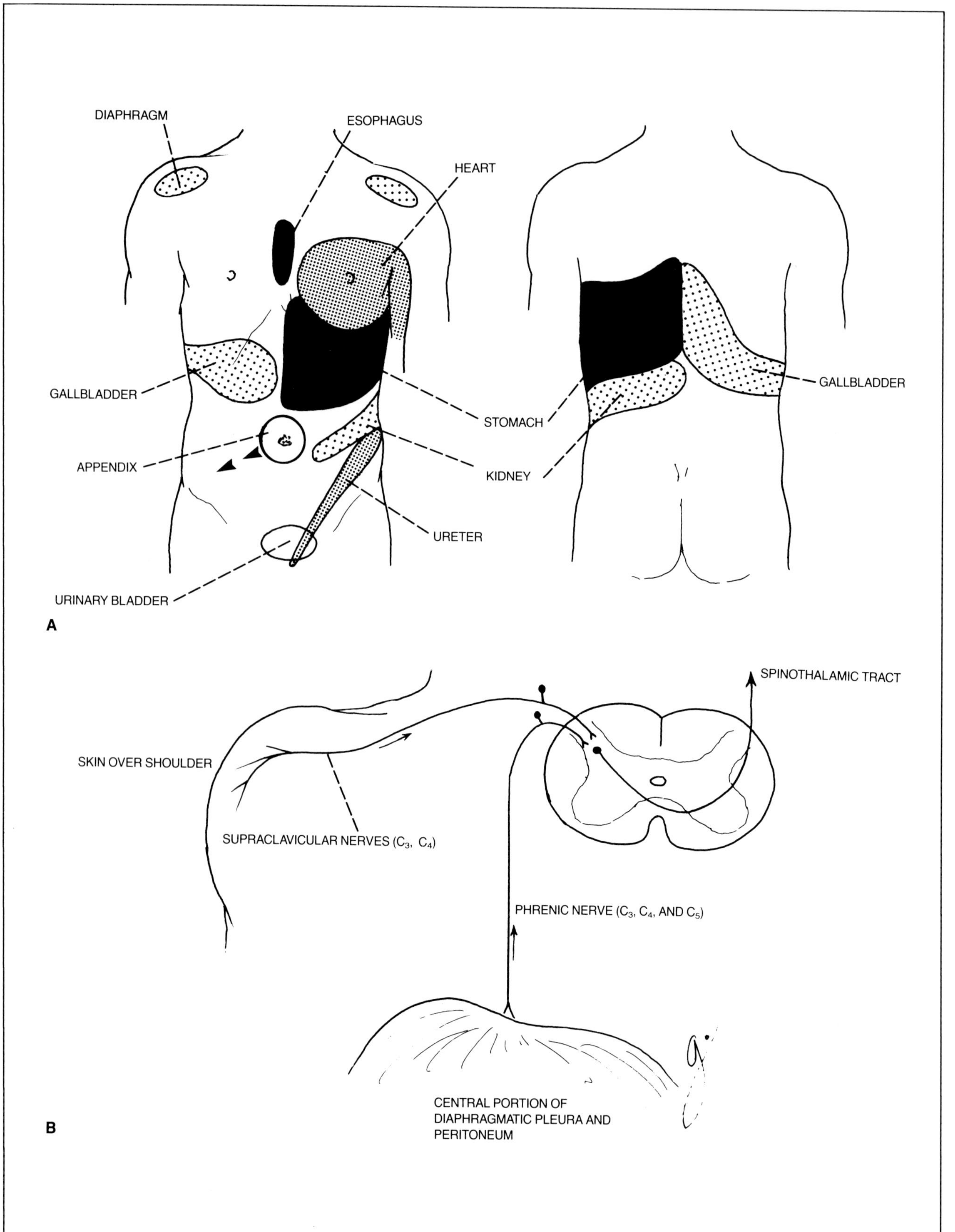

Figure 9–6. A. Some important skin areas involved in referred visceral pain. **B.** Convergence of somatic and visceral afferent pain fibers in the posterior gray column.

cus, which is innervated by the tenth intercostal nerve. Later, the pain shifts to the right iliac fossa, where the inflamed appendix irritates the parietal peritoneum, which is innervated by the 12th thoracic and first lumbar spinal nerves. Here, the pain is conducted along fast delta type A fibers and is precise, severe, and localized.

CONTROL OF MUSCLE ACTIVITY

Innervation of Skeletal Muscle

The nerve trunk to a muscle is a mixed nerve, about 60 percent being motor and 40 percent sensory, and it also contains some sympathetic autonomic fibers. The nerve enters the muscle along with the arterial supply; the place of entrance is known as the **neurovascular hilus.** Segmentally innervated muscles, such as those of the anterior abdominal wall, may be innervated by several nerve trunks. Muscles innervated by spinal nerves commonly receive motor and sensory nerve fibers from several segments of the spinal cord but some muscles, for example, the small muscles of the hand, are supplied from a single segment.

The motor fibers are of two types: (1) the large **alpha fibers** derived from large cells in the anterior gray horn of the spinal cord or the motor nuclei of cranial nerves and (2) the smaller **gamma fibers** derived from smaller cells in the spinal cord and brainstem. Each fiber is myelinated and ends by dividing into many branches, each of which ends on a muscle fiber at a **motor end plate.**

The sensory fibers are myelinated. The majority arise from specialized sensory endings lying within the muscle or the tendons and are called **neuromuscular spindles** or **neurotendinous spindles,** respectively. These endings are stimulated by elongation and tension in the muscle, which may occur during active contraction or by passive stretching. The function of these sensory fibers is to convey to the central nervous system information regarding the degree of tension of the muscles, which is essential for the maintenance of muscle tone and body posture and for carrying out coordinated voluntary movements. Other sensory fibers arise from a variety of sensory endings in the connective tissue of the muscle.

Sympathetic fibers are postganglionic nonmyelinated fibers and pass to the smooth muscle in the walls of the blood vessels supplying the muscle. The fibers are vasodilator as well as vasoconstrictor and regulate the blood flow to the muscles.

Motor Unit

A motor unit may be defined as a single alpha motor neuron and the muscle fibers that it innervates. The muscle fibers of a single motor unit are widely scattered throughout the muscle. In the small muscles of the hand or the extraocular muscles, where fine precision movements are required, the motor units possess only a few muscle fibers. In muscles, such as the gluteus maximus, where fine control is not required, a single motor nerve may innervate many hundreds of muscle fibers.

Motor End Plate

The large alpha myelinated nerve fiber on entering a skeletal muscle branches many times. A single branch then terminates on a muscle fiber at the motor end plate (*Fig. 9–7*). The nerve loses its myelin sheath and breaks up into a number of subsidiary branches. Here, the naked axon is expanded slightly and contains numerous mitochondria and vesicles containing acetylcholine. At the site of the motor end plate, the surface of the muscle fiber is elevated to form what is referred to as the **sole plate.** The elevation is caused by the local accumulation of granular sarcoplasm beneath the sarcolemma and the presence of numerous nuclei and mitochondria.

The naked axon lies in a groove on the surface of the muscle fiber. Each groove is formed by the infolding of the sarcolemma and is capped or roofed over by a flattened Schwann cell. The groove branches many times, each branch containing a division of the axon. The floor of the groove is formed of sarcolemma, a site of numerous folds called **junctional folds,** which serve to increase the surface area of the sarcolemma that lies close to the naked axon. The plasma membrane of the axon (presynaptic membrane) is separated by a space about 20 to 50 nm wide from the plasma membrane of the muscle fiber (the sarcolemma or postsynaptic membrane). This space, the **synaptic cleft,** is filled with the basal laminae of the axons and the muscle fiber.

The protein molecules necessary for the synthesis and storage of acetylcholine at the nerve ending are made in the nerve cell body and are transported along the axon to the nerve terminal. The small molecule choline is absorbed locally from the tissue fluid, whereas the acetate molecule is produced by the mitochondria. Acetylcholine is synthesized by the enzyme choline acetyltransferase.

A nerve impulse, on reaching the motor end plate, allows calcium ions to enter the nerve ending. The calcium causes the presynaptic vesicles to fuse with the presynaptic membrane so that acetylcholine is released into the synaptic cleft. The acetylcholine diffuses through the basal laminae to reach the receptors on the postsynaptic membrane.

The receptors on the postsynaptic membrane are tubular structures formed of protein subunits that pass completely through the muscle cell membrane. In the resting state the tubes or channels are closed. If acetylcholine becomes attached to the receptor, however, the channel opens allowing sodium ions and calcium ions to flow inside the muscle fiber and potassium ions to flow outside, therefore, a local potential is created, called the **end plate potential.** If the end plate potential is large enough, an action potential will be initiated, which will spread along

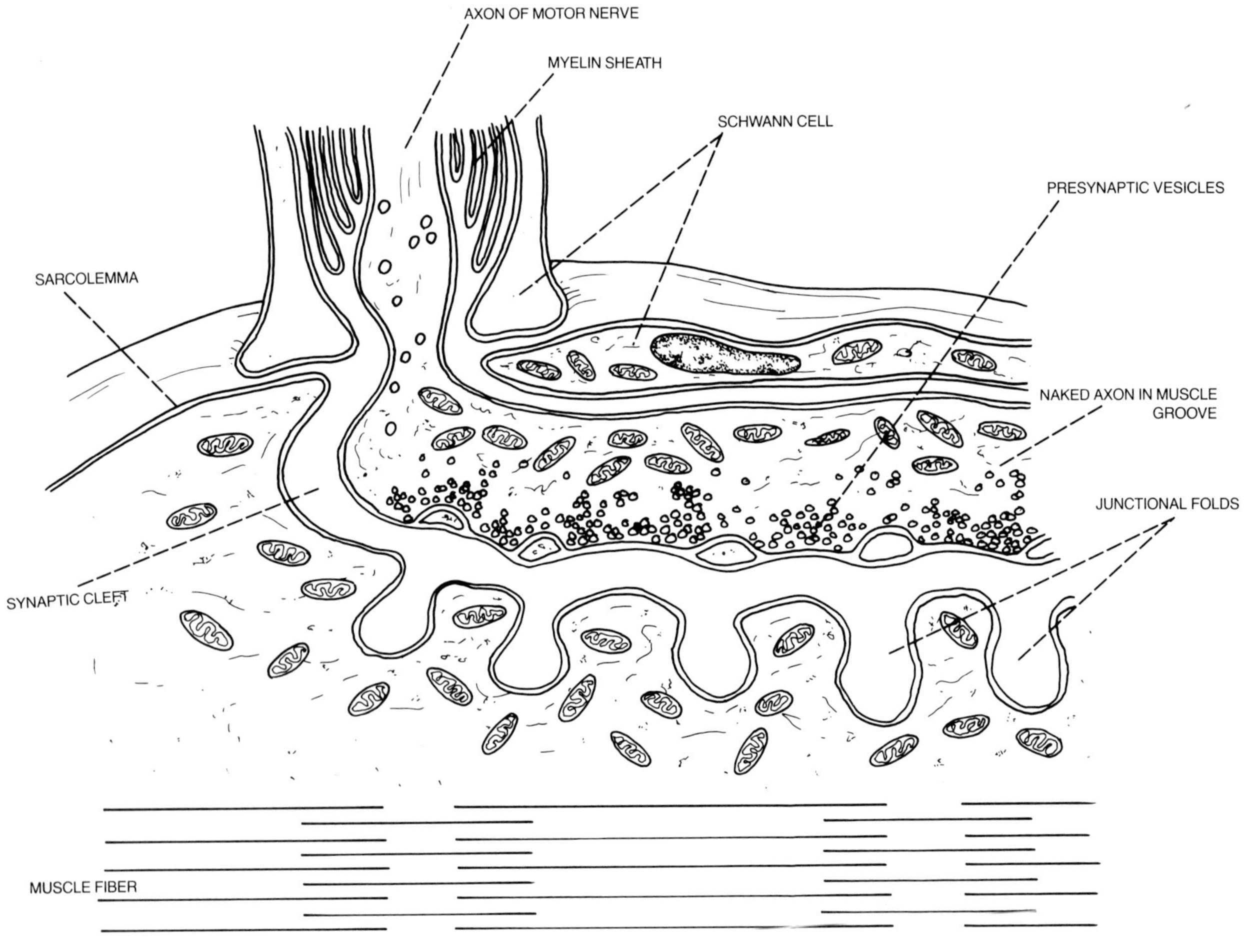

Figure 9–7. Longitudinal section of a neuromuscular junction.

the surface of the sarcolemma. The wave of depolarization is carried into the muscle fiber to the contractile myofibrils through the system of T-tubules. This leads to the release of calcium ions from the sarcoplasmic reticulum, which in turn causes the muscle to contract.

The receptors on the postsynaptic membrane are composed of five subunits—two alpha, one beta, one gamma, and one delta units. The binding sites for acetylcholine on the alpha units are the sites for competition between cholinergic agonists and antagonists. Thus, tubocurarine can bind to both alpha units and prevent acetylcholine from opening up the receptor channel.

Once the acetylcholine crosses the synaptic cleft it immediately undergoes hydrolysis due to the presence of **acetylcholinesterase** (*Fig. 9–7*) in the basal laminae in the synaptic cleft. The acetylcholine remains for about 1 msec in contact with the postsynaptic membrane and it is rapidly destroyed to prevent re-excitation of the muscle fiber.

Neuromuscular Spindles

Muscular spindles (*Fig. 9–8*) are found in skeletal muscle and are most numerous toward the tendinous attachment of the muscle. Each spindle measures about 1 to 4 mm in length and is surrounded by a fusiform capsule of connective tissue. Within the capsule are 6 to 14 slender **intrafusal muscle fibers;** the ordinary muscle fibers situated outside the spindles are referred to as **extrafusal fibers.** The intrafusal fibers of the spindles are of two types: the **nuclear bag fibers** and the **nuclear chain fibers.** The nuclear bag fibers are recognized by the presence of numerous nuclei in the equatorial region, which consequently is expanded; also, cross-striations are absent in this region. In the nuclear chain fibers, the nuclei form a single longitudinal chain in the center of each fiber at the equatorial region. The nuclear bag fibers are larger in diameter than the nuclear chain fibers, and they extend beyond the capsule at each end to be attached to the endomysium of the extrafusal fibers.

There are two types of sensory innervation of muscle spindles: the annulospiral and the flower spray. The **annulospiral endings** are situated at the equator of the intrafusal fibers. As the large myelinated nerve fiber pierces the capsule, it loses its myelin sheath and the naked axon winds spirally around the nuclear bag or chain portions of the intrafusal fibers.

The **flower spray endings** are situated mainly on the nuclear chain fibers, some distance away from the equatorial region. A slightly smaller myelinated nerve fiber than that of the annulospiral ending, it pierces the capsule and loses its myelin sheath, and the naked axon branches terminally and ends as varicosities; it resembles a spray of flowers.

Stretching (elongation) of the intrafusal fibers results in stimulation of the annulospiral and flower spray endings, and nerve impulses pass to the spinal cord in the afferent neurons.

Motor innervation of the intrafusal fibers is provided by fine gamma motor fibers. The nerves terminate in small motor end plates situated at both ends of the intrafusal fibers. Stimulation of the motor nerves causes both ends of the intrafusal fibers to contract and activate the sensory endings. The equatorial region, which is without cross-striations, is noncontractile. The extrafusal fibers of the remainder of the muscle receive their innervation in the usual way from large alpha-size axons.

Function of the Neuromuscular Spindle

Under resting conditions the muscle spindles are continually giving rise to afferent nerve impulses, and most of this information is not consciously perceived. When muscle activity occurs, either actively or passively, the intrafusal fibers are stretched and there is an increase in the rate of passage of nerve impulses to the spinal cord in the afferent neurons. Similarly, if the intrafusal fibers are relaxed because of the cessation of muscle activity, there is a decrease in the rate of passage of nerve impulses to the spinal cord. The neuromuscular spindle, thus, plays a very important role in keeping the central nervous system informed about muscle activity, and thus, indirectly influences the control of voluntary muscle.

Control of the Intrafusal Fibers of the Neuromuscular Spindle

In the brain and the spinal cord there are centers that give rise to tracts that synapse directly or indirectly with gamma motor neurons in the spinal cord. The reticular formation, the basal ganglia, and the cerebellum are examples of such centers. It is by this means that these centers can greatly influence voluntary muscle activity. The gamma efferent motor fibers cause shortening of the intrafusal fibers, stretching the equatorial regions, and stimulating the annulospiral and flower spray endings. This in turn initiates the reflex contraction of the extrafusal fibers described previously.

It is estimated that about one-third of all motor fibers passing to a muscle are gamma efferents, the remaining two-thirds are large alpha motor fibers. It is believed that the nuclear bag fibers are concerned with dynamic responses and are associated more with position and velocity of contraction, whereas the nuclear chain fibers are associated with slow static contractions of voluntary muscle.

Neurotendinous Spindles (Golgi Tendon Organs)

Neurotendinous spindles are present in tendons and are located near the junctions of tendons with muscles (*Fig. 9–9*). Each spindle consists of a fibrous capsule that surrounds a small bundle of loosely arranged tendon fibers

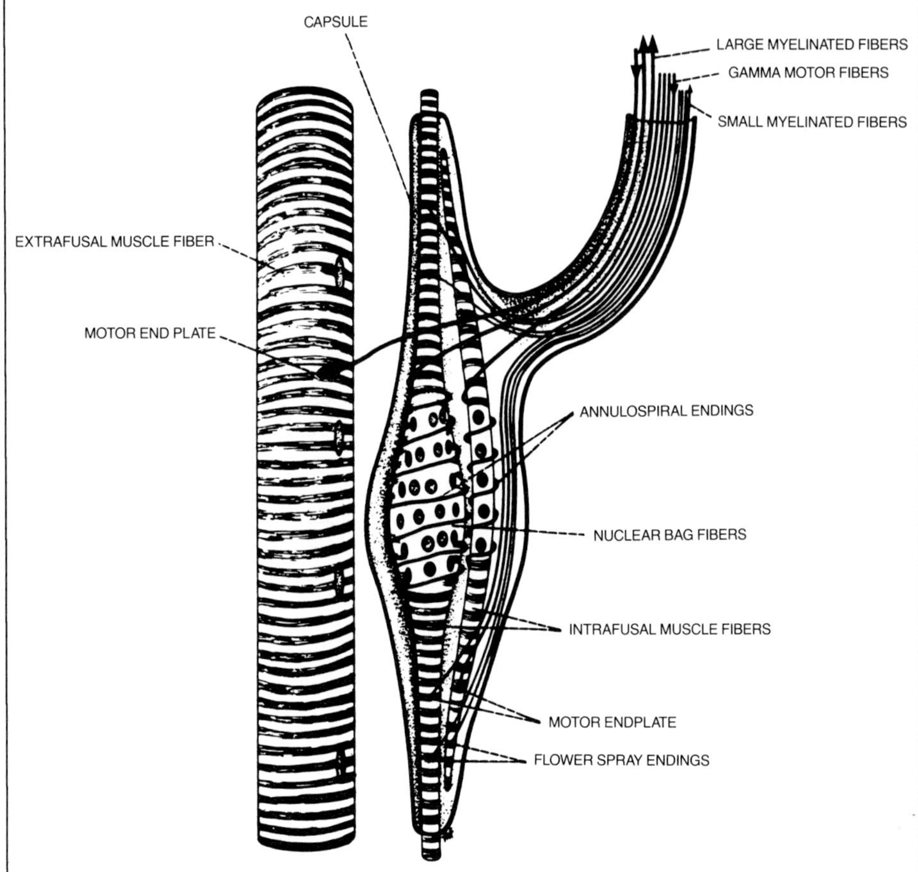

Figure 9–8. A neuromuscular spindle showing two types of intrafusal fibers: the nuclear bag and nuclear chain fibers.

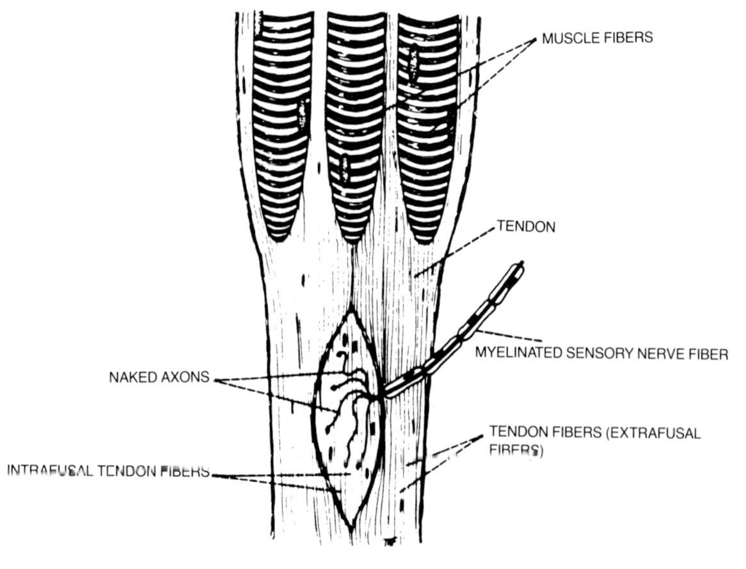

Figure 9–9. A neurotendinous spindle.

(intrafusal fibers). These tendon cells are larger and more numerous than those found elsewhere in the tendon. One or more myelinated sensory nerve fibers pierce the capsule, lose their myelin sheath, branch, and terminate in club-shaped endings.

The nerve endings are activated by being squeezed by the adjacent fibers within the spindle when tension develops in the tendon. Unlike the neuromuscular spindle, which is sensitive to changes in muscle length, the neurotendinous organ detects changes in muscle tension.

Function of the Neurotendinous Spindle

Increased muscle tension stimulates the neurotendinous spindles and an increased number of nerve impulses reach the spinal cord through the afferent nerve fibers. These fibers synapse with the large alpha motor neurons situated in the anterior gray horns of the spinal cord. Unlike the muscle spindle reflex, this reflex is inhibitory and limits muscle contraction. In this manner the tendon reflex prevents the development of too much tension in the muscle. Although this function is probably important as a protective mechanism, its main function is to provide the central nervous system with information that can influence voluntary muscle activity.

Muscle Tone

Muscles at rest exhibit a certain degree of tautness, which can be felt as a firmness, with a degree of elasticity, on applying pressure. Although some of these physical qualities can be attributed to the viscosity of the tissue, the condition is mainly due to the fact that the resting muscle is always in a state of partial contraction. Because there is no intermediate phase (muscle fibers are either fully contracted or fully relaxed), it follows that a few muscle fibers within a muscle are fully contracted all the time. To avoid fatigue, different groups of motor units, and thus different groups of muscle fibers, are brought into action at different times. This is accomplished by the asynchronous discharge of nervous impulses in the alpha motor neurons in the anterior gray horn of the spinal cord.

Basically, muscle tone is dependent on the activity of the alpha motor neurons, which in turn depends on input from other neurons: (1) afferent fibers arising from the muscle spindles and tendon spindles that synapse with the alpha motor neurons, (2) reflex input from the ligaments and the joints to the alpha motor neurons, (3) impulses arising in the brain and descending to influence the alpha motor neurons, and (4) input from the Renshaw cells. The degree of muscle tone depends on the summation of the nerve impulses received by the alpha motor neurons from other neurons of the central nervous system.

Segmental Innervation of Muscles

A spinal nerve supplies the muscle derived from the myotome of the same segment. In many parts of the body muscles are formed by the fusion of separate myotomes and in the limbs the muscles are formed in situ from the mesenchyme in the developing limbs. All this makes the learning of the segmental innervation of muscles extremely difficult. It should be remembered, however, that most muscles are innervated by two, three, or four spinal nerves, and therefore, by the same number of segments of the spinal cord. Thus, to paralyze a muscle completely it is necessary to block several spinal nerves.

Although it is impossible to remember the segmental innervation of all the muscles in the body, the following should be known as it is possible to test them by eliciting simple muscle reflexes (*Fig. 9–10*).

- *Biceps brachii tendon reflex* C_5 and C_6 (flexion of the elbow joint by tapping the biceps tendon)
- *Triceps tendon reflex* C_6, C_7, and C_8 (extension of the elbow joint by tapping the triceps tendon)
- *Brachioradialis tendon reflex* C_5, C_6, and C_7 (supination of the radioulnar joints by tapping the insertion of the brachioradialis tendon)
- *Abdominal superficial reflexes* (contraction of underlying abdominal muscles by stroking the skin). Upper abdominal skin T_6 to T_7; middle abdominal skin T_8 to T_9; lower abdominal skin T_{10} to T_{12}
- *Patellar tendon reflex* (knee jerk) L_2, L_3, and L_4 (extension of knee joint on tapping the patellar tendon)
- *Achilles tendon reflex* (ankle jerk) S_1 and S_2 (plantar flexion of ankle joint on tapping the Achilles tendon)

Stretch Reflex

The neurons involved in the simple stretch reflex are as follows: Stretching a muscle results in the elongation of the intrafusal fibers of the muscle spindle and stimulation of the annulospiral and flower spray endings. The nerve impulses reach the spinal cord in the afferent neurons and synapse with the large alpha motor neurons situated in the anterior gray horns of the spinal cord. Nerve impulses now pass via the efferent motor nerves and stimulate the extrafusal muscle fibers and the muscle contracts. This simple stretch reflex depends on a two-neuron arc, an afferent neuron, and an efferent neuron. It is interesting to note that the muscle spindle afferent impulses inhibit the alpha motor neurons supplying the antagonist muscles. This effect is called **reciprocal inhibition.**

Cerebral Cortex and Muscle Activity

Above the level of the spinal cord, the cerebral cortex, the basal ganglia, the cerebellum, and the brainstem all play a vital and an integrated role in the performance of muscular movement. The old concept of the simple division of cortical areas into motor and sensory is erroneous, for many of the sensory areas are far more extensive than originally described, and it is known that motor responses can be obtained on stimulation of sensory areas.

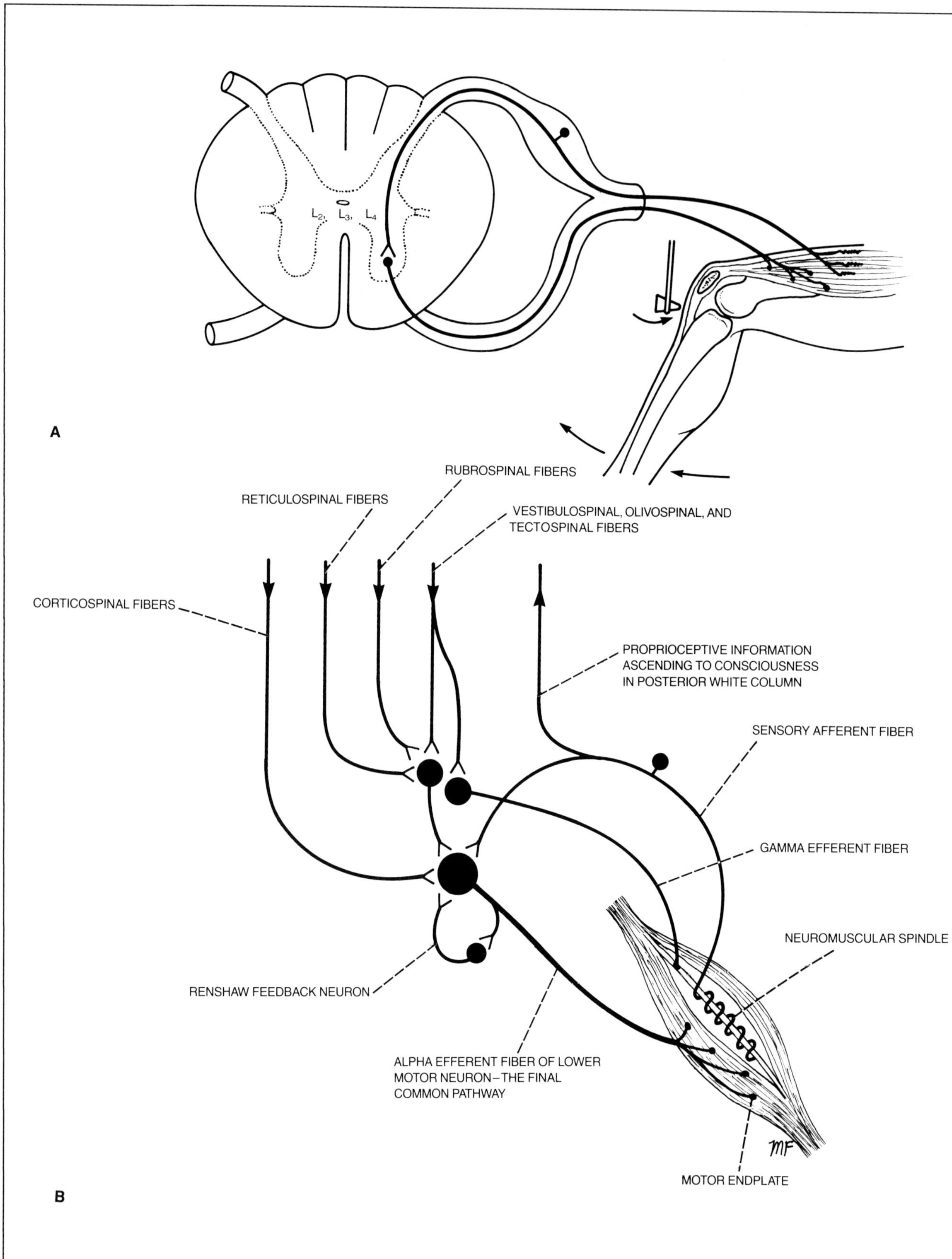

Figure 9–10. A. A monosynaptic reflex arc. **B.** Multiple neurons synapsing with the lower motor neuron.

It is now believed that the sensory cortex provides the motor cortex with information that is processed by the motor cortex and integrated with other supraspinal centers.

Motor Cortex

The motor cortex is divided into two separate areas, the primary motor area and the premotor area (*Fig. 9–11*). The primary motor area, or Brodmann's area 4, occupies the precentral gyrus extending over the superior border of the cerebral hemisphere into the paracentral lobule. The secondary motor area, or Brodmann's area 6 and parts of areas 8, 44, and 45, occupies the anterior part of the precentral gyrus and the posterior parts of the superior, middle, and inferior frontal gyri.

The primary motor area, if electrically stimulated, produces isolated movements on the opposite side of the body and contraction of muscle groups concerned with the performance of a specific movement. Although isolated ipsilateral movements do not occur, bilateral movements of the extraocular muscles, the muscles of the upper part of the face, the tongue and the mandible, and the larynx and the pharynx, do occur.

The movement areas of the body are represented in an inverted form in the precentral gyrus (*Fig. 9–12*). Starting from below and passing superiorly are structures involved in swallowing and in the movement of the tongue, the jaw, the lips, the larynx, the eyelid, and the eyebrow. The next area is an extensive region for movements of the fingers, especially the thumb, the hand, the wrist, the elbow, the shoulder, and the trunk. The movements of the hip, the knee, and the ankle are represented in the highest areas of the precentral gyrus and the toes are situated on the medial surface of the cerebral hemisphere in the paracentral lobule. The anal and vesical sphincters are also located in the paracentral lobule. The area of the cortex controlling a particular movement is proportional to the skill involved in performing the movement and is not related to the mass of muscle participating in the movement. In this respect it is interesting to note that more than one-half of the motor cortex is involved in the control of the hands and the muscles of speech.

The function of the primary motor area is, therefore, to carry out the individual movements of different parts of the body. To assist in this function, it receives numerous afferent fibers from the premotor area, the sensory cortex, the thalamus, the cerebellum, and the basal ganglia. The primary motor cortex is not responsible for the design of the pattern of movement but is the final station for the conversion of the design into the execution of the movement.

The premotor area receives numerous inputs from the sensory cortex, the thalamus, and the basal ganglia. The function of the premotor area is to store programs of motor activity assembled as the result of past experience. The premotor area, therefore, programs the activity of the primary motor area. It is particularly involved in the control of coarse postural movements through its connections with the basal ganglia.

Efferent Fibers from the Motor Cortex

Corticospinal Tract. Fibers of the corticospinal tract arise as axons of pyramidal cells situated in the fifth layer of the cerebral cortex. About one-third of the fibers originate from the primary motor cortex (area 4), one-third from the secondary motor cortex (area 6), and one-third from the parietal lobe (areas 3, 1, and 2). The descending fibers converge in the corona radiata and then pass through the posterior limb of the internal capsule (*Fig. 9–12*). The further course of these important fibers has been described on page 166.

The majority of the corticospinal fibers synapse with internuncial neurons, which in turn synapse with alpha motor neurons and some gamma motor neurons. Only the largest corticospinal fibers synapse directly with the motor neurons.

Fibers to the Caudate Nucleus and Lentiform Nucleus. Descending fibers pass to the caudate nucleus and the putamen of the lentiform nucleus. New connections arise that traverse to the globus pallidus with further connections to the thalamus and feed back to the cerebral cortex (*Fig. 9–13*).

Fibers to the Reticular Formation. Descending fibers from the motor cortex end in the reticular formation of the brainstem (*Fig. 9–13*). The reticular formation then gives rise to the reticulospinal tracts, which influence the activity of the anterior gray horn cells.

Fibers to the Red Nucleus. Descending fibers from the motor cortex end in the red nucleus in the midbrain. From here the rubrospinal tract descends to influence the activity of the anterior gray horn cells.

Fibers to the Cerebellum. Descending fibers from the frontal and temporal lobes descend to the pontine nuclei. Here, they synapse and give rise to the pontocerebellar fibers that cross the midline and influence the opposite cerebellar hemisphere (*Fig. 9–14*).

It is thus seen that the motor cortex not only controls the lower motor neurons, but keeps in communication with the basal ganglia, the reticular formation, red nucleus, and cerebellum, which by their activity can influence skeletal muscle activity.

Basal Ganglia and Motor Activity

The basal ganglia consist of the caudate nucleus and the lentiform nucleus (putamen and globus pallidus). Together with the substantia nigra and the reticular formation they can influence muscle activity. In the past it

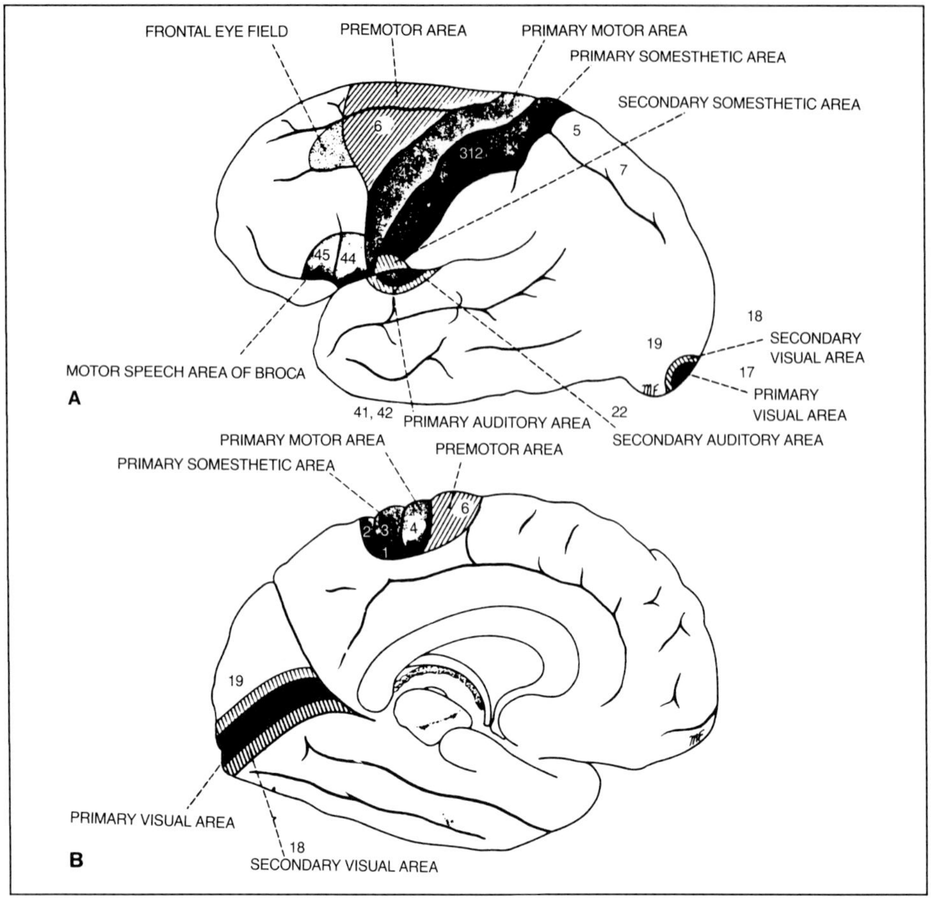

Figure 9–11. Functional localization of the cerebral cortex. **A.** Lateral view of the left cerebral hemisphere. **B.** Medial view of the left cerebral hemisphere.

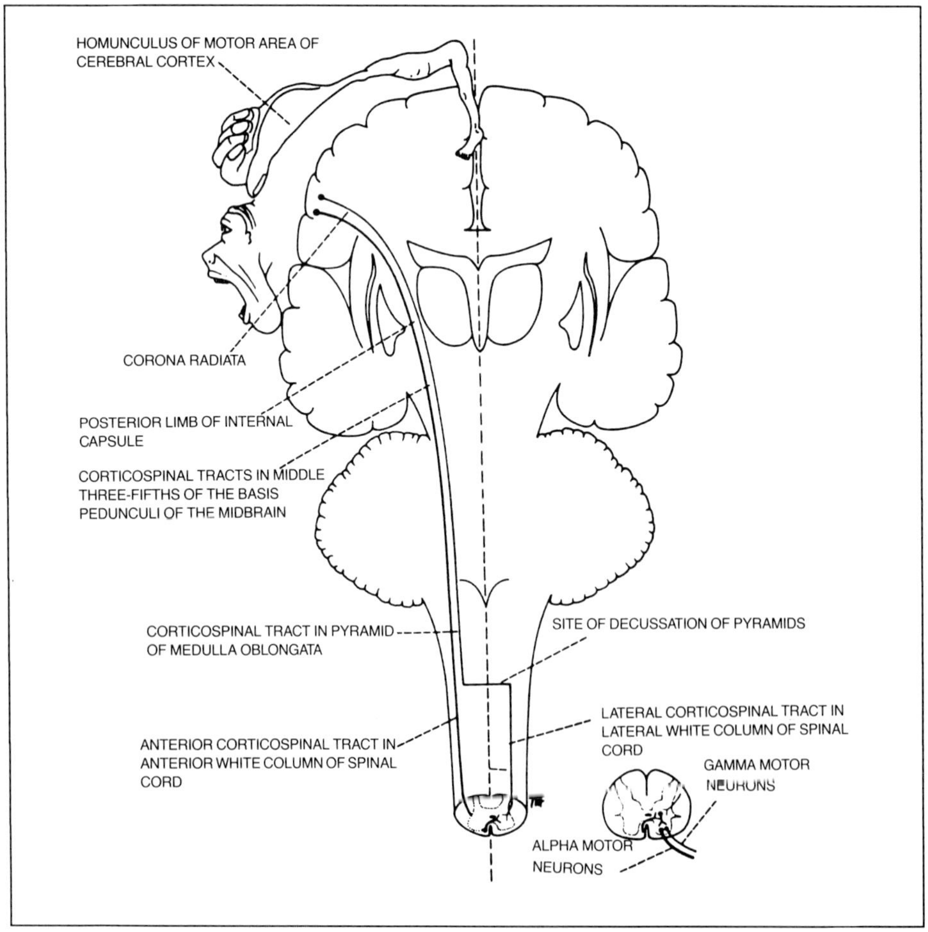

Figure 9–12. Corticospinal tracts.

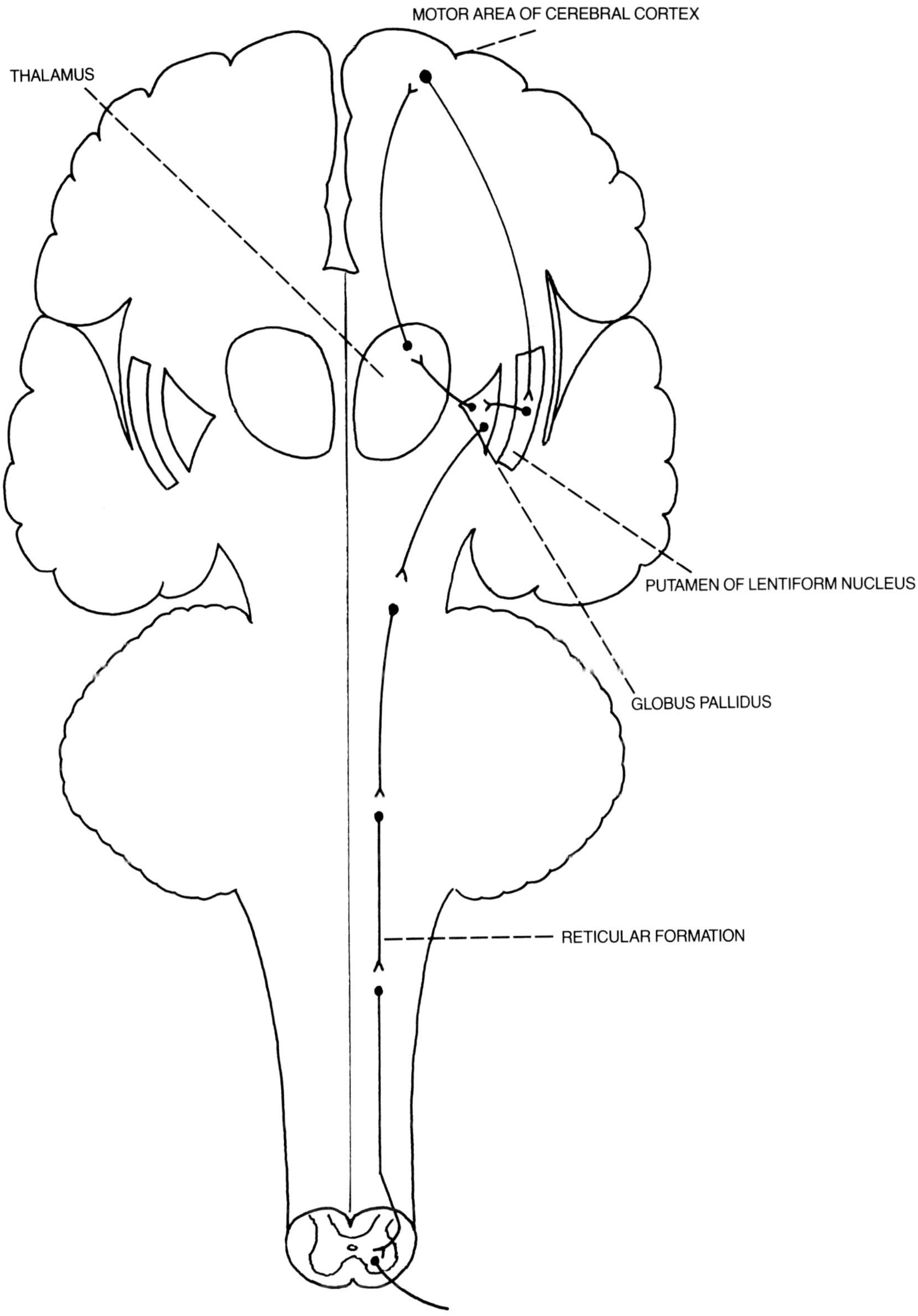

Figure 9–13. Some neuronal connections of the basal ganglia and how they may influence the activities of the cerebral cortex and the lower motor neurons.

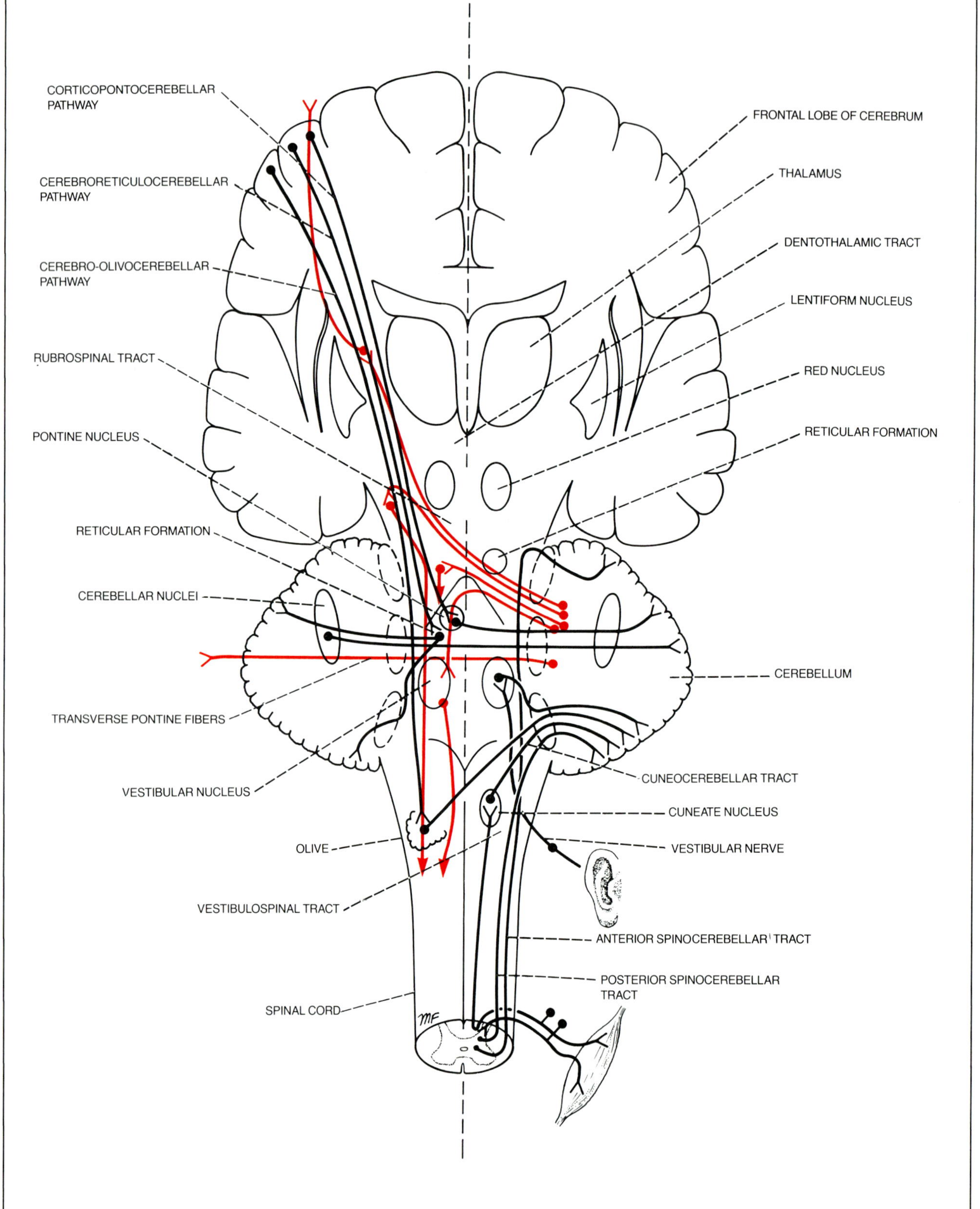

Figure 9–14. Some of the main connections of the cerebellum. The cerebellar peduncles are shown as ovoid dashed lines.

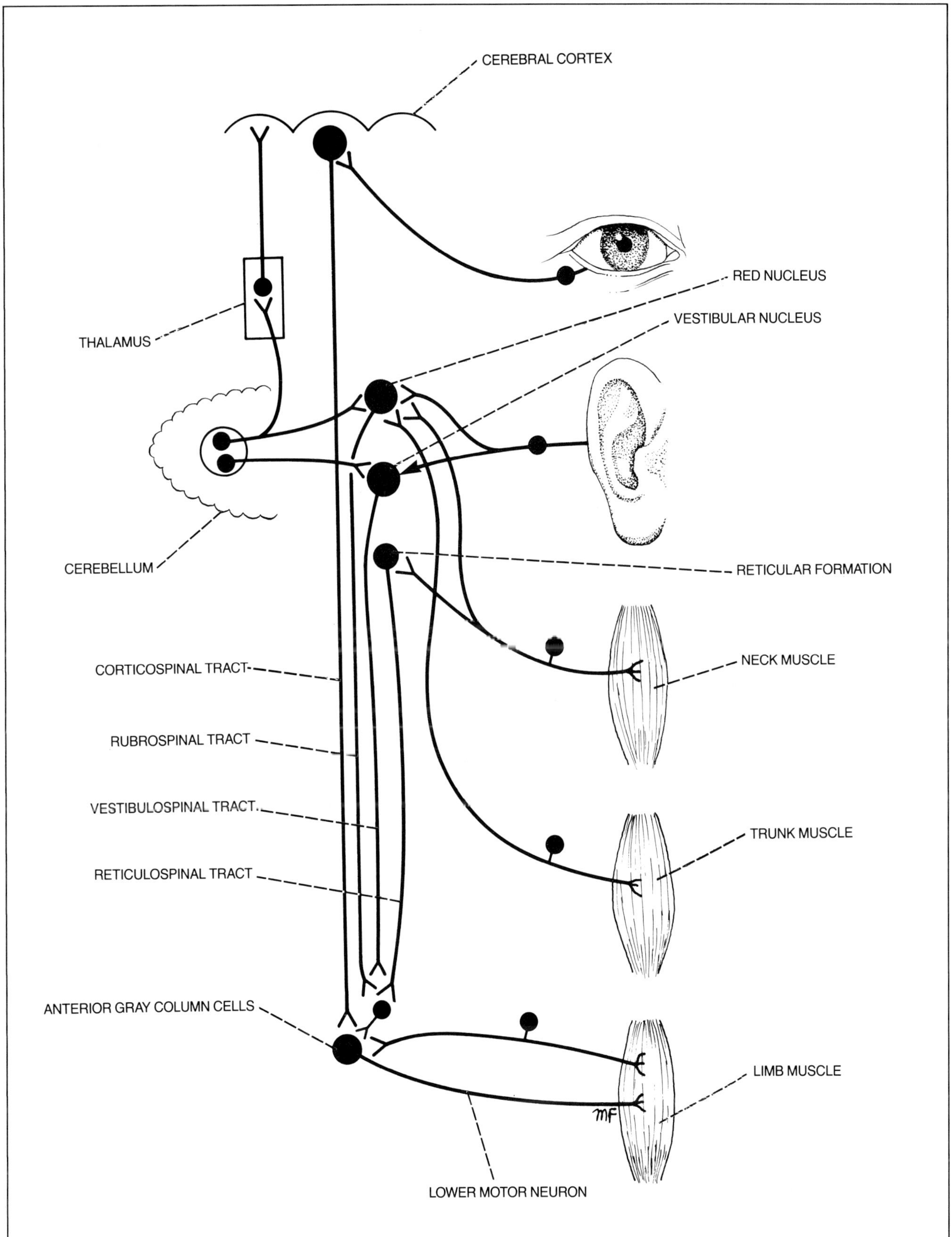

Figure 9–15. Nervous input from higher levels of the central nervous system that influence the activity of the anterior gray column cells of the spinal cord.

was believed that these areas of the nervous system were activated by the motor cortex, but now it is thought that afferent information received from the sensory cortex and from other areas of the nervous system activates the basal ganglia, which in turn influences the lower motor neurons and modulates the activities of the motor cerebral cortex. In general it may be said that excitation of the basal ganglia inhibits muscle tone throughout the body. It follows, therefore, that destruction of the basal ganglia causes increased muscle tone or rigidity.

Cerebellum and Motor Activity

The cerebellum receives afferent information concerning voluntary movement from the motor areas of the cerebral cortex and from muscle spindles and tendon organs. It also receives information concerning balance from the vestibular nerve and possibly sight through the tectocerebellar tract. The following main afferent pathways are involved: (1) corticopontocerebellar, (2) cerebro-olivocerebellar, (3) cerebroreticulocerebellar, (4) anterior spinocerebellar, (5) posterior spinocerebellar, and (6) the cuneocerebellar. These pathways are summarized in Figure 9–14.

After several synaptic relays in the cerebellar cortex, afferent impulses converge on the Purkinje cells. Axons of most Purkinje cells project onto the deep cerebellar nuclei exerting an inhibitory influence upon them. The cerebellar output is not passed directly to the lower motor neurons as one would expect, but is conducted to the motor cortex, the red nucleus, the reticular formation, and the vestibular nucleus. Thus the cerebellum exerts its influence indirectly through the cerebral cortex and brain stem.

The efferent pathways of the cerebellum are summarized in Figure 9–14. Note that each cerebellar hemisphere is connected principally with the same side of the body by nervous pathways, therefore, a lesion of one cerebellar hemisphere gives rise to signs and symptoms that are limited to the same side of the body.

The cerebellum functions as a coordinator of precise movements by continually comparing the output of the motor area of the cerebral cortex with proprioceptive information received from the site of muscle action. It is then able to bring about the necessary adjustments by influencing the activity of the lower motor neurons. This is accomplished by controlling the sequence of firing of the alpha and gamma motor neurons. Loss of cerebellar function leads to hypotonia owing to the absence of the cerebellar influence on the simple stretch reflex mechanism.

Voluntary Movement

Voluntary movement is initiated by the individual. A series of different muscles are made to contract for the purpose of reaching a goal. This would suggest that the descending tracts that influence the activity of the lower motor neurons are driven by information received by the sensory systems (such as eyes and ears) and the muscles themselves. In addition, they are affected further by past afferent information that has been stored in the memory and may be colored by prior or current emotional input. The limbic structures appear to play a role in emotion, motivation, and memory, and may influence the initiation process of voluntary movement by their projections to the cerebral cortex.

The descending pathways from the cerebral cortex and the brainstem, i.e., upper motor neurons, influence the activity of lower motor neurons either directly or through internuncial neurons (*Fig. 9–15*). Most of the tracts originating in the brainstem that descend to the spinal cord also receive input from the cerebral cortex.

The corticospinal tracts are believed to control the prime mover muscles, especially those responsible for highly skilled movements of distal limb parts. Other supraspinal descending tracts play a major role in the simple basic voluntary movements, and in addition bring about an adjustment of muscle tone, so that easy and rapid movements of joints can take place.

CLINICAL NOTES

DISEASES AFFECTING THE MONOSYNAPTIC REFLEX ARC AND MUSCLE TONE

A disease process that blocks part of the reflex arc controlling muscle tone will abolish it. Some examples are syphilitic infection of the posterior root (tabes dorsalis); destruction of the motor anterior horn cells, as in poliomyelitis or syringomyelia; destruction of a segment of the spinal cord by trauma, or pressure from a tumor; section of an anterior root; pressure on a spinal nerve by a prolapsed intervertebral disc; section of a peripheral nerve, as in a stab wound; and block of a peripheral nerve supplying a muscle by a local anesthetic. All these conditions cause loss of muscle tone.

Muscle Wasting

After section of a motor nerve, the muscle undergoes wasting within 2 or 3 weeks. The volumes of the muscle mass in the limbs is easily tested by measuring the diameter of the limbs at a given point over the paralyzed muscle and comparing that measurement with the one at the same site on the opposite limb.

Muscular Fasciculation

Twitching of groups of muscle fibers is seen in chronic diseases affecting the anterior horn cells, for example, progressive muscular atrophy.

Muscle Fibrillation

Three to 5 days after a skeletal muscle has been completely denervated, spontaneous contractions appear in the muscle fibers. Once muscle atrophy commences the fibrillation gradually disappears.

Lesions of the Lower Motor Neuron

Trauma, infection (poliomyelitis), vascular disorders, degenerative diseases, and neoplasms may all produce a lesion of the lower motor neuron by destroying the nerve cell body in the anterior gray horn or its axon in the anterior root or spinal nerve. The clinical signs of flaccid paralysis, atrophy of the involved muscles, loss of muscle reflexes, and muscular fasciculation may all be present.

PYRAMIDAL AND EXTRAPYRAMIDAL TRACTS

Pyramidal tract is a term commonly used clinically and refers specifically to the corticospinal tract (does not include corticonuclear tracts). It came into use when it was found that the corticospinal tracts are situated in an area in the anterior part of the medulla oblongata known as the pyramids. Extrapyramidal tracts is a loose term referring to all the other descending tracts other than the corticospinal tracts.

Lesions of the Upper Motor Neurons

Corticospinal Tracts. Lesions of the corticospinal tracts produce a positive Babinski sign. A positive Babinski sign is present during the first year of life because the corticospinal tract is not myelinated until then. The superficial abdominal reflexes are absent. Why this reflex is dependent on the integrity of the corticospinal tracts is not understood. There is a decrease in fine skilled voluntary movements especially in the hands and the feet.

Other Descending Tracts. Lesions of descending tracts other than the corticospinal tract produce clinical signs that include severe paralysis with little or no muscle atrophy, spasticity or hypertonicity, exaggerated deep muscle reflexes, clonus, and clasp-knife reaction.

Lesions of the Cerebellum

The following symptoms and signs are characteristic of cerebellar dysfunction and are caused by the loss of cerebellar influence on the motor cortex and brainstem nuclei: hypotonia, ataxia, past pointing, intention tremor, nystagmus, and dysarthria. There is a tendency to fall to the same side when walking. A lesion of the cerebellum gives rise to ipsilateral symptoms and signs.

CLINICAL PROBLEMS

1. A 26-year-old woman was walking along a dusty road when she suddenly became aware of a foreign body in her right eye. Because the cornea is extremely sensitive, she suffered considerable pain and discomfort. What sensory endings are present in the cornea? Is the cornea sensitive to stimuli other than pain? Trace the pathway taken by pain impulses as they travel from the cornea to the central nervous system.

2. Different adjectives are used by patients to describe the different qualities of pain that they are experiencing. Terms such as sharp, dull, aching, burning, stinging, are commonly used. Are there different receptors involved in detecting different types of pain? Are different types of pain conducted along different pathways? Are the pain pathways faster in their conduction rates than pathways concerned with other sensory modalities?

3. Thermocoagulation is a method of treating severe intractable trigeminal neuralgia. Trace the pain pathway from the skin over the chin to the cerebral cortex and indicate the site where this form of treatment should be directed.

4. An 18-year-old woman woke up one morning nauseated with a vague dull aching pain in the umbilical region. During the morning the quality of the pain changed, became more intense and sharp, and was felt most severely in the right iliac region. On examination, the muscles of the lower part of the anterior abdominal wall in the right iliac region showed guarding and rigidity. There was acute rebound tenderness. The diagnosis of acute appendicitis with possible perforation was made. Can you explain in neuroanatomic terms why cases of acute appendicitis commonly experience initially a dull, vague pain in the umbilical region that later becomes an acute localized pain in the right iliac region? Can you explain the rigidity of the abdominal muscles?

5. Although definite pain pathways exist for the transmission of noxious painful impulses, once the information reaches the posterior gray column of the spinal cord it can be modulated. Explain the various systems that exist in the spinal cord and brainstem for modifying the pain input. Explain how a soldier on a battlefield can continue to fight even though he has been gravely wounded.

6. A 59-year-old man stated that he experiences a dull aching pain down the inner side of his left upper limb when he becomes excited or when he goes for a long walk. Can you explain the pathway taken by the pain fibers from the heart to the central nervous system. Why is the pain referred to the arm?

7. It is common practice to give a patient an anticholinesterase drug at the end of an operation in which a nondepolarizing muscle relaxant has been used. Where does a nondepolarizing muscle relaxant act at the motor end plate? The anticholinesterase drug acts on the enzyme acetylcholinesterase. Where is this enzyme located at the motor end plate?

8. Muscle relaxants adhere to the postsynaptic membrane at a motor end plate. What is the postsynaptic membrane? What is a receptor? What is a receptor composed of? What is a receptor channel? What ions normally pass through a receptor channel?

9. Acetylcholine is the chemical transmitter at a motor end plate. Where is it produced? What is responsible for its release? How is it stored?

10. A 45-year-old woman visited her physician because she had noticed an increasing clumsiness of her left arm. She also noticed that her left hand had a tremor when she attempted to perform fine movements. When she walked she said she had a tendency to fall over to the left. On physical examination she was found to hold her left shoulder lower than her right. There was also hypotonia

of the muscles of the left upper and lower limbs. The heel to toe walking test along a straight line showed that the patient tended to sway over to the left side. The finger to nose test revealed intention tremor and past pointing of the left index finger. There was no dysarthria and no evidence of nystagmus. In neuroanatomic terms explain the signs and symptoms of this patient.

11. When examining a patient with a hemiplegia caused by a cerebrovascular accident it is important to determine which clinical signs can be attributed to a lesion of the corticospinal tracts and which signs can be caused by damage to the other decending tracts. In neuroanatomic terms group the signs caused by the different tract lesions.

12. A 35-year-old man was diagnosed as having a tumor arising within the vertebral canal. On physical examination he was found to have weakness and diminished tone of the deltoid and biceps brachii muscles on both sides of the body. The biceps tendon jerk was absent on the left side and diminished on the right side. The muscles of the trunk and lower limbs showed increased tone and spastic paralysis. Both knee jerks were exaggerated. Using your knowledge of anatomy (a) identify the segments of the spinal cord that are being pressed upon by the tumor. (b) Which segments of the spinal cord are involved in the reflex arcs responsible for the biceps brachii tendon jerks and the knee jerks? (c) What is responsible for the diminished tone of the deltoid and biceps brachii muscles? (d) What is responsible for the increased tone of the muscles of the lower limb?

ANSWERS FOR CLINICAL PROBLEMS

1. The only sensory receptors present in the cornea are free nerve endings. The cornea is sensitive to light touch and temperature changes in addition to pain. The question now arises—How does a single type of sensory receptor detect different types of sensory stimuli? The answer is that each free nerve ending has a nerve fiber that terminates at a specific site in the brain and the type of sensation felt is determined by the area in the brain at which the nerve terminates.

   The cornea is innervated by sensory fibers from the ophthalmic division of the trigeminal nerve. The fibers leave the eyeball as the long ciliary nerves and travel in the nasociliary branch of the ophthalmic nerve. The cell bodies reside in the trigeminal ganglion (see page 289) and the central processes of these cells enter the brain on the anterior surface of the pons.

2. All types of pain stimuli are now believed to be received at free nerve endings. Two main types of pain are recognized (1) the sharp pricking type of pain of short duration that is localized and (2) the dull aching, burning pain that may persist for some time and is poorly localized. The first type is conducted along the delta type A fibers at relatively fast velocities, whereas the second type is conducted along the slower type C fibers. Sharp pain passes to the reticular formation, the thalamus, the hypothalamus, and the sensory cortex. Dull aching pain goes to the reticular formation and the intralaminar nuclei of the thalamus. The lateral spinothalamic tracts are the ascending pathways. As illogical as it may seem the conduction rate in the pain pathways is slower than for those conducting other sensory modalities.

3. The skin over the chin is innervated by the mental nerve, a branch of the inferior alveolar branch of the mandibular division of the trigeminal nerve. The peripheral nerve fibers within the mandibular nerve now pass through the foramen ovale to join the convex surface of the trigeminal ganglion. The central processes of the ganglion cells then conduct the pain impulses into the pons via the large sensory root of the trigeminal nerve. Pain fibers terminate in the superior part

of the spinal nucleus of the fifth nerve. Axons from the spinal nucleus now cross the median plane and ascend as the trigeminal lemniscus. The fast traveling pain fibers then travel to the reticular formation, the thalamus, and the hypothalamus. The thalamic fibers then pass to the postcentral gyrus of the cerebral cortex. The slow traveling pain fibers terminate in the reticular formation and the intralaminar nuclei of the thalamus.

The thermocoagulation technique has been used to interrupt pain conducting nerve fibers in trigeminal neuralgia. The electrode is introduced into the trigeminal (gasserian) ganglion by passage through the skin and the infratemporal fossa and into the skull through the foramen ovale.

4. In the early stages of acute appendicitis the visceral pain in the appendix is produced by the distention of its lumen, or spasm of its muscle. Afferent fibers enter the spinal cord at the level of the tenth thoracic segment, having ascended through the superior mesenteric plexus and lesser splanchnic nerve. A diffuse aching pain is referred to the region of the umbilicus, which is innervated by the tenth intercostal nerve. Later in this case the pain changed its character and became sharp and localized in the right iliac region. Here, the inflamed appendix and later the pus from the ruptured appendix irritated the parietal peritoneum, which is innervated by the 12th thoracic and 1st lumbar spinal nerves. The irritation of the parietal peritoneum reflexly produced an increase in the tone of the muscles in this region, which are also supplied by the 12th thoracic and 1st lumbar spinal nerves. The increase in muscle tone is interpreted as rigidity by the examiners hand.

5. Once the pain impulses have reached the posterior gray column of the spinal cord their further progress can be modulated by (a) the descending analgesic system of fibers that can suppress the pain being conducted by the delta type A fibers and the type C fibers (see also page 289). The suppression also probably takes place at higher levels in the brainstem. (b) The gating mechanism by which afferent nonpainful stimuli reaching the posterior gray column can inhibit, i.e., close the gate, for the further passage of the painful impulses. Should the nonpainful impulses cease, the gate would open so that pain conduction would continue unimpeded. (c) The presence of naturally occurring enkephalins and endorphins as well as other chemical inhibitors within the central nervous system.

Soldiers who have been known to continue fighting after being severely wounded usually are conscious of receiving the wound and experience some pain, but it is of short duration. This can be explained on the basis that the pain is felt while the tissue is being damaged. Later, other nonpainful afferent sensory impulses reach the nervous system and suppress the further conduction of pain by the gating mechanism.

6. The pain impulses from the heart pass along afferent nerve fibers within the cardiac branches of the sympathetic trunk and enter the spinal cord through the posterior roots of the upper four thoracic spinal nerves. The pain is referred to the skin areas supplied by the corresponding spinal nerves. The pain is referred to the medial side of the arm via the intercostobrachial nerve (T_2).

7. A nondepolarizing muscle relaxant prevents the access of acetylcholine to the receptors on the postsynaptic membrane. The enzyme acetylcholinesterase is situated in the basal laminae in the synaptic cleft.

8. The postsynaptic membrane at the motor end plate is the sacrolemma or cell membrane of the muscle fiber. A receptor is a tubular structure formed of protein subunits that pass completely through the postsynaptic membrane. Receptors are formed by the muscle cell. Receptors have been analyzed biochemically and are composed of five subunits, namely two alpha, one beta, one gamma, and one delta units. A receptor channel is the name given to the lumen within the tubular receptor. Normally the channel is closed, but it is opened by the reaction of acetylcholine on the receptor. When the channel is open sodium and calcium ions pass from the extracellular space into the muscle cell and potassium ions flow from the inside of the cell to the outside.

9. Acetylcholine is produced at the nerve terminal of the motor end plate. The choline molecule is absorbed from the tissue fluid and the acetate molecule is produced by the mitochondria. Choline acetyltransferase is the enzyme responsible for forming the acetylcholine. The release of acetylcholine is initiated by the arrival of a nerve impulse at the nerve terminal. Calcium ions then enter the terminal axon and cause the presynaptic vesicles containing the acetylcholine to fuse with the presynaptic membrane. The acetylcholine is then released into the synaptic cleft by exocytosis. Acetylcholine is stored at the axon terminal in presynaptic vesicles.

10. This patient had an astrocytoma of the left cerebellar hemisphere. The clumsiness, the tremor, the muscle incoordination, and the hypotonia occurred on the left, the same side of the body as the lesion. The hypotonia was caused by the removal of the influence of the cerebellum on the simple stretch reflex involving the muscle spindles and tendon organs. The clumsiness, tremor, and overshooting on the finger to nose test were caused by the absence of cerebellar influence on muscle coordination. The falling to the left side and the drooping of the left shoulder were caused by loss of muscle tone.

11. Interruption of the corticospinal tracts produces the following clinical signs: (a) positive Babinski sign, (b) loss of superficial abdominal reflexes, and (c) loss of performance of fine, skilled voluntary movements, especially at the distal ends of the limbs.

Interruption of descending tracts other than the corticospinal tracts produces the following clinical signs: (a) severe paralysis on the opposite side of the body, (b) spasticity of the paralyzed muscles, (c) exaggerated deep muscle reflexes on the opposite side of the body to the lesion (clonus may be demonstrated), and (d) clasp-knife reaction, which may be felt in the paralyzed muscles.

12. (a) The deltoid and the biceps brachii muscles are innervated by the fifth and sixth cervical segments of the spinal cord. These spinal cord segments lie within the vertebral foramina of the fourth and fifth cervical vertebrae, respectively. (b) The biceps brachii reflex arc involves the fifth and sixth cervical segments of the spinal cord. The patellar tendon reflex (knee jerk) involves the second, third, and fourth lumbar segments of the spinal cord. (c) The diminished tone of the deltoid and biceps brachii muscles is caused by the interruption of the reflex arcs upon which their tone depends. (d) The increased tone of the muscles of the lower limbs is caused by the interruption of the descending tracts from supraspinal levels other than the corticospinal tracts. The tract lesion was produced by the pressure of the expanding tumor in the vertebral canal in the cervical region.

Index